Addison-Wesley Essentials of Mathematics

Randall I. Charles
Stanley A. Smith
Penelope P. Booth
John A. Dossey

Addison-Wesley Publishing Company
Menlo Park, California • Reading, Massachusetts
New York • Don Mills, Ontario • Wokingham, England
Amsterdam • Bonn • Sydney • Singapore • Tokyo
Madrid • Bogotá • Santiago • San Juan

Copyright © 1989 by Addison-Wesley Publishing Company, Inc. All rights reserved. No part of this publication may be reproduced, stored in a retrieval system, or transmitted, in any form or by any means, electronic, mechanical, photocopying, recording, or otherwise, without the prior written permission of the publisher. Printed in the United States of America. Published simultaneously in Canada.

ISBN 0-201-21989-1

5 6 7 8 9 10 - VH - 95 94 93 92 91

Acknowledgements

Illustration
Art by AYXA
Christensen and Son Design
Marilyn Hill
Marcia McGetrick
Phyllis Rockne
George Samuelson
Sally Shimizu

Typesetting
Typothetae

Design
Design Office

Photography

page	
3	Richard Steedman/The Stock Market
6	© David Madison
20	Raoul Hacker/Stock, Boston
31	Bob Daemmrich/Stock, Boston
44	Philip Jon Bailey/Stock, Boston
50	Larry Lee
59	Chuck Fishman/Woodfin Camp & Associates
62	Shepard Sherbell/Stock, Boston
89	Jon Feingersh/Tom Stack & Associates
98	David Madison/Duomo
104	Gabe Palmer/The Stock Market
106	Stephen Frisch*
119	Dan McCoy/Rainbow
120	© David Madison
145	Charles Krebs/The Stock Market
150	Ed Simpson/After-Image
158	UPI/Bettman Newsphotos
164	Stephen Frisch*
172	Kevin Schafer/Tom Stack & Associates
174	Mark Tuschman*
181	Julie Houck/Stock, Boston
190	© The Academy of Motion Pictures Arts & Sciences ®
192	Stuart Cohen/Stock, Boston
206	Julie Habel/West Light
207	David Burnett/Contact-Woodfin Camp
213	Jose Azel/Woodfin Camp & Associates
239	Robert Jureit/The Stock Market
240	Norman Owen Tomalin/Bruce Coleman Inc.
252	UPI/Bettman Newsphotos
267	Bill Gallery/Stock, Boston
268	Charles Gupton/Stock, Boston
272	Larry Dech/Tom Stack & Associates
276	Frank P. Rossotto/The Stock Market
286	Udo Schreiber/Shooting Star
303	Dan McCoy/Rainbow
318	Will Hart/Focus West
331	Brian Parker/Tom Stack & Associates
350	Kevin Schafer/Tom Stack & Associates
363	John Marmaras/Woodfin Camp & Associates
364	Michal Heron/Woodfin Camp & Associates
368	Bill Gillette/Stock, Boston
374	© David Madison
397	Michael Grecco/Stock, Boston
431	Pete Saloutos/After-Image
457	Gary Milburn/Tom Stack & Associates
458	James Wilson/Woodfin Camp & Associates
464	Tom Stack/Tom Stack & Associates
468	Michal Heron/Woodfin Camp & Associates
472	Steve Solum/Bruce Coleman Inc.

Wayland Lee*/Addison-Wesley Publishing Company: Pages 14, 146, 156, 175, 182, 290, 340, 370, 378

*Photographs taken expressly for the publisher

CONTENTS

1 EXPLORING THE FUNDAMENTALS

Chapter 1 Overview **2**
1-1 Reading and Writing Standard Numerals **4**
1-2 Comparing and Ordering Whole Numbers and Decimals **6**
1-3 Rounding **8**
1-4 Understanding the Operations: Addition and Subtraction **10**
1-5 Estimating Sums and Differences Using Rounding **12**
1-6 Understanding the Operations: Multiplication and Division **14**
1-7 Estimating Products Using Rounding **16**
1-8 Estimating Quotients Using Compatible Numbers **18**
1-9 Deciding When to Estimate **20**
1-10 An Introduction to Problem Solving **22**
1-11 Enrichment: Writing Checks **24**
1-12 Computer: Flowcharts **25**

Chapter 1 Review **26**
Chapter 1 Test **27**
Chapter 1 Cumulative Review **28**

2 ADDITION AND SUBTRACTION OF WHOLE NUMBERS AND DECIMALS

Chapter 2 Overview **30**
2-1 Problem Solving: Learning to Use Strategies **32**
2-2 Adding Whole Numbers and Decimals **34**
2-3 Subtracting Whole Numbers and Decimals **36**
2-4 Subtracting Across Zero **38**
Mixed Practice **39**
2-5 Estimating Sums and Differences Using Front-End Estimation **40**
2-6 Consumer Math: Filling Out Deposit Slips **42**
2-7 Consumer Math: Keeping a Running Balance **43**
2-8 Mental Math: Counting **44**
2-9 Choosing a Calculation Method **46**
2-10 Problem Solving: Determining Reasonable Answers **48**
2-11 Problem Solving: Developing Thinking Skills **50**
2-12 Practice Through Problem Solving **51**
2-13 Enrichment: Reconciling a Bank Statement **52**
2-14 Calculator: Maintaining a Checking Account **53**

Chapter 2 Review **54**
Chapter 2 Test **55**
Chapter 2 Cumulative Review **56**

3 MULTIPLICATION OF WHOLE NUMBERS AND DECIMALS

Chapter 3 Overview **58**
- **3-1** Problem Solving: Learning to Use Strategies **60**
- **3-2** Multiplying by a 1-Digit Number **62**
- **3-3** Multiplying by a 2-Digit Number **64**
- **3-4** Multiplying by Larger Factors **66**
- **3-5** Estimating Sums Using Clustering **68**
  Mixed Practice **69**
- **3-6** Problem Solving: Multiple-Step Problems **70**
- **3-7** Multiplying Decimals **72**
- **3-8** Multiplying by 10, 100, or 1000 **74**
- **3-9** Consumer Math: Reading a Paycheck **76**
- **3-10** Exponents **78**
- **3-11** Mental Math: Break Apart Numbers **79**
- **3-12** Problem Solving: Data Collection and Analysis **80**
- **3-13** Practice Through Problem Solving **81**
- **3-14** Enrichment: Prime Factorization **82**
- **3-15** Computer: Multiplying Decimals **83**

Chapter 3 Review **84**
Chapter 3 Test **85**
Chapter 3 Cumulative Review **86**

4 DIVISION OF WHOLE NUMBERS AND DECIMALS

Chapter 4 Overview **88**
- **4-1** Problem Solving: Learning to Use Strategies **90**
- **4-2** Dividing by a 1-Digit Number **92**
- **4-3** Dividing by a 2-Digit Number **94**
- **4-4** Dividing with Zeros in the Quotient **96**
- **4-5** Dividing by a 3-Digit Number **98**
- **4-6** Consumer Math: Reading a Telephone Bill **100**
- **4-7** Problem Solving: Using Data from an Advertisement **102**
  Mixed Practice **103**
- **4-8** Dividing a Decimal by a Whole Number **104**
- **4-9** Dividing by 10, 100, or 1000 **106**
- **4-10** Dividing a Decimal by a Decimal **108**
- **4-11** Problem Solving: Group Decisions **110**
- **4-12** Practice Through Problem Solving **111**
- **4-13** Enrichment: Order of Operations **112**
- **4-14** Calculator: Inverse Operations **113**

Chapter 4 Review **114**
Chapter 4 Test **115**
Chapter 4 Cumulative Review **116**

5 TABLES, CHARTS, AND GRAPHS

Chapter 5 Overview 118
- **5-1** Problem Solving: Learning to Use Strategies 120
- **5-2** Bar Graphs 122
- **5-3** Pictographs 124
- **5-4** Line Graphs 126
- **5-5** Reading Circle Graphs 128
- **5-6** Estimating Sums Using Compatible Numbers 130
- **5-7** Mental Math: Compensation 131
- **5-8** Consumer Math: Evaluating Graphs 132
- **5-9** Problem Solving: Understanding the Situation 134
- **5-10** Problem Solving: Data Collection and Analysis 136
- **5-11** Enrichment: Double Bar and Line Graphs 138
- **5-12** Computer: Graphs 139

Chapter 5 Review 140
Chapter 5 Test 141
Chapter 5 Cumulative Review 142

6 MEASUREMENT

Chapter 6 Overview 144
- **6-1** Problem Solving: Learning to Use Strategies 146
- **6-2** Customary Units of Length 148
- **6-3** Accuracy of Measurement 150
- **6-4** Customary Capacity 152
- **6-5** Customary Weight 154
- **6-6** Consumer Math: Reading Customary Scales 156
- **6-7** Metric Units of Length 158
- **6-8** Changing Metric Units of Length 160
- **6-9** Metric Capacity 162
- **6-10** Metric Mass 164
- **6-11** Reading Metric Measures 166
- **6-12** Temperature 168
- **6-13** Problem Solving: Using Formulas 170
- **6-14** Problem Solving: Group Decisions 172
- **6-15** Practice Through Problem Solving 173
- **6-16** Enrichment: Calculating Customary Measures 174
- **6-17** Calculator: Working with Measurement 175

Chapter 6 Review 176
Chapter 6 Test 177
Chapter 6 Cumulative Review 178

7 FRACTIONS

Chapter 7 Overview 180
- **7-1** Problem Solving: Learning to Use Strategies 182
- **7-2** Using Fractions 184
- **7-3** Equivalent Fractions 186
- **7-4** Writing Fractions in Lowest Terms 188
- **7-5** Improper Fractions and Mixed Numbers 190
- **7-6** Least Common Multiple 192
- **7-7** Least Common Denominator 193
- **7-8** Comparing Fractions and Mixed Numbers 194
- **7-9** Consumer Math: Estimating Gauge Readings 196
- **7-10** Estimating Fractions Using Compatible Numbers 197
- **7-11** Writing Fractions as Decimals 198
- **7-12** Writing Decimals as Fractions 200
- **7-13** Problem Solving: Using Data from a Table 202
- **7-14** Problem Solving: Developing Thinking Skills 204
- **7-15** Enrichment: Elapsed Time 206
- **7-16** Computer: Telecommunications 207

Chapter 7 Review 208
Chapter 7 Test 209
Chapter 7 Cumulative Review 210

8 MULTIPLICATION AND DIVISION OF FRACTIONS

Chapter 8 Overview 212
- **8-1** Problem Solving: Learning to Use Strategies 214
- **8-2** Multiplying Fractions 216
- **8-3** Multiplying with Mixed Numbers 218
- **8-4** Consumer Math: Installment Buying 220
- **8-5** Reciprocals 221
- **8-6** Dividing Fractions 222
- **8-7** Dividing with Mixed Numbers 224
- **8-8** Estimating Products and Quotients Using Compatible Numbers 226
- **8-9** Problem Solving: Developing a Plan 227
- **8-10** Problem Solving: Using Data from a Chart 228
- **8-11** Mental Math: Break Apart Numbers 229
- **8-12** Problem Solving: Developing Thinking Skills 230
- **8-13** Enrichment: Federal Income Tax Form 1040EZ 232
- **8-14** Calculator: Multiplying and Dividing Fractions 233

Chapter 8 Review 234
Chapter 8 Test 235
Chapter 8 Cumulative Review 236

9 ADDITION AND SUBTRACTION OF FRACTIONS

Chapter 9 Overview **238**
- **9-1** Problem Solving: Learning to Use Strategies **240**
- **9-2** Adding and Subtracting Fractions with Like Denominators **242**
- **9-3** Adding and Subtracting Mixed Numbers with Like Denominators **244**
- **9-4** Adding and Subtracting Fractions with Unlike Denominators **246**
- **9-5** Adding and Subtracting Mixed Numbers with Unlike Denominators **248**
- **9-6** Mental Math: Compatible Numbers **250**
- **9-7** Regrouping Mixed Numbers **251**
- **9-8** Subtracting Mixed Numbers with Regrouping **252**
- **9-9** Estimating Sums and Differences Using Rounding **254**
- **9-10** Consumer Math: Framing Pictures **256**
- **9-11** Problem Solving: Understanding the Situation **258**
- **9-12** Problem Solving: Group Decisions **259**
- **9-13** Enrichment: Time Zones **260**
- **9-14** Computer: Loops **261**

Chapter 9 Review **262**
Chapter 9 Test **263**
Chapter 9 Cumulative Review **264**

10 RATIO, PROPORTION, AND PERCENT

Chapter 10 Overview **266**
- **10-1** Problem Solving: Learning to Use Strategies **268**
- **10-2** Ratios **270**
- **10-3** Proportions **272**
- **10-4** Solving Proportions **274**
- **10-5** Rates **276**
- **10-6** Consumer Math: Better Buy **278**
- **10-7** Similar Figures and Proportions **280**
- **10-8** Scale Drawings **282**
- **10-9** Percents **284**
- **10-10** Percents and Decimals **286**
- **10-11** Percents and Fractions **288**
- **10-12** Comparing Percents, Decimals, and Fractions **290**
- **10-13** Problem Solving: Determining Reasonable Answers **292**
- **10-14** Problem Solving: Data Collection and Analysis **294**
- **10-15** Enrichment: Telephone Costs **296**
- **10-16** Calculator: Scale Drawings **297**

Chapter 10 Review **298**
Chapter 10 Test **299**
Chapter 10 Cumulative Review **300**

11 USING PERCENT

Chapter 11 Overview **302**
- **11-1** Problem Solving: Learning to Use Strategies **304**
- **11-2** Finding a Percent of a Number **306**
- **11-3** Finding What Percent One Number Is of Another **308**
- **11-4** Finding a Number Given a Percent **310**
- **11-5** Mental Math: Percents **312**
  Mixed Practice **313**
- **11-6** Estimating Percents Using Compatible Numbers **314**
- **11-7** Consumer Math: Simple Interest **316**
- **11-8** Percent of Increase or Decrease **318**
- **11-9** Problem Solving: Using Data from a Table **320**
- **11-10** Problem Solving: Developing Thinking Skills **322**
- **11-11** Enrichment: Commission **324**
- **11-12** Computer: Percents **325**

Chapter 11 Review **326**
Chapter 11 Test **327**
Chapter 11 Cumulative Review **328**

12 GEOMETRY

Chapter 12 Overview **330**
- **12-1** Problem Solving: Learning to Use Strategies **332**
- **12-2** Points, Lines, and Segments **334**
- **12-3** Rays and Angles **336**
- **12-4** Drawing and Naming Angles **338**
- **12-5** Perpendicular and Parallel Lines **340**
- **12-6** Polygons **342**
- **12-7** Triangles **344**
- **12-8** Quadrilaterals **346**
- **12-9** Perimeter **348**
- **12-10** Circles and Circumferences **350**
- **12-11** Consumer Math: Budgets and Circle Graphs **352**
- **12-12** Problem Solving: Determining Reasonable Answers **354**
- **12-13** Problem Solving: Group Decisions **355**
- **12-14** Enrichment: Constructing Regular Polygons **356**
- **12-15** Calculator: Finding Missing Measures **357**

Chapter 12 Review **358**
Chapter 12 Test **359**
Chapter 12 Cumulative Review **360**

13 PROBABILITY AND STATISTICS

Chapter 13 Overview **362**
13-1 Problem Solving: Learning to Use Strategies **364**
13-2 Counting Problems **366**
13-3 Permutations **368**
13-4 Simple Probability **370**
13-5 Experimental Probability **372**
13-6 Compound Probability **374**
13-7 Odds **376**
13-8 Expectations **378**
13-9 Consumer Math: Using Samples **380**
13-10 Problem Solving: Working with Data **382**
13-11 Mean and Mode **384**
13-12 Range and Median **386**
13-13 Problem Solving: Data Collection and Analysis **388**
13-14 Enrichment: Stem-and-Leaf Diagrams **390**
13-15 Computer: Arrays **391**

Chapter 13 Review **392**
Chapter 13 Test **393**
Chapter 13 Cumulative Review **394**

14 AREA AND VOLUME

Chapter 14 Overview **396**
14-1 Problem Solving: Learning to Use Strategies **398**
14-2 Area of Squares and Rectangles **400**
14-3 Area of Parallelograms **402**
14-4 Area of Triangles **404**
14-5 Area of Circles **406**
14-6 Consumer Math: Buying Carpet **408**
14-7 Space Figures **410**
14-8 Surface Area of Rectangular Prisms **412**
14-9 Volume of Rectangular Prisms **414**
14-10 Volume of Cylinders **416**
14-11 Volume of Pyramids and Cones **418**
14-12 Problem Solving: Using Logic **420**
14-13 Problem Solving: Developing Thinking Skills **422**
14-14 Enrichment: Spatial Visualization **424**
14-15 Calculator: Irregular Areas **425**

Chapter 14 Review **426**
Chapter 14 Test **427**
Chapter 14 Cumulative Review **428**

15 INTEGERS

Chapter 15 Overview **430**
15-1 Problem Solving: Learning to Use Strategies **432**
15-2 Positive and Negative Numbers **434**
15-3 Adding Integers **436**
15-4 Subtracting Integers **438**
15-5 Consumer Math: Checking a Budget **440**
15-6 Estimating Sums **441**
15-7 Multiplying Integers **442**
15-8 Dividing Integers **444**
15-9 Exploring Operations with Integers **446**
15-10 Problem Solving: Missing Data **448**
15-11 Problem Solving: Group Decisions **449**
15-12 Enrichment: Wind Chill Factor **450**
15-13 Calculator: Integers **451**

Chapter 15 Review **452**
Chapter 15 Test **453**
Chapter 15 Cumulative Review **454**

16 ALGEBRA

Chapter 16 Overview **456**
16-1 Problem Solving: Learning to Use Strategies **458**
16-2 Writing Algebraic Expressions Using Variables **460**
16-3 Solving Equations with Addition **462**
16-4 Solving Equations with Subtraction **464**
16-5 Solving Equations with Multiplication or Division **466**
16-6 Solving Multiple-Step Equations **468**
16-7 Consumer Math: Solving Problems Using Formulas **470**
16-8 Squares and Square Roots **472**
16-9 Pythagorean Theorem **474**
16-10 Graphing Ordered Pairs **476**
16-11 Problem Solving: Determining Reasonable Answers **478**
16-12 Problem Solving: Data Collection and Analysis **480**
16-13 Enrichment: Reading a Grid Map **482**
16-14 Computer: Squares and Square Roots **483**

Chapter 16 Review **484**
Chapter 16 Test **485**
Chapter 16 Cumulative Review **486**

DATA BANK **488**
EXTRA MIXED SKILLS REVIEW **497**
GLOSSARY **513**
TRY THIS ANSWERS **519**
INDEX **527**

Addison-Wesley
Essentials of Mathematics

Chapter 1 Overview

Key Ideas

- Use whole number and decimal place value.
- Round whole numbers and decimals.
- Choose the operation needed to solve a problem.
- Estimate sums, differences, products, and quotients.
- Decide if a problem needs an exact answer or an estimate.
- Review the problem-solving strategies and the problem-solving checklist.

Key Terms

- standard numeral
- is less than (<)
- difference
- compatible numbers
- place value
- round
- quotient
- problem solving
- notation
- sum
- estimate
- strategies
- is greater than (>)
- product
- operations

Key Skills

Add.

1. $6 + 5$
2. $5 + 7$
3. $1 + 2$
4. $7 + 9$
5. $5 + 4$
6. $4 + 2$
7. $7 + 6$
8. $7 + 4$
9. $4 + 6$
10. $3 + 7$
11. $4 + 8$
12. $7 + 2$
13. $7 + 7$
14. $8 + 9$
15. $7 + 9$
16. $8 + 5$
17. $3 + 9$
18. $6 + 8$

Subtract.

19. $11 - 2$
20. $8 - 6$
21. $6 - 6$
22. $10 - 2$
23. $12 - 8$
24. $11 - 8$
25. $7 - 4$
26. $11 - 7$
27. $12 - 4$
28. $12 - 3$
29. $9 - 4$
30. $17 - 8$
31. $7 - 7$
32. $11 - 9$
33. $13 - 6$
34. $16 - 7$
35. $17 - 9$
36. $13 - 5$

Multiply.

37. 4×1
38. 3×2
39. 3×4
40. 5×8
41. 6×0
42. 7×6
43. 2×9
44. 4×7
45. 6×8
46. 5×3
47. 5×5
48. 5×4
49. 6×9
50. 7×6
51. 8×6
52. 9×4
53. 5×8
54. 9×9

Divide.

55. $12 \div 4$
56. $30 \div 5$
57. $18 \div 9$
58. $28 \div 7$
59. $56 \div 8$
60. $64 \div 8$
61. $36 \div 4$
62. $45 \div 9$
63. $72 \div 9$
64. $63 \div 9$
65. $42 \div 6$
66. $49 \div 7$
67. $72 \div 8$
68. $81 \div 9$
69. $35 \div 7$
70. $24 \div 4$
71. $28 \div 4$
72. $56 \div 7$

EXPLORING THE FUNDAMENTALS 1

1-1 Reading and Writing Standard Numerals

Objective
Read and write whole numbers and decimals.

House numbers, page numbers, numbers in newspaper headlines—most of the numbers you see every day are called **standard numerals.** You can read and write standard numerals if you know the **place-value** system. When you write a number with a $ or a ¢, you are using **money notation.**

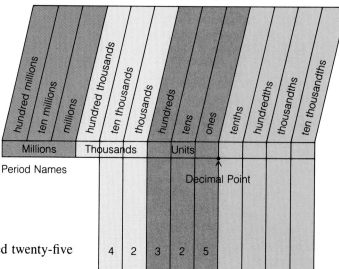

Examples Write the word name.

1. 42,325

forty-two thousand, three hundred twenty-five

2. 0.603

six hundred three thousandths

3. $460.89

four hundred sixty dollars and eighty-nine cents

Try This Write the word name.
    a. 87,076   **b.** 24.34   **c.** 0.54   **d.** $565   **e.** $8.68

Examples Write as a standard numeral or in money notation.
    4. three hundred twenty-five thousand, seven hundred six   325,706
    5. seven and seventy-four thousandths   7.074
    6. seventy-five dollars and thirty-four cents $75.34

Try This Write as a standard numeral or in money notation.
    f. four hundred seventy thousand, two hundred forty-three
    g. five and two hundred eight thousandths   **h.** one dollar and forty-nine cents

Exercises

Write the word name.
1. 540
2. 186
3. 2345
4. 1276
5. 4.75
6. 1.03
7. 0.46
8. 65,040
9. 24,253
10. 0.064
11. 0.0075
12. 425,000
13. 0.5675
14. 125,500
15. $0.42
16. $0.06
17. $4.35
18. $13.59
19. $530
20. $3208.00

Write as a standard numeral or in money notation.
21. six hundred fifty-seven
22. eight hundred four
23. nine thousand, eighty-three
24. two thousand, four hundred seventy
25. seven and thirty-four hundredths
26. three hundred sixty-three thousandths
27. ninety-five hundredths
28. sixty-one thousandths
29. seventeen dollars and forty cents
30. two hundred dollars and ten cents
31. eighty-four cents
32. twenty-nine dollars and nine cents
33. 3 in the dimes place, 8 in the dollars place, 1 in the pennies place
34. 1 in the tens place, 5 in the tenths place, 6 in the hundredths place, 2 in the ones place
35. 9 in the hundreds place, 4 in the thousands place, 7 in the tens place, 6 in the ten thousands place, 2 in the millions place
36. 4 in the dimes place, 3 in the hundred dollars place, 7 in the dollars place

Calculator Write commas between the periods in the calculator display. Use the chart on p. 4 to help you.

37. 56900000
38. 97432.769
39. 248600.52
40. 17345
41. 300245.5
42. 62987351

Problem Solving Write in money notation.
43. 2 ten-dollar bills, 14 one-dollar bills, 8 dimes, 12 pennies
44. 2 twenty-dollar bills, 1 one-dollar bill, 6 quarters, 5 pennies
45. 1 ten-dollar bill, 23 one-dollar bills, 5 nickels, 2 pennies
46. 6 five-dollar bills, 7 one-dollar bills, 5 quarters, 2 dimes, 1 nickel
47. One way to make $1111 is shown here. Find three other ways to make $1111 using $1, $10, $100, and $1000 bills. You do not have to use each type of bill.

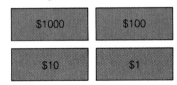

EXPLORING THE FUNDAMENTALS

1-2 Comparing and Ordering Whole Numbers and Decimals

Objective
Compare and order whole numbers and decimals.

Track and field records for many running events are timed to the nearest hundredth of a second. Judges for these events compare and order the runners' times.

Examples Compare. Write < (is less than), > (is greater than), or = (is equal to).

1. 12.17 ☐ 12.42

| Compare the place values from left to right. Find the first place where the digits are different. | → | The numbers compare the same way these digits compare. |

12.17
12.42

1 < 4   so   12.17 < 12.42

Compare. Write <, >, or =.

2. 0.045 ☐ 0.0438   0.045 > 0.0438   (5 > 3 in the thousandths place.)

3. 2.06 ☐ 2.6   2.06 < 2.6   (0 < 6 in the tenths place.)

4. 1354 ☐ 1345   1354 > 1345   (5 > 4 in the tens place.)

5. 24.8 ☐ 24.80   24.8 = 24.80   (0 in the hundredths place does not change the value.)

Try This Compare. Write <, >, or =.

a. 538 ☐ 583   **b.** 0.056 ☐ 0.0578   **c.** 1.006 ☐ 1.06   **d.** 1076 ☐ 1070

Example Arrange the numbers in order from least to greatest.

6. 8.03, 7.8, 8, 8.35, 7.25, 8.3

To arrange numbers in order, compare them two at a time.

7.25, 7.8, 8, 8.03, 8.3, 8.35

Try This Arrange the numbers in order from least to greatest.

e. 12.3, 12.42, 12.03, 12, 12.045   **f.** 0.032, 0.003, 0.02, 0.23, 0.03
g. 0.8, 1.02, 0.99, 0.9, 9.0, 1   **h.** 1834, 1842, 1839, 1983, 1883

Exercises

Compare. Write <, >, or =.

1. 3.08 ☐ 3.80
2. 0.09 ☐ 0.009
3. 3240 ☐ 3420
4. 2.605 ☐ 2.6050
5. 34,567 ☐ 34,389
6. 1450.6 ☐ 14,506
7. 12.50 ☐ 1.250
8. 1.045 ☐ 1.450
9. 74,560 ☐ 74,559
10. 0.56 ☐ 0.560
11. 2.78 ☐ 2.708
12. 0.099 ☐ 0.01

Arrange the numbers in order from least to greatest.

13. 456, 546, 465, 446, 565
14. 12,308; 12,383; 12,843; 12,830
15. 2.8, 2.43, 2.2, 2.34, 2.04
16. 1.09, 1.009, 1.010, 1.091, 1.190
17. 24.09, 24, 24.9, 24.89, 24.98
18. 32, 33.08, 31.2, 31.09, 31.990
19. 76, 84, 86.23, 79.9, 75, 0.90
20. 0.01, 0.011, 0.001, 0.101, 0.111

Practice Through Problem Solving

Place the decimal point in the middle numeral so that the numerals are in order from least to greatest. Write zeros if needed.

21. 2, 234, 3
22. 0.1, 473, 1
23. 34, 4503, 50
24. 0.03, 768, 0.3
25. 0.04, 543, 0.3
26. 0.001, 18, 0.002

Place one digit in each space so that the numerals are in order from least to greatest. Use two different digits in each problem.

27. 0.3, 0._ _, 0.4
28. 7, 7._ _6, 7.1
29. 0.99, _._, 1.1
30. 0.40, 0.4_ _, 0.41
31. 6.01, 6._ _, 6.1
32. 0.01, 0._1_, 0.02

Calculator

33. What number could you add to 2357 so that only the 3 changes to a 4? Use a calculator to check.

34. What number could you add to 12.57 so that only the 7 changes to an 8? Use a calculator to check.

Working with Variables

The list shows all of the whole numbers that you can write in the box so that these numerals are in order from least to greatest.

24 . . . ☐ . . . 28
24 . . . 25 . . . 28
24 . . . 26 . . . 28
24 . . . 27 . . . 28

In mathematics, instead of using boxes to stand for numbers, you will use letters. You could write: 24 . . . a . . . 28. Letters used this way to represent numbers are called **variables.**

List all of the whole numbers that you can write for each variable so that these numerals are in order from least to greatest.

35. 8 . . . d . . . 15
36. 97 . . . x . . . 104
37. 1203 . . . y . . . 1210

1-3 Rounding

30-Second Spots Reach Almost $400,000
New York—Today network marketing departments released these figures on the costs of commercials. The graph gives some sample costs for well-known shows.

Objective
Round numbers to a given place.

Sometimes you can **round** numbers instead of using exact numbers.

Understand the Situation

- What does each mark on the top scale of the graph represent?
- Which show's commercials cost the most?
- Which show's commercials cost the least?

You can round the cost of the *Cosby Show* commercials shown in the bar graph.

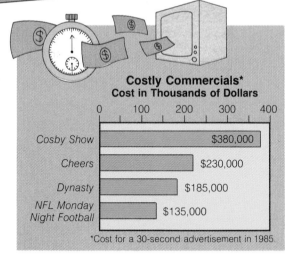

Costly Commercials*
Cost in Thousands of Dollars

Cosby Show	$380,000
Cheers	$230,000
Dynasty	$185,000
NFL Monday Night Football	$135,000

*Cost for a 30-second advertisement in 1985.

Examples

1. Round $380,000 to the nearest hundred thousand dollars.

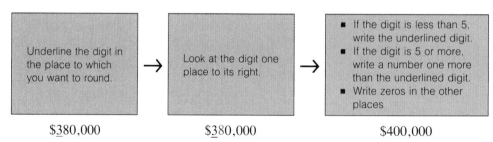

$380,000 $3̲80,000 $400,000

$380,000 rounded to the nearest hundred thousand dollars is $400,000.

2. Round 4538 to the nearest hundred.   4538 rounds to 4500.
3. Round 6509 to the highest place.   6509 rounds to 7000.
4. Round 2.046 to the nearest hundredth. 2.046 rounds to 2.05.
5. Round $0.28 to the nearest dollar. $0.28 rounds to $0.
6. Round $0.98 to the nearest dime. $0.98 rounds to $1.00.

Try This   Round to the given place.

a. 46   b. 1.53   c. 3438   d. 3.546   e. $7.45   f. $1.89
   tens   tenths   hundreds   hundredths   dollar   dime

Exercises

Round to the nearest ten.
1. 87
2. 34
3. 25
4. 3
5. 624
6. 478

Round to the nearest hundred.
7. 463
8. 232
9. 999
10. 4555
11. 23,096

Round to the highest place.
12. 472
13. 2833
14. 6604
15. 58,218
16. 114,909

Round to the nearest hundredth.
17. 0.86
18. 8.055
19. 0.0025
20. 99.986
21. 4.666

Round to the nearest tenth.
22. 0.56
23. 3.41
24. 6.97
25. 0.081
26. 0.6355

Round to the nearest dollar.
27. $12.35
28. $8.50
29. $119.82
30. $19.95
31. $42.49

Round to the nearest dime.
32. $4.56
33. $0.75
34. $15.54
35. $1.80
36. $62.39

You are an editor. Write a headline for each entry. Use rounded numbers.

Advertising Costs for September
37. TV: $456,789
38. Radio: $35,904
39. Newspaper: $2808
40. Magazine: $4599

Number Sense

41. The least whole number that rounds to 40 is 35. What is the greatest whole number that rounds to 40?

42. The greatest whole number that rounds to 300 is 349. What is the least whole number that rounds to 300?

Mixed Skills Review

Find the answer.
43. 14 − 5
44. 7 + 7
45. 8 × 8
46. 54 ÷ 9
47. 8 + 6
48. 56 ÷ 8
49. 5 × 8
50. 8 + 5
51. 16 − 8
52. 7 × 7

Write the word name.
53. $67.29
54. 3,606,000
55. 579,014
56. 21.98
57. 42,309
58. Write in money notation: 4 five-dollar bills, 9 one-dollar bills, 12 dimes

1-4 Understanding the Operations: Addition and Subtraction

Objective
Choose the correct addition or subtraction action.

You can solve many problems by choosing the correct operation. The action in a problem, not the numbers, tells you which operation you need. You can cover the numbers and still tell if you need to add, subtract, multiply, or divide.

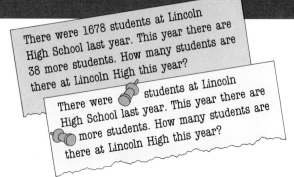

Examples Describe the action. Should you use addition or subtraction to solve?

1. At the first stop, ☐ students got on the bus. At the second stop, △ more students got on the bus. Now how many students are on the bus?

    Put together groups.
    Use **addition**.

2. There were ○ students on Bus 35. ☐ of these students got off the bus. How many students were still on the bus?

    Take away part of the total group.
    Use **subtraction**.

3. There are △ students on Bus 35. There are ☐ students on Bus 12. How many fewer students are on Bus 35?

    Compare the number in one group to the number in the other group.
    Use **subtraction**.

    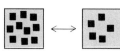

4. A large bus can hold a total of ○ students. There are ☐ on the bus now. How many more students can the bus hold?

    The total and one part are known.
    Find the **missing part**.
    Use **subtraction**.

Try This Describe the action. Should you use addition or subtraction to solve?

a. The drama club has a total of △ members. The car club has only ☐ members. How many more members are in the drama club?

b. The lunch room can hold ○ students. There are ☐ students in the lunch room now. How many more students can the lunch room hold?

Exercises

Describe the action. Should you use addition or subtraction to solve the problem?

> **Addition Action**
> - Put together.
>
> **Subtraction Actions**
> - Take away.
> - Compare.
> - Find the missing part.

1. There are ☐ students that ride a motorcycle to school. There are △ students that drive a car to school. How many more students drive cars to school than ride motorcycles?

2. A school jacket costs ☐. Tanya has saved ○. How much more does she need to buy the jacket?

3. After high school, △ students went to college and ○ went into the armed services. What was the total number of students who went to college or the armed services?

4. Ramon gave a cafeteria clerk ☐ for his lunch. He got △ in change. How much did he pay for lunch?

5. A parking permit for a car at school costs ○. A permit for a motorcycle costs ☐. How much more does a car permit cost than a motorcycle permit?

6. There are ☐ spaces for motorcycles in the school parking lot. The school sold spaces for △ motorcycles. How many more parking spaces can the school sell?

7. Hot lunch at Fallbrook High School costs ○. The cold lunch costs ☐. How much less does the cold lunch cost than the hot lunch?

8. A drink at the school store costs ☐. Popcorn costs △ and a bag of peanuts costs ○. What does it cost in all to buy one of each?

Write Your Own Problem

Use the table for Exercises 9–11. Write problems that you could solve using the given statements.

9. $23 + $15
10. $45 − $35
11. $15 + $30 + $35
12. Write an addition problem with $65 as the answer.
13. Write a subtraction problem with $8 as the answer.

> **School Costs**
> - Parking permit—car $23
> - Parking permit—motorcycle $15
> - Monthly lunch ticket—$30
> - Gym clothes for 1 year—$45
> - Books—$35

Find the answer.

14. 7×9
15. $15 - 7$
16. $5 + 5$
17. $36 \div 4$
18. $12 - 9$
19. $42 \div 6$
20. 8×6
21. $14 - 6$
22. $9 + 8$
23. $35 \div 5$

EXPLORING THE FUNDAMENTALS

1-5 Estimating Sums and Differences Using Rounding

Objective
Estimate sums and differences using rounding.

Kahla and Art used their calculators to find the total number of lightning strikes on trees and telephone lines. Whose answer is most reasonable?

Kahla's Answer

Art's Answer

Lightning Strikes

Location	Number of Strikes
Trees	262
Water	185
Telephone lines	118
Houses	95
Others	61

Understand the Situation
- How many trees were struck by lightning?
- How many telephone lines were struck by lightning?
- Which operation do you need to find the total? Tell why.

You can use estimation to check the reasonableness of answers. An **estimate** gives an approximation of the exact answer.

Examples Estimate. Round to the highest place.

1. 262 + 118

   262   rounds to 300
   +118 rounds to 100   262 + 118 is about 400.
                   400

   Art's answer is closer to the estimate. His answer is most reasonable.

2. 975 − 588

   1000 − 600 = 400   975 − 588 is about 400.

3. $8.28 + $4.95   Round to the nearest dollar.
   $8 + $5 = $13 $8.28 + $4.95 is about $13.

Try This Estimate. Round to the highest place or to the nearest dollar.

a. 67 + 34   b. 45 − 23   c. 55 + 12   d. 234 + 448
e. 645 − 334   f. 951 − 544   g. 7681 − 4579   h. 2560 + 9357
i. $7.24 − $4.99   j. $8.25 + $3.89   k. $13.50 − $6.99   l. $8.61 + $7.70

Exercises

Estimate. Round to the highest place.

1. 34 + 62
2. 28 + 39
3. 66 − 26
4. 81 − 43
5. 75 − 42
6. 54 + 58
7. 21 + 19
8. 35 − 9
9. 97 + 84
10. 89 − 24
11. 235 + 467
12. 267 − 105
13. 496 − 250
14. 989 + 264
15. 602 + 449
16. 734 + 255
17. 827 + 457
18. 624 − 470
19. 820 − 366
20. 919 − 729
21. 5680 + 2542
22. 1267 + 7340
23. 5500 − 2344
24. 8230 − 6780
25. 3660 + 8904
26. 2789 − 1889
27. 7643 − 1290
28. 5380 + 3850

Estimate. Round to the nearest dollar.

29. $8.63 − $2.75
30. $12.66 − $4.56
31. $7.25 + $4.45
32. $3.35 + $9.10
33. $14.25 − $8.50
34. $11.10 − $7.05
35. $6.99 + $9.75
36. $17.50 − $8.89
37. $18.13 − $8.59

Estimate. Use rounding.

38. 345 + 223 + 145 + 298
39. 1236 + 5780 + 1678 + 990
40. 3562 + 2109 + 3099 + 1508
41. 529 + 972 + 183 + 116

Number Sense

Find the missing number.

42. 46 + □ is about 80.
43. 67 − □ is about 30.
44. 34 + □ is about 50.
45. 88 − □ is about 40.

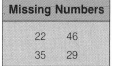

Missing Numbers

| 22 | 46 |
| 35 | 29 |

Data Bank

46. Estimate the difference between the record snowfall for one season and the record snowfall for 24 hours. Use the data on page 488.

Mixed Applications

Solve. Use data from the graph.

47. Which two months have a difference of about 70?
48. Which two months have a sum of about 140?
49. Which three months had a total of about 160 storms?

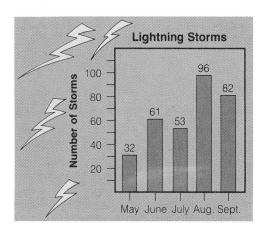

EXPLORING THE FUNDAMENTALS

1-6 Understanding the Operations: Multiplication and Division

Objective
Choose the correct multiplication or division action.

Some actions suggest that you should multiply to solve a problem. Others suggest that you should divide.

Examples Describe the action. Should you use multiplication or division to solve the problem?

1. Jana worked ☐ weeks last year at The Hamburger Hut. Each week she earned ○. How much did she earn in all?

   **Put together same-sized groups.
   Use multiplication.**

   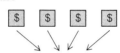

2. There is a total of △ pickle slices to a jar. Each hamburger order gets ☐ slices. How many hamburger orders will one jar serve?

   **Separate a total amount into groups of a given size.
   Use division.**

   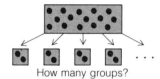
   How many groups?

3. Jana worked a total of ○ hours last month. She worked △ days. She worked the same number of hours each day. How many hours did she work each day?

   **Separate a total amount into a given number of groups of equal size.
   Use division.**

   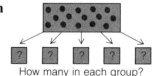
   How many in each group?

Try This Describe the action. Should you use multiplication or division to solve the problem?

a. The store manager gave a teacher a total of ☐ hamburger boxes for a class project. ○ students shared the boxes equally. How many boxes did each student get?

b. Carl works △ hours a week washing dishes. Each summer he works ☐ weeks. How many hours does he work in the summer?

Exercises

Describe the action. Should you use multiplication or division to solve the problem?

> **Multiplication Action**
> - Put together same-sized groups.
>
> **Division Actions**
> - Separate into groups of a given size.
> - Separate into a given number of groups of equal size.

1. From Monday through Friday, The Hamburger Hut is open a total of △ hours. It is open the same number of hours each day. How many hours is it open each day?

2. The Hamburger Hut is open ○ hours each day. The owner estimates that they serve about △ customers an hour. About how many customers do they serve each day?

3. One catsup container holds □ ounces. How many containers can Roberta fill from a △-ounce jar?

4. There are ○ rolls in a bag. A box contains △ bags. How many rolls are in a box?

5. A cook can make □ hamburgers from each pound of ground beef. A box of ground beef contains △ pounds. How many hamburgers can the cook make from a box of ground beef?

6. Charles needs to pour a △-quart jar of salad dressing into ○ bottles with the same amount in each. How much salad dressing should Charles pour into each bottle?

7. An average of ○ customers come to the drive-in window each hour. About how many customers would come to the window in □ hours?

8. The Hamburger Hut uses about □ paper napkins a day. How many days would a carton of △ paper napkins last?

9. Masako earns □ a weekend at the restaurant. Last month she worked △ weekends. How much money did she earn last month working on weekends?

10. Jonathan earned △ for ○ days of work. About how much is this each day?

Write Your Own Problem

Use the table for Exercises 11–13. Write problems that you could solve using the given statements.

11. $15 × 3
12. $45 ÷ 12
13. $11 × 12

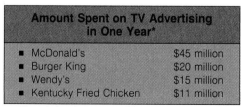

Amount Spent on TV Advertising in One Year*	
■ McDonald's	$45 million
■ Burger King	$20 million
■ Wendy's	$15 million
■ Kentucky Fried Chicken	$11 million

*Rounded to the nearest million.

14. Write a multiplication problem with $36 as the answer.
15. Write a division problem with $7 as the answer.

Suppose

16. Suppose in Example 1 on p. 14, □ = 3 and ○ = $50. Solve the problem.
17. Suppose in Example 3 on p. 14, ○ = 40 and △ = 10. Solve the problem.

EXPLORING THE FUNDAMENTALS **15**

1-7 Estimating Products Using Rounding

Objective
Estimate products using rounding.

Suppose that every high school performance of *You Can't Take It With You* was sold out. If there were 185 seats for each performance, about how many people saw this play?

Popular High School Plays	
Play	Number of Performances in One Year
You Can't Take It With You	37
Bye Bye Birdie	32
The Miracle Worker	32
Our Town	28

Understand the Situation

- How many times did high schools perform *You Can't Take It With You* in one year?
- How many people attended each performance?
- The same number of people attended each performance of *You Can't Take It With You*. Would it be easier to add or to multiply to find the total? Tell why.

Examples   Estimate. Use rounding.

1. 185 × 37

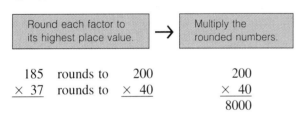

Shortcut: Multiply the leading digits. Write the total number of zeros to the right of this product.

```
 185   rounds to    200         200           200      2 zeros
× 37   rounds to   × 40        × 40          × 40     +1 zero
                               8000          8000      3 zeros
```

185 × 37 is about 8000. About 8000 people saw *You Can't Take It With You*.

2. Estimate 382 × 8. 382 × 8 is about 400 × 8, or 3200.

3. Estimate 2375 × 48. 2375 × 48 is about 2000 × 50, or 100,000.

4. Estimate 974 × 226. 974 × 226 is about 1000 × 200, or 200,000.

5. Estimate 43 × 19 × 28. 43 × 19 × 28 is about 40 × 20 × 30, or 24,000.

Try This   Estimate. Use rounding.

a. 521 × 4   **b.** 68 × 32   **c.** 134 × 51   **d.** 275 × 586
e. 3462 × 63   **f.** 1820 × 642   **g.** 28 × 12 × 46   **h.** 18 × 33 × 65

Exercises

Estimate. Use rounding.

1. 24 × 37
2. 35 × 18
3. 51 × 32
4. 67 × 81
5. 134 × 52
6. 243 × 61
7. 435 × 28
8. 246 × 67
9. 775 × 80
10. 381 × 63
11. 672 × 66
12. 564 × 27
13. 235 × 626
14. 714 × 422
15. 266 × 287
16. 780 × 257
17. 345 × 27
18. 67 × 258
19. 980 × 786
20. 1340 × 70
21. 1245 × 54
22. 2476 × 18
23. 7544 × 8124
24. 30 × 63
25. 4350 × 320
26. 3288 × 740
27. 1987 × 364
28. 6650 × 87
29. 4532 × 2450
30. 7843 × 4552
31. 24 × 38 × 51
32. 16 × 9 × 75
33. 124 × 7 × 33
34. 85 × 99 × 20
35. 405 × 760 × 94
36. 230 × 120 × 350

Use the table on p. 16 for Exercises 37–38.

37. About 115 people saw each performance of *Bye Bye Birdie*. About how many people saw this play?
38. About 215 people saw each performance of *Our Town*. About how many people saw this play?

Write Math

39. Write a rule for estimating products with three factors.

Number Sense

40. A student estimates 4254 × 32 to be about 1200. What did this student do wrong?
41. A student estimates 552 × 17 to be about 6000. What did this student do wrong?

Mixed Applications

Solve. Use the table on Broadway plays.

42. Which play had the greatest number of performances?
43. Which play had the least number of performances?
44. A ticket to *Annie* costs $28. About 220 people saw the Saturday evening performance. About how much money did *Annie* make in ticket sales that night?
45. If about 535 people attended each performance of *South Pacific* during its run, about how many people saw the play?

Broadway Plays	
Plays	**Number of Performances**
Fiddler on the Roof	3242
Annie	2377
South Pacific	1925
Grease	3388
The Wiz	1672

EXPLORING THE FUNDAMENTALS

1-8 Estimating Quotients Using Compatible Numbers

Objective
Estimate quotients using compatible numbers.

What was the average number of black mayors in each state in 1984?

Understand the Situation

- How many black mayors were there in 1984?
- How many states are in the United States?
- Which operation would you use to find the average number of black mayors for each state? Tell why.

Black Mayors in the US	
Year	Number of Black Mayors
1980	182
1982	223
1984	255
1986	289

You can estimate quotients using compatible numbers. **Compatible numbers** are pairs of numbers that make the problem easier for you to solve using mental math. They are close in value to the numbers in the problem. Change each number to make a pair of compatible numbers.

Examples Choose the best pair of compatible numbers for each estimate.

1. 255 ÷ 50   **A.** 260 ÷ 50   **B.** 260 ÷ 60   **C.** 250 ÷ 50

   Answer **c** is the best choice. Think of the basic fact: 25 ÷ 5.

2. 231 ÷ 32   **A.** 230 ÷ 30   **B.** 210 ÷ 30   **C.** 240 ÷ 30

   Answers **B** and **c** are both easy to solve using mental math. The numbers in answer **c** are closer to the real values. Answer **c** is the best choice.

Try This Choose the best pair of compatible numbers for each estimate.

a. 157 ÷ 8   **A.** 160 ÷ 8   **B.** 160 ÷ 10   **C.** 150 ÷ 8
b. 4328 ÷ 64   **A.** 4300 ÷ 60   **B.** 4200 ÷ 60   **C.** 4300 ÷ 65

Examples Estimate. Use compatible numbers.

3. 187 ÷ 6

   $6\overline{)187}$ → $6\overline{)180}$ (30)

   187 ÷ 6 is about 30.

4. 5235 ÷ 46

   $46\overline{)5325}$ → $50\overline{)5000}$ (100)

   5235 ÷ 46 is close to 100.

Try This Estimate. Use compatible numbers.

c. 272 ÷ 4   d. 318 ÷ 60   e. 6434 ÷ 79   f. 4702 ÷ 81

CHAPTER 1

Exercises

Choose the best pair of compatible numbers for each estimate.

1. 724 ÷ 9   A. 700 ÷ 9   B. 720 ÷ 9   C. 700 ÷ 10
2. 2826 ÷ 12   A. 2800 ÷ 12   B. 2400 ÷ 12   C. 2830 ÷ 10
3. 5248 ÷ 24   A. 5000 ÷ 25   B. 5000 ÷ 20   C. 5200 ÷ 20
4. 3754 ÷ 75   A. 3800 ÷ 80   B. 3500 ÷ 70   C. 3700 ÷ 70

Estimate. Use compatible numbers.

5. 146 ÷ 8
6. 231 ÷ 6
7. 618 ÷ 9
8. 469 ÷ 7
9. 447 ÷ 5
10. 317 ÷ 8
11. 271 ÷ 3
12. 281 ÷ 3
13. 324 ÷ 50
14. 288 ÷ 70
15. 372 ÷ 90
16. 135 ÷ 40
17. 254 ÷ 51
18. 335 ÷ 82
19. 276 ÷ 39
20. 633 ÷ 76
21. 192 ÷ 19
22. 185 ÷ 24
23. 417 ÷ 69
24. 804 ÷ 86
25. 3123 ÷ 64
26. 1338 ÷ 42
27. 2760 ÷ 67
28. 2218 ÷ 72

Use the table on p. 18 for Exercises 29–30.

29. Estimate the average number of black mayors for each state in 1982.
30. Estimate the average number of black mayors for each state in 1986.

Talk Math

31. Estimate 347 ÷ 60 and 347 ÷ 50 using compatible numbers. Did you choose the same number for 347 both times? Explain your answer.

Number Sense

Find the missing number.

32. 268 ÷ □ is about 30.
33. 429 ÷ □ is about 60.
34. 170 ÷ □ is about 20.
35. 520 ÷ □ is about 90.

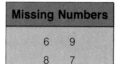

Missing Numbers: 6, 9, 8, 7

Mixed Skills Review

Estimate.

36. 89 × 31
37. $33 + $16
38. 98 − 27
39. $24 ÷ 5
40. $402 − $291
41. 225 + 649
42. 552 × 28
43. $56 − $49

Round to the given place.

44. 99 tens
45. 13.45 tenths
46. $67.02 dime
47. 4.3936 hundredths
48. $19.39 dollar
49. 20,364 hundreds
50. 0.12 tenths
51. 56,511 thousands

1-9 Deciding When to Estimate

Objective
Decide if you need an exact answer or an estimate.

When you solve some problems, you will need to decide if you will need an **estimate** or an **exact** answer.

Examples Tell if you need an exact answer or an estimate. Tell why.

1. The cashier needs to figure the 6% tax on your meal.

   It is against the law to charge an incorrect amount for tax. The cashier needs to know the exact amount of tax.

2. You want to check the amount of tax on your bill.

   You may want to figure the exact amount. However, most people choose to make a quick check for a large error. You would probably estimate the amount of tax.

3. A newspaper editor is writing a headline about the budget for a local school district.

   The exact amount of a school district's budget might be $1,345,758. Most newspaper headlines would give an estimate of the budget to make the headline short and clear. The headline might read $1.3 million.

4. You have saved $75. You want to know if you can buy two blouses and a sweater.

   All you want to know is if you have enough money. An estimate is all you need.

5. You want to know the amount you will have to pay each month for a car loan. You ask several banks what their rates are.

   You need to compare the rates of all the banks. You want an exact answer.

Try This Tell if you need an exact answer or an estimate. Tell why.

a. You want to know how much money you will need to take a friend to a movie and then out for a snack.

b. You want to know the hourly wage you will earn selling clothes at a department store.

Exercises

Tell if you need an exact answer or an estimate. Tell why.

1. You want to make a report to your class about the area of your state.
2. Tapes are 3 for $10. You want to know the cost of 1 tape.
3. You want to know how long it takes to get from your house to the airport.
4. You have to buy one glass of lemonade for each student in your class. The class budget is low.
5. You paid for a sandwich with a $5 bill. You want to know how much change you should get back.
6. Your school has 210 students to place in homerooms. At most, 25 students can be in one homeroom. You want to know how many homerooms the school needs.
7. You want to tell a friend the likely temperature in your home town in July.
8. You tell the gas station attendant to fill up the tank. You have to pay for it.
9. You want to know how many points you need on the next test to get an A in the class. You ask your teacher.
10. A sweater costs $45. It is on sale for $\frac{1}{2}$ off. You want to know about how much you will save.
11. You buy a new car. You want to know how many miles you get per gallon of gas.
12. As you look at the clock, you remember that you forgot to put more money in the parking meter. You want to know when the time on the meter ran out.

Group Activity

Take turns. Give an example of a situation where you would need an exact answer and an example where all you would need is an estimate. Use the topics shown.

13. a rock concert
14. an aerobics class
15. a school cafeteria

Data Hunt

16. Use an almanac to find the heights of the three tallest buildings in the U.S. Estimate how high they would be if they were stacked on top of each other.

Mixed Skills Review

Write each word name.

17. 8692
18. $12.69
19. 833
20. $64,039

Estimate.

21. 3752 − 3234
22. 2614 + 6014
23. 96 + 21
24. 54 − 39

Compare. Write <, >, or =.

25. 1.10 ☐ 11
26. 0.504 ☐ 0.54
27. 101.01 ☐ 10.101
28. 0.30 ☐ 0.3
29. 6.9274 ☐ 6.9724
30. $701.00 ☐ $701
31. 90,894 ☐ 90,884
32. $2345 ☐ $23.45

Objective
Name problem-solving strategies.

You can use these **problem-solving strategies** to understand, plan, and solve problems.

The checklist shows the *process* that you can use to help you solve problems.

Problem-Solving Strategies
- Guess and check.
- Choose an operation.
- Make an organized list.
- Act it out.
- Find a pattern.
- Write an equation.
- Use logical reasoning.
- Draw a picture or diagram.
- Make a table.
- Use objects.
- Work backward.
- Solve a simpler problem.

5-Point Checklist
1. Understand the *Situation*.
2. Find the needed *Data*.
3. *Plan* what to do.
4. Find the *Answer*.
5. *Check* back.

This box is used throughout the book to remind you that you can use this process to help you solve problems.

Study the solution to each problem. Answer the questions that follow.

Problem 1

Janine needs to seat 14 people for a meeting. She is going to make 1 long table by pushing together small tables. Each small table can seat 1 person on a side. What is the least number of tables that Janine needs?

Janine's solution:

#tables	#people
1	4
2	6
3	8
4	10

$2 \times 6 = 12$
$12 + 2 = 14$

(6 tables)

↑ 2 times #tables plus 2

1. What do the small squares and x's represent?
2. What do you see about the numbers in the column *# people*?
3. Name a strategy that Janine used to solve this problem.
4. Write the correct answer to this problem in a complete sentence.

Problem 2

There are 18 animals in a barnyard. Some are chickens and some are cows. All together there are 50 legs. How many of the animals are chickens and how many are cows?

Scotty's solution:

```
|| || || || || ||     12 c and 6 ch  too much      5 c and 13 ch  too low      7 c and 11 ch

|||| |||| ||||         12        6                  5        13                 7        11
                      × 4       × 2                ×4        × 2               ×4       × 2
|||| |||| ||||        ----      ----               ----      ----              ----     ----
                       48        12                 20        26                28       22
|||| |||| ||||

|||| |||| ||||         48 + 12 = 60                 20 + 26 = 46                28 + 22 = 50
                                                                                 (7 and 11)
```

5. What do the small vertical marks represent?
6. Why did Scotty write *too much* after *12c* and *6ch*?
7. Name a strategy that Scotty used to solve this problem.
8. Write the correct answer to this problem in a complete sentence.

Problem 3

In how many ways can the class pick 3 out of 5 students to be homeroom representatives?

Kelly's solution:

```
A B C              B C D   2          C D E   1
A B D   3          B C E   2
A B E                                          10
                   B D E   1
A C D   2
A C E
A D E   1
```

9. What do the letters represent?
10. Name a strategy that Kelly used to solve this problem.
11. Write the answer to this problem in a complete sentence.

Summary Questions

12. For which of the above problems might you use objects to solve the problem? What objects might you use?
13. Can you use more than one strategy to solve some problems?
14. Can you solve some problems in more than one way? Give another way of solving one of the above problems.

Objective
Write a check.

Mr. Olvera received a utility bill for $31.48 for the month of August. Mr. Olvera wrote a check on September 10 to City of Harris Utilities to pay for the bill.

Example

Write a check for the total amount of the utility bill.

Pay to the Order of is the person/place to whom you are paying the money.

Write today's date.

Write the amount in money notation. Write the first digit as close to the dollar sign as possible.

Write what the check was for.

Write the amount in words. Write *and* for the decimal point. Write the cents as a fraction over 100. Start at the far left. Fill the remaining space on the right with a line.

Sign your name.

Try This

Write a check for the amount. Use today's date. Copy the blank check on p. 496.

a. Shoe World $75.50   b. The Ticket Outlet $50.00

Exercises

Write a check for the amount. Use today's date. Copy the blank check on p. 496.

1. Sports City
   $66.67
2. *Teen Magazine*
   $19.92
3. The Clothes Horse
   $109.97

You Decide

Decide if you need an estimate or an exact amount for the situation.

4. You need to pay $23.84 for the telephone bill. You write a check for the bill.

1-12 Computer: Flowcharts

Objective
Make and read a flowchart.

Imagine a process or task such as washing dishes. Here are three steps in this process.

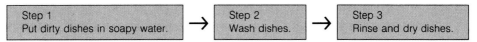

There is a **sequence of operations.** You can represent a sequence of operations in a **flowchart.** A computer programmer sometimes uses a flowchart to write a computer program.

You can make a flowchart of any process.

Example Make a flowchart for feeding a cat.

Each shape around an instruction means something different.

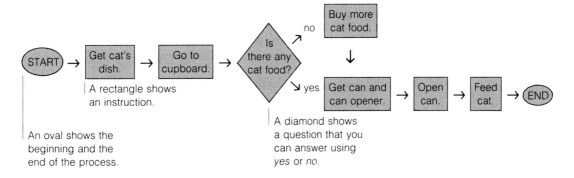

Exercises

Make a flowchart for the process.

1. making a sandwich
2. making your bed
3. brushing your teeth
4. rounding a 3-digit number to the nearest ten
5. comparing two 3-digit numbers

6. Rearrange the following to make a flowchart for washing a car.
   - Rinse car.
   - END
   - Are the windows rolled up?
   - START
   - Dry car.
   - Roll up windows.
   - Fill bucket with soap and water.
   - Rinse soap.

Data Hunt 7. Find a recipe. Write the recipe in the form of a flowchart.

CHAPTER 1 REVIEW

Choose the correct money notation for each.

1. two $5 bills, twelve $1 bills, 6 quarters, 4 nickels
2. three $10 bills, three $1 bills, 3 quarters, 3 pennies
3. one $20 bill, two $1 bills, 8 dimes, 2 nickels
4. five $5 bills, 10 quarters, 7 dimes, 5 nickels

$33.78 $29.95
$28.45 $23.70
$22.90 $31.55

Choose the number that makes the statement true.

5. 34.2 < ___   A. 34.1   B. 33.4   C. 34.3
6. 1.1 > ___   A. 1.2   B. 1.01   C. 1.11
7. ___ < 0.012   A. 0.02   B. 0.1   C. 0.011
8. 0.3 < ___ < 0.5   A. 0.04   B. 0.31   C. 0.51
9. 1.9 < ___ < 2   A. 1.99   B. 2.01   C. 1.9

Choose the one that does not belong in the group. Explain why not.

10. A. 213 rounds to 200.
    B. 5543 rounds to 5500.
    C. 3658 rounds to 4000.
    D. 10 rounds to 0.
11. A. Draw a picture.
    B. Use objects.
    C. Estimate.
    D. Guess and check.
12. A. 1258
    B. 57
    C. eleven and six tenths
    D. 10.98
13. A. put together
    B. find the missing part
    C. take away
    D. compare
14. A. 39, 42, 43, 50
    B. 125, 222, 300, 305
    C. 41, 29, 18, 7, 3
    D. 625, 733, 802, 1000
15. A. situation
    B. take away
    C. plan
    D. data

Match.

16. 1111
17. addition
18. take away
19. estimate
20. division
21. put together same-sized groups
22. compatible numbers
23. Draw a picture.
24. situation, data, plan, answer, check
25. >
26. hundredths

A. separate into groups of a given size
B. approximation
C. estimation strategy
D. subtraction
E. put together
F. standard numeral
G. multiplication
H. problem-solving strategy
I. place value
J. exact
K. is greater than
L. is less than
M. problem-solving process

CHAPTER 1 TEST

Write the word name. **1.** 37,392 **2.** 23.083 **3.** 0.64 **4.** $68.07

Write as a standard numeral or in money notation.

5. eighty thousand, six hundred fifty-seven
6. one hundred sixty dollars and thirty-four cents
7. two and one hundred seven thousandths

Compare. Use <, >, or =. **8.** 8450 □ 845.0 **9.** 0.0307 □ 0.0317 **10.** $2.82 □ $12.82

Arrange in order from least to greatest. **11.** 64.02, 6.402, 0.64, 64.002, 64.2

Round to the given place.

12. 3 ten
13. 243 highest place
14. 4601 hundred
15. 7.4929 hundredth
16. $1.50 dollar

Describe the action. Should you use addition or subtraction to solve?
17. There are □ ninth-graders taking math. ∆ are boys. How many are girls?

Estimate. Round to the highest place or to the nearest dollar.
18. 423 + 392 **19.** 3741 − 1435 **20.** $7.15 + $2.69 **21.** $11.49 − $5.88

Describe the action. Should you use multiplication or division to solve?
22. Al earns □ an hour. He worked ∆ hours. How much money did he earn?

Estimate.
23. 34 × 79 **24.** 22 × 389 **25.** 741 × 92 **26.** 572 × 823 **27.** 421 × 5
28. 238 ÷ 6 **29.** 352 ÷ 61 **30.** 943 ÷ 89 **31.** 4744 ÷ 81 **32.** 6925 ÷ 69

Decide if you need an exact answer or an estimate. Explain why.
33. You bought a pen. How much change should you get from a $5 bill?

34. Name a strategy that Sonia used to solve this problem.
Problem: There are 12 animals in a barn. Some are chickens and some are horses. All together there are 32 legs. How many of the animals are chickens?

4 chickens
6 horses
10 animals
too low

8 chickens
4 horses
12 animals

EXPLORING THE FUNDAMENTALS

CHAPTER 1 CUMULATIVE REVIEW

1. Arrange in order from least to greatest.
   6.04, 6, 5.7, 6.25, 5.35, 6.2
   A. 6.25, 6, 6.04, 6.2, 5.35, 5.7
   B. 5.35, 5.7, 6, 6.2, 6.04, 6.25
   C. 5.35, 5.7, 6, 6.04, 6.2, 6.25
   D. 5.35, 6.04, 6.2, 5.7, 6, 6.25

2. Decide which operation you would use to solve this problem.
   Problem: Liza worked □ hours on Tuesday. She worked △ hours on Friday. How many more hours did she work on Friday?
   A. addition
   B. subtraction
   C. multiplication
   D. division

3. Decide which operation you would use to solve this problem.
   Problem: Gene planted □ trees. He will plant △ more. How many trees will he plant all together?
   A. addition
   B. subtraction
   C. multiplication
   D. division

4. Arrange in order from least to greatest.
   7309, 7390, 7039, 7930, 7903, 7093
   A. 7930, 7903, 7390, 7309, 7093, 7039
   B. 7039, 7093, 7390, 7309, 7903, 7930
   C. 7039, 7093, 7309, 7390, 7930, 7903
   D. 7039, 7093, 7309, 7390, 7903, 7930

5. Name a strategy that Riva used to solve this problem.
   Problem: Oliver needs to seat 24 people for dinner. He is going to make one long table by pushing together smaller tables. Each smaller table can seat 2 people on a side. What is the least number of tables Oliver needs?
   A. Guess and check.
   B. Write an equation.
   C. Draw a picture.
   D. Work backward.

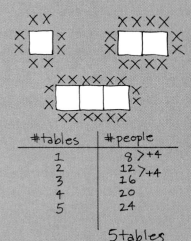

6. Round to the nearest tenth.
   0.039
   A. 0.04
   B. 0.039
   C. 0
   D. 0.1

7. Write as a standard numeral.
   eight hundred sixty-two thousand, one hundred fifty-two
   A. 86,152
   B. 0.862152
   C. 86.2152
   D. 862,152

8. Decide if you need an exact answer or an estimate.
   You bought 3 sweaters for $64.77. You need to write a check for the amount.
   A. exact answer
   B. estimate

9. Compare. Use <, >, or =.
   6.09 ☐ 6.9
   A. <
   B. >
   C. =

10. Write in money notation.
    six dollars and two cents
    A. $6.02
    B. $60.02
    C. $6.20
    D. $60.20

11. Estimate.
    418 ÷ 8
    A. 500
    B. 60
    C. 50
    D. 5

12. Write the word name.
    0.86
    A. eighty-six
    B. eighty-six tenths
    C. eighty-six hundredths
    D. eight hundredths

13. Estimate.
    372 × 220
    A. 8000
    B. 60,000
    C. 90,000
    D. 80,000

14. Round to the nearest dime.
    $4.65
    A. $5.00
    B. $4.50
    C. $4.70
    D. $4.00

15. Round to the nearest hundredth.
    16.837
    A. 16.8
    B. 16.84
    C. 16.838
    D. 17

16. Estimate.
    1209 ÷ 62
    A. 20
    B. 10
    C. 40
    D. 2

17. Estimate. Round to the highest place.
    4061 − 2865
    A. 2000
    B. 3000
    C. 1000
    D. 1900

18. Estimate. Round to the nearest dollar.
    $12.62 + $15.09
    A. $17.70
    B. $27
    C. $28
    D. $27.70

19. Estimate.
    3892 × 27
    A. 12,000
    B. 120,000
    C. 100,000
    D. 80,000

20. Estimate. Round to the nearest dollar.
    $9.62 − $3.50
    A. $6.10
    B. $7.00
    C. $6.00
    D. $6.50

21. Estimate.
    2801 ÷ 42
    A. 75
    B. 80
    C. 60
    D. 70

22. Decide which operation you would use to solve this problem.
    Problem: Tao works ☐ hours a week. Each summer he works for Δ weeks. How many hours does he work in the summer?
    A. addition
    B. subtraction
    C. multiplication
    D. division

23. Compare. Use <, >, or =.
    1094 ☐ 1090
    A. <
    B. >
    C. =

EXPLORING THE FUNDAMENTALS

Chapter 2 Overview

Key Ideas

- Use the problem-solving strategies *draw a picture or diagram* and *use objects*.
- Add and subtract whole numbers and decimals.
- Use front-end estimation.
- Fill out deposit slips for a checking account.
- Use the mental math strategy *counting*.
- Choose a calculation method.
- Evaluate the reasonableness of answers.

Key Terms

- regroup
- running balance
- reasonable answer
- front-end estimation
- mental math
- sum
- deposit slip
- calculation method
- addend

Key Skills

Add.

1. $4 + 5$
2. $3 + 7$
3. $6 + 3$
4. $8 + 2$
5. $7 + 5$
6. $8 + 6$
7. $4 + 9$
8. $3 + 8$
9. $5 + 5$
10. $9 + 5$
11. $6 + 6$
12. $8 + 8$
13. $7 + 8$
14. $9 + 6$
15. $9 + 8$
16. $7 + 6$
17. $1 + 9$
18. $4 + 7$

Subtract.

19. $14 - 7$
20. $18 - 9$
21. $11 - 4$
22. $13 - 6$
23. $8 - 6$
24. $9 - 3$
25. $12 - 5$
26. $12 - 8$
27. $10 - 5$
28. $13 - 7$
29. $10 - 3$
30. $9 - 4$
31. $7 - 4$
32. $14 - 8$
33. $15 - 7$
34. $17 - 8$
35. $16 - 7$
36. $12 - 9$

Tell the place value of the underlined digit.

37. 2<u>5</u>
38. 4<u>6</u>
39. 12<u>5</u>
40. <u>2</u>24
41. 5<u>6</u>7
42. 3<u>0</u>0
43. 2<u>0</u>8
44. 1<u>2</u>77
45. <u>3</u>466
46. 66<u>0</u>8
47. 0.<u>3</u>4
48. 0.0<u>4</u>
49. <u>1</u>.86
50. 0.03<u>5</u>
51. 0.<u>7</u>05
52. 1.04<u>6</u>
53. 2.<u>8</u>86
54. 0.87<u>5</u>

ADDITION AND SUBTRACTION OF WHOLE NUMBERS AND DECIMALS

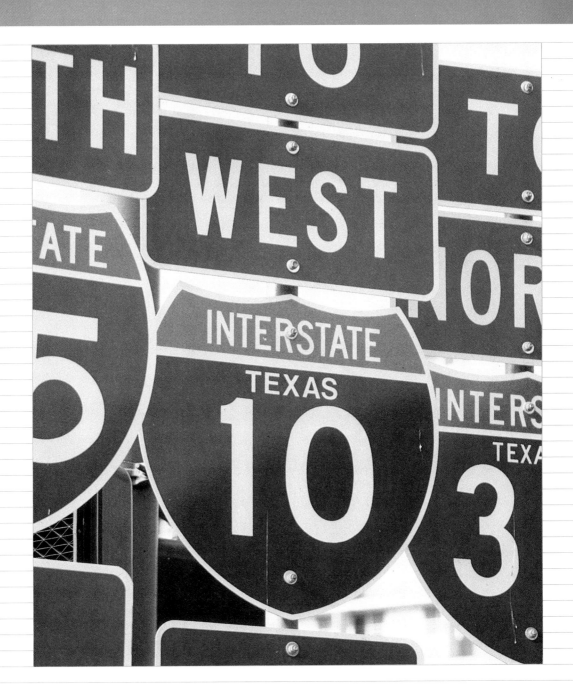

2-1 Problem Solving: Learning to Use Strategies

Objective
Use the strategies draw a picture or diagram *and* use objects.

- **SITUATION**
- **DATA**
- **PLAN**
- **ANSWER**
- **CHECK**

You can solve many problems in more than one way. In this example, Ben used the strategy **draw a picture or diagram.** Elena **used objects** to solve this problem.

Problem-Solving Strategies
- Guess and check.
- Choose an operation.
- Make an organized list.
- Act it out.
- Find a pattern.
- Write an equation.
- Use logical reasoning.
- **Draw a picture or diagram.**
- Make a table.
- **Use objects.**
- Work backward.
- Solve a simpler problem.

Example Lincoln High School held a car show to raise money. Students roped off the 200-foot by 250-foot parking lot. They placed posts at each corner and every 50 feet in between each corner. How many posts did they use?

Ben's solution:

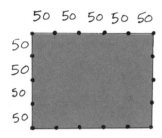

The students used 18 posts.

Elena's solution:

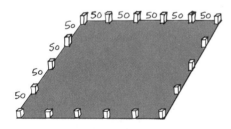

The students used 18 posts.

Try This Solve.

a. There are 5 towns on the same road. Two of the towns are at each end of the road. Each town is 8 miles apart. How long is the road?

b. Ridge, Grove, Knoll, and Marble are towns on the same road. Grove is 10 miles east of Ridge. Knoll is 5 miles west of Grove. Marble is 2 miles east of Ridge. How far apart are Knoll and Marble?

Exercises

Solve. Use one or more of the problem-solving strategies.

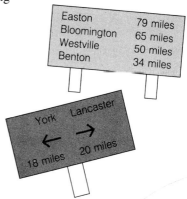

1. You live between Benton and Westville. Your house is 10 miles from Westville. How far is your house from Bloomington?

2. You live between York and Lancaster. You live 5 miles from York. How far do you live from Lancaster?

3. Bus B got to the park first. Bus A got there last. Bus C got there just after Bus D. Bus E got there just after Bus C. Which bus got there second?

4. A city flower garden is in the shape of a rectangle 30 meters long and 20 meters wide. A night light is placed at each corner and every 5 meters in between corners. How many lights are in the garden?

5. An elevator takes 10 seconds to go from the first floor to the third floor. How many seconds would it take this elevator to go from the first floor to the sixth floor?

6. How many blocks do you need to build the fifth staircase?

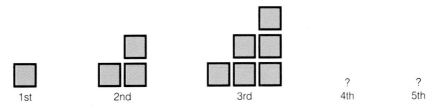

7. The length of a rectangular lot is 12 meters longer than the width. The length is 28 meters. What is the distance around, or perimeter of, the lot?

Write Your Own Problem

8. Write three problems that you could answer using data from this map.

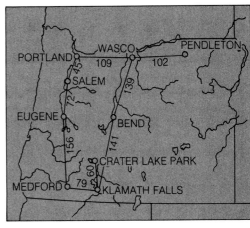

ADDITION AND SUBTRACTION OF WHOLE NUMBERS AND DECIMALS

2-2 Adding Whole Numbers and Decimals

Objective
Add whole numbers and decimals.

In LANDLORD, a game like MONOPOLY® Keiko owns Cherry Avenue with 1 house. She also owns Brook Place with 4 houses. Suppose you landed on these properties. What would you have to pay Keiko in all?

Understand the Situation

- What would you pay if you landed on Cherry Avenue with 1 house?
- What would you pay if you landed on Brook Place with 4 houses?
- Are you asked to compare the two amounts or to find the total?

To find the total amount, add $175 and $875. You can estimate the sum. $175 + $875 is about $200 + $900, or $1100. The sum is about $1100.

Examples Add.

1. $175 + $875

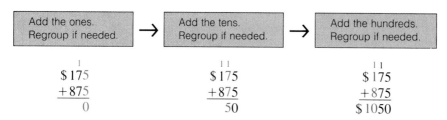

The answer seems reasonable. You would have to pay $1050.

```
       1           1 1           1 1
2.   38      3.  1336     4.  $24.59    5.   0.074    6.   0.190  ←
    +44         + 827        + 2.53        +0.388        +0.838
     82          2163         $27.12        0.462         1.028
                              Line up the place values.         Write a zero
                              Add as with whole numbers.        if needed.
```

Try This Add.

```
a.   38      b.   529     c.  2346    d.   48.26    e.   0.037    f.   1.38
    +26         +557        + 185        +14.83        +0.86        +2.472
```

Exercises

Add.

1. 780 + 36
2. 237 + 438
3. 529 + 625
4. 1247 + 826
5. 7255 + 1388
6. $18.76 + 0.36
7. 0.708 + 0.777
8. 3.5 + 2.838
9. 0.58 + 6.95 + 10.8
10. 16,079 + 45,322 + 84,809
11. 65 + 7
12. 57 + 24
13. 72 + 65
14. 66 + 34
15. $29 + $46
16. 34.83 + 2.47
17. 65.08 + 8.74
18. 0.54 + 0.876
19. $84.99 + $3.88
20. $135.75 + $10.88
21. $12.80 + $32.65
22. 124 + 453 + 5961
23. 3.24 + 6.01 + 2.5
24. 42.6 + 2.35 + 11.8
25. $3.708 + 4.66 = n$
26. $x = 409 + 1388$
27. $\$6.75 + \$8.95 = a$
28. Find the sum of 68 and 83.
29. Find the total of 2.37 and 38.

Practice Through Problem Solving

30. Use the digits 1, 2, 3, 4, 5, and 6. Write a problem with a sum that is close to 100.

31. Make a true equation. Place two addition signs between the digits to the left of the equal sign.
    3 4 5 6 7 8 9 = 690

32. Make a true equation. Place one decimal point in each addend.
    3.5 6 + 5 2.3 = 40.83

Show Math

33. Use or make game money to show how to add $5600 + $3700. Use only $1000 bills and $100 bills. Regroup as needed.

Mixed Applications

Solve. Use data from the chart for Exercises 34 and 35.

34. Suppose you paid the property tax on one turn and bought Motor Avenue on the next. How much did you spend?

35. Suppose you have $1950. After passing RENT, how much money would you have?

36. You own 8 pieces of property. You have 3 houses on each of 2 of your properties. You have 2 houses on each of the rest of your properties. How many houses do you have all together?

LANDLORD Facts
- Collect $200 when passing RENT!
- Four subways cost $200 each.
- Property tax is $75.
- Motor Avenue and Lake Avenue cost the least: $60 each.
- Beachwalk costs the most: $400. Brook Place is next: $350.

ADDITION AND SUBTRACTION OF WHOLE NUMBERS AND DECIMALS

2-3 Subtracting Whole Numbers and Decimals

Objective
Subtract whole numbers and decimals.

How many more parking meters worked properly and had all parts than meters that gave some time but not enough time?

Understand the Situation

- How many meters gave some time but not enough time?
- How many meters worked properly and had all parts?
- What is the action in the problem?

A Survey of Parking Meters	
Worked properly and had all parts.	313
Worked but were missing parts.	385
Would not accept coins.	144
Took coins, but gave no time.	34
Took coins, gave some time but not enough time.	59
Took coins, but gave too much time.	104

To compare the numbers of parking meters, subtract 59 from 313.

You can estimate the difference. 313 − 59 is about 300 − 60, or 240.

Examples Subtract.

1. 313 − 59

Subtract the ones. Regroup if needed.	→	Subtract the tens. Regroup if needed.	→	Subtract the hundreds.

$$\begin{array}{r} 0\,13 \\ 3\cancel{1}\cancel{3} \\ -\;59 \\ \hline 4 \end{array} \qquad \begin{array}{r} 10 \\ 2\,\cancel{0}\,13 \\ \cancel{3}\cancel{1}\cancel{3} \\ -\;59 \\ \hline 54 \end{array} \qquad \begin{array}{r} 10 \\ 2\,\cancel{0}\,13 \\ \cancel{3}\cancel{1}\cancel{3} \\ -\;59 \\ \hline 254 \end{array}$$

The answer seems reasonable. 254 is about 240. There were 254 more meters.

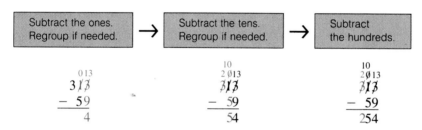

2. $\begin{array}{r} 3\,12 \\ \cancel{4}\cancel{2}7 \\ -\;83 \\ \hline 344 \end{array}$
3. $\begin{array}{r} 4\,9\,13 \\ \cancel{5}\cancel{0}\cancel{3}8 \\ -\;264 \\ \hline 4774 \end{array}$
4. $\begin{array}{r} 12\;5\,15 \\ \cancel{1}\cancel{2}.\cancel{6}\cancel{5} \\ -\;4.38 \\ \hline 8.27 \end{array}$ Line up the place values. Subtract as with whole numbers.
5. $\begin{array}{r} 2\;12 \\ \cancel{3}.\cancel{2}47 \\ -0.332 \\ \hline 2.915 \end{array}$
6. $\begin{array}{r} 5\,10 \\ 1.5\cancel{6}\,0 \\ -0.423 \\ \hline 1.137 \end{array}$ Write a zero if needed.

Try This Subtract.

a. $\begin{array}{r} 385 \\ -\;68 \end{array}$
b. $\begin{array}{r} 607 \\ -233 \end{array}$
c. $\begin{array}{r} 3275 \\ -\;428 \end{array}$
d. $\begin{array}{r} 25.06 \\ -\;2.84 \end{array}$
e. $\begin{array}{r} 235.1 \\ -\;61.8 \end{array}$
f. $\begin{array}{r} 4.26 \\ -1.032 \end{array}$

Exercises

Subtract.

1. 273 − 66
2. 437 − 264
3. 806 − 414
4. 3455 − 186
5. 23,456 − 7,128
6. 44,980 − 23,977
7. 23.54 − 2.26
8. 64.08 − 2.64
9. 0.05 − 0.038
10. 2.516 − 0.688
11. 63 − 7
12. 78 − 9
13. 54 − 26
14. 70 − 55
15. $94 − $55
16. 136 − 18
17. 224 − 43
18. 546 − 28
19. 602 − 11
20. $131 − $78
21. $35.67 − $8.28
22. 0.04 − 0.017
23. $8.98 − $6.89
24. $0.634 − 0.225 = v$
25. $81.49 − $32.68 = t$
26. $b = 65{,}503 − 7{,}943$
27. 43.7 less 6.4 equals what number?
28. Find the difference between 806 and 14.
29. 7 less than 63 equals what number?

Calculator

30. Copy and fill in the calculator keystrokes to show two different ways to find this difference: 32 − 12 − 8.

   A. ☐☐☐☐☐☐☐

   B. ☐☐☐☐☐☐☐

Number Sense

What number would you write in each box to make a true equation?

31. 235 + ☐ = 768
32. ☐ + 522 = 866
33. 156 + ☐ = 652

Working with Variables

A **numerical expression,** like 24 + 35, is a name for a number. 24 + 35 is another name for 59. An expression like $x + 35$ that has a variable is called an **algebraic expression.** To evaluate an algebraic expression, **substitute** a numerical value for the variable and compute. For example, to evaluate $x + 35$ for $x = 10$, substitute 10 for x: $10 + 35 = 45$.

Evaluate.

34. $y + 82$ for $y = 16$ and for $y = 35$
35. $5.63 − m$ for $m = 1.4$ and for $m = 0.9$
36. $12.5 + t$ for $t = 7.8$ and for $t = 34$
37. $a + b + 54$ for $a = 12$ and $b = 28$
38. $p − 24 − q$ for $p = 130$ and $q = 12$
39. $8.59 − t$ for $t = 3.99 and for $t = 8.29
40. $45 + c + d$ for $c = 39$ and $d = 82$
41. $181 − v − w$ for $v = 59$ and $w = 102$

2-4 Subtracting Across Zero

Objective
Subtract across zero.

Carrie and Kim each worked the same problem on the chalkboard. Study their work. Then answer Exercises 1 and 2.

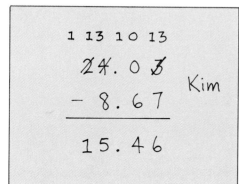

Exercises

1. Whose answer is correct?
2. Explain what mistake the other student made. How could you tell a classmate to avoid this mistake?
3. The answer to A. is correct. Without solving, how can you tell that the answer to B. is wrong?

   A. $\;\;2.03$
       -0.462
       $\;\;1.568$

   B. $\;\;2.03$
       -1.568
       $\;\;0.362$

Practice Through Problem Solving

4. Use each of the digits given one time. Write a problem with a difference that is greater than 1.2 and less than 1.3.

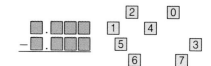

5. Write a subtraction problem where the difference is 12.042.
6. What is the greatest number with four decimal places that you can add to 2.04 and have a sum that is less than 3?

Use Objects

Toss a cube numbered 1 to 6 eight times. Record your tosses. Use your numbers to fill in the boxes.

7. Write a problem with the least possible difference.

8. Write a problem with the greatest possible difference.

Mixed Practice

Add or subtract.

1. $62 - 8$
2. $\$7.64 + \$2.29$
3. $92 - 23$
4. $27.34 - 8.3$
5. $54 + 8$
6. $0.03 - 0.002$
7. $273 - 148$
8. $36 + 57$
9. $442 + 605 + 828$
10. $90 + 22$
11. $\$45.61 - \$32.85$
12. $\$25.32 - \$12.04$
13. $28 + 92$
14. $0.603 - 0.017$
15. $\$36.20 - \$18.10$
16. $148 + 353 + 920$
17. $5697 - 3299$
18. $0.62 + 0.867$
19. $\$17.86 + 0.45$
20. $6.4 - 0.42$
21. $\$44.77 - 8.23$
22. $702 - 358$
23. $766 + 136$
24. $\$14.09 + 3.87$
25. $3955 + 826$
26. $1402 - 187$
27. $0.129 + 0.29$
28. $67{,}028 - 24{,}096$
29. $\$63.02 - 7.08$
30. $65{,}329 + 82{,}754$
31. $0.05 + 0.699$
32. $6.909 - 5.12$
33. $n = 1400 - 56$
34. $1204 + 88 = x$
35. $0.1 + 0.091 = y$
36. $12{,}000 - 462 = d$
37. $m = 14.5 + 0.39$
38. $682.9 - 107 = s$
39. $\$19.24 - \$1.50 = t$
40. $29 - 13.4 = w$
41. $v = \$25.09 + \$0.82$

42. What is the total of 79 and 479?
43. 3217 minus 294 equals ☐.
44. $724 plus $53 equals ☐.
45. What is 78.4 less 2.05?
46. Add 607 to 129.
47. Subtract $34.95 from $62.
48. Take 0.048 from 0.2.
49. 66 is how much less than 120?
50. Add $12 to $45, then subtract $4.02. What is the result?
51. Subtract 0.26 from 1.45, then add 12. What is the result?

Which is the closest estimate?

52. $39 + 72 + 150$ A. 240 B. 260 C. 280
53. $63 - 19.5$ A. 43 B. 46 C. 39
54. $402 + 96 + 12$ A. 510 B. 500 C. 490
55. $12.84 - 10.76$ A. 12 B. 2 C. 0.2

Evaluate.

56. $258 + k$ for $k = 511$
57. $5.96 - b$ for $b = 1.29$
58. $d + \$56$ for $d = \$77$
59. $n + 15.06$ for $n = 366.9$
60. $x - \$28$ for $x = \$41$
61. $6000 - f$ for $f = 633$

ADDITION AND SUBTRACTION OF WHOLE NUMBERS AND DECIMALS

2-5 Estimating Sums and Differences Using Front-End Estimation

Objective
Estimate sums and differences using front-end estimation.

Marcus received a $3000 scholarship for college. Is this enough money to cover the costs of one year's tuition and room and board at the local college?

An estimate can quickly give you the answer.

You can estimate the total cost for tuition and room and board for a resident using **front-end** estimation.

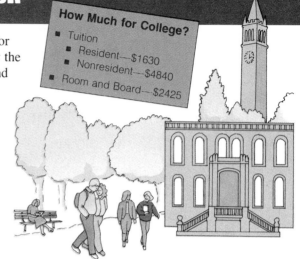

Examples Estimate. Use front-end estimation.

1. $1630 + $2425

Add the front-end, or lead, digits.

$1630 + $2425
1 + 2 = 3, or 3000

Adjust this estimate. Use compatible numbers.

630 + 425 is about 600 + 400, or 1000. 3000 + 1000 = 4000

$1630 + $2425 is about $4000. The total cost for tuition and room and board for a resident is about $4000. Marcus' scholarship does not cover the costs.

2. $4840 − $1630

$4840 − $1630 = 4 − 1 = 3, or 3000
$840 − $630 is about $800 − $600, or $200.
$3000 + $200 = $3200
$4840 − $1630 is about $3200.

3. 421 + 332 + 549

421 + 332 + 549 = 4 + 3 + 5 = 12, or 1200
21 + 32 + 49 is about 20 + 30 + 50, or 100.
1200 + 100 = 1300
421 + 332 + 549 is about 1300.

Try This Estimate. Use front-end estimation.

a. 563 + 144   **b.** 831 + 424 + 247   **c.** 754 + 209 + 618
d. 245 + 890   **e.** 764 − 208   **f.** 2328 − 1225   **g.** 3211 + 7689

Exercises

Which numbers would you use to adjust the front-end estimate?

1. 646 + 249   A. 50 + 50   B. 46 + 50   C. 45 + 49

2. 446 − 142   A. 45 − 42   B. 40 − 40   C. 46 − 40

3. 768 − 254   A. 65 − 54   B. 70 − 55   C. 70 − 50

4. 854 + 222 + 479   A. 55 + 22 + 80   B. 50 + 20 + 80   C. 50 + 25 + 80

Estimate. Use front-end estimation.

5. 435 + 228   6. 389 + 576   7. 762 − 358

8. 687 − 129   9. 548 + 775   10. 134 + 643

11. 563 − 234   12. 978 − 623   13. 3429 + 6124

14. 7181 + 7865   15. 2247 − 1110   16. 4793 − 2840

17. 2135 + 1780   18. 4560 − 3250   19. 8778 − 5634

20. 845 + 346 + 751   21. 164 + 986 + 896   22. 256 + 125 + 349

Choose the best estimate.

23. 783 − 218   A. 500   B. 560   C. 580

24. 573 + 729   A. 1300   B. 1200   C. 1270

25. 345 + 238 + 262   A. 800   B. 770   C. 840

Decide if each estimate is too high, too low, or if it is too hard to tell.

26. 435 + 228 is about 650.   high   low   ?

27. 2247 − 1111 is about 1100.   high   low   ?

28. 8578 − 5634 is about 3000.   high   low   ?

29. 2981 + 7465 is about 10,500.   high   low   ?

Talk Math   30. Estimate 4478 + 2561 first by using front-end estimation and then by using compatible numbers. Which strategy would you rather use here? Tell why.

You Decide   Estimate each sum or difference. Use one of the estimation strategies shown so far. Tell which strategy you used.

Estimation Strategies
- Rounding
- Compatible numbers
- Front-end

31. 134 + 596   32. 405 + 598   33. 783 − 479

34. 554 − 121   35. 272 + 130   36. 795 + 288

ADDITION AND SUBTRACTION OF WHOLE NUMBERS AND DECIMALS

2-6 Consumer Math: Filling Out Deposit Slips

Objective
Fill out deposit slips for checking and savings accounts.

- SITUATION
- DATA
- PLAN
- ANSWER
- CHECK

Jeff deposited one $5 roll of dimes and paycheck 26-143 for $127.35. He asked for $50 in cash. Jeff used a deposit slip to list the items for the bank.

		Dollars	Cents	
Cash	Currency	0	00	1.
	Coin	5	00	2.
Checks	26-143	127	35	3.
	Total	132	35	4.
	Less Cash Received	50	00	5.
	Net Deposit	82	35	6.

DEPOSIT Jeffrey Skeritt
236 Crockett Drive
Lufkin, TX 75901

Date *July 16, 1988*

Jeffrey Skeritt
Sign here for cash received

Arrow Bank of Texas
392 Main Street
Lufkin, TX 75901

095101340:...1649:52149:

Examples Write the amount that should appear on each line of Jeff's deposit slip.

1. **Currency**   Jeff did not deposit any bills. Write zero.
2. **Coin**   Jeff had one $5 roll of dimes. Write $5.
3. **Checks**   Write each bank number and its amount on a separate line.
4. **Total**   Add the amounts for cash and checks. Write the total of $132.35.
5. **Less cash received**   Jeff received $50 of the total in cash. Write $50.
6. **Net deposit**   Subtract cash received from the total. Write $82.35.

Try This Fill out a deposit slip for this deposit. Copy the form on p. 496.

Savings account deposit of one $10 roll of quarters, eight $1 bills, and two checks: 24-953 for $25 and 24-871 for $11.75. No cash received.

Exercise

Fill out a deposit slip for this deposit. Copy the form on p. 496.

One 50¢ roll of pennies, check 23-146 for $48.50, check 45-1590 for $35, and $25.50 cash received.

2-7 Consumer Math: Keeping a Running Balance

Objective
Keep a running balance for a checking account.

- SITUATION
- DATA
- PLAN
- ANSWER
- CHECK

Use a checkbook register to keep a record of your checking account. Fill in the amount of each check or withdrawal and the amount of each deposit. Add or subtract each entry to keep a **running balance** of the account.

NUMBER	DATE	DESCRIPTION OF TRANSACTION	(−) PAYMENT	T	(−) FEE (IF ANY)	(+) DEPOSIT	BALANCE	
							139	23
100	9/29	City Feet Shoes	42	46			42	46
							96	77
	10/1	Gardening money				15 00	15	00
							111	77

1.
2.
3.

Examples Fill in the balances for the register.

1. Margaret opened her account with $139.23.
   Begin each register with the amount of money already in the account.

2. On September 29, Margaret wrote check #100 for $42.46 to City Feet.
   For each check, record the check number, the date it was written, the name of the payee, and the amount. You may choose to add a note about how you used the money. Subtract the amount from the running balance.

3. On October 1, Margaret deposited the $15 she earned gardening.
   For each deposit, record the date and the amount. Add the amount to the last balance. Write the new amount.

Try This Fill in the balances for the register. Copy the form on p. 496.

At the start of a new register, Peter's account balance was $205.67.
On September 24, Peter deposited the $26.80 he earned busing dishes.
On September 27, Peter wrote check #349 for $21.24 to Stuffed Shirts.

Exercise

Fill in the balances for the register. Copy the form on p. 496.

At the start of a new register, Julio's checking account balance was $359.66. On October 3, Julio wrote check # 238 for $31.89 to Sports World for sweats. On October 4, Julio wrote check # 239 for $7.67 to City Feet for socks. On October 7, Julio deposited a $25 birthday check.

2-8 Mental Math: Counting

Objective
Use the mental math strategy counting.

What is your change if you pay for a four-day world passport ticket with a $100 bill?

You can solve some problems easily in your head using **mental math.** The mental math strategy **counting** can sometimes help you to find answers.

Ticket Costs
Three-Day World Passport—$66
Child (3–11)—$50
Four-Day World Passport—$78
Child (3–11)—$60

Examples Find each sum or difference using mental math. Use counting.

1. $100 − $78   *"$78. How much more to make $100? Count up 2. "$78 . . . $79, $80." Count up by 10s. "$80 . . . $90, $100. 2 and 2 tens are 22."*

The change is $22.

2. 89 + 3   *Count up 3. "89 . . . 90, 91, 92."*
89 + 3 = 92

3. 40 − 2   *Count back 2. "40 . . . 39, 38."*
40 − 2 = 38

Try This Find each sum or difference using mental math. Use counting.

a. $7 − $3.50   **b.** $20 − $15.75   **c.** $150 − $85   **d.** 112 − 3
e. 2 + 49   **f.** 51 − 2   **g.** $24.50 − $19   **h.** $62.75 − $48

CHAPTER 2

Exercises

Find each sum or difference using mental math. Use counting.

1. $4 − $1.50
2. $5 − $2.75
3. 68 + 3
4. 39 + 3
5. $40 − $25.50
6. $50 − $34.50
7. 21 − 2
8. 30 − 3
9. $10 − $3.45
10. $20 − $9.25
11. 72 − 3
12. 81 − 2
13. $150 − $115
14. $200 − $135
15. 119 + 2
16. 237 + 3
17. 3 + 78
18. 2 + 69
19. $49 + $3
20. $2 + $19
21. $35.50 − $27
22. $24.75 − $15
23. $45.75 − $24
24. $75.50 − $62
25. $65.50 − $44.75
26. $86.25 − $75.50
27. $82.45 − $16.30

Mixed Applications

28. How much warmer is the average temperature in April than in March?

29. How much farther is it from Ft. Myers to Disney World than from Jacksonville to Disney World?

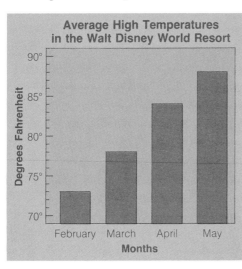

Average High Temperatures in the Walt Disney World Resort

Driving Distances to Walt Disney World

Daytona Beach	65 miles
Tampa	82 miles
St. Petersburg	104 miles
Ft. Myers/Sanibel Island	157 miles
Jacksonville	140 miles
Ft. Lauderdale	207 miles
Miami	229 miles

Talk Math

30. Tell why most people think counting is an easy way to find $20.50 − $15.90 but a difficult way to find $20.63 − $15.36.

You Decide

Which answers would be easy for you to find using the mental math strategy *counting?* Find those answers.

31. 68 − 39
32. 68 + 3
33. 81 − 2
34. $34.50 − $28
35. 2339 − 1269
36. $7.50 − $2.75
37. 506 − 428
38. 80 − 2
39. $5 − $3.25
40. 349 + 3
41. $45.03 − $3.19
42. $3 + $29

2-9 Choosing a Calculation Method

Objective
Choose a calculation method.

One of the decisions you must make when solving problems is to choose a calculation method.

Even though you may have a calculator, there are certain times when it would be better not to use it.

Calculation Method
- Paper and pencil
- Mental math
- Calculator

Examples Choose the calculation method you would use to find the answer. Tell why.

1. 24×35

   If you had your calculator and you needed an answer quickly, you might use your calculator. Since the numbers are small, you might choose to use paper and pencil instead of getting out your calculator.

2. 150×30

   Both numbers are a multiple of 10. Numbers that are multiples of 10 are usually easy to multiply using mental math. It would probably be faster to use mental math.

3. $\$34 + \$46 + \$52 + \$38$

   There are many numbers to add. You might use a calculator. Look at the numbers before choosing to use a calculator. The numbers 34 and 46 and the numbers 52 and 38 are compatible. You can add these numbers using mental math.

4. $\$365.25 - \$48.89$

   There are many computations needed to find the answer. In this case, it is often faster if you use your calculator.

Try This Choose the calculation method you would use to find the answer. Tell why.

a. $775 - 150$ b. 135×68 c. 122×50 d. $\$140 + \$62.50$

Exercises

Choose the calculation method you would use to find the answer. Tell why.

1. $75.36 + $24.95
2. 6 × 126
3. 32 + 53 + 28
4. 49 + 24 + 35 + 166
5. $54 − $32.50
6. $350 × 20
7. 5643 − 138
8. 23 × 51
9. 125 + 250 + 650
10. 6 × 2 × 30
11. 3002 − 356
12. 700 − 203
13. 3004 − 800
14. $75 − $24.25
15. $23.50 × 8
16. 56 × 235
17. 0.54 − 0.35
18. $649.95 − $23

Choose the number that would make it reasonable to compute each problem with the given calculation method.

19. mental math: 124 + ___   A. 47   B. 27   C. 16
20. calculator: $34 × ___   A. 128   B. 5   C. 20
21. paper and pencil: $165 − ___   A. 120   B. 127   C. 25.43
22. calculator: 45.07 − ___   A. 0.07   B. 25.07   C. 8.09
23. mental math: 72 × ___   A. 3.6   B. 400   C. 85
24. paper and pencil: 7.4 + ___   A. 12   B. 32.56   C. 0.6

Suppose your calculator is in the other room. For which of the following exercises would you go get your calculator?

25. You are balancing your checkbook. There are three pages of calculations.
26. A television ad says that each tire costs $49.99. You want to know the approximate cost for 4 tires.
27. Interest on your credit card is 1.5% (0.015) per month. You want to know how much interest you would pay on a balance of $175.50 for one month.
28. You just received your first paycheck. You want to check that the amount is correct.

Use Objects 29. Use paper squares to build the fourth design. Guess how many blocks you would need for the fifth design. Check your guess by building the fifth design.

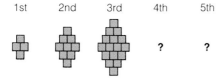

ADDITION AND SUBTRACTION OF WHOLE NUMBERS AND DECIMALS

2-10 Problem Solving: Determining Reasonable Answers

Objective
Decide if an answer is reasonable.

- **SITUATION**
- **DATA**
- **PLAN**
- **ANSWER**
- **CHECK**

The last part of the 5-point checklist for problem solving is to **check your work.**

This chart shows you some ways you can do this.

Check Your Work
- Is the arithmetic correct?
- Did you use the strategies correctly?
- Is the answer reasonable?

Examples Decide if the answer given is reasonable without solving the problem. If it is not reasonable, tell why not.

1. *Problem:* Teenage females watch about 23 hours of television each week. About how much do they watch each day?

   Answer: They watch about 6 hours each day.

   Six hours seems like a high number for one day. If they watch 6 hours each day, they would watch 7 × 6, or 42 hours a week. Six hours a day is not a reasonable answer.

2. *Problem:* At Ed's TV & Stereo, the total cost of a large-screen television is $485, with the finance charge. How much would you have to pay each month to pay off the TV in one year?

   Answer: You would have to pay $40.42 each month.

   Use multiplication to check division. $40.42 × 12 is about $40 × 12, or $480. The answer seems reasonable.

Try This Decide if the answer given is reasonable without solving the problem. If it is not reasonable, tell why not.

a. *Problem:* Children between 2 and 5 years of age watch about 32 complete television programs a week. About how many complete programs do they watch each day?

   Answer: They watch about 5 complete programs a day.

b. *Problem:* In 1989, Fred's TV was 18 years old. In what year did Fred buy his TV?

   Answer: Fred bought his TV in 1960.

Exercises

Decide if the answer given is reasonable without solving the problem. If it is not reasonable, tell why not.

Average Number of Hours of Weekly TV Viewing by Age

Women 18–34	32 hours
Women 35–54	34 hours
Women 55 and over	42 hours
Men 18–34	26 hours
Men 35–54	29 hours
Men 55 and over	38 hours
Female teens	23 hours
Male teens	24 hours
Children 6–11	27 hours
Children 2–5	28 hours

1. *Problem:* About how much television do male teenagers watch in one month?
   Answer: They watch about 96 hours in one month.

2. *Problem:* About how much more television do women over age 55 watch per week than men over age 55?
   Answer: They watch about 42 hours more per week.

3. *Problem:* In a one-hour television program, a viewer watches about 12 minutes of commercials. About how many minutes of commercials does a child under the age of five watch per week?
   Answer: They watch about 11 minutes of commercials.

4. *Problem:* A small color television can be bought for 12 monthly payments of $31.25 with interest. What is the total cost for the television?
   Answer: The total cost is $375.

5. *Problem:* A 30-second commercial on a television show 15 years ago cost $12,750. Today, that same commercial costs about 12 times more. How much does the commercial cost today?
   Answer: A commercial today costs $1,530,000.

6. *Problem:* The father on a television show in 1975 is 55 years old today. How old was he in 1975?
   Answer: The father was 12 years old.

Mixed Skills Review

Estimate.

7. 282 ÷ 4
8. 506 × 5
9. 386 + 291
10. 627 ÷ 9
11. 472 ÷ 7
12. 9 × 297
13. 358 ÷ 9
14. 414 − 211
15. 418 ÷ 6
16. 663 − 59

Add or subtract.

17. 73.1 + 89
18. 203 − 77
19. 4.59 − 1.6
20. 395 + 833

Write in order from least to greatest.

21. 5.02, 4.8, 5, 5.45, 4.56, 5.2
22. 0.088, 0.08, 0.87, 0.078, 0.8

2-11 Problem Solving: Developing Thinking Skills

Objective
Use problem-solving strategies.

- SITUATION
- DATA
- PLAN
- ANSWER
- CHECK

Workers built a 100-yard tunnel. During daylight, they tunneled 30 yards. Each night, 10 yards of dirt caved in. How many days did it take to finish the tunnel?

Exercises

Use the *Thinking Actions*. Answer the question.

Try these **before** starting to write.

Thinking Actions Before
- Read the problem.
- Ask yourself questions to understand it.
- Think of possible strategies.

1. What was the total length of the tunnel?
2. How far did they tunnel during the day?
3. What happened to the tunnel at night?
4. What is the problem asking you to find?
5. What strategy, or strategies, might help you to solve this problem?

Try these **during** your work.

Thinking Actions During
- Try your strategies.
- Stumped? Try answering these questions.
- Check your work.

6. Can you draw a picture showing a 100-yard tunnel?
7. At the end of the first day, how far did they tunnel?
8. At the end of the second day, how far did they tunnel?

9. Solve the problem. Write the answer in a complete sentence.
10. Name the strategy, or strategies, that you used to solve this problem.

Suppose

11. Suppose the tunnel was 200 yards long. How many days would it take to finish the tunnel?
12. Suppose 20 yards of dirt caved in in the tunnel at night. How many days would it take to finish a 100-yard tunnel?

CHAPTER 2

2-12 Practice Through Problem Solving

Objective
Practice addition. Use problem-solving strategies.

Suppose that each letter of the alphabet is given a specific value. These values are represented by the numbers 1 through 9, as shown in the table.

Values								
1	2	3	4	5	6	7	8	9
A	B	C	D	E	F	G	H	I
J	K	L	M	N	O	P	Q	R
S	T	U	V	W	X	Y	Z	

Here is how you would find the value of the word *Amelia*.

Use the table. Find the value of each letter. Add the values.

$A + M + E + L + I + A$
$1 + 4 + 5 + 3 + 9 + 1$
$= 23$

The value of the word *Amelia* is 23.

Exercises

Find each value. Use the table.

1. What is the value of your first name? Of your last name?
2. What is the value of your best friend's name?
3. What is the value of *The United States*?
4. What is the value of the name of your school?
5. What is the value of your teacher's name?
6. What is the value of *The Beatles*?
7. Find a friend's name with a value of 25.
8. Find the name of a city with a value of 24.
9. Find the last name of an actor with a value of 23.
10. Can the value of a name be 0?

Write Your Own Problem

Write one addition and one subtraction problem that have the given number as the answer.

11. $54
12. 6.3
13. 228
14. 7.06
15. $33.12
16. 302

Write an addition problem with the given answer. Use five addends.

17. 24
18. 46
19. $57
20. 100
21. 13.44
22. $36.30

2-13 Enrichment: Reconciling a Bank Statement

Objective
Reconcile bank statements with checkbook balances.

Leslie received a bank statement for her checking account. The balance in her checkbook is $235.31.

Leslie needs to **reconcile** the bank statement and her checkbook. This means that she needs to adjust the balances to make them agree. Leslie had written checks that the bank had not received when the bank statement was calculated. These are called **outstanding checks**. Leslie made deposits that did not appear on the bank statement. These are called **outstanding deposits**.

FIRST AMERICAN BANK
Checking Account Number: 1043216545
Statement Date: 7-20-1987

Previous Balance	Total Deposits	Total Checks	Service Charge	New Balance
$115.90	$531.32	$420.16	$2.50	$224.56

Check Number	Date	Amount
529	7-02	15.30
530	7-03	25.00
531	7-07	13.98
533	7-11	47.63
534	7-12	59.61
535	7-12	115.73
536	7-15	17.56
539	7-16	53.20
540	7-19	72.15

Deposits Date	Amount
7-05	256.92
7-15	215.70
7-19	58.70

Example Reconcile Leslie's bank statement with her checkbook balance.

outstanding checks $35.42 + $23.51 + $50 + $7.52 = $116.45
outstanding deposits $124.70

bank statement balance	$224.56	checkbook balance	$235.31
outstanding deposits	+$124.70	service charge	−$2.50
total	$349.26	total	$232.81
outstanding checks	−$116.45		
total	$232.81		

The adjusted balance, $232.81, is the true amount in her account.

Exercises

Reconcile the bank statement information with the checkbook balance.

1. bank statement balance $225.67
   outstanding checks $52.31, $26.80
   outstanding deposits $88.90
   service charge $1.75
   checkbook balance $237.21

2. bank statement balance $126.75
   outstanding checks $56.71, $39.50, $25
   outstanding deposits $104.70, $231.75
   service charge $2.75
   checkbook balance $344.74

3. bank statement balance $45.90
   outstanding checks $132.67, $56.20
   outstanding deposits $300
   service charge $1.50
   checkbook balance $158.53

4. bank statement balance $56.89
   outstanding checks $35.91, $20, $41.70
   outstanding deposits $123.91, $30, $71.65
   service charge $2
   checkbook balance $186.84

2-14 Calculator: Maintaining a Checking Account

Objective
Use a calculator to maintain a checkbook register.

Sandy wrote checks for two weeks in September without recording them in his checkbook. The calendar shows his checks, bank fees, and deposits. Use the data and your calculator to complete the checkbook register. Find the balance after each transaction. 100 checks cost $4.95.

Monday	Tuesday	Wednesday	Thursday	Friday	Saturday
8 Balance $89.24	9 Phone bill $19.60 #391	10 200 new checks	11 Team jacket $46.12 #392	12 Monthly bank fee $5	13 Birthday check deposit $60
15 Records $11.49 #393	16 Jeans $18.36 #394	17 Shirt $26.74 #395	18 Transfer to savings account $10	19 Paycheck deposit $154.19	20 New calculator $10.75 #396

NUMBER	DATE	DESCRIPTION OF TRANSACTION	(−) PAYMENT	T	(−) FEE (IF ANY)	(+) DEPOSIT	BALANCE $89.24
1.							
2.							
3.							
4.							
5.							
6.							
7.							
8.							
9.							
10.							
11.							

ADDITION AND SUBTRACTION OF WHOLE NUMBERS AND DECIMALS

CHAPTER 2 REVIEW

Choose the one that does not belong in the group. Explain why not.

1. Estimate.
  A. 503 × 5
  B. 1997 + 507
  C. 4110 − 1608
  D. 5003 + 507

2. Estimate.
  A. 45.9 − 16.1
  B. 3.07 × 25.1
  C. 5.01 × 5.9
  D. 19.921 + 10.03

3.
  A. 32
     +46
  B. 201
     +123
  C. 48.16
     +10.53
  D. 1.47
     +3.75

4.
  A. 28
     −19
  B. 3.72
     −1.58
  C. 4.62
     −3.21
  D. 473
     −295

5.
  A. 10.5 + 3.2
  B. 5.356 − 5.3
  C. 4.13 + 1.47
  D. 63.2 − 7.2

6.
  A. $y + 14$ for $y = 7$
  B. $23.5 - x$ for $x = 1.5$
  C. $p + 12.3$ for $p = 7.7$
  D. $38 - q$ for $q = 25$

Find the errors. Write the correct balance.

7.

NUMBER	DATE	DESCRIPTION OF TRANSACTION	(−) PAYMENT	T	(−) FEE (IF ANY)	(+) DEPOSIT	BALANCE
							347 85
510	10/21	San Jose Shoes running shoes	43 62				43 62
							391 47
	10/21	paycheck				203 40	203 40
							188 07
511	10/25	Alvarez & Sons auto repair	134 82				134 82
							53 25
	10/28	paycheck				203 40	203 40
							256 65

Match.

8. Name three calculation methods.
9. Name one mental math strategy.
10. Name two problem-solving strategies.
11. Name one estimation strategy.

  A. front-end
  B. Draw a picture.
  C. paper and pencil
  D. Use objects.
  E. calculator
  F. counting
  G. mental math

Determine which answer is reasonable without finding the answer.

12. 483 + 177
  A. 550
  B. 560
  C. 660
  D. 600

13. $25.88 − $16.24
  A. $8.74
  B. $9.64
  C. $12.84
  D. $11.94

14. 2979 + 1465
  A. 3334
  B. 5004
  C. 3844
  D. 4444

CHAPTER 2

CHAPTER 2 TEST

1. A sign shows Springdale is 35 miles to the west and Emery is 25 miles to the east. You live between Springdale and Emery. You live 6 miles from Springdale. How far do you live from Emery?

Add or subtract.

2. 558 + 765
3. 6.72 + 40.3
4. $4.65 + $32.76
5. 45 + 439 + 295
6. 87 − 48
7. 509 − 71
8. 48.2 − 3.23
9. $54.08 − $2.64

Estimate. Use front-end estimation.

10. 458 + 793
11. 2452 − 1133
12. 771 − 369
13. 858 + 244 + 302

14. Write the amount that should appear on each line of the deposit slip.

    one 50¢ roll of pennies, check 44-234 for $45.32, check 87-342 for $78.34, and $25 cash received

	Dollars	Cents
Currency		
Coin		
Total		
Less Cash Received		
Net Deposit		

15. Fill in the balances for the register.

    The opening balance was $473.54. On May 5, Cassie wrote check #256 to Sound Waves for $57.64. On May 7, she deposited $254.39. On May 9, she wrote check #257 to the Pant Station for $38.22.

NUMBER	DATE	DESCRIPTION OF TRANSACTION	(−) PAYMENT	T	(−) FEE (IF ANY)	(+) DEPOSIT	BALANCE
		Sound Waves					
		Deposit					
		Pant Station					

Find the answer. Use mental math.

16. 59 + 3
17. $47.55 − $3
18. 80 − 3
19. $25.25 − $0.75

Choose the calculation method you would use to solve each problem. Tell why.

20. $100 + $55.75
21. 133 × 60
22. 298 × 86
23. 120 × 40

Decide if the answer given is reasonable without solving the problem. If it is not reasonable, tell why not.

24. *Problem:* A plane ticket costs $189 for adults. It costs $95 for children. How much more is the adult's ticket than the child's ticket?

    Answer: The adult's ticket is $284 more than the child's ticket.

CHAPTER 2 CUMULATIVE REVIEW

1. There are 8 towns on the same road. There is a town at each end of the road. Each town is 6 miles from the next town. How long is the road?
   A. 54 miles
   B. 48 miles
   C. 42 miles
   D. 36 miles

2. Find the net deposit: check 32-197 for $65, one $10 roll of quarters, check 16-153 for $129.35, and $35.50 cash received.
   A. $239.85
   B. $167.95
   C. $168.95
   D. $168.85

3. Subtract. Use mental math.
   $14.50 − $12.75
   A. $2.85
   B. $1.75
   C. $2.25
   D. $2.75

4. Estimate.
   1360 + 2775
   A. 3500
   B. 4000
   C. 3000
   D. 5000

5. Add.
   74.86
   +22.75
   A. 107.61
   B. 98.51
   C. 98.61
   D. 97.61

6. Subtract.
   $62.04 − $6.87
   A. $55.27
   B. $56.27
   C. $55.17
   D. $56.17

7. Estimate.
   426 ÷ 40
   A. 15
   B. 12
   C. 8
   D. 10

8. Name a strategy that Rajit used to solve this problem.
   Problem: In how many ways can Shana pick 2 out of 4 museums to visit tomorrow?
   A. Make a table.
   B. Work backward.
   C. Guess and check.
   D. Make an organized list.

9. Add.
   3495 + 2099
   A. 5584
   B. 5494
   C. 5484
   D. 5594

10. Estimate. Round to the highest place.
    498 + 549
    A. 900
    B. 1000
    C. 1100
    D. 950

11. Write in money notation. seventeen dollars and four cents
    A. $17.04
    B. $17.40
    C. $174.00
    D. $70.04

12. Round to the nearest ten.
    584
    A. 600
    B. 585
    C. 590
    D. 580

13. Find the new balance. At the start of a new register, Kaya's checking account balance was $489.46. On July 22, Kaya wrote check #324 for $26.75. On July 27, Kaya wrote check #325 for $38.91. On August 1, Kaya deposited $40.
    A. $463.80
    B. $423.80
    C. $463.81
    D. $399.70

14. Decide which operation you would use to solve this problem.
    Problem: Barry worked △ days last month. He worked a total of □ hours. How many hours did he work each day?
    A. addition
    B. subtraction
    C. multiplication
    D. division

15. Compare. Write <, >, or =.
    29.034 □ 29.30
    A. <
    B. >
    C. =

16. Decide which operation you would use to solve this problem.
    Problem: Glennis played soccer □ hours a day for △ weeks. How many hours did she play all together?
    A. addition
    B. subtraction
    C. multiplication
    D. addition

17. Subtract. Use mental math.
    $90 − $2
    A. $92
    B. $88
    C. $87
    D. $89

18. Subtract.
    4003 − 595
    A. 3518
    B. 4408
    C. 3408
    D. 3508

19. Write in standard notation.
    seventy-two thousandths
    A. 72,000
    B. 0.72
    C. 0.072
    D. 7200

20. Round to the nearest hundredth.
    0.0349
    A. 0.03
    B. 0.035
    C. 0.04
    D. 0.034

21. Estimate.
    344 × 52
    A. 150
    B. 1500
    C. 2000
    D. 15,000

22. Choose the calculation method you would use to solve this problem.
    30,008 − 377
    A. mental math
    B. calculator
    C. paper and pencil

23. Subtract.
    5907.2 − 38.51
    A. 5940.71
    B. 5868.69
    C. 5517.1
    D. 5864.69

Chapter 3 Overview

Key Ideas

- Use the problem-solving strategy *make an organized list*.
- Multiply whole numbers and decimals.
- Estimate sums using clustering.
- Solve multiple-step problems.
- Read a paycheck stub.
- Read and write exponents.
- Use the mental math strategy *break apart numbers*.
- Collect and analyze data.

Key Terms

- product
- deduction
- factor
- exponent
- clustering
- break apart numbers
- paycheck

Key Skills

Multiply.

1. 3×5
2. 8×6
3. 7×4
4. 9×2
5. 6×7
6. 1×6
7. 4×2
8. 4×3
9. 8×4
10. 0×7
11. 6×6
12. 7×9
13. 9×9
14. 8×5
15. 9×4
16. 7×7
17. 4×5
18. 9×6

Add.

19. 23 + 31
20. 34 + 18
21. 62 + 25
22. 18 + 16
23. 45 + 35
24. 33 + 58
25. 74 + 58
26. 65 + 62
27. 90 + 30
28. 44 + 57
29. 81 + 25
30. 98 + 41
31. 60 + 120
32. 54 + 140
33. 34 + 230
34. 75 + 540
35. 80 + 670
36. 92 + 720
37. 123 + 2450
38. 500 + 3400
39. 625 + 3460
40. 2346 + 1260
41. 3402 + 5420
42. 6538 + 1250

MULTIPLICATION OF WHOLE NUMBERS AND DECIMALS

3

3-1 Problem Solving: Learning to Use Strategies

Objective
Use the strategy make an organized list.

- SITUATION
- DATA
- **PLAN**
- ANSWER
- CHECK

This ballot shows one possible outcome for a school election. When you work with probability, you often need to find how many outcomes are possible.

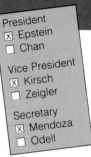

Example How many outcomes are possible for the student council election?

One way to solve this problem is to **draw a picture**.

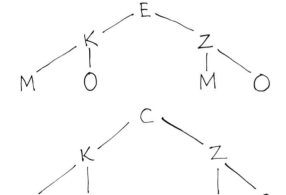

Another way to solve this problem is to **make an organized list**.

E–K–M   E–Z–M
E–K–O   E–Z–O

C–K–M   C–Z–M
C–K–O   C–Z–O

The solutions show that there are 8 possible outcomes for the student council election.

Try This Solve.

a. Toothpaste comes in three sizes—small, medium, and large—and two flavors—mint and regular. How many choices are possible for one size of toothpaste in one flavor?

b. Ambassadors from four countries are to be seated at tables arranged as shown. How many arrangements are possible?

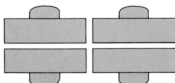

60 CHAPTER 3

Exercises

Solve. Use one or more of the problem-solving strategies.

Problem-Solving Strategies
- Guess and check.
- Choose an operation.
- **Make an organized list.**
- Act it out.
- Find a pattern.
- Write an equation.
- Use logical reasoning.
- **Draw a picture or diagram.**
- Make a table.
- Use objects.
- Work backward.
- Solve a simpler problem.

1. Khuon and Wendy have part-time typing jobs. Each is paid $2.50 per page. One evening Khuon typed 15 pages and Wendy typed 8 pages. How much less money than Khuon did Wendy make that evening?

2. How many combinations of 2 toppings can you choose for your hamburger?

Big Burgers
Your Choice of Any Two Toppings
- Mushrooms
- Swiss cheese
- Bacon
- Cheddar cheese

3. A high school football team won its first game by a score of 12–0. What are the different ways the team could have scored 12 points?

- Touchdown — 6 pts.
- Point after touchdown — 1 pt.
- Field goal — 3 pts.
- Safety — 2 pts.

4. The distance from Jennifer's house to school is 2.5 miles. She bikes the round trip to school 20 times each month. How many miles will she bike in 9 months going to and from the school?

5. What are the different prices you could pay for one model of car with one option package?

Otto's Autos

Models	Options		
■ Standard $8000	**A.** Air conditioning	**B.** Air conditioning Power windows	**C.** Air conditioning Power windows Cruise control
■ Deluxe $10,000	$850	$1200	$2000

6. A club has a telephone tree to send messages to members. First, the president of the club calls 3 committee heads. Each of them calls 4 members. Each of these people calls 3 more members to reach everyone in the club. How many members are in the club?

Finishing a Solution

7. Five students ran in a race: Andy, Bert, Carlos, Dan, and Earl. How many different ways can they win the first- and second-place ribbons? Finish the organized list to solve the problem. What patterns do you see?

```
A-B    B-A    C-A
A-C    B-C    C-B
A-D
```

MULTIPLICATION OF WHOLE NUMBERS AND DECIMALS

3-2 Multiplying by a 1-Digit Number

Objective
Multiply by a 1-digit number.

The walkway at a shopping mall is covered with one-foot-square tiles. How many tiles are there? What is the **area** of the walkway?

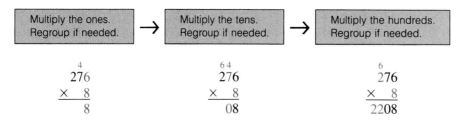

8 feet wide

276 feet long

Area = length × width
$A = l \times w$

Understand the Situation

- How do you find the area of a rectangle?
- What is the length of the walkway?
- What is the width of the walkway?

You can estimate the product. 276 × 8 is about 300 × 8, or 2400.

Examples Multiply.

1. 276 × 8

| Multiply the ones. Regroup if needed. | → | Multiply the tens. Regroup if needed. | → | Multiply the hundreds. Regroup if needed. |

```
    4              6 4               6
   276             276              276
 ×   8           ×   8            ×   8
     8              08             2208
```

The product 2208 seems reasonable, since it is close to the estimate 2400. There is a total of 2208 tiles. The area of the walkway is 2208 square feet.

```
       3            4                        1         2 1 3
2.    89    3.    381    4.   507    5.   1242   6.   3426
    ×  4        ×   6       ×   8       ×    3       ×    5
      356        2286        4056        3726        17,130
```

Try This Multiply.

a. 36   b. 1803   c. 124   d. 2429   e. 85   f. 305
  × 4   ×   6   × 5   ×   8   × 3   × 7

Exercises

Multiply.

1. 32 × 3	2. 174 × 4	3. 26 × 5	4. 1324 × 2	5. 168 × 3
6. 41 × 2	7. 225 × 6	8. 505 × 7	9. 37 × 4	10. 1461 × 8
11. 6026 × 9	12. 827 × 6	13. 97 × 8	14. 7262 × 5	15. 6008 × 7

Find each product.

16. $3 \times 73 = c$
17. $2343 \times 6 = p$
18. $q = 226 \times 4$
19. $u = 5 \times 318$
20. $3006 \times 9 = k$
21. $x = 3 \times 402$
22. $4 \times 1649 = y$
23. $m = 6070 \times 4$
24. $4 \times 627 = b$

Find the area of the rectangle.

25. $l = 456$ meters, $w = 6$ meters
26. $l = 508$ meters, $w = 8$ meters
27. $w = 7$ feet, $l = 1432$ feet
28. $w = 8$ yards, $l = 5280$ yards

Practice Through Problem Solving

29. Find three ways to place one digit in each box to get a product between 200 and 400.

    ☐☐ × ☐

30. Find three ways to place one digit in each box to get a product between 5000 and 8000.

    ☐☐☐☐ × ☐

Estimation

Estimate. The exact answer is between which two numbers? Check using a calculator.

31. 122×7: 700, 800, 900, 1000
32. 2342×3: 6000, 6600, 7200, 7800
33. 4009×4: 15,000; 15,500; 16,000; 16,500

Show Math

34. Draw three different rectangles each with an area of 48 square units.

Working with Variables

You can write multiplication *expressions* different ways. You can write *4 times r* as:

$4 \times r$	$4 \cdot r$	$4(r)$	$4r$
Write an × between the numeral and the variable.	Write a dot between the numeral and the variable.	Write parentheses around the variable.	Use nothing between the numeral and the variable.

Evaluate each expression.

35. $5 \times t$ for $t = 23$
36. $6 \cdot a$ for $a = 45$
37. $245(r)$ for $r = 7$
38. $5ab$ for $a = 7$ and $b = 3$

3-3 Multiplying by a 2-Digit Number

Objective
Multiply by a 2-digit number.

How fast is a chicken? The top rate of speed for a running chicken is estimated to be about 792 feet in 1 minute. How far might this chicken travel if it could run at this rate for 15 minutes?

distance = rate × time
$d = r \times t$

Understand the Situation

- What is the formula that uses distance, rate, and time?
- Is the problem asking you to find distance, rate, or time?

You can estimate the product. 792 × 15 is about 800 × 15. 800 × 10 = 8000 and 800 × 20 = 16,000. Since 15 is halfway between 10 and 20, 800 × 15 should be halfway between 8000 and 16,000. 800 × 15 = 12,000

Examples Multiply.

1. 792 × 15

| Multiply by the ones. | → | Multiply by the tens. | → | Add the products. |

```
   4 1              4 1                4 1
   792              792                7 9 2
 × 15             × 15              ×   1 5
  3960             3960              3 9 6 0
                   792⊙   Write a zero   7 9 2 0
                          as a place holder.  1 1,8 8 0
```

The product 11,880 seems reasonable since it is close to the estimate, 12,000. The chicken might run 11,880 feet in 15 minutes.

```
                           2               1 2           2 2 4
                         1 4                 1             1
2.   321       3.    4 2 8      4.    2 0 4 5    5.    8 3 2 6
   × 24            ×   3 5          ×     4 2        ×     8 3
    1284            2 1 4 0          4 0 9 0          2 4 9 7 8
    6420           1 2 8 4 0         8 1 8 0 0        6 6 6 0 8 0
    7704           1 4,9 8 0         8 5,8 9 0        6 9 1,0 5 8
```

Try This Multiply.

| a. 421 × 23 | b. 635 × 19 | c. 704 × 35 | d. 2518 × 46 | e. 4926 × 74 |

Exercises

Multiply.

1. 23 × 12
2. 34 × 16
3. 312 × 21
4. 103 × 32
5. 264 × 18
6. 325 × 17
7. 418 × 32
8. 677 × 41
9. 423 × 75
10. 1231 × 34
11. 7042 × 63
12. 4218 × 71
13. 246 × 86
14. 682 × 93
15. 780 × 88

Find each product.

16. $332 \times 23 = h$
17. $41 \times 450 = p$
18. $y = 78 \times 603$
19. $z = 874 \times 64$
20. $v = 1234 \times 42$
21. $2003 \times 31 = r$
22. $t = 3475 \times 52$
23. $u = 69 \times 4582$

Solve.

24. The top rate of speed for an elephant is 2200 feet in 1 minute. At this rate, how far could an elephant run in 15 minutes?
25. The top rate of speed for a pig is 968 feet in 1 minute. At this rate, how far could a pig run in 15 minutes?

Calculator Write the calculator keystrokes you used to find the answer.

26. Find the product of 1604 and 37.
27. Find the product of 40 and 365.

Estimation Decide if the exact answer is greater or less than the given number. Check using a calculator.

28. 234×20 < 400 or > 400
29. 388×30 < 1200 or > 1200
30. 508×29 $< 15{,}000$ or $> 15{,}000$
31. 688×38 $< 28{,}000$ or $> 28{,}000$
32. 78×61 < 4800 or > 4800
33. 1299×42 $< 50{,}000$ or $> 50{,}000$

Mixed Skills Review Decide if each person needs an exact answer or an estimate.

34. a cashier counting money in a cash register
35. the gas station attendant figuring change for a $10 bill
36. a shopper choosing what to spend for lunch
37. a family figuring costs for a trip

Add.

38. $5640 + 365$
39. $243 + 692$
40. $7024 + 99$
41. $45 + 67 + 93 + 5$

MULTIPLICATION OF WHOLE NUMBERS AND DECIMALS

3-4 Multiplying by Larger Factors

Objective
Multiply by a 3-digit number.

A survey compared households' median incomes to their years of education.

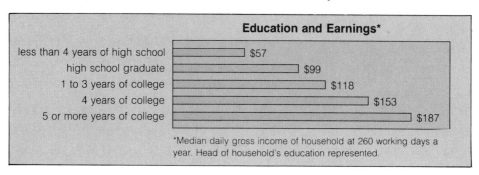

Education and Earnings*

- less than 4 years of high school — $57
- high school graduate — $99
- 1 to 3 years of college — $118
- 4 years of college — $153
- 5 or more years of college — $187

*Median daily gross income of household at 260 working days a year. Head of household's education represented.

Exercises

Use the graph to answer Exercises 1–4.

Problem: How much might a household earn in 1 year if the head of the household has had 4 years of college?

1. Lonell multiplied 153 by 260 to solve this problem. Why is multiplication a correct operation to use?

2. Yolanda multiplied 260 by 153 to solve this problem. Should she find the same answer as Lonell? How do you know?

3. Would you use a calculator, paper and pencil, or mental math to solve this problem? Why?

4. Joan used her calculator to solve this problem. Use estimation to decide if her answer is reasonable.

Joan's answer: 3975.

Tim's paper-and-pencil solution is correct. Use his solution to answer Exercises 5–7.

5. Which numbers did Tim multiply to get the 1300?

6. Tim multiplied 260 × 1 to get 260. Does the 1 stand for 1 ten or 1 hundred in the problem?

7. Tim said the 1300 is really 13,000. Is he correct? Explain.

8. About how much money does a household earn in one year if the head of the household has graduated from high school?

Tim's answer:
```
      3
      1
    2 6 0
  × 1 5 3
  -------
    7 8 0
  1 3 0 0
  2 6 0
  -------
  3 9,7 8 0
```

Decide if each product is reasonable or not reasonable. Tell how you decided.

9. 235 × 188 =   **10.** 255 × 530 =   **11.** 2341 × 322 =

`44180.`   `179240.`   `887902.`

12. 365 × 425 =   **13.** 7582 × 501 =   **14.** 6322 × 1144 =

`15125.`   `3798582.`   `623268.`

Estimate each product. Without finding the answer, decide if your estimate is greater or less than the exact answer. Tell how you decided.

15. 236
    × 102

16. 293
    × 475

17. 406
    × 644

18. 3802
    × 610

19. 7293
    × 610

Problem Solving

Look for a pattern. Complete each sequence.

20. 176 × 2 = 352
176 × 20 = 3520
1760 × 20 = 35,200
1760 × 200 = _____

21. 263 × 4 = 1052
263 × 40 = 10,520
2630 × 40 = 105,200
26,300 × 40 = _____

22. 523 × 4 = 2092
5230 × 400 = _____

23. 126 × 324 = 40,824
12,600 × 32,400 = _____

Number Sense

24. Is 64 × 250 the same as 640 × 25? How can you tell without multiplying?

Mental Math

Find each product. Use mental math.

25. 200 × 100   **26.** 300 × 400   **27.** 20 × 400 × 200

28. 300 × 100 × 200   **29.** 250 × 100   **30.** 425 × 200

31. 231 × 400   **32.** 500 × 210   **33.** 200 × 630

Calculator

Solve. Write the calculator keys you used to find the answer.

34. The manager of the Corner Restaurant finds that they serve about 230 customers every day. How many customers would the restaurant serve in a year?

35. The restaurant owes $485 each week in sales tax. What is the yearly total?

Estimation

Find a partner. Each person picks digits to place in the squares. The person with a product closest to the one given gets one point. Use a calculator to check.

36. 405 × □□ = 4860   **37.** 254 × □□ = 9906

38. 325 × □□□ = 81,250   **39.** □□□ × □□□ = 110,835

MULTIPLICATION OF WHOLE NUMBERS AND DECIMALS

3-5 Estimating Sums Using Clustering

Objective
Estimate sums using clustering.

Another strategy for estimating sums is **clustering.** This strategy is useful when there are several addends that are close in value.

When the addends cluster near the same number, use that number to estimate the sum. Replace each addend with the cluster number. Then multiply by the number of addends.

Examples Estimate. Use clustering.

1. 23 + 27 + 26 + 24 + 26
   Each addend is near 25. There are 5 addends.

   5 × 25 = 125
   23 + 27 + 26 + 24 + 26 is about 125.

2. 19 + 22 + 42 + 21
   19, 22, and 21 cluster near 20.

   19 + 22 + 21 is about 3 × 20, or 60.
   60 + 42 = 102
   19 + 22 + 42 + 21 is about 102.

Try This Estimate. Use clustering.

a. 41 + 42 + 37 + 38   b. 97 + 105 + 186 + 101 + 103

Exercises

Estimate. Use clustering.

1. 18 + 19 + 21
2. 73 + 68 + 69
3. 29 + 31 + 30 + 28
4. 42 + 41 + 40 + 37
5. 52 + 47 + 48 + 51
6. 38 + 39 + 43 + 40
7. 57 + 58 + 60 + 63
8. 39 + 42 + 41 + 39 + 43
9. 84 + 78 + 79 + 81 + 80
10. 24 + 17 + 19 + 20 + 22
11. 194 + 203 + 201 + 195 + 200
12. 610 + 590 + 597 + 603 + 605
13. 23 + 19 + 21 + 45
14. 47 + 49 + 80 + 51
15. 35 + 61 + 58 + 60
16. 87 + 18 + 19 + 22 + 21
17. 42 + 39 + 21 + 18
18. 51 + 49 + 18 + 23
19. 69 + 68 + 33 + 28 + 72
20. 18 + 33 + 21 + 28 + 31
21. 80 + 80 + 21 + 19 + 78
22. 71 + 61 + 58 + 69 + 73
23. 104 + 23 + 98 + 19
24. 280 + 312 + 295 + 64
25. 19 + 58 + 60 + 21
26. 409 + 387 + 416 + 390
27. 32 + 21 + 19 + 29 + 23
28. 500 + 102 + 497 + 98 + 504

Mixed Practice

Write the word name for the standard numeral or the money notation.
1. 56,043
2. 81.26
3. 0.9
4. $158
5. $2.49
6. 316,290
7. 47.388
8. 0.17
9. $5960
10. $38.52

Compare. Write <, >, or =.
11. 2557 ☐ 2575
12. 36.2 ☐ 36.20
13. $4086 ☐ $4080
14. 7.09 ☐ 7.9
15. 0.01 ☐ 0.001
16. 34.26 ☐ 3.426
17. $83,420 ☐ $83,419
18. 9.052 ☐ 9.520
19. 0.09 ☐ 0.10

Round to the given place.
20. 2.56 tenth
21. 313 hundred
22. $89.49 dollar
23. 99 ten
24. $5.70 dime
25. 0.0087 hundredth
26. 1.648 tenth
27. $29.51 dollar

Estimate.
28. 23 × 36
29. 634 ÷ 70
30. 77 + 32
31. 7655 − 7028
32. 2414 ÷ 80
33. 905 ÷ 94
34. $118 + $509
35. 670 × 317
36. 632 − 328
37. 642 × 194

Evaluate.
38. $m + 87$ for $m = 26$
39. $7.94 - x$ for $x = 2.26$
40. $c + d + 50$ for $c = 75$ and $d = 120$
41. $r - 36 - s$ for $r = 95$ and $s = 11$

Add or subtract.
42. 4027 − 275
43. 3.158 − 0.226
44. 60.9 + 72.84
45. 0.05 − 0.048
46. 2983 + 6615
47. 9.624 + 12.7

Add or subtract. Use mental math.
48. 60 − 4
49. $65.50 − $52
50. 3 + 107
51. 37 − 2
52. $140 − $115
53. 298 + 3
54. 2 + 69
55. $76.25 − $65.50

Find the area of the rectangle.
56. $l = 12$ yards, $w = 8$ yards
57. $l = 17$ inches, $w = 9$ inches
58. $w = 8.6$ miles, $l = 24$ miles
59. $w = 4$ feet, $l = 31$ feet

3-6 Problem Solving: Multiple-Step Problems

Objective
Solve multiple-step problems.

- **SITUATION**
- **DATA**
- **PLAN**
- **ANSWER**
- **CHECK**

You can solve many problems using two or more of the operations: addition, subtraction, multiplication, or division. These problems are called **multiple-step** problems.

TRACY'S SEMI-ANNUAL CLOTHING SALE TODAY ONLY

skirts—all sizes, $25
blouses—were $18, now $12
slacks—$35 cotton, $45 wool
top name tennis shoes—$28
sweaters—all styles, $32
running outfits—$12
summer shorts—2 for $14
dresses—clearance, $65–$95

Example

Kara bought 2 blouses and 3 running outfits at the sale. How much did she pay all together, without tax?

Sean's solution:

A. Find the cost of the blouses.
$12
× 2
$24

B. Find the cost of the running outfits.
$12
× 3
$36

C. Find the total cost.
$24
+$36
$60

Tina's solution:

A. Blouses and running outfits cost the same. Find the total number of blouses and running outfits.
2 + 3 = 5

B. Find the total cost.
$12
× 5
$60

The total cost for 2 blouses and 3 running outfits is $60.

You can solve some multiple-step problems in more than one way. Sean and Tina found different ways to solve this problem. Both solutions are correct.

Try This

Solve. Use the sale ad.

What is the total cost, without tax, for 3 skirts and 2 pairs of wool slacks?

Exercises

Pick one or more of the statements that give the steps you could use to solve the problem. Use data from the sale ad on p. 70 for Exercise 2.

1. Last week 24 girls and 35 boys bought tickets for a class trip. Each ticket cost $5. How much money did they pay all together for the tickets?
   A. Multiply $5 times 24. Add 35.
   B. Add 24 and 35. Multiply the sum by $5.
   C. Add 24, 35, and $5.
   D. Multiply 24 by $5. Multiply 35 by $5. Add the two products.

2. Karen bought 2 pairs of cotton slacks. The total tax was $4.20. How much did Karen pay for the slacks, with tax?
   A. Add $35 and $4.20.
   B. Multiply $4.20 by 2. Add $35.
   C. Multiply $35 by 2. Add $4.20.
   D. Add $35 and $4.20. Multiply the sum by 2.

3. At Munro's Store, Denise bought 3 pairs of tennis shoes for $28 each, with tax. She paid with a $100 bill. How much was her change?
   A. Subtract $28 from $100.
   B. Multiply $28 by 3. Subtract the product from $100.
   C. Subtract $28 from $100. Multiply the difference by 3.
   D. Add 3 to $28. Subtract the sum from $100.

4. Movie tickets for adults cost $6 each and children's tickets cost $4 each. What is the cost for 4 adult tickets and 5 children's tickets?
   A. Multiply 4 and $6. Multiply $4 and 5. Add the products.
   B. Multiply 4 and $4. Multiply 5 and $6. Add the two products.
   C. Add $6 and 4. Add $4 and 5. Multiply the two sums.
   D. Multiply 4 and $6. Multiply 4 and $5. Multiply the products.

Solve.

5. Rafer bought 2 CDs for $12 each and 2 tapes for $8 each. Tax for these items was $2.40. How much did he pay all together?

6. A school bus rents for $125 a day plus 10¢ a mile. How much would it cost to rent the bus one day for a 300-mile trip?

Suppose

7. Use the example on p. 70. Suppose Tracy's Store marked down all the sale clothing an additional $3 per item. Find the cost of the clothing Kara bought.

Write Your Own Problem

8. Write a word problem that you could solve using these steps.
   A. Multiply $5 by 15.
   B. Multiply $4 by 25.
   C. Add the two products.

3-7 Multiplying Decimals

Objective
Multiply decimals.

Lynn would like to work 6.5 hours on the weekends to earn extra money. Her guidance counselor gave her a list of jobs. Suppose Lynn takes one of these jobs. What is the most Lynn could make on a weekend?

Job and Hourly Wage
- sales clerk — $5.25
- fast-food server — $4.25
- garage helper — $4.75
- waiter/waitress — $3.50
- grocery clerk — $6.25

Understand the Situation
- How many hours does Lynn want to work on weekends?
- Why is multiplication an operation you can use to find the answer?

You can estimate the product. Round each decimal to the nearest whole number and multiply. 6.25×6.5 is about 6×7, or 42.

Examples Multiply.

1. 6.25×6.5

| Multiply as with whole numbers. | → | Find how many places are to the right of the decimal points. | → | Place the decimal point so that there is this number of places to the right. |

```
   6.2 5              6.2 5                        6.2 5
 × 6.5              × 6.5    three places        × 6.5
   3 1 2 5            3 1 2 5                      3 1 2 5
   3 7 5 0 0          3 7 5 0 0                    3 7 5 0 0
   4 0 6 2 5          4 0 6 2 5   three places     4 0.6 2 5
```

The answer 40.625 seems reasonable since it is close to the estimate 42. When computing wages, fractions of cents are usually rounded down. Round $40.625 down to the nearest cent. Lynn could make $40.62.

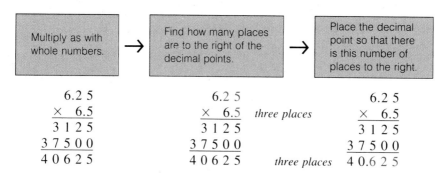

2. 2.4
      ×0.6
      1.44

3. 3.47
      ×0.02
      0.0694
      Write a zero to show place value.

4. 23.6
      × 24
      944
      4720
      566.4

5. 0.605
      × 0.18
      4840
      6050
      0.10890

Try This Multiply.

a. 8.23
      × 1.4

b. 4.5
      ×2.4

c. 0.78
      ×0.02

d. 4.5
      ×6.9

e. $4.85
      × 15

Exercises

Write each product by placing the decimal point in the correct place. Write zeros as needed.

1. 7.4 × 0.6 = 444	2. 5.31 × 0.2 = 1062	3. 0.85 × 6.3 = 5355	4. 0.02 × 0.8 = 16	5. 0.025 × 0.4 = 1

Multiply.

6. 2.5 × 0.3	7. 4.1 × 0.8	8. 0.75 × 1.2	9. 0.84 × 2.4	10. $3.60 × 6.2
11. 0.17 × 8.2	12. 0.55 × 4.6	13. $1.24 × 0.5	14. 4.06 × 0.8	15. 0.42 × 0.7
16. 0.48 × 0.06	17. $1.25 × 0.04	18. 0.065 × 0.18	19. 0.108 × 0.26	20. 0.776 × 4.2

21. $4.5 \times 2.8 = h$
22. $c = 8.2 \times 1.7$
23. $d = 75 \times 3.4$
24. $64 \times 8.2 = p$
25. $\$3.45 \times 0.6 = x$
26. $\$8.04 \times 4.5 = y$
27. $b = \$2.65 \times 18$
28. $m = \$5.75 \times 24$

Mixed Applications

29. Ralph worked at the garage for 5 hours. His brother, Jack, worked 5 hours at the fast-food restaurant. How much more did Ralph earn than Jack for these 5 hours? Use the table on p. 72.

Problem Solving

Pick one or more of the statements that give the steps you could use to solve this problem. Use the table on p. 72.

30. Masuo worked 5.5 hours as a waitress. She also earned $12.50 in tips. How much did she earn all together?
    A. Multiply 5.5 and $12.50. Add $3.50.
    B. Add $12.50 and $3.50. Multiply by 5.5.
    C. Multiply 5.5 and $3.50. Add $12.50.

Estimation

Estimate. The exact answer is between which two numbers? Check using a calculator.

31. $7.89 × 8: $56, $60, $66, $72
32. $4.35 × 6: $24, $28, $32, $36
33. $8.10 × 4.5: $35, $40, $45, $50
34. $9.50 × 0.5: $3.50, $4.00, $4.50, $5.00

Talk Math

35. Tell how you can use addition to find 6 × 0.25.

Suppose

Use the table on p. 72. Suppose Lynn wanted to work 8.5 hours on the weekends.

36. What is the most Lynn could make on a weekend working as a sales clerk?
37. What is the most Lynn could make on a weekend working as a waitress?

3-8 Multiplying by 10, 100, or 1000

Objective
Multiply by 10, 100, or 1000.

To get data for this graph, a consumer group questioned 1000 people between the ages of 45 and 54. What was the total number of cars owned by people in this age range who were questioned?

Understand the Situation

- What does *head of the household* mean?
- What was the average number of cars per household for ages 45 to 54?
- Can a family own 2.7 cars?

You can estimate using rounding.
2.7 × 1000 is about 3 × 1000, or 3000.

Who Owns the Cars

Age of head of household	Number of vehicles per household
65 and over	1.1
55–64	2.1
45–54	2.7
35–44	2.3
25–34	1.9
Under 25	1.2

Examples Multiply.

1. 2.7 × 10 = 27.0
Multiply by 10.
Move the decimal point one place to the right.

2. 2.7 × 100 = 270.0
Multiply by 100.
Move the decimal point two places to the right.

3. 2.7 × 1000 = 2700.0
Multiply by 1000.
Move the decimal point three places to the right.

The product 2700 in Example 3 seems reasonable because it is close to the estimate 3000. People in the 45 to 54 age range owned a total of 2700 cars.

4. 4.56 × 10 = 45.6   **5.** 3.264 × 100 = 326.4   **6.** 1.53 × 1000 = 1530

Try This Multiply.

a. 9.34 × 10   **b.** 0.0456 × 100   **c.** 1.758 × 1000   **d.** 14 × 100
e. 5.02 × 10   **f.** 789.4 × 100   **g.** 21.6 × 1000   **h.** 9.012 × 1000

Exercises

Multiply.

1. 4.81 × 10
2. 0.72 × 10
3. 80.5 × 100
4. 0.712 × 100
5. 0.204 × 1000
6. 100 × 1.24
7. 0.778 × 1000
8. 1000 × 4.223
9. 10 × 100
10. 100 × 100
11. 100 × 24
12. 0.7 × 1000
13. 23.6 × 10
14. 7.8 × 10
15. 86.03 × 1000

Evaluate.

16. 10d for $d = 12.55$
17. 100a for $a = 3.42$
18. 10xy for $x = 10$ and $y = 0.527$
19. 100mn for $m = 2.2$ and $n = 3.5$

Practice Through Problem Solving

20. Find three ways to place the same digit in each box to get a product less than 100. You can place a decimal point between any two of the boxes.
    □ □ □ □ × 10

21. Find three ways to place the same digit in each box to get a product less than 1000. You can place a decimal point between any two of the boxes.
    □ □ □ □ × 100

Calculator

Use a calculator to find each product. Look for a pattern. Write a statement telling how to multiply a decimal by 10,000 or 100,000.

22. 2.4056 × 10,000
23. 0.00604 × 10,000
24. 12.04 × 10,000
25. 0.468 × 100,000
26. 0.072031 × 100,000
27. 32.089 × 100,000

Mixed Applications

Solve. Use the Car Cost per Mile graph.

28. What did it cost in 1986 to own and run a car for 1000 miles?
29. About how much did it cost in 1980 to own and run a car for 100 miles?
30. How much more did it cost to own and run a car for 100 miles in 1986 than it did in 1976?

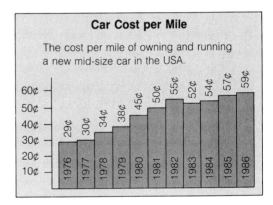

Car Cost per Mile

The cost per mile of owning and running a new mid-size car in the USA.

Mixed Skills Review

Find each answer.

31. 296 × 7
32. 12,004 − 765
33. $23.02 + $9.89
34. 4286 × 35
35. Gwen's checking account balance on Monday was $115.23. She deposited a check for $25 on Tuesday. On Friday, she wrote a check for $37.54. What was her new balance?

3-9 Consumer Math: Reading a Paycheck

Objective
Read a paycheck.

- SITUATION
- DATA
- PLAN
- ANSWER
- CHECK

With a paycheck, you receive a statement of your pay and deductions. A **deduction** is an amount subtracted from your total earnings.

SOC. SEC. NO. 000-11-0000 MARTINSON, LESLIE ANN			PERIOD ENDING 03-31-88		**ACE GRAPHICS**	CHECK NO. 546663
RATE	HOURS	GROSS	TYPE OF DEDUCTION	AMOUNT	YEAR TO DATE	AMOUNT
12.38	40	495.20	FICA	34.66	Total Gross	1485.60
			Federal Tax	89.14	FICA	103.98
			State Tax	10.27	Fed. Tax	267.42
					State Tax	30.81
CURRENT GROSS		495.20	CURRENT DEDUCTIONS	134.07	CURRENT NET	361.13

Examples

1. The **gross** amount is your total pay before any deductions are taken. To find the gross amount, multiply the hourly rate by the number of hours worked. Is the current gross amount given on the statement correct?

    $12.38 \times 40 = 495.20$ The amount is correct.

2. **Current deductions** gives the sum of all deductions for the pay period. Is this sum on the statement correct?

    $34.66 + 89.14 + 10.27 = 134.07$ The amount is correct.

3. The **net** amount is the amount of money you take home. It is sometimes called **take-home pay.** Find this amount by subtracting the current deductions from the current gross. Is the current net amount on the statement correct?

    $495.20 - 134.07 = 361.13$ The amount is correct.

4. The **year-to-date** data gives a total of your pay and deductions for the calendar year. The year-to-date data is given for how many months?

    The pay period ended 3-31-88. The year-to-date data is for three months.

Try This   What is the total amount of federal and state tax she has paid so far this year?

Exercises

Solve. Use Leslie Martinson's statement on p. 76 for Exercises 1 and 2. (Each month's paycheck was the same.)

1. Check if the year-to-date federal tax amount is correct.
2. Check if the year-to-date state tax amount is correct.

Use this monthly paycheck statement for Dee Marshal to do Exercises 3–8. (Each month's paycheck was the same.)

SOC. SEC. NO. 000-22-0000 MARSHAL, DEE DIANNE		PERIOD ENDING 06-30-88		**S.F.** **PRESS**	CHECK NO. 1244900	
RATE	HOURS	GROSS	TYPE OF DEDUCTION	AMOUNT	YEAR TO DATE	AMOUNT
7.75	40	? ? ?	FICA	21.70	Total Gross	1860.00
			Federal Tax	55.36	FICA	130.20
			State Tax	6.43	Fed. Tax	? ? ?
					State Tax	? ? ?
CURRENT GROSS	? ? ?	CURRENT DEDUCTIONS	? ? ?		CURRENT NET	? ? ?

3. What was Dee's gross income for the month?
4. What was the total amount of her deductions for this pay period?
5. How much federal tax has Dee paid so far this year?
6. How much state tax has Dee paid so far this year?
7. How much money did Dee take home for this pay period?
8. How much money has Dee taken home for the year to date?

Solve It Another Way

9. Raul solved this checkbook problem using the strategy *draw a picture*. Find another way to solve this problem.

Problem: A bank has 3 choices for checkbook covers: dark blue, black, or white. There are 5 choices for check colors: green, blue, yellow, tan, or purple. How many different ways can you pick the colors for a checkbook cover and checks?

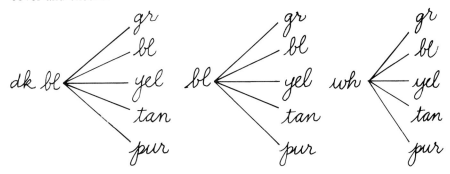

MULTIPLICATION OF WHOLE NUMBERS AND DECIMALS

3-10 Exponents

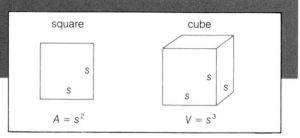

Objective
Write numbers using exponents.

Exponents show how many times the same factor is repeated. The repeated factor is called the **base**.

$16 = \underbrace{2 \times 2 \times 2 \times 2}_{\text{factors}} = 2^4 \leftarrow \text{exponent}$

↑ standard numeral ↑ base

Exponents show **powers** of the base.

Examples Write using exponents. Then write as a standard numeral.

1. $3 \times 3 \times 3 \times 3 \times 3 = 3^5 = 243$ — 3^5 is 3 to the **fifth power**.
2. $10 \times 10 = 10^2 = 100$ — 10^2 is 10 to the **second power**, or 10 **squared**.
3. $(0.4)(0.4)(0.4) = (0.4)^3 = 0.064$ — $(0.4)^3$ is 0.4 to the **third power**, or 0.4 **cubed**.

Try This Write using exponents. Then write as a standard numeral.

a. 5×5 b. $4 \times 4 \times 4$ c. $(2.1)(2.1)(2.1)$ d. $7 \times 7 \times 7 \times 7 \times 7$

Exercises

Write using exponents.

1. $10 \times 10 \times 10$
2. $3 \times 3 \times 3 \times 3 \times 3 \times 3$
3. $2 \times 2 \times 2 \times 2 \times 2 \times 2 \times 2$
4. $(0.1)(0.1)(0.1)(0.1)$
5. 12×12
6. $10 \times 10 \times 10 \times 10$
7. 5 to the fourth power
8. 7 squared
9. 6.9 cubed
10. 20.6 squared

Write as a standard numeral.

11. 2^3 12. 3^4 13. 8^2 14. 9^3 15. $(0.2)^3$
16. $(0.5)^2$ 17. 4^4 18. 10^5 19. $(1.5)^3$ 20. $(2.3)^2$

Data Bank

21. Find the volume of a cylinder with a radius $r = 4$ centimeters and a height $h = 12$ centimeters. Use the formula table on p. 495.

Write Math

22. Write a statement telling how to write $2^3 \times 2^2$ using the base one time with only one exponent.

CHAPTER 3

3-11 Mental Math: Break Apart Numbers

Objective
Use the mental math strategy break apart numbers.

Sometimes the exact answer to a problem can be found using mental math. Break apart numbers to make smaller, easier calculations. Think about place values.

Examples Find each product or sum. Use mental math. Break apart numbers to help.

1. 42×3 *Think: Break apart 42 into 40 and 2.*
   $40 \times 3 = 120$ and $2 \times 3 = 6$
   42×3 is $120 + 6 = 126$

2. $325 + 440$ *Think: Break apart 325 into 300 and 25.*
   Break apart 440 into 400 and 40.
   $400 + 300 = 700$ and $25 + 40 = 65$
   $325 + 440$ is $700 + 65 = 765$

Try This Find each product or sum. Use mental math. Break apart numbers to help.

a. 32×4 b. 21×3 c. $160 + 332$ d. $38 + 7$ e. 32×5

Exercises

Find each product or sum. Use mental math. Break apart numbers to help.

1. 51×4
2. 43×3
3. $34 + 25$
4. $82 + 26$
5. $215 + 60$
6. $440 + 26$
7. 35×3
8. 72×5
9. $65 + 15$
10. $25 + 35$
11. $123 + 220$
12. $457 + 220$
13. $78 + 9$
14. 210×5
15. 201×4
16. $53 + 20 + 21$
17. $435 + 220 + 5$
18. $23 + 160 + 360$

You Decide Find each answer using one of the mental math strategies shown so far.

19. $68 + 3$
20. 53×3
21. $72 - 3$
22. 62×5
23. $\$4 - \$2.50$
24. $2 + 219$
25. $200 - 78$
26. $134 + 263$

Mental Math Strategies
- Counting
- Break apart numbers

These are the strategies presented so far.

MULTIPLICATION OF WHOLE NUMBERS AND DECIMALS

3-12 Problem Solving: Data Collection and Analysis

Objective
Collect and analyze data.

- SITUATION
- DATA
- PLAN
- ANSWER
- CHECK

Theatre advertisements in the newspaper give a lot of information in a small space. Read carefully to find the information you need to make a decision.

FAVORITE FAMILY MUSICAL
THE KING AND I
Mon.-Thurs. Evgs. at 8 & Sat. Mat. at 2: $22.50, 18.50, 16. Fri. & Sat. Evgs. at 8: $25, 21, 18.50. Wed. Mat. at 2: $18.50, 16, 13.50.
Charge by phone: **555-8330**
•• WATERFRONT THEATRE ••
395 West Street.   Free Parking.

FINAL 12 PERFORMANCES!
FIDDLER ON THE ROOF
Wed., Thurs., Fri., Sat. at 8:00 p.m.; Sun. at 7:30 p.m. Tickets: $22.50, 19.50, 17.50, 13.50; 11.50 for Wed. & Thurs. mat.; $35, 30, 22.50, 18, 16 for Fri., Sat., and Sun. eve.
THE STAR THEATRE • 555-9246
1428 South Street

STARTS THIS FRIDAY!
GREASE
Tue.-Thur. 8 PM, Fri. & Sat. 8:30 PM. Wed. & Sat. Mats. 2:30 PM, Sun. Mats. 3 PM. Tickets $11-$32.50. Tickets at Circle Theater Box Office and major agencies.
CALL 555-6229
CIRCLE THEATRE
249 6th St. (between Lark and Howard)

1976 TONY & PULITZER PRIZE WINNER!
A CHORUS LINE
Tue.-Thur at 8 p.m.: $22.50, 19, 17, 15, 13
Fri. & Sat. Evgs. at 8: $30, 27, 25, 23, 20
Mats. Sat. at 2, Sun. at 2:30: $20, 17, 15, 13, 11; Wed Mats. at 2: $18, 15, 13, 12, 9
62 Main Street   Call: **555-8181**
or call any major ticket agency
THEATRE ON THE SQUARE

Situation

Suppose you and a friend want to see *A Chorus Line* and *The King and I*.
Use the newspaper listings to make a chart of the information for both plays.
Use one column for each play.

- List days when no performances are given.
- List days when two performances are given.
- List the cost for the highest-priced seat on the least expensive day.
- List the cost for the highest-priced seat on the most expensive days.
- List the cost for the highest-priced tickets for two people.
- List the cost for the lowest-priced tickets for two people.
- List the days when matinees are given.
- Tell which performance time and price you choose to see *A Chorus Line*. Give your reasons.

3-13 Practice Through Problem Solving

Objective
Practice multiplication using problem-solving strategies.

Multiplication Games

1. *That's Really Great*

   Material: number cube, marked 1–6
   Players: two to four individual players or teams of two
   Directions:
   Copy the multiplication problem boxes (one for each player or team). Take turns tossing the cube. Each time, write the digit in one of the boxes. The person with the greatest product is the winner. Repeat five times.

   ☐ ☐ ☐
   ×   ☐

   Change the game: Make the winner the one with the least product.

2. *Hit the Target*

   Material: number cube, marked 1–6
   Players: two to four individual players or teams of two
   Directions:
   Toss the cube four times and record the digits in order. This number is the target number. Copy the multiplication problem boxes (one for each player or team). Take turns tossing the cube. Each time write the digit in one of the boxes. The person with the product closest to the target number is the winner. Repeat five times.

      ☐ ☐
   ×☐ ☐

   Change the game: Change the target number to a 3-digit number.

Multiplication Riddles

3. If you multiply the two of us together, you get 120. If you subtract us, you get 2. Who are we?

4. If you multiply me by myself 3 times, you get 15,625. Who am I?

5. I am less than 1. If you double me and add 0.2, you get 2. Who am I?

Get It in the Ballpark

6. Will 7 × $1.98 fall between $7 and $10, $10 and $13, or $13 and $15?

7. Will 12 × $3.05 fall between $32 and $36, $36 and $40, or $40 and $44?

8. Will 4 × $2.47 fall between $8 and $10, $10 and $12, or $12 and $14?

3-14 Enrichment: Prime Factorization

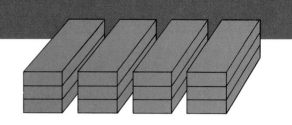

Objective
Find the prime factorization of a number.

Disc jockey Wild Willie has 12 songs to play in the next hour. He wants to divide the music into sets with the same number of songs in each set. He decides to play 4 sets with 3 songs each.

Understand the Situation

- How else could Wild Willie divide the songs into sets?
- If Willie has 12 songs, how many different sets of an equal number of songs could he play?
- If Willie has 13 songs, how many different sets of an equal number of songs could he play?

The factors of 12 are 1, 2, 3, 4, 6, and 12. The factors of 13 are 1 and 13.
A **prime number** is a number with exactly two factors. Some prime numbers are 2, 3, 5, 7, 11, 13, 17, and 19.

A **composite number** is a number that has factors other than 1 and itself. An example of a composite number is 12. Zero and 1 are neither prime nor composite.

The **prime factorization** of a number is the number written as a product of prime numbers.

Example

Find the prime factorization of 24.

Make a factor tree until each branch is a prime number.

The prime factorization of 24 is $3 \times 2 \times 2 \times 2$.
You can write this as $24 = 3 \times 2^3$.

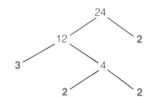

Exercises

Find the prime factorization of the number.

1. 100
2. 42
3. 66
4. 63
5. 36
6. 195

7. 60
8. 144
9. 64
10. 75
11. 72
12. 300

13. 108
14. 909
15. 250
16. 25
17. 81
18. 1000

19. Find the sum of the three least prime numbers. Find the prime factorization of the sum.

20. Find the sum of the five least prime numbers. Find the prime factorization of the sum.

CHAPTER 3

3-15 Computer: Multiplying Decimals

Objective
Use a BASIC program to multiply decimals.

Your computer understands a computer language called BASIC. BASIC programs can do addition using the [+] key, subtraction using the [−] key, multiplication using the [*] key, and division using the [/] key.

Example Coach Vitale ordered 17 baseball caps for the team. Each cap cost $4.97. What was the total cost of the caps?

10 REM FINDING THE TOTAL COST	REM *is a remark.*
20 PRINT "HOW MANY ITEMS?"	PRINT *prints everything inside the quotes.*
30 INPUT N	INPUT *waits for you to enter a number* N.
40 PRINT "WHAT IS THE COST OF EACH ITEM?"	
50 INPUT C	INPUT *waits for you to enter a number* C.
60 LET T = N*C	LET *sets the value of* T *equal to the product of* N *and* C. *The* * *symbol is used for multiplication.*
70 PRINT "THE TOTAL COST IS $";T	*This line prints everything in quotes and the value* T.
80 END	*This line tells the program to stop.*

Run the program by typing RUN and pressing Return.

HOW MANY ITEMS?	*Line* 20 *prints this question.*
?17	*The* INPUT *in line* 30 *prints a* "?" *and waits for you to type 17.*
WHAT IS THE COST OF EACH ITEM?	*Line* 40 *prints this question.*
?4.97	*The* INPUT *in line* 50 *prints a* "?" *and waits for you to type 4.97.*
THE TOTAL COST IS $84.49	*Line* 60 *calculates the product. Line* 70 *prints it.*

Exercises

1. Type the program above. Then type LIST and press Return. The program should appear on your screen. Correct any mistakes. Type RUN to use the program.
2. Use the program to find the cost of 174 tapes at $3.87 each.
3. Use the program to find the cost of 136 T-shirts at $6.47 each.

MULTIPLICATION OF WHOLE NUMBERS AND DECIMALS

CHAPTER 3 REVIEW

Decide if the answers will be <, >, or = for the choices given.

1. 73×5
   - ☐ 300
   - ☐ 350
   - ☐ 400
   - ☐ 500

2. 325×20
   - ☐ 700
   - ☐ 5000
   - ☐ 7000
   - ☐ 10,000

3. 240×105
   - ☐ 2405
   - ☐ 20,000
   - ☐ 25,000
   - ☐ 30,000

4. 32×19
   - ☐ 30×20
   - ☐ 640
   - ☐ 6400
   - ☐ 35×20

5. 8.03×1.5
   - ☐ 0.9
   - ☐ 10
   - ☐ 16
   - ☐ 80

6. 0.48×0.02
   - ☐ 0.00096
   - ☐ 0.0096
   - ☐ 0.096
   - ☐ 0.96

7. $0.032 \times 10,000$
   - ☐ 0.000032
   - ☐ 0.00032
   - ☐ 32
   - ☐ 32,000

8. 3^4
   - ☐ $3 + 3 + 3 + 3$
   - ☐ 3×4
   - ☐ $4 \times 4 \times 4$
   - ☐ 81

9. $2^2 \times 3^2$
   - ☐ 4×6
   - ☐ 5^2
   - ☐ 36
   - ☐ 6^4

10. 2.13×10
    - ☐ 0.213
    - ☐ 2.13
    - ☐ 21.3
    - ☐ 213

Choose the correct answer.

11. 54×25
    - A. 378
    - B. 1250
    - C. 1339
    - D. 1350

12. 146×103
    - A. 1898
    - B. 14,038
    - C. 15,028
    - D. 15,038

13. 54.2×1.5
    - A. 58.91
    - B. 81.3
    - C. 589.10
    - D. 813.00

14. 6.93×100
    - A. 69.3
    - B. 6.93
    - C. 0.693
    - D. 693

15. 0.035×1000
    - A. 0.35
    - B. 35
    - C. 3.5
    - D. 350

Estimate.

16. $18 + 21 + 17 + 19 + 20$
    - A. 90
    - B. 100
    - C. 95
    - D. 105

17. $0.00502 \times 10,000$
    - A. 5.02
    - B. 50.2
    - C. 502
    - D. 5020

18. $(1.5)^3$
    - A. 3.375
    - B. 4.5
    - C. 33.75
    - D. 0.045

Match.

19. amount subtracted from total earnings
20. total earnings minus deductions
21. total earnings for the calendar year
22. total earnings before any money is taken out

A. net amount
B. deduction
C. gross amount
D. year-to-date amount

CHAPTER 3 TEST

Pizza Choices
Choose Any 2 Toppings
Anchovies Onions
Green Peppers Sausage
Pepperoni Mushrooms

1. How many different kinds of pizza can you make?

Multiply.

2. 27 × 8
3. 217 × 6
4. 709 × 4
5. 378 × 7
6. 1475 × 3
7. 9583 × 8

8. 37 × 63
9. 48 × 39
10. 593 × 77
11. 739 × 84
12. 4963 × 52
13. 9037 × 38

14. 822 × 430
15. 9764 × 125
16. 2903 × 477
17. 8001 × 1008
18. 4624 × 3299
19. 3605 × 1525

Estimate. Use clustering. 20. 149 + 153 + 147 + 203 21. 54 + 53 + 49 + 47 + 52

22. Rosita bought 2 bags of peanuts that cost $0.60 each. She bought 3 large apples that cost $0.50 each. How much did she pay for everything?

Multiply. Round answers in money notation to the nearest cent.

23. 4.7 × 0.7
24. 0.62 × 7.9
25. $3.18 × 0.75
26. $37.20 × 100
27. 10.6 × 1000
28. 0.56 × 10

29. Arthur received his November paycheck. What was his net pay?

SOC. SEC. NO. 000-11-6934 GONZALES, ARTHUR M.			PERIOD ENDING 11-30-88		**ACE HARDWARE**		CHECK NO. 593201
RATE	HOURS	GROSS	TYPE OF DEDUCTION	AMOUNT	YEAR TO DATE		AMOUNT
6.40	40	256.00	FICA	17.92	Total Gross		12,288.00
			Federal Tax	45.57	FICA		860.16
			State Tax	5.12	Fed. Tax		2187.36
					State Tax		245.76
CURRENT GROSS		256.00	CURRENT DEDUCTIONS	68.61	CURRENT NET		

Write using exponents. Then write as a standard numeral.

30. 11 × 11 × 11 × 11
31. 4 × 4 × 4 × 4 × 4
32. eight cubed

Find each product or sum. Use mental math. Break apart numbers to help.

33. 65 × 4
34. 320 × 6
35. 730 + 255
36. 41 + 320 + 504

CHAPTER 3 CUMULATIVE REVIEW

1. Multiply.
   2.368 × 100
   A. 236.8
   B. 23.68
   C. 2368
   D. 236

2. Decide if the answer given is reasonable without solving the problem.
   Problem: In 1988, Justin's radio was 12 years old. In what year did Justin buy his radio?
   Answer: 1946
   A. reasonable
   B. unreasonable

3. Decide which operation you would use to solve this problem.
   Problem: Mubina sold △ carwash tickets a day for ☐ days. How many tickets did she sell in all?
   A. addition
   B. subtraction
   C. multiplication
   D. division

4. Estimate.
   358 ÷ 6
   A. 30
   B. 50
   C. 6
   D. 60

5. Write the net deposit: one $0.50 roll of pennies, check 26-189 for $36.75, check 38-2974 for $84.32, less $25 cash received.
   A. $96.57
   B. $101.07
   C. $146.57
   D. $121.57

6. A tent comes in 4 sizes (1, 2, 4, or 8 persons) and 3 colors (red, blue, or green). How many possible combinations of sizes and colors are there?
   A. 12
   B. 6
   C. 8
   D. 7

7. Write using exponents.
   (0.6)(0.6)(0.6)
   A. 6^3
   B. $(0.6)^3$
   C. $(0.6)^6$
   D. (0.6)(3)

8. Multiply.
   $4.45
   × 17
   A. $47.45
   B. $75.35
   C. $72.65
   D. $75.65

9. Multiply.
   1096 × 7
   A. 7672
   B. 7096
   C. 7696
   D. 7632

10. Compare. Use <, >, or =.
    3.008 ☐ 3.80
    A. <
    B. >
    C. =

11. Round to the nearest dime.
    $1.83
    A. $1.90
    B. $2.00
    C. $1.80
    D. $1.85

12. Estimate.
    481 − 368
    A. 200
    B. 110
    C. 130
    D. 90

13. Choose the calculation method that you would use to solve this problem.
    790 − 190
    A. mental math
    B. calculator
    C. paper and pencil

14. Multiply.
    56 × 82
    A. 4592
    B. 4192
    C. 4582
    D. 560

15. Multiply.
    342 × 360
    A. 3078
    B. 123,120
    C. 121,120
    D. 12,312

16. Carey, Trent, Caspar, and Barns are towns on the same road. Trent is 5 miles east of Carey. Caspar is 10 miles west of Trent. Barns is 3 miles east of Carey. How far apart are Caspar and Barns?
    A. 13 miles
    B. 10 miles
    C. 8 miles
    D. 3 miles

17. Write in money notation. seventy-four dollars and thirty-two cents
    A. $704.32
    B. $74.302
    C. $74.32
    D. $7.432

18. Estimate. Round to the highest place.
    222 + 189
    A. 320
    B. 400
    C. 300
    D. 500

19. Subtract.
    28.06
    − 9.37
    A. 18.79
    B. 19.69
    C. 18.89
    D. 18.69

20. Multiply. Use mental math.
    62 × 8
    A. 496
    B. 416
    C. 486
    D. 506

21. Add.
    56.24 + 16.97
    A. 73.11
    B. 72.21
    C. 73.21
    D. 63.21

22. Subtract. Use mental math.
    $165 − $20.50
    A. $145
    B. $144.50
    C. $145.50
    D. $144

23. Estimate.
    47 + 48 + 38 + 51
    A. 170
    B. 150
    C. 190
    D. 200

24. Estimate.
    684 × 3
    A. 1800
    B. 1850
    C. 2200
    D. 2100

25. Subtract.
    4.56 − 0.9
    A. 4.66
    B. 4.47
    C. 3.66
    D. 4.57

26. Decide which operation you would use to solve this problem.
    Problem: There are □ sportscars in the parking lot. There are △ trucks. How many more trucks are there?
    A. addition
    B. subtraction
    C. multiplication
    D. division

27. Farrell bought 2 shirts for $23 each. The total tax was $3.22. How much did Farrell pay in all?
    A. $49.22
    B. $29.44
    C. $42.78
    D. $52.44

Chapter 4 Overview

Key Ideas

- Use the problem-solving strategy *guess and check*.
- Divide whole numbers and decimals.
- Read a phone bill.
- Analyze data from an advertisement.
- Make group decisions.

Key Terms

- divisor
- dividend
- quotient
- remainder

Key Skills

Find the answer.

1. $12 \div 3$
2. $24 \div 6$
3. $54 \div 6$
4. $28 \div 7$
5. $36 \div 6$
6. $32 \div 8$
7. $48 \div 8$
8. $49 \div 7$
9. $42 \div 6$
10. $35 \div 7$
11. 4×5
12. 7×6
13. 8×3
14. 4×9
15. 5×4
16. 7×8
17. 3×9
18. 7×9
19. 5×3
20. 4×6
21. $12 - 5$
22. $9 - 5$
23. $13 - 6$
24. $11 - 4$
25. $15 - 8$
26. $16 - 7$
27. $17 - 8$
28. $14 - 9$
29. $12 - 8$
30. $13 - 8$

Subtract.

31. 46 − 42
32. 75 − 72
33. 26 − 18
34. 62 − 56
35. 89 − 81
36. 75 − 38
37. 36 − 32
38. 50 − 42
39. 41 − 36
40. 76 − 64
41. 60 − 48
42. 80 − 63

Multiply, then add.

43. $5 \times 3 =$ ___ $+ 4 =$ ___
44. $2 \times 7 =$ ___ $+ 6 =$ ___
45. $6 \times 8 =$ ___ $+ 6 =$ ___
46. $5 \times 7 =$ ___ $+ 4 =$ ___
47. $8 \times 5 =$ ___ $+ 9 =$ ___
48. $6 \times 0 =$ ___ $+ 8 =$ ___
49. $9 \times 3 =$ ___ $+ 4 =$ ___
50. $8 \times 8 =$ ___ $+ 6 =$ ___
51. $3 \times 2 =$ ___ $+ 8 =$ ___
52. $7 \times 7 =$ ___ $+ 8 =$ ___
53. $4 \times 8 =$ ___ $+ 3 =$ ___
54. $6 \times 7 =$ ___ $+ 6 =$ ___

DIVISION OF WHOLE NUMBERS AND DECIMALS 4

4-1 Problem Solving: Learning to Use Strategies

Objective
Use the strategy guess and check.

- SITUATION
- DATA
- **PLAN**
- ANSWER
- CHECK

The shipping department forgot to label this week's shipping graph. The manager wanted to know how many boxes of each kind the department shipped.

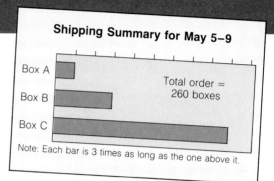

Example How many boxes of each kind did the department ship?

One way to understand the situation is to **draw a picture**.

□ —→ □□□ —→ □□□ Add to
  3 times   3 times □□□ find the
                     □□□ total.

You might use the strategy **guess and check** to solve the problem.

Try 10:   Try 30:   Try 20:
10 + 30 + 90 = 130   30 + 90 + 270 = 390   20 + 60 + 180 = 260
   too low   too high   Correct!

The department shipped 20 A boxes, 60 B boxes, and 180 C boxes.

Try This Solve.

a. What is the cost of the bike and the cost of the helmet? The bike is worth four times more than the price of the helmet.

For Sale
Bike and helmet for $100
Call Roy: 555-3842

b. Helen paid for a 50¢ pineapple juice with 5 coins. She did not use all dimes. What coins did she use?

Exercises

Solve. Use one or more of the problem-solving strategies.

Problem-Solving Strategies
- Guess and check.
- Choose an operation.
- Make an organized list.
- Act it out.
- Find a pattern.
- Write an equation.
- Use logical reasoning.
- Draw a picture or diagram.
- Make a table.
- Use objects.
- Work backward.
- Solve a simpler problem.

1. Jeff bought 3 record albums. Each album cost the same. The tax was $1 and the total cost was $16. What was the price of each album?

2. The Album Attic sold a total of 225 jazz and rock albums. How many albums does each ⊙ represent?

The Album Attic
Album Sales for November

3. Tickets to the county fair cost $6.75 for adults and $4.25 for children. Children under the age of 3 can enter free. What would tickets cost for two adults, their twin 6-year-old sons, and their newborn daughter?

Which two VCRs have a total cost of about:

4. $600
5. $900
6. $400
7. $800
8. $1000
9. $700

VCR Sale	
Model #	Price
2345	$105
5440	$275
A-235	$310
70-89	$480
E104	$575

Which three VCRs have a total cost of about:

10. $1000
11. $1400

12. Sabrina worked two jobs on the weekend. She made a total of $120. One job paid twice as much as the other. How much did she earn on each job?

13. Khalil lives between Bloomington and Joliet. He lives 20 miles from Bloomington. How far does he live from Joliet?

Bloomington 45	Pontiac 38
Springfield 105	Joliet 90

Determining Reasonable Answers

14. Ralph worked this problem on his calculator. Without finding the correct answer, tell how you know that the answer he found is wrong.

    Problem: A box of albums weighs 24 pounds. A box of tapes weighs 18 pounds. What would a shipment of 10 album boxes and 20 tape boxes weigh?

    Ralph's answer: The shipment would weigh 360 pounds.

Use Objects

15. Make $5 and $7 tickets from paper. Use your paper tickets to help you solve this problem.

    Problem: Melinda spent $50 on tickets. She bought some $5 tickets and some $7 tickets. How many tickets of each kind did she buy?

4-2 Dividing by a 1-Digit Number

Objective
Divide whole numbers by 1-digit divisors.

Aretha's father lent her money, without interest, to buy a motorcycle. She paid him back in 6 equal monthly payments. How much was each payment?

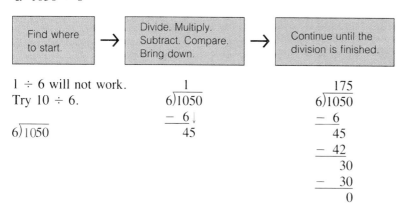

Understand the Situation

- What was the price of the motorcycle?
- Was each payment the same amount?
- What tells you that division is needed?

You can estimate the quotient. $1050 \div 6$ is about $1000 \div 5$, or 200.

Examples Divide.

1. $1050 \div 6$

| Find where to start. | → | Divide. Multiply. Subtract. Compare. Bring down. | → | Continue until the division is finished. |

$1 \div 6$ will not work.
Try $10 \div 6$.

$$6\overline{)1050}$$

$$\begin{array}{r} 1 \\ 6\overline{)1050} \\ -\,6\downarrow \\ \hline 45 \end{array}$$

$$\begin{array}{r} 175 \\ 6\overline{)1050} \\ -\,6 \\ \hline 45 \\ -\,42 \\ \hline 30 \\ -\,30 \\ \hline 0 \end{array}$$

You can check division using multiplication. Check: $175 \times 6 = 1050$.

175×6 is a little less than 200×6, or 1200. 1050 is a little less than 1200. The answer seems reasonable. Each monthly payment will be $175.

2.
$$\begin{array}{r} 184 \\ 4\overline{)736} \\ -\,4 \\ \hline 33 \\ -\,32 \\ \hline 16 \\ -\,16 \\ \hline 0 \end{array}$$

3.
$$\begin{array}{r} 54\text{ R}2 \\ 5\overline{)272} \\ -\,25 \\ \hline 22 \\ -\,20 \\ \hline 2 \end{array}$$ *The remainder is 2.*

4.
$$\begin{array}{r} 1163\text{ R}2 \\ 7\overline{)8143} \\ -\,7 \\ \hline 11 \\ -\,7 \\ \hline 44 \\ -\,42 \\ \hline 23 \\ -\,21 \\ \hline 2 \end{array}$$

Try This

Divide.

a. 2)84   b. 3)144   c. 8)928   d. 434 ÷ 5   e. 2936 ÷ 6

Exercises

Finish each problem.

1.
```
    1
6)834
 -6
  2
```

2.
```
    6
5)340
 -30
   4
```

3.
```
    7
2)1542
 -14
   1
```

4.
```
    1
4)5326
 -4
  13
```

5.
```
    2
3)6340
 -6
  3
```

Divide.

6. 7)98   7. 4)93   8. 6)90   9. 3)97   10. 4)644

11. 5)625   12. 6)359   13. 4)915   14. 3)852   15. 8)526

16. 7)324   17. 3)1566   18. 6)1254   19. 9)2307   20. 5)1972

21. 8)2328   22. 6)2865   23. 9)5589   24. 2)3908   25. 8)1883

26. $345 \div 5 = n$ 27. $t = 174 \div 4$ 28. $m = 286 \div 6$

29. $754 \div 9 = c$ 30. $1055 \div 5 = d$ 31. $3452 \div 2 = p$

32. Divide 742 by 6.
33. What is 363 divided by 9?
34. Find the quotient when 4496 is divided by 8.
35. Find the quotient when the divisor is 3 and the dividend is 143.

Estimation

Estimate.

36. 4)783   37. 5)184   38. 3)2387   39. 9)2044

Estimate to check each quotient. If you think the quotient is wrong, find the correct quotient.

40.
```
   49 R3
5)248
```

41.
```
   350 R1
2)631
```

42.
```
   993 R6
8)7150
```

43.
```
   862 R2
4)3450
```

Write Math

44. Finish each sentence to tell how to check your work on a division problem.

    Multiply the _____ by the _____. Add the _____ to this product. This number should be the _____.

```
              quotient  remainder
                  ↓        ↓
                   649 R1
divisor →    5)3246      ← dividend
              -30
               24
              -20
               46
              -45
                1
```

DIVISION OF WHOLE NUMBERS AND DECIMALS

4-3 Dividing by a 2-Digit Number

Los Angeles ⟷ New York
2849 miles

Objective
Divide whole numbers by 2-digit divisors.

A truck driver can travel at an average rate of 52 miles per hour. About how many hours of driving time will it take to travel from Los Angeles to New York City?

time = distance ÷ rate

Understand the Situation

- What is the distance from Los Angeles to New York City?
- How many miles will the truck driver travel in one hour?
- What should you divide the distance by to find the travel

You can estimate the quotient. 2849 ÷ 52 is about 3000 ÷ 50, or 60.

Examples Divide.

1. 2849 ÷ 52

Find where to start. → Divide.

```
                        54 R41
52)2849            52)2849
28 ÷ 52 will not work.  -260    Think: 280 ÷ 50. Try 5.
Try 284 ÷ 52.            249
                        -208
                          41
```

Estimate to check. Round 54 × 52 down to 50 × 50, or 2500. The dividend should be greater than 2500. 2849 is greater than 2500, so the answer seems reasonable. At an average rate of 52 miles per hour, it would take about 55 hours of driving to travel from Los Angeles to New York City.

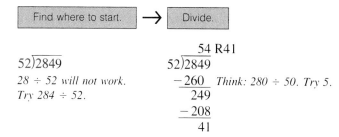

Try This Divide.

a. 30)185 **b.** 42)320 **c.** 418 ÷ 58 **d.** 2356 ÷ 34 **e.** 4502 ÷ 21

Exercises

Finish each problem.

1. $\begin{array}{r}1\\20\overline{)360}\\-20\end{array}$
2. $\begin{array}{r}8\\81\overline{)7285}\\-648\end{array}$
3. $\begin{array}{r}2\\37\overline{)1036}\\-74\end{array}$
4. $\begin{array}{r}1\\55\overline{)7260}\\-55\\\hline 17\end{array}$

Divide.

5. $12\overline{)492}$
6. $30\overline{)750}$
7. $19\overline{)694}$
8. $27\overline{)902}$

9. $52\overline{)468}$
10. $36\overline{)288}$
11. $72\overline{)442}$
12. $46\overline{)552}$

13. $26\overline{)8476}$
14. $56\overline{)3578}$
15. $24\overline{)8448}$
16. $39\overline{)2285}$

17. $62\overline{)2170}$
18. $36\overline{)1872}$
19. $42\overline{)3910}$
20. $26\overline{)2106}$

21. $42\overline{)6573}$
22. $28\overline{)1862}$
23. $13\overline{)27{,}482}$
24. $25\overline{)21{,}100}$

25. $u = 432 \div 24$
26. $658 \div 12 = w$
27. $705 \div 25 = b$

28. $x = 664 \div 31$
29. $2235 \div 50 = y$
30. $z = 4068 \div 28$

31. $d = 7440 \div 62$
32. $r = 9285 \div 47$

33. Evaluate $636 \div d$ for $d = 53$.
34. Evaluate $r \div 72$ for $r = 6049$.

Estimation Tell if the quotient shown on the calculator is reasonable. Find the correct answer if it is not reasonable.

35. $72\overline{)23{,}400}$
36. $58\overline{)35{,}960}$
37. $43\overline{)96{,}535}$

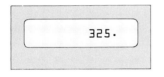
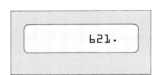
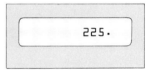

Data Bank 38. Use the mileage chart on p. 490. How much longer would it take you to drive from Chicago to San Francisco at 50 miles per hour than at 55 miles per hour?

Show Math 39. Four friends found $1420 in an old chest. The bills they found are shown. The police said that they could divide the money evenly. How many bills of each type did each person get? What bills did they have to trade for smaller bills so that they could divide the money evenly? Use paper money to show.

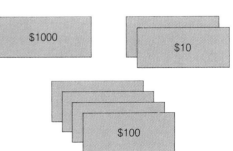

DIVISION OF WHOLE NUMBERS AND DECIMALS

4-4 Dividing with Zeros in the Quotient

Objective
Divide whole numbers with zeros in the quotient.

Rachel thought her answer was wrong because it was not close to her estimate. She worked the problem again and found her mistake.

Estimate: 1830 ÷ 6 is about 1800 ÷ 6, or 300.

Incorrect answer:

```
       35
    _____
  6)1830
   -18
    ___
     30
    -30
    ___
      0
```

Correct answer:

```
      305
    _____
  6)1830
   -18
    ___
      3
     -0
     ___
      30
     -30
     ___
       0
```

Remember:
Every time you bring down a digit and divide, write a digit in the quotient. You may need to write a 0.

Examples Divide.

```
      30 R2
1. 4)122
   -12
    ___
      2
     -0
    ___
      2
```

```
      600 R4
2. 8)4804
   -48
    ___
      0
     -0
    ___
      4
     -0
    ___
      4
```

```
       108 R2
3. 23)2486
    -23
    ____
      18
     - 0
    ____
     186
    -184
    ____
       2
```

Try This Divide.

a. 5)253   b. 3)1513   c. 8)4803   d. 8195 ÷ 27   e. 27,212 ÷ 34

Exercises

Divide.

1. 6)242
2. 7)427
3. 9)270
4. 4)123
5. 8)2416
6. 4)2044
7. 9)1814
8. 8)6472
9. 14)2804
10. 21)4373
11. 33)6732
12. 26)15,730

13. $321 \div 3 = m$
14. $2021 \div 4 = z$
15. $s = 9624 \div 12$
16. $h = 4406 \div 11$
17. $2655 \div 26 = e$
18. $a = 9696 \div 16$
19. $11,088 \div 36 = b$
20. $505 \div 33 = w$

Evaluate.

21. $1640 \div k$ for $k = 15$
22. $l \div 16$ for $l = 6496$
23. $7000 \div n =$ for $n = 23$
24. $j \div 13$ for $j = 10,400$

Mental Math

Find each quotient using mental math.

25. 4)82
26. 5)201
27. 2)605
28. 34)6800
29. 25)1004

Number Sense

30. Tell what numbers you should place in the boxes to make the sentences true. Do not divide.

    $285 \div 15 = \square$ $285 \div \square = 19$

Problem Solving

31. Push each of the digits shown on the calculator one time to get the quotient shown.

    ```
           3 0 7
    2□)7□□□
    ```

Understanding Division

32. Show that you understand the division steps. Change each ● to a numeral to finish the division.

    ```
         1●●9 R5
    6)74●9
      -●
       1●
      -12
        ●3
       -18
         5●
        -●●
          ●
    ```

4-5 Dividing by a 3-Digit Number

Objective
Divide whole numbers by 3-digit divisors.

Gerald Paige videotaped 30 hours (1800 minutes) of the Olympics. Each tape can record 240 minutes. How many tapes did he use?

When you have a large divisor, it is often better to use a calculator. Be sure to estimate the quotient.

Example 1800 ÷ 240

Estimate: 1800 ÷ 200 = 9. The quotient must be less than 9 because the divisor is greater than 200.

Calculator:

```
7.5
```

Paper-and-Pencil:

```
         7 R120
240)1800
    -1680
      120
```

Exercises

1. Write an answer to the example in a complete sentence.
2. Suppose each tape could record 120 minutes. How many tapes would Gerald need to use?

Use Ted's paper-and-pencil solution below to answer Exercises 3–6.

3. What numbers did Ted multiply to get 864?
4. Why did Ted subtract 0 from 216?
5. What numbers can you multiply to check the solution to this problem?
6. 88,560 ÷ 205 equals what number?

```
           205
432)88,560        Ted's
   -864           Work
    216
    - 0
    2160
   -2160
       0
```

Estimation Estimate. Tell which number is the exact answer.

7. 127)635   5 or 50
8. 232)3712   6 or 16
9. 405)12,150   30 or 300
10. 325)65,000 20 or 200

Number Sense

Without finding the answer, tell if the quotient for the first problem is greater or less than the given quotient. Use < or >.

11. 944 ÷ 8 ☐ 944 ÷ 4 = 236
12. 2340 ÷ 40 ☐ 2340 ÷ 45 = 52
13. 5000 ÷ 250 ☐ 5000 ÷ 200 = 25
14. 8275 ÷ 300 ☐ 6000 ÷ 300 = 20

Calculator

15. Use your calculator to fill in the value for each symbol. What pattern do you find?

   A. 177,177 ÷ 13 = ☐
      ☐ ÷ 11 = ○
      ○ ÷ 7 = △
   B. 386,386 ÷ 13 = ☐
      ☐ ÷ 11 = ○
      ○ ÷ 7 = △
   C. 529,529 ÷ 13 = ☐
      ☐ ÷ 11 = ○
      ○ ÷ 7 = △

Mixed Skills Review

Find each answer.
16. 3.6 × 70
17. $245 + $864
18. 902 − 57.4
19. $248 ÷ 8
20. 5.07 − 2.98
21. 752 × 33
22. 9)‾289
23. 9654 + 7523

Estimate.
24. 452 + 348
25. 610 × 52
26. 1972 − 988
27. 3621 ÷ 42

Find each answer using mental math.
28. $48 + $2
29. 62 × 3
30. 425 + 540
31. $5.75 − $1.25

Round to the given place.
32. 4.96 hundredths
33. $8.50 dollar
34. 48,523,602 thousands
35. 0.04 tenths

Arrange in order from least to greatest.
36. 8031, 80.31, 803.1, 8.031, 8013
37. 1.45, 1.4, 1.9, 1.405, 1.451

Evaluate.
38. $v + \$486$ for $v = \$1932$
39. $7.032 - r$ for $r = 0.89$
40. $48.02 \times y$ for $y = 1000$
41. $206t$ for $t = 57$
42. $100s$ for $s = 0.829$
43. $w - \$546$ for $w = \$6000$
44. $38.09 + n$ for $n = 7.543$
45. $b \div 4$ for $b = 64$

Decide if the situation needs an exact answer or an estimate.

46. Ricardo bought 3 shirts for $13.39 each. He wrote a check for the total amount. For how much money did he write the check?

47. Emilia bought 2 belts for $42.39 all together. Both belts cost the same. How much money did she spend on each belt?

DIVISION OF WHOLE NUMBERS AND DECIMALS

4-6 Consumer Math: Reading a Telephone Bill

Objective
Read a telephone bill.

- SITUATION
- DATA
- PLAN
- ANSWER
- CHECK

Eva received her monthly phone bill. Most phone bills contain two parts, a bill summary and a detailed list of charges.

The bill summary shows the total amount due to your local carrier and sometimes to your long distance carrier. Your long distance carrier may bill you separately.

The detailed list of charges breaks down the costs of services provided by your local carrier and your long distance carrier. Your local carrier lists your monthly service charge, calls to places outside your immediate area, and other charges. Your long distance carrier lists itemized calls to places outside the local area.

```
*D=Day      N=Night/Weekend     ** DETAIL OF CHARGES **    214 555 0583
 E=Evening  #=state taxable              JUL 09 87
                                        AMOUNT    U.S. TAX   ST/LOC TAX    TOTAL
 CURRENT LONG DISTANCE CO.  CHARGES      43.96      2.34        1.01       47.31
 ─────────────────── LONG DISTANCE CO. ITEMIZED CALLS ───────────────────
```

NO	DATE	TIME	PLACE CALLED		AREA	NUMBER	*	MIN	AMOUNT
1	6 14	340PM	STAUNTON	IL	618	555 3302	N	39	5.52
2	6 18	307PM	TULSA	OK	918	555 5000	D	8	2.15
3	6 25	826AM	AIRLINE	TX	713	555 0101	D	2	.75#
4	6 29	1140AM	STAUNTON	IL	618	555 3302	D	28	8.46
5	6 29	830PM	STAUNTON	IL	618	555 3302	E	31	5.80
6	7 01	1044AM	BENLD	IL	217	555 4585	D	15	4.40
7	7 03	814PM	STAUNTON	IL	618	555 3302	E	10	1.82
8	7 04	1229PM	STAUNTON	IL	618	555 3302	N	13	1.79
9	7 06	1142AM	AUSTIN	TX	512	555 9585	D	5	1.86#
10	7 06	1148AM	BABCOCK	TX	512	555 7054	D	1	.38#
11	7 06	1150AM	BRITTON	OK	405	555 1212	D	1	.31
12	7 06	1151AM	TEMPE	AZ	602	555 8511	D	1	.34
13	7 06	253PM	HELOTES	TX	512	555 2135	D	4	1.49#
14	7 06	259PM	HOUSTON	TX	713	555 5376	D	24	8.89#

ITEMIZED CALLS EXCLUDING TAX   43.96

See Reverse   THANK YOU FOR PAYING BY MAIL

Example Find the cost above of the call to Austin, Texas, on July 6, at 11:42 a.m.

Read the DATE column until you find 7 06. Look for the Austin call at 11:42 a.m. Then read across to the AMOUNT column. The cost of the call was $1.86.

Try This Find the length of the call to Airline, Texas, on June 25.

Exercises

Solve. Use the telephone bill shown on p. 100.

1. How many long distance calls did Eva make?
2. On which long distance call did Eva spend the longest time on the phone?
3. Which long distance call cost the most?
4. How many long distance calls did Eva make at the day rate?
5. How many long distance calls were taxed by the state?
6. How many different states did Eva call?
7. What was the charge for U.S. tax on long distance calls?
8. What was the total charge for long distance calls?
9. Find the cost of the call to Helotes, Texas, on July 6.
10. Find the length of the call to Tempe, Arizona, on July 6.
11. To which city did Eva make a call that lasted 13 minutes?
12. On what date did Eva call Staunton, Illinois, for 31 minutes?

Working with Variables

13. The expression $\$0.15 + \$0.07m$ gives the cost for a weekday phone call less than 10 miles away. This cost varies based on the number of minutes for the call.

    $0.15 + 0.07m$
    - 15¢ for the initial minute
    - 7¢ for each additional minute
    - Multiply 0.07 by the number of additional minutes (m). Add 0.15 to this product.

Complete the table to show how the total cost of a call varies based on the number of minutes.

Number of minutes	1	2	3	4	5	6
Additional minutes (m)	0		2		4	
Total cost (0.15 + 0.07m)	0.15	0.22			0.43	

Mixed Skills Review Find the answer.

14. $12.62 - 10.4$
15. $238 + 52$
16. 38×5.4
17. $504 \div 42$
18. $7654 + 6105$
19. $\$17.25 - \$6.27$
20. $\$42 \times 1.25$
21. $676 \div 13$

Evaluate.

22. $789 + z$ for $z = 4018$
23. $x - \$799$ for $x = \$1605$
24. $23r$ for $r = 48$
25. $s \div 19$ for $s = 285$

DIVISION OF WHOLE NUMBERS AND DECIMALS

4-7 Problem Solving: Using Data from an Advertisement

Objective
Read data from an advertisement.

- **SITUATION**
- **DATA**
- **PLAN**
- **ANSWER**
- **CHECK**

Use data in the advertisement to answer these questions.

Fall Sale

Pocket Radio (with earphones) was $45
Special Savings Today

CD Player $395 today SAVE $50!

VCR—Half-Price Sale	
Model #	Original Cost
0134	$275
0135	$450
0140	$575

Videotape Sale were $7 Special Sale Prices

Record Albums 2 for $10

1. Is the original price of the CD player more or less than $400?
2. Is the sale price of the pocket radio more or less than $45?
3. Would you most likely pay more or less than $5 for one record album?
4. Suppose you had $200. Which VCR could you buy?

Use the advertisement for the audiotape recorders to decide what each person bought.

5. Elena paid more than $200. She paid less than $300. She paid an odd number of dollars.
6. Mickey paid less than $300. Both 9 and 3 divide the amount she paid with no remainder.
7. Kevin bought 2 machines. He paid $650 all together.
8. Estella bought one machine. The number 50 divides the amount she paid with no remainder.
9. Akio bought 3 machines. He paid $895 all together.
10. Cliff bought 2 machines for $500.

Audiotape Recorders	
Model#	Price
1180	$180
1150	$210
2000	$235
17A	$290
17B	$360
164-C	$425
165-C	$450

Find a Related Problem

11. *Problem:* Maurice bought 3 main-floor seats for $8 each. He bought 7 balcony seats for $5 each. How much did he pay all together for the 10 tickets?

Tell which of these problems you could solve in the same way as the above problem.

A. Small boxes of 3 golf balls cost $5 each. Large boxes of 7 golf balls cost $8 each. What would you pay all together if you bought 10 large boxes?

B. A large van holds 8 people. A small van holds 5 people. How many people can 3 large vans and 7 small vans carry all together?

Mixed Practice

Find each answer.

1. 4.65×100
2. $1032 \div 8$
3. $\$601 \times 23$
4. $1301 - 178$
5. $0.048 + 0.96$
6. 9×275
7. 3.48×0.03
8. $\$123 + \$22 + \$84$
9. 41×2175
10. $\$624 \div 4$
11. $235.2 - 71.7$
12. 7.25×7.5
13. $5.808 - 4.11$
14. $2213 \div 11$
15. $\$2518 \times 7$
16. $4844 + 936$

Estimate.

17. $83 + 82 + 79 + 78 + 84$
18. $32 + 29 + 94 + 31 + 89$
19. $345 + 790$
20. 38×6
21. $4209 \div 61$
22. $763 - 459$
23. 32×41
24. $5511 \div 92$
25. $2237 - 1135$
26. $6399 \div 82$

Evaluate.

27. $836 + t$ for $t = 9227$
28. $37r$ for $r = \$29$
29. $456 \div w$ for $w = 4$
30. $z - 548$ for $z = 737$

Write using exponents. Then write the standard numeral.

31. $3 \times 3 \times 3$
32. 10×10
33. $9 \times 9 \times 9 \times 9$
34. 8 squared
35. 1.5 cubed
36. 6 to the fifth power

Find each answer using mental math.

37. 62×5
38. $2 + 59$
39. $211 - 2$
40. 202×6
41. $\$23.50 - \$19$
42. 7×33
43. $425 + 225$
44. $29 + 3$
45. $223 + 220$
46. 4×310
47. $\$121 - \$3$
48. $\$6 - \$4.50$

Choose the calculation method you would use to find each answer.

49. 232×40
50. $\$130 + \$53.50$
51. 672×99
52. 20×659

53. You are in a department store. You want to know the total cost of 2 beach towels that cost $17.39 each.
54. You are a sales clerk. You need to add 6.5% sales tax to a sale of $24.87.
55. Your yard measures 12 feet by 18 feet. You want to find the number of feet of fencing to put around your yard.
56. You only have $10. You want to buy 2 novels that cost $4.75 each without tax.
57. You are a cashier at a restaurant. You need to find the total amount of a bill.
58. You are planning to paint your room. You need to decide how much paint to buy to cover four walls and the ceiling.

4-8 Dividing a Decimal by a Whole Number

Objective
Divide a decimal by a whole number.

How much would each monthly payment be for the video/audio system?

Understand the Situation

- What is the total price?
- Payments will be made for how many months?
- Will each payment be the same amount?

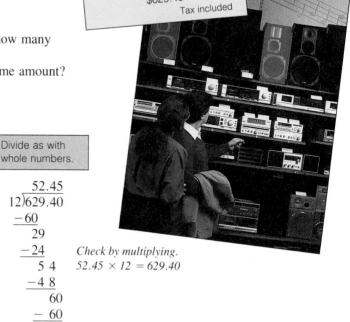

Video/Audio System
Cash-and-Carry or
12 Equal Monthly Payments
Today Only
$629.40
Tax included

Examples Divide.

1. 629.40 ÷ 12

| Write the decimal point in the quotient. | → | Divide as with whole numbers. |

```
         .                    52.45
    12)629.40             12)629.40
                            -60
                             29
                            -24
                              5 4      Check by multiplying.
                             -4 8      52.45 × 12 = 629.40
                               60
                             - 60
                                0
```

Check by estimating: 52.45 × 12 is a bit more than 50 × 12, or 600. 629.40 is a bit more than 600. The answer seems reasonable. Each monthly payment would be $52.45.

```
       0.028                    0.48                    3.06
2.  6)0.168            3.  15)7.20            4.  8)24.48
      -0                      -6 0                   -24
       16    Remember          1 20   You can          4
      -12    to write a       -1 20   write a zero    -0
       48    zero.                0   and keep        48
      -48                             dividing.      -48
        0                                              0
```

Try This   Divide.

a. 6)18.72   **b.** 24)60.72   **c.** 0.585 ÷ 3   **d.** 8.376 ÷ 8   **e.** 58.8 ÷ 35

Exercises

Place the decimal point in each quotient.

1. 4)9.44 quotient 236
2. 8)2.928 quotient 366
3. 6)141 quotient 235
4. 25)12.5 quotient 5
5. 35)0.875 quotient 25

Divide.

6. 5)16.8
7. 8)20.64
8. 9)32.85
9. 6)14.1
10. 4)0.128
11. 3)0.2091
12. 45)104.85
13. 36)185.04
14. 41)360.39
15. 19)0.475
16. 62)1990.2
17. 33)1.8612

18. $n = 0.17 \div 5$
19. $7.77 \div 3 = y$
20. $c = 313.8 \div 6$
21. $0.264 \div 4 = b$
22. $183.6 \div 36 = v$
23. $d = 1.058 \div 23$
24. $a = 312.52 \div 52$
25. $93.38 \div 46 = p$

Number Sense

Match each divisor with its division problem.

26. ☐)188 quotient 23.5
27. ☐)18.2 quotient 3.64
28. ☐)0.0156 quotient 0.0026
29. ☐)21.12 quotient 5.28

Divisors: 4 5 6 8

Mixed Applications

Solve. Use these ads for Exercises 30–31.

30. What is the monthly payment if you buy the TV at Romano's TV World?
31. Which dealer has the least expensive 12-month plan?
32. How much more money would you pay on a 3-year, monthly-payment plan at $125.95 per month than paying cash for a $4200 motorcycle?

Romano's TV World
Color TV
Model 360-x   Sale!
You pay only $495.60 total in 12 monthly payments.

Kazuo's Electronics
Model 360-x Color TV
Pay just $40 a month for 12 months.

Mixed Skills Review

Write as a standard numeral or in money notation.

33. six thousand, seven hundred eight
34. seventeen dollars and ten cents
35. nine hundred dollars and two cents
36. thirty-two and three hundredths

Tell if you need an exact answer or an estimate.

37. You want to compare the interest rates that banks give for savings accounts.
38. You want to know if you have enough money to buy your groceries.

DIVISION OF WHOLE NUMBERS AND DECIMALS

4-9 Dividing by 10, 100, or 1000

Objective
Divide by 10, 100, or 1000.

The **meter** is the standard unit of length in the metric system. **Deka**meter, **hecto**meter, and **kilo**meter are units of metric length greater than one meter.

Some Units of Metric Length
- Dekameter   10 meters
- Hectometer 100 meters
- Kilometer   1000 meters

If you ran a 5000-meter race, how many dekameters would you have run? How many hectometers? How many kilometers? You can find how many dekameters, hectometers, and kilometers are in 5000 meters by dividing.

Examples   Divide.

1. 5000 ÷ 10 = 500.0 , or 500.   *Divide by 10. Move the decimal point one place to the left.*

2. 5000 ÷ 100 = 50.00 , or 50.   *Divide by 100. Move the decimal point two places to the left.*

3. 5000 ÷ 1000 = 5.000 , or 5.   *Divide by 1000. Move the decimal point three places to the left.*

If you ran 5000 meters, you would have run 500 dekameters, 50 hectometers, or 5 kilometers.

4. 28.45 ÷ 10 = 2.845   **5.** 120.3 ÷ 100 = 1.203

6. 2895 ÷ 1000 = 2.895   **7.** 34.2 ÷ 1000 = 0.0342
↑
Sometimes you need to write extra zeros to show the place value.

Try This   Divide.

a. 45.6 ÷ 10   **b.** 78.4 ÷ 100   **c.** 3000 ÷ 1000   **d.** 0.45 ÷ 10

Exercises

Place the decimal point in each quotient. Write zeros as needed.

1. $34.5 \div 10 = 345$
2. $75.6 \div 100 = 756$
3. $3456.7 \div 1000 = 34567$
4. $3.67 \div 100 = 367$
5. $0.5 \div 10 = 5$
6. $400 \div 100 = 400$
7. $4.78 \div 100 = 478$
8. $100.1 \div 10 = 1001$
9. $0.07 \div 100 = 7$

Divide.

10. $10\overline{)3.89}$
11. $100\overline{)6.52}$
12. $10\overline{)0.73}$
13. $1000\overline{)8422}$
14. $100\overline{)87}$
15. $1000\overline{)0.92}$
16. $10\overline{)46.7}$
17. $100\overline{)78.1}$
18. $1000\overline{)10.3}$
19. $10\overline{)0.7}$
20. $100\overline{)7.2}$
21. $1000\overline{)145.6}$
22. $0.44 \div 100 = e$
23. $8.71 \div 1000 = g$
24. $a = 92.6 \div 1000$
25. $n = 0.82 \div 10$
26. $l = 8.3 \div 100$
27. $r = 3.3 \div 10$
28. $4013 \div 1000 = x$
29. $6.72 \div 100 = q$

Write Math

Finish each statement.

30. To divide a decimal number by 100, move the decimal point _____ places to the _____ .

31. To multiply a decimal number by 100, move the decimal point _____ places to the _____ .

Understanding Division

32. Without dividing, tell what the quotient is. (Hint: Compare the decimal points.)

$2.45\overline{)24.5}$

Problem Solving

33. A 200-meter racecourse had a flag at the starting line, a flag at the finishing line, and a flag every 10 meters in between the starting and finishing lines. How many flags did they use on the course?

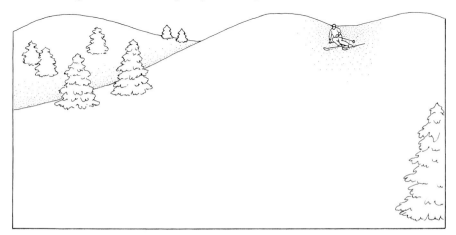

4-10 Dividing a Decimal by a Decimal

Objective
Divide decimals by decimals.

Ellen has a package that needs $4.50 in postage. She has some 25¢ stamps. How many stamps does she need to use to mail the package?

Math Fact

$$2\overline{)8} 20\overline{)80}$$
$$ \uparrow \uparrow$$
$$2 \times 10 8 \times 10$$

If you multiply the divisor and the dividend by the same number, the quotient does not change.

Understand the Situation

- How much postage does the package need?
- What is the value of each stamp that Ellen has?
- What operation could you use to find the number of stamps Ellen needs?

You can use the given Math Fact to make the divisor a whole number. Since you want to find how many 0.25 are in 4.50, you should divide.

Estimate. Use compatible numbers. 4.50 ÷ 0.25 is about 5.00 ÷ 0.25, or 20.

Examples Divide. Round each quotient to the nearest hundredth.

1. 4.50 ÷ 0.25

To make the divisor and the dividend into whole numbers, multiply both by a power of 10. → Divide as with whole numbers.

$$0.25\overline{)4.50}$$
Move each decimal point two places to the right.

$$\begin{array}{r} 18 \\ 25\overline{)450} \\ -25 \\ \hline 200 \\ -200 \\ \hline 0 \end{array}$$

Ellen needs eighteen 25-cent stamps for postage totaling $4.50.

2. $1.5\overline{)3.630}$
$$\begin{array}{r} 2.42 \\ -30 \\ \hline 63 \\ -60 \\ \hline 30 \\ -30 \\ \hline 0 \end{array}$$
Write zeros as needed.

3. $0.08\overline{)0.02600}$ → 0.33
$$\begin{array}{r} 0.325 \\ -24 \\ \hline 20 \\ -16 \\ \hline 40 \\ -40 \\ \hline 0 \end{array}$$
0.325 rounds to 0.33.

4. $5.3\overline{)1.3400}$ → 0.25
$$\begin{array}{r} 0.252 \\ -106 \\ \hline 280 \\ -265 \\ \hline 150 \\ -106 \\ \hline 44 \end{array}$$
0.252 rounds to 0.25.

Try This Divide.
a. $0.4\overline{)1.28}$ b. $2.5\overline{)0.65}$ c. $3.12 \div 0.24$ d. $0.0273 \div 0.42$

Exercises

Place the decimal point in each quotient. Write zeros as needed.

1. $0.6\overline{)1.44}^{\,24}$
2. $0.8\overline{)0.4336}^{\,542}$
3. $0.45\overline{)11.25}^{\,25}$
4. $2.1\overline{)0.0504}^{\,24}$

Divide. Round to the nearest hundredth.

5. $1.4\overline{)3.57}$
6. $0.7\overline{)0.2352}$
7. $0.3\overline{)0.0768}$
8. $1.3\overline{)9}$
9. $2.1\overline{)0.69}$
10. $0.35\overline{)12.35}$
11. $6.2\overline{)0.0312}$
12. $0.04\overline{)0.432}$
13. $0.17\overline{)0.09}$
14. $4.2\overline{)9.5}$
15. $1.05\overline{)0.68}$
16. $2.8\overline{)6}$

17. $4.5 \div 0.26 = b$
18. $v = 0.225 \div 0.4$
19. $8 \div 12 = h$
20. $s = 5.05 \div 0.44$
21. $x = 12.4 \div 0.24$
22. $0.06 \div 1.4 = k$
23. $0.05 \div 0.8 = d$
24. $n = 0.15 \div 0.08$
25. $2.7 \div 0.59 = f$
26. $n = 1.06 \div 0.45$
27. $z = 6.02 \div 0.8$
28. $0.062 \div 0.006 = j$

Estimation Estimate. Decide if the quotient shown on the calculator is reasonable.

29. $0.78 \div 0.5$
30. $8.4 \div 1.5$
31. $4.2 \div 0.016$

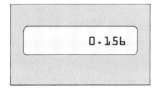
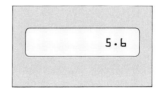
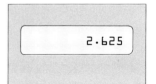

Group Activity

32. Suppose you have 25-cent stamps and 20-cent stamps. What amounts of postage less than $1 can you make with these stamps? (Hint: Make paper stamps to help.)

Mixed Applications

33. Suppose you have 22-cent stamps and 39-cent stamps. How many of each could you put on a package needing $2.10 postage?

34. Draw and label the length, width, and thickness of two packages that meet the minimum mail size. Draw two that do not meet the minimum sizes.

Mail Fact

Minimum Mail Sizes
All mail must be at least 0.007 in. thick. Mail less than 0.25 in. thick must be at least 3.5 in. in height, at least 5 in. in length, and rectangular in shape. (Note: You can mail pieces greater than 0.25 in. thick if they are less than 3.5 by 5 in.)

4-11 Problem Solving: Group Decisions

Objective
Make group decisions.

- **SITUATION**
- **DATA**
- **PLAN**
- **ANSWER**
- **CHECK**

Your class is collecting money to buy a remote control car for a third grader in the hospital. The car is already picked out. Now you need to decide which of three types of batteries to buy. You could buy regular batteries, alkaline batteries, or rechargeable batteries along with a battery recharger.

Facts to Consider
- The remote control car costs $45.50.
- The remote control car requires six C batteries and ten AA batteries.
- Cost of batteries:

Battery	C	AA
Regular	$1.05/pair	$0.78/pair
Alkaline	$2.49/pair	$1.49/pair
Rechargeable	$5.99/each	$3.99/each

- Alkaline batteries last 40 to 60 minutes of play. Alkaline batteries last 3 to 4 times longer than regular batteries.
- The battery recharger for rechargeable batteries costs $12.95. Batteries will recharge up to 300 times.

Plan and Make a Decision
- Describe what you are being asked to do.
- What are the advantages of each choice? How might you make a list or chart to help you compare?
- What data will you use when comparing costs? How could a calculator help you?
- How will you collect money?
- How much money do you need to collect?
- How will you come to a group decision?

Share Your Group's Decision
- Explain your decision to the class.
- Be ready to back up your decision with facts and reasons.

Suppose
- Suppose there was a two-for-one sale on alkaline batteries. How would this affect your choice?
- Suppose you wanted to buy two remote control cars. How would this affect your choice?

4-12 Practice Through Problem Solving

Objective
Practice operations with decimals in problem-solving situations.

Write a decimal number in each blank that will give an answer that is about equal to the one given. ≈ means *about equal to*.

1. _____ ÷ 4.3 ≈ 28
2. _____ ÷ 2.1 ≈ 6.8
3. _____ × 0.85 ≈ 12
4. _____ × 1.6 ≈ 20
5. _____ ÷ 0.5 ≈ 2
6. _____ ÷ 0.9 ≈ 1.4

On Your Own

Write your own problem with an answer that is about equal to the one given. Write one digit in each box. Write decimal points where needed.

7. ☐☐☐ ÷ ☐ ≈ 1.5
9. ☐ ÷ ☐☐ ≈ 5
8. ☐☐ ÷ ☐☐ ≈ 4.2
10. ☐☐☐☐ ÷ ☐☐☐ ≈ 0.24

All Four It

11. Use 4 fours and any of the operations (addition, subtraction, multiplication, and division) to make as many of the numbers between 10 and 20 as you can.

(For example: 4 × 4 + (4 − 4) = 16)

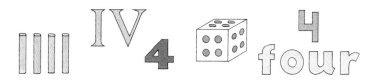

Choose the Sign

Place + or − in each box to get the final answer.

12. 0.08 ☐ 0.05 ☐ 0.09 ☐ 0.01 ☐ 0.11 ☐ 0.06 = 0.08
13. 1.1 ☐ 2.3 ☐ 6.6 ☐ 5.7 ☐ 0.9 ☐ 1.3 = 4.7
14. 36.5 ☐ 29.4 ☐ 4.9 ☐ 5.5 ☐ 9.9 ☐ 7.3 = 0.3

Estimation

Choose the better answer.

15. $18.25 − $8.96
    A. less than $10
    B. greater than $10
16. $5.79 ÷ 6
    A. less than $1
    B. greater than $1
17. $25.18 − $10.23
    A. less than $15
    B. greater than $15
18. $19.89 × 9
    A. less than $180
    B. greater than $180

DIVISION OF WHOLE NUMBERS AND DECIMALS

4-13 Enrichment: Order of Operations

Objective
Use the order of operations to evaluate expressions.

Erik and Marta found two different answers to this problem using their calculators. Whose answer is correct?

35 + 55 × 45   *Erik's Answer*   *Marta's Answer*

The answer you get depends on which operation is done first.

Understand the Situation
- How did Erik get 4050 for the answer?
- How did Marta get 2510 for the answer?

You can use the **order of operations** to find the answer. First do the operations inside parentheses. Then do all multiplication and division from left to right. Then do all addition and subtraction from left to right.

Examples Find the answer.

1. 35 + 55 × 45   35 + 55 × 45 = 35 + 2475   Multiply first. Then add.
                                  = 2510   Marta's answer is correct.

2. 36 ÷ (3 + 3) × (15 − 5) ÷ 5   Do operations in parentheses.
   36 ÷ (3 + 3) × (15 − 5) ÷ 5 = 36 ÷ 6 × 10 ÷ 5   Divide.
                              = 6 × 10 ÷ 5   Multiply.
                              = 60 ÷ 5   Divide.
                              = 12

Try This Find the answer.
a. 5 + 8 × 2   b. 2 × 3 + 4 × 5   c. 12 ÷ 3 − 1   d. 36 ÷ (18 − 9) ÷ 6

Exercises

Find the answer.

1. 6 + 3 × 9
2. 8 × 5 + 3
3. 14 − 3 × 2
4. 3 × (4 + 8) × 1
5. 6 × 5 − 7 × 4
6. 1 × (10 − 4) × 2
7. 24 ÷ (6 − 2)
8. 36 ÷ 9 + 3
9. (28 − 4) ÷ 2
10. 48 ÷ 8 − 6 ÷ 3
11. 16 ÷ 2 + 8 ÷ 4
12. 72 ÷ 9 − 36 ÷ 12
13. 21 ÷ (7 − 4) × 3 − 27 ÷ 9
14. 81 ÷ 3 − 24 ÷ 8 + 5 × 3

4-14 Calculator: Inverse Operations

Objective
Use inverse operations to check answers.

You can use inverse operations to check your work. **Inverse operations** are operations that undo each other. Multiplication is the inverse operation of division. Division is the inverse operation of multiplication.

Inverse Operations
Addition ↔ Subtraction
Multiplication ↔ Division

Examples Use the inverse operation to check the answer.

1. $19{,}188 \div 52 = 369$

   $19{,}188 \div 52 = 369$

   Use the inverse operation. Multiply the divisor and quotient. The product should equal the dividend.

   ON/C 5 2 × 3 6 9 =   ⌐ 19188. ⌐
   The answer is correct.

2. $492 \times 562 = 276{,}503$

   $492 \times 562 = 276{,}503$

   Use the inverse operation. Divide the product by either factor. The quotient should equal the other factor.

   ON/C 2 7 6 5 0 3 ÷ 5 6 2 =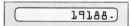
   The answer is wrong.

Try This Use the inverse operation to check the answer. If the answer is wrong, find the correct answer.

a. $41{,}370 \div 42 = 985$ b. $845 \times 33 = 37{,}885$ c. $132 \times 158 = 2856$

Exercises

Use the inverse operation to check the answer. If the answer is wrong, find the correct answer.

1. $869 \times 497 = 516{,}251$
2. $122 \times 367 = 44{,}764$
3. $89{,}987 \div 841 = 107$
4. $40{,}194 \div 77 = 522$
5. $929 \times 344 = 318{,}576$
6. $99{,}684 \div 468 = 113$

Write the number to make a true sentence.

7. $\square \times 516 = 219{,}816$
8. $\square \div 899 = 690$
9. $45 \times \square = 27{,}045$

Write Your Own Problem
10. Write three multiplication problems. Cover up one factor in each problem. Have a classmate find the missing number.

CHAPTER 4 REVIEW

Choose the one that does not belong in each group. Explain why not. **Answers may vary.**

1. A. Make an organized list.
   B. Guess and check.
   C. Multiply first.
   D. Draw a picture.
   is not a problem-solving strategy

2. A. dividend
   B. quotient
   C. divisor
   D. sum

3. A. 50 ÷ 10
   B. 49 ÷ 7
   C. 65 ÷ 6
   D. 63 ÷ 9

Match.

4. 3144 ÷ 6   A. 507
5. 1424 ÷ 7   B. 756
6. 7020 ÷ 65   C. 203 R3
7. 11,154 ÷ 22   D. 524
8. 42,054 ÷ 6   E. 7009
9. 152,712 ÷ 202   F. 108

10. These answers have been scrambled. Rearrange the answers to make each a true sentence.
    A. 52.5 ÷ 10 = 525
    B. 5.25 ÷ 100 = 5.25
    C. 52,500 ÷ 100 = 0.0525

11. On Saturday Elliot worked at Simmons Grocery Store. On Sunday he earned 3 times as much money tutoring math students. He earned a total of $36 on Sunday. How much did he earn on Saturday?
    A. $30   B. $18   C. $9   D. $27   E. $12

Place the decimal point in the quotient.

12. 4)5.32 → 133
13. 2)1.452 → 726
14. 2.5)390 → 1560
15. 0.35)23.87 → 682

Complete the sentence.

16. To divide a decimal number by 1000, move the decimal point _____ places to the _____.

17. To multiply a decimal number by 1000, move the decimal point _____ places to the _____.

Match.

18. divisor
19. dividend
20. quotient
21. remainder

```
      (A) 39
   (B) 4)158 (D)
      -12
       38
      -36
        2  (C)
```

CHAPTER 4 TEST

1. Bill is a clerk in Millard's Drugstore. He gave Mrs. Appleby $0.65 in change. He used 6 coins. What coins and how many of each kind of coin did he use?

Divide.

2. 6)96
3. 8)35
4. 4)956
5. 9)5232
6. 14)98
7. 27)1301
8. 61)4951
9. 89)46,013
10. 37)3959
11. 23)7597
12. 18)18,018
13. 28)56,844
14. 230)58,880
15. 421)95,998
16. 702)87,054
17. 632)353,288

18. How much was the phone call to Tulsa, Oklahoma?

```
*D=Day      N=Night/Weekend    ** DETAIL OF CHARGES **   214 555 0583
 E=Evening  #=state taxable                               JUL 09 87
                                        AMOUNT    U.S. TAX  ST/LOC TAX      TOTAL
 CURRENT LONG DISTANCE CO.  CHARGES     43.96       2.34       1.01         47.31
 ─────────────────── LONG DISTANCE CO. ITEMIZED CALLS ───────────────────
   NO    DATE    TIME     PLACE CALLED     AREA   NUMBER     *    MIN   AMOUNT
    1    6  14   340PM    STAUNTON    IL   618    555 3302   N    39     5.52
    2    6  18   307PM    TULSA       OK   918    555 5000   D     8     2.15
    3    6  25   826AM    AIRLINE     TX   713    555 0101   D     2      .75#
    4    6  29   1140AM   STAUNTON    IL   618    555 3302   D    28     8.46
    5    6  29   830PM    STAUNTON    IL   618    555 3302   E    31     5.80
    6    7  01   1044AM   BENLD       IL   217    555 4585   D    15     4.40
    7    7  03   814PM    STAUNTON    IL   618    555 3302   E    10     1.82
```

Use the advertisement to answer items 19 and 20.

19. Can Marsha buy a toothbrush and a tube of toothpaste with $2.25?
20. Pam bought one tube of toothpaste, one roll of dental floss, and a toothbrush on sale. How much more money would she have spent for these items before the sale?

Dental Items Sale
$\frac{1}{2}$ **Off All Items**

Regular Price with Tax
Toothpaste.......... $1.80
Dental floss.......... $1.30
Toothbrush.......... $2.46

Divide.

21. 7)9.52
22. 9)407.43
23. 27)1.5633
24. 48)144.096
25. 10)9.87
26. 100)0.0654
27. 1000)53.06
28. 10)0.009

Divide. Round to the nearest hundredth.

29. 5.7)13.96
30. 0.06)0.529
31. $0.732 \div 0.35$
32. $0.00438 \div 0.7$

DIVISION OF WHOLE NUMBERS AND DECIMALS

CHAPTER 4 CUMULATIVE REVIEW

1. Multiply.
   62.93 × 10
   A. 62.930
   B. 620.93
   C. 629.3
   D. 6293

2. Subtract.
   134.2
   − 51.6
   A. 72.4
   B. 82.4
   C. 83.6
   D. 82.6

3. Round to the hundreds place.
   1684
   A. 1700
   B. 1680
   C. 1600
   D. 2000

4. Ira bought 2 tennis rackets. He paid $200 all together. Which models did he buy?
   A. #12A and #1650
   B. #12A and #12B
   C. #3000 and #2350
   D. #128-C and #127-C

Tennis Rackets	Half-Price Sale
Model #	Regular Price
12A	$250
12B	$150
127-C	$120
128-C	$175
1650	$90
2350	$50
3000	$200

5. Divide.
   6)192
   A. 33
   B. 35
   C. 32
   D. 31 R5

6. Which long distance call cost the most?
   A. Austin, TX   B. Staunton, IL   C. Houston, TX   D. Oklahoma City, OK

```
*D=Day      N=Night/Weekend    ** DETAIL OF CHARGES **    214 555 6583
 E=Evening  #=state taxable        JUL 09 87                    PAGE 5
                               AMOUNT    U.S. TAX   ST/LOC TAX    TOTAL
CURRENT LONG DISTANCE CO.       63.45      2.34       1.01        66.80
─────────────── LONG DISTANCE CO. ITEMIZED CALLS ───────────────
 NO  DATE    TIME    PLACE CALLED      AREA  NUMBER    *   MIN   AMOUNT
  1  6  14  1151AM   TEMPE       AZ     602  555 8511  D    1     .34
  2  6  18   253PM   HELOTES     TX     512  555 2135  D    7    2.60#
  3  6  25   259PM   HOUSTON     TX     713  555 5376  D   19    7.04#
  4  6  29  1140AM   STAUNTON    IL     618  555 3322  D   28    8.46
  5  6  29   830PM   STAUNTON    IL     618  555 3322  E   31    5.80
  6  7  01  1044AM   BENLD       IL     217  555 4585  D   15    4.40
  7  7  03   814PM   STAUNTON    IL     618  555 3322  E   10    1.82
  8  7  04  1229PM   STAUNTON    IL     618  555 3322  N   13    1.79
  9  7  06  1122AM   AUSTIN      TX     512  555 9585  D   14    5.19#
 10  7  06  1141AM   BABCOCK     TX     512  555 7054  D    6    2.23#
 11  7  06  1150AM   BRITTON     OK     405  555 1212  D    1     .31
 12  7  06  1151AM   TEMPE       AZ     602  555 8511  D    1     .34
 13  7  06   126PM   OKLA CITY   OK     405  555 5537  D   52   13.00
 14  7  06   237PM   AUSTIN      TX     512  555 9585  D   27   10.00#
                           ITEMIZED CALLS EXCLUDING TAX          63.32
See Reverse              THANK YOU FOR PAYING BY MAIL
```

7. Multiply.
   4326 × 7
   A. 30,282
   B. 29,382
   C. 30,196
   D. 30,292

8. Subtract.
   29.01
   −10.67
   A. 18.38
   B. 19.34
   C. 18.44
   D. 18.34

9. Divide.
   7635 ÷ 4
   A. 1908 R3
   B. 198 R3
   C. 1907 R3
   D. 1908 R4

10. Decide which operation you need to solve this problem.
    Problem: The Farmer's Market is open every Saturday for ☐ hours. It is open for △ weeks of the year. How many hours is the market open all together?
    A. addition
    B. subtraction
    C. multiplication
    D. division

11. Estimate. Round to the highest place.
    62 − 38
    A. 30
    B. 40
    C. 20
    D. 35

12. Add. Use mental math.
    $35 + $12.75
    A. $47.75
    B. $45.75
    C. $57.75
    D. $48.50

13. Multiply.
    286
    ×122
    A. 34,692
    B. 8692
    C. 34,892
    D. 25,892

14. At the start of a new register, Juanita's checking account balance was $432.56. On October 2, Juanita wrote check #104 for $56.79. On October 5, Juanita deposited $30. On October 7, Juanita wrote check #105 for $19.57. What is her current balance?
    A. $326.20
    B. $356.20
    C. $386.20
    D. $376.20

15. Julie, David, Amelia, and Ollie are friends who live on the same road. David lives 5 miles west of Julie. Amelia lives 2 miles east of David. Ollie lives 6 miles west of Julie. How far apart do Amelia and Ollie live?
    A. 3 miles
    B. 8 miles
    C. 6 miles
    D. 12 miles

16. Divide.
    0.735 ÷ 3.5
    A. 0.21
    B. 2.1
    C. 21
    D. 0.021

17. Compare. Use <, >, or =.
    9.6070 ☐ 9.608
    A. <
    B. >
    C. =

18. Write as a standard numeral.
    16^4
    A. 4096
    B. 64
    C. 65,636
    D. 65,536

Chapter 5 Overview

Key Ideas

- Use the problem-solving strategies *make a table* and *find a pattern*.
- Read and make bar graphs, pictographs, and line graphs.
- Read circle graphs.
- Estimate using compatible numbers.
- Use the mental math strategy *compensation*.
- Evaluate graphs.
- Understand the situation.
- Collect and analyze data to solve a problem.

Key Terms

- bar graph
- horizontal
- pictograph
- vertical
- line graph
- compatible numbers
- circle graph
- compensation
- scale

Key Skills

Find the answer.

1. 23 + 31
2. 74 + 58
3. 82 + 25
4. 28 + 16
5. 76 + 35
6. 31 + 85
7. 146 + 56
8. 547 + 66
9. 246 + 248
10. 3508 + 776
11. 2485 + 325
12. 9712 + 1382
13. 46 − 42
14. 60 − 48
15. 26 − 18
16. 62 − 56
17. 89 − 81
18. 75 − 46
19. 135 − 74
20. 236 − 83
21. 405 − 29
22. 707 − 248
23. 600 − 83
24. 1035 − 348
25. 42 × 3
26. 64 × 6
27. 76 × 4
28. 38 × 8
29. 89 × 7
30. 52 × 9
31. 125 × 6
32. 308 × 7
33. 58 × 34
34. 75 × 80
35. 450 × 34
36. 360 × 25
37. 8)69
38. 4)35
39. 3)26
40. 7)149
41. 6)188
42. 4)523
43. 2)726
44. 8)186
45. 5)807
46. 4)764

TABLES, CHARTS, AND GRAPHS 5

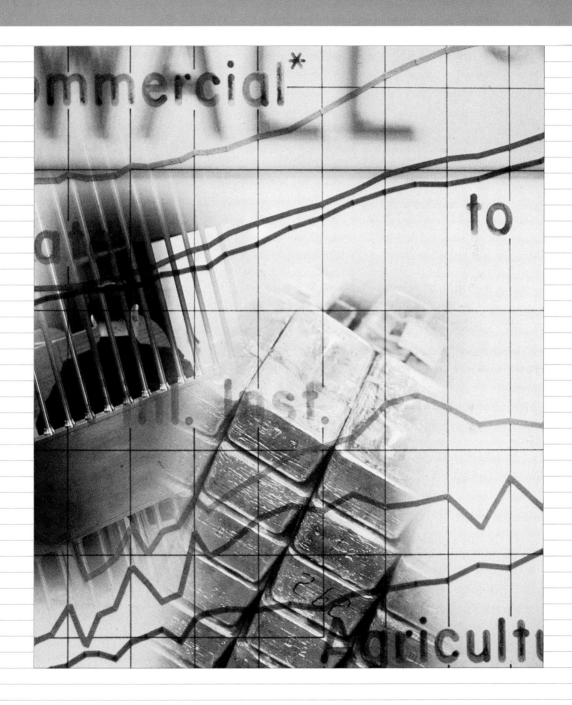

5-1 Problem Solving: Learning to Use Strategies

Objective
Use the strategies *make a table* and *find a pattern*.

- **SITUATION**
- **DATA**
- **PLAN**
- **ANSWER**
- **CHECK**

At each track meet, the girls' mile relay team improved their time. In the first meet, their time improved 0.2 s. In the second meet, their time improved 0.4 s. In the third meet, their time improved 0.7 s. At this rate, by how much will they improve their time in the eighth track meet?

You can use the strategies **make a table** and **find a pattern** to solve this problem.

Example Solve.

At this rate, by how much will the girls' mile relay team improve their time in the eighth track meet?

Make a table. Use the data given in the problem.

Track meet	1	2	3
Seconds	0.2	0.4	0.7

Find the pattern. Fill in the table until you find the solution.

Track meet	1	2	3	4	5	6	7	8
Seconds	0.2	0.4	0.7	1.1	1.6	2.2	2.9	3.7

+0.2 +0.4 +0.6 +0.8
   +0.3 +0.5 +0.7

At this rate, the girls' mile relay team will improve their time by 3.7 s in the eighth track meet.

Try This Solve.

a. Alberto wants to improve his tennis serve. He practices serves every other day. On the first day, he served 50 balls. On the third day, he served 75 balls. On the fifth day, he served 100 balls. If he continues practicing this way, on what day will he serve 225 balls?

b. Sui-Lan saved a penny, a dime, and a quarter every day. How many days did it take her to have an exact number of dollars? How much money did she save?

Exercises

Solve. Use one or more of the problem-solving strategies.

Problem-Solving Strategies
- Guess and check.
- Choose an operation.
- Make an organized list.
- Act it out.
- **Find a pattern.**
- Write an equation.
- Use logical reasoning.
- Draw a picture or diagram.
- **Make a table.**
- Use objects.
- Work backward.
- Solve a simpler problem.

1. Julie is in a swimming program. On any given day, she receives 1.5 points for every 6 laps she swims. On Julie's best day in the first week, she got 9 points. How many laps did she swim?

2. At a 4-week sports camp, students play in 4 different sports: basketball, soccer, baseball, and track. Students pick one sport per week. How many different ways can Beth choose to arrange the sports?

3. Brad was in charge of buying 9 trophies for the inter-school diving meet. Large trophies were $7. Small trophies were $4. If he spent $51, how many of each size trophy did he buy?

4. Four friends ran in a 10K race. They finished the race one right after the other. David had one person ahead of him. Juri was not last. Michael was not first. Katie was ahead of Juri. In what order did they finish the race?

5. Laura and Tina have been jogging together for 4 years. Tina is now 16. At what age will she have been jogging for half of her life with Laura?

6. Kendra's gymnastics coach is twice as old as Kendra. The sum of their ages is 42. How old are they?

7. Tony bought a new baseball mitt for $45. He had one $20 bill and some $5 and $1 bills. He used 14 bills to pay for the mitt. Tell how many of each kind of bill Tony used to pay for his mitt.

8. Victoria wants to buy a snack in a snack machine that will take only quarters, dimes, and nickels. Each snack costs 45¢. She must use exact change. How many different combinations of coins can Victoria use to buy a snack?

9. On a camping trip, Mr. Hasegawa and his 2 sons canoed across a river. Their canoe held only 200 pounds at a time. The father weighed 200 pounds. His sons each weighed 100 pounds. How did they cross the river in the canoe?

Suppose

10. Suppose in the example on p. 120 that the girls' relay team improved their practice time by 0.3 s in the first meet. They improved their time 0.6 s in the second meet. They improved their time 0.9 s in the third meet. At this rate, by how much will they improve their practice time in the eighth meet?

Write Your Own Problem

11. Write a problem that you could solve using the strategy *make a table*. Use the information in the sign.

> **Bring back your rental skis on time!**
> Remember—The late charge increases for every day you are late.

TABLES, CHARTS, AND GRAPHS

5-2 Bar Graphs

Objective
Read and make bar graphs.

A **bar graph** compares quantities using vertical or horizontal bars.

Understand the Situation
- What are the units on the horizontal scale?
- What does the mark between 2000 and 3000 mean?

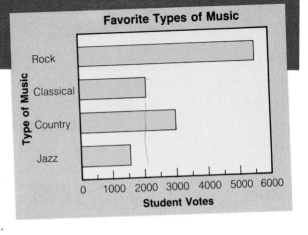

Example

1. Estimate the number of votes for rock.

   To estimate, follow the bar to its end. Read the number directly across on the horizontal scale. The bar showing votes for rock is about halfway between 5000 and 6000. A good estimate is about 5500 votes.

Try This

Estimate the number of votes for each type of music.
   a. classical   **b.** country   **c.** jazz

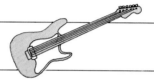

Example

2. Make a vertical bar graph of this data. Student votes for favorite rock group: A-ha 37, U2 30, Huey Lewis 45, The Police 19.

   Ⓐ Find the greatest and least values for the data. Choose units for the scale.

   The greatest number of votes was 45. The least number of votes was 19. Use a scale from 0 to 50 with marks every 5 units. Label every 10 units.

   Ⓑ Use graph paper. Draw and label the horizontal and vertical sides of the graph.

   Ⓒ Draw bars to show the data.

   Ⓓ Write a clear title.

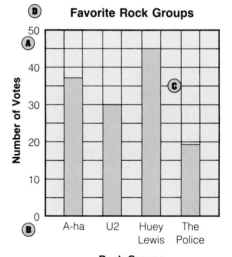

Try This

d. Make a horizontal bar graph of this data. Student votes for favorite Huey Lewis hits: Jacob's Ladder 56, Heart and Soul 23, Hip to be Square 48, Power of Love 42.

122 CHAPTER 5

Exercises

Use the Favorite Types of Music bar graph on p. 122.

1. Which type of music was in second place?
2. Which two types of music received a total of about 9000 votes?
3. Which type of music received about half as many votes as country?
4. Estimate the difference between the number of votes for country and the number of votes for classical.
5. Estimate the difference between the number of votes for rock and the number of votes for country.

Use this bar graph to answer Exercises 6–10.

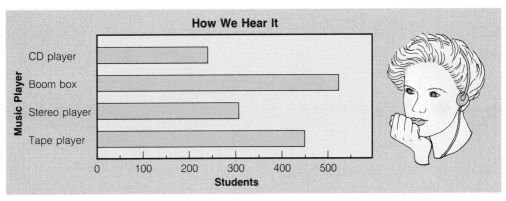

6. Estimate the number of students who use CD players.
7. Estimate the number of students who use tape players.
8. Estimate the difference between the number of students who use stereo players and the number of students who use boom boxes.
9. Is it possible that some students use more than one music player?
10. Why might CD players be used by the fewest students?
11. Make a horizontal bar graph of this data.

Favorite Country Stars

Singer	Votes
George Strait	37
Alabama	26
The Judds	52
Willie Nelson	46
Ronnie Milsap	31

12. Make a vertical bar graph of this data.

Attendance at Music Performances

Concerts	Attending
Jazz	25,000
Classical	37,000
Opera	7,500
Musical play	48,000

Mixed Skills Review Find each answer.

13. $n = 768 \div 32$
14. $94.1 - 29.04 = x$
15. $t = 52 \times 37.12$

TABLES, CHARTS, AND GRAPHS

5-3 Pictographs

Objective
Read and make pictographs.

A **pictograph** uses symbols to compare data.

Understand the Situation
- What does each TV stand for?
- What does half a TV stand for?
- Which type of video received the least votes?
- Is it likely that the number of rentals is exact or rounded?

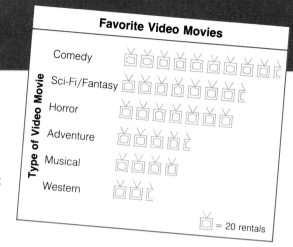

Example

1. Estimate the number of students who rented adventure videos.

   4 full symbols   *Count the number of whole symbols.*
   4 × 20 = 80   *Multiply by the value of one symbol.*
   $\frac{1}{2}$ symbol = 10 votes   *Add the value of any half symbols.*

   About 90 students rented adventure videos.

Try This

Estimate the number of students who rented each type of video.

   a. comedy   **b.** sci-fi/fantasy   **c.** horror   **d.** musical   **e.** western

Example

2. Make a pictograph for this data. A Star Trek fan club ran a survey to find members' favorite episodes.

   Ⓐ Draw and label the sides of the graph.
   Ⓑ Draw symbols to show the data. Round amounts where needed.
   Ⓒ Write a clear title.

Trekkies' Choice	
Title	Votes
Where No Man Has Gone Before	64
City on the Edge of Forever	45
Menagerie	78
What Are Little Girls Made Of?	34
The Corbomite Maneuver	55

Ⓒ **Trekkies' Choice**

Episodes:
- Where No Man Has Gone Before
- City on the Edge of Forever
- Menagerie
- What Are Little Girls Made Of?
- The Corbomite Maneuver

Ⓐ 🖖 = 10 votes

CHAPTER 5

Try This

f. Make a pictograph for this data. Votes for science fiction classics: *2001* 34, *Star Wars* 45, *Alien* 21, *Bladerunner* 15.

Exercises

Use the Favorite Video Movies pictograph on p. 124.

1. Which types of videos were about 10 rentals apart?
2. Find the difference between the number of students renting sci-fi/fantasy videos and adventure videos.
3. What was the combined rental for the top two types of video movies?

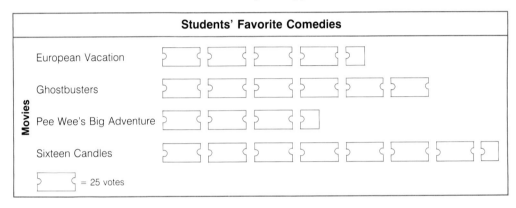

Use the pictograph of Students' Favorite Comedies.

4. Find the number of students voting for *Ghostbusters*.
5. What was the combined vote for the two most popular comedies?
6. What is the difference in the number of votes for *Sixteen Candles* and *European Vacation*?

Make a pictograph for the data in these tables.

7.

Favorite Types of TV Shows	
Type of Show	**Votes**
Soaps	56
Comedy shows	49
Quiz shows	19
Detective shows	30
News programs	12

8.

Favorite TV Sports	
Sport	**Votes**
Soccer	27
Football	30
Baseball	45
Tennis	22
Basketball	42

Problem Solving

9. Alicia, Brenda, and Cassie want to find an afternoon to watch a video of *2001*. Brenda has band practice every other afternoon beginning this Monday. Alicia plays volleyball the first and third Tuesday of each month. Cassie has to go to the dentist every Friday. The month starts on Sunday. What is the first weekday they could all watch the video?

5-4 Line Graphs

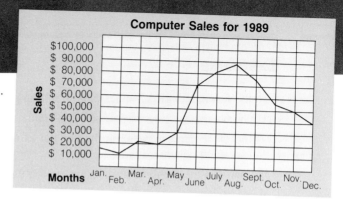

Objective
Read and make line graphs.

Denise Delgado sells computers. Denise uses a line graph to show changes in monthly sales for her region. A **line graph** uses line segments to show change in data.

Understand the Situation
- What is the direction of change in sales between June and July?
- What is the direction of change in sales between January and June?

Example 1. Estimate the dollar amount of sales for October.

Find the month on the horizontal scale. Find that month's dot on the graphed line. Read across to the number scale. The dot falls about halfway between $50,000 and $60,000. The sales for October 1989 were about $55,000.

Try This
a. Estimate the dollar amount of sales for March.
b. Estimate the difference between the sales for October and April.

Example 2. Make a line graph of this data.
Computer games sold: July 32, August 29, September 39, October 45, November 53, December 76.

Ⓐ Choose the units.
Ⓑ Draw, mark, and label the scales.
Since sales change over time, put months on the horizontal axis. The greatest number of sales is 76. The vertical scale can go from 0 to 80.
Ⓒ Mark the dots for the data.
On each month line, make a dot opposite the number of games sold. For July the dot is placed slightly above the 30 line to show 32.
Ⓓ Connect the dots with line segments.
Ⓔ Write a clear title.

Try This
c. Make a line graph of this data on regional computer sales.

Week	1	2	3	4	5	6	7	8	9	10
Number of Sales	176	170	167	158	150	148	145	140	145	140

Exercises

Use the Computer Sales graph on p. 126.
1. Estimate the dollar sales for December.
2. Estimate the total sales for November and December.
3. Estimate the difference between sales for May and January.
4. Which month had the least sales in 1989?
5. Which month had the greatest sales in 1989?
6. In which month did sales increase the most?
7. In which month did sales decrease the most?
8. Estimate the total sales for the year.

Make a line graph for the data.

9. **Computer System Sales**

Period	1	2	3	4	5	6	7	8	9	10
Number of Sales	110	105	98	95	90	90	88	85	80	85

10. **Computer Sales to Small Business**

Year	Number of Sales
1983	12
1984	19
1985	29
1986	41
1987	73
1988	88
1989	126

Estimation
11. Estimate the total number of sales for all sales periods in Exercise 9.
12. Estimate the total computer sales in Exercise 10.

Use the Computer Games Sales graph in Example 2 to solve these exercises.
13. What is the direction of change between July and August?
14. What is the overall direction of change from July to December?

Talk Math
15. Explain to the class why line graphs are used to show changes in data over time.
16. Find a line graph in a newspaper or a magazine. Show the graph to the class and explain what it shows.

Group Activity
17. Find the daily attendance record for your class for the last five weeks. Arrange the data in a table. Make a line graph to show the data. Was there an increase in absences on a certain day?

Mixed Skills Review

Find the answer.
18. $407 - 68$
19. 68×407
20. $408 \div 68$
21. $407 + 86 + 2.72$

Find the answer. Use mental math.
22. 200×15
23. $21 + 460$
24. $\$7.47 - \$0.50$
25. $96.8 \div 100$

TABLES, CHARTS, AND GRAPHS

5-5 Reading Circle Graphs

Objective
Read circle graphs.

Brandi asked the 30 members of the Cycle Club to pick their favorite pizza topping. This circle graph shows the results of her survey. A **circle graph** shows parts of a whole. Each part of the whole can be represented as a percent of the whole.

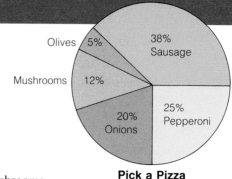

Pick a Pizza

Understand the Situation

- Which part of the graph is the largest?
- What percent of the Cycle Club chose mushrooms as their favorite topping?
- How many people does the whole circle represent?

You can add or subtract percents in the graph. The sum of all the percents is 100%.

Examples

1. Find the total percent of the members voting for either pepperoni or olives.

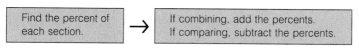

Pepperoni 25%
Olives 5%

$25\% + 5\% = 30\%$

A total of 30% of the members voted for either pepperoni or olives.

2. What percent more of the members chose sausage than chose onions?

38% chose sausage.   *Find the percent of each section.*
20% chose onions.

$38\% - 20\% = 18\%$ *To compare, subtract the percents.*

18% more of the members chose sausage than chose onions.

Try This

a. Find the total percent of the members voting for either olives or mushrooms.

b. What percent more of the members chose pepperoni than chose olives?

c. Find the total percent of the members voting for either sausage or pepperoni.

d. What percent more of the members voted for onions than voted for mushrooms?

128   CHAPTER 5

Exercises

Use the Pick a Pizza circle graph on p. 128.

1. What percent more of the members chose sausage than chose olives?
2. Find the total percent of the members voting for sausage, onions, or pepperoni.
3. What percent of the members voted for toppings other than sausage?
4. Later, 5% of the members changed their votes from mushrooms to sausage. What are the new percents for mushrooms and sausage?
5. Brandi can buy only two kinds of pizza for the club. What kinds should she buy to please the most members?

The Cycle Club has saved $100 for a party. This circle graph shows how they will spend the money.

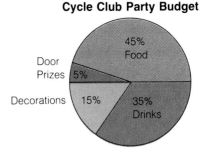

Cycle Club Party Budget

6. How much money does the whole circle represent?
7. What percent more of the money will they spend on drinks than on prizes?
8. Find the total percent they will spend for items other than food and drink.
9. Find the total percent they will spend on food, drinks, and door prizes.

Number Sense

10. What part of each dollar will the Cycle Club spend for food?
    A. $45   B. $0.45   C. $4.50   D. $.045

11. What part of each dollar will they spend for drinks and decorations?
    A. $0.25   B. $0.75   C. $0.50   D. $0.05

Mental Math

Use the Pick a Pizza graph on p. 128.

12. Find three choices that total 50%, or half the members.
13. Suppose the five toppings were all chosen by the same percent of members. What would each percent be?

Working with Variables

Solve and check.

$x + 2 = 10$ You can undo the action of adding by subtracting.
                     To get the variable alone on one side, subtract 2.

$x + 2 - 2 = 10 - 2$ Remember to use the same action on both sides of an equation
$x = 8$ to keep the sides equal.

$8 + 2 = 10$ Check by substituting your answer for the variable.
                     The solution is 8.

Solve and check.

14. $y + 5 = 19$ 15. $t + 12 = 30$ 16. $x + 4.25 = 12$ 17. $r + 3.26 = 7$

TABLES, CHARTS, AND GRAPHS

5-6 Estimating Sums Using Compatible Numbers

Objective
Estimate sums using compatible numbers.

Understand the Situation

- How many members joined the Boosters Club during the first week?
- In which week did about 200 members join?
- Is the membership drive over?

Boosters Club Membership Drive

Week	1	2	3	4	5	6
New Member	46	209	54			

Examples Estimate. Use compatible numbers.

1. 46 + 209 + 54

46 and 54 are compatible. Their sum is 100.
209 is about 200.
200 + 100 = 300
Estimate: 300

2. 55 + 24 + 44 + 276

There are two pairs of compatible numbers.
55 + 44 is about 100
and 24 + 276 = 300
100 + 300 = 400
Estimate: 400

3. $8 + $6.32 + $12.03

$8.00 and $12.03 are compatible. Their sum is about $20.
$6.32 is about $6.
$6 + $20 = $26
Estimate: $26

Try This Estimate. Use compatible numbers.

a. 31 + 23 + 126   **b.** 24 + 18 + 27 + 32   **c.** $19.13 + $4.05 + $21.09

Exercises

Estimate. Use compatible numbers.

1. 50 + 412 + 46
2. 116 + 600 + 402
3. 1099 + 82 + 20
4. 221 + 770 + 32 + 180
5. 300 + 93 + 1702 + 110
6. 2604 + 352 + 401 + 150
7. $4.04 + $15.28 + $1.11
8. $26.06 + $2.18 + $4.08
9. $130.09 + $48.14 + $70.05
10. The Hereford High Sports Boosters raised $2356 at a fundraiser. They plan to divide the money equally among 11 different sports programs. Estimate the amount of money each program will receive.

Determining Reasonable Answers Decide if each answer is reasonable. Tell how you reached your answer.

11. Shanda added the following club expenses on her calculator: postage $32.18, duplicating $19.16, certificates $24.98. Her calculator showed this answer: 176.32.
12. During the final three weeks of the membership drive, the following numbers of members joined: 120, 76, 180. Shanda added the numbers on her calculator. Her calculator showed this answer: 276.

5-7 Mental Math: Compensation

Objective
Use the mental math strategy compensation.

The circle graph shows the results of a student survey at Austin High School.

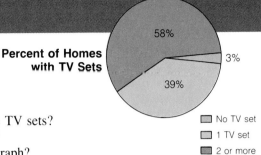

Percent of Homes with TV Sets

- No TV set
- 1 TV set
- 2 or more TV sets

Understand the Situation
- What percent of homes had two or more TV sets?
- What percent of homes had no TV sets?
- What is the total of the percents in the graph?

To find the difference between the number of homes having two or more TV sets and the number having one TV set, Chen used the mental math strategy called **compensation**.

Examples Find the answer. Use mental math.

1. 58 − 39
Adjust to a simpler fact.
58 − 40 = 18   *This subtracts 1 too many.*
Compensate by adding 1 to the difference.
18 + 1 = 19
58 − 39 = 19

2. 48 × 7
Adjust to a simpler fact.
50 × 7 = 350   *This adds 2 more 7s.*
Compensate by subtracting two 7s, or 14, from the product.
350 − 14 = 336
48 × 7 = 336

Try This Find the answer. Use mental math.

a. 23 × 9   **b.** 369 + 225   **c.** 279 − 98   **d.** 49 × 4   **e.** $198 − $69

Exercises

Find the answer. Use mental math.

1. 9 × 33
2. 8 × 52
3. 102 − 49
4. 227 + 125
5. 9 × 53
6. 348 − 51
7. 99 × 32
8. 677 + 150
9. $3.99 × 9
10. 4 × 19
11. $9.95 × 3
12. $49.95 × 7
13. Jessie bought 3 CDs for $14.99 each. What was the total cost?

Calculator Add. Use mental math to decrease the number of addends that you need to enter into the calculator.

14. 55 + 62 + 45 + 32 + 58 + 48
15. 8 + 80 + 15 + 82 + 5 + 99

5-8 Consumer Math: Evaluating Graphs

Objective
Evaluate graphs.

Graphs can sometimes mislead readers. It is important to read a graph carefully. Here are two different graphs that show the popularity ratings of a political candidate.

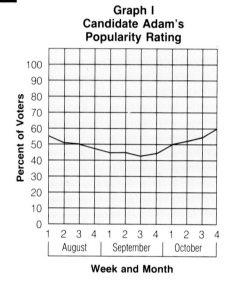

- What is shown on the vertical scale of each graph?
- What is shown on the horizontal scale of each graph?
- Which graph gives the most positive view of Adam's popularity?

Example Compare the data in each graph for the first week of October. What are the differences in the graphs?

Both graphs show a rating of 50 percent for that week. The difference is in how the data is shown. Graph II shows only part of the scale and uses larger spaces.

Try This
a. Compare the data in each graph for the last two weeks in October. What are the differences in the two graphs?

b. Suppose Graph II showed the full month of September. How would the graphed line change?

132 CHAPTER 5

Exercises

Study Graph III and Graph IV. Then answer the questions.

**Graph III
Political Contributions**

**Graph IV
Political Contributions**

1. Which graph gives the more positive view of the rate of political contributions? Why?
2. How do the contributions compare monthly on each graph?
3. What does each space on the vertical scale represent in Graph III? In Graph IV?
4. On which graph does the increase from October to February seem greatest? Why?

Suppose

5. Suppose that you changed Graph I and II on p. 132 into bar graphs. Would you still get the same impression from the two graphs?
6. Suppose that you switched the scales on Graph I and II. Would you still get the same impression from the two graphs?

Data Hunt

7. Newspapers often tell the results of polls in many fields. Find an article that gives this kind of data. Display the data using different graphs so that it may be pictured in more than one way.

Mixed Skills Review

8. Reggie's checkbook balance is $47.15. He writes a check for $15 and deposits $23.18. What is the new balance?
9. Show how Reggie writes $26.49 in checkbook notation.

Write using exponents.

10. $4 \times 4 \times 4 \times 4$
11. 16×16
12. $(0.3)(0.3)(0.3)$
13. 9 squared

5-9 Problem Solving: Understanding the Situation

Objective
State the question in the problem.

- **SITUATION**
- **DATA**
- **PLAN**
- **ANSWER**
- **CHECK**

How many more people surveyed listen to their favorite radio station because of the music than because of the newscasts?

Reasons for Listening to Your Favorite Radio Station

Reasons	Number of People
Music	945
D.J.	225
Loyalty	135
Newscasts	75
Do not know	60
No favorite station	60

One of the first things you must do when you solve a problem is **understand the situation**.

Example Read the question. Then decide which question asks the same thing.

Question: How many more people surveyed listen to their favorite radio station because of the music than because of the newscasts?

A. How many people surveyed listen to their favorite radio station because of the music and because of the newscasts?

B. How many fewer people listen to their favorite radio station because of newscasts than because of the music?

Decide what the original question asks.
The original question asks for the difference between the number of people who listen because of the music and people who listen because of the newscasts.

Decide what questions **A** and **B** ask.

*Question **A** asks for the total number of people who listen because of the music and people who listen because of the newscasts. Question **B** asks for the difference between the number of people who listen because of the newscasts and because of the music.*

Question **B** asks the same thing as the original question.

Try This Read the question. Then decide which question asks the same thing. Use the table.

Question: How many people surveyed listen to their favorite radio station for reasons other than the music, other than the newscasts, for no reason, or just do not have a favorite radio station?

A. How many more people listen to their favorite radio station because of loyalty than because of the D.J.?

B. What is the total number of people who listen to their favorite radio station because of loyalty and because of the D.J.?

Exercises

Read each question. Then decide which question asks the same thing. Use the table.

When You Get Extra Money, What Is the First Thing You Buy?

Clothes	39
Records/tapes	52
Things for the car	24
Food	16
Movies	13
Shoes	8
Do not know	22

1. *Question:* How many more students buy records and tapes than buy clothes?
   A. How many students buy records, tapes, or clothes?
   B. How many fewer students buy clothes than buy records and tapes?

2. *Question:* How many students knew what they would buy if they had extra money?
   A. What is the difference between the total number of students surveyed and the number that did not know what they would buy?
   B. What is the difference between the total number of students surveyed and the number who knew what they would buy?

3. *Question:* What is the difference between the number of students choosing records, tapes, and movies and the number choosing things for the car?
   A. How many students buy things for the car?
   B. How much is 13 plus 52 minus 24?

Write a question for each statement. Then solve the problem.

4. Clothes, records, and tapes were the most popular items.

5. Suppose all of the students that said *do not know* would have said *clothes*.

6. Of the 39 students who buy clothes with extra money, 16 students are females.

7. In a survey at another school, 25 students said that they would buy clothes with their extra money.

Evaluating Strategies

8. Tell why this picture is not correct for the given question. Draw a correct picture. Solve the problem.

How far is it from Paisley to La Pine?

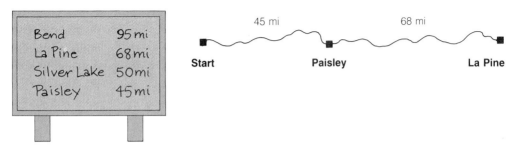

Finding All Solutions

9. How many different ways can you make change for a 50¢ piece without using pennies?

TABLES, CHARTS, AND GRAPHS

5-10 Problem Solving: Data Collection and Analysis

Objective
Collect and analyze data.

- SITUATION
- DATA
- PLAN
- ANSWER
- CHECK

Consumer magazines compare different brands of the same product. Here is a sample table that compares television sets.

Brand and Model	Price	Overall Score	Picture Clarity	Screen Reflectance	Color Fidelity	Airplane Flutter	Fringe Reception	Tone Quality	Extra Remote Control Features	Advantages	Disadvantages
Beston B29	$300	◐	○	◐	◐	○	●	–	E	l
Perfectavox S2	$330	⊙	○	◐	◐	◐	●	–	E	–
PreSet 007A	$412	◐	○	◐	◐	○	○	–	D,E	l
Magnitube 800	$480	⊙	○	⊙	○	◐	◑	–	–	–
Electrobox XX5	$299	⊙	○	◐	◐	○	○	✓	D,E	b
Zeno 006T	$250	⊙	◑	◑	◐	◑	◑	–	E	m
Jackson 2333A	$550	⊙	◑	◐	◑	◑	●	–	A	g,l
Kawatachi 725x	$449	⊙	◑	⊙	◐	◐	◑	✓	–	l
Silver Tone 4000	$339	⊙	○	⊙	○	○	◑	–	–	a
Spirit View 250	$420	⊙	◐	◐	◐	◐	●	–	E	b,d

⊙ ◐ ○ ◑ ●
Best ⟷ Worst

Advantages:

A - numeric keypad for selecting channels
D - channel number displayed on screen
E - can scan through channels

Disadvantages:

a - transformer needed to hook up to VHF
b - picture sharpness control in back of set
d - must use remote control to reinstate a channel already scanned by
g - background noise worse than usual
l - picture shrinkage worse than most
m - channel numbers harder to read than most

Work in groups. Complete the following page.

Situation

Consumer magazines usually rank the different brands by their quality. The brands of television sets in the table on p. 136 are not in the correct order. Give a score to each brand. Rank the brands by their quality.

Data Collection

Write your own questionnaire. Survey your classmates about the quality of the television sets they have at home. Make a table of the results. Rank the brands by their quality.

Facts to Consider

- Price may not reflect quality.
- *Picture clarity* measures how crisp and sharp the images are. Fuzziness and outlines will bring down the score.
- *Color fidelity* measures how well the set shows colors.
- *Screen reflectance* measures how well the screen does not reflect room light.
- *Airplane flutter* measures how well the set resists interference when airplanes fly overhead.
- *Fringe reception* measures how well the set pulls in weak signals.
- *Tone quality* measures how well the set sounds.
- *Extra remote control features* tells if the brand has more than these standard features: power, volume, mute, off timer, quick view, channel selector.

Share Your Group's Plan

- Explain how you decided on the rankings for the various brands. Tell how you gathered data for your own consumer table.
- Compare your results to other groups.
- Use point values to rate the television sets in your survey. Make a bar graph to compare the scores of the brands and models.

Suppose

- Suppose the brands on p. 136 all had the best rating for picture clarity. Would this change your ranking?

Mixed Skills Review

Find the answer.

1. $1205 - 349$
2. 16×702
3. $3368.4 \div 8.4$
4. $520.7 - 46.93$
5. $4628.2 \div 73$
6. 0.42×0.067
7. $381 + 45.8$
8. $6.617 \div 13$
9. 708×39
10. $1136.8 \div 56$
11. $8.601 - 0.978$
12. $\$57.16 \times 0.35$

Estimate.

13. $24 + 106 + 77$
14. $113 \div 11$
15. 4×39
16. $874 - 309$

Find the answer. Use mental math.

17. $\$10.25 + \$2.50$
18. 100×278
19. 4×39
20. $394 \div 10$

5-11 Enrichment: Double Bar and Line Graphs

Objective
Read double bar and line graphs.

Drew and Tip drove from Danville to the mountains for a vacation. Drew left 3 h before Tip. The **double line graph** and **double bar graph** show the time and the number of miles Drew and Tip drove. After 1 h of driving, who had driven the farthest?

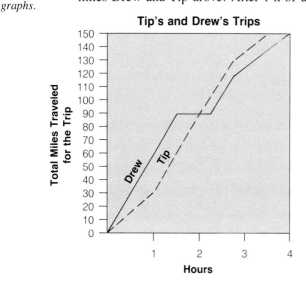
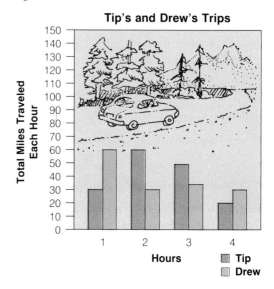

Double graphs compare two sets of data. The key to the graph tells how to find each set of data. Read double line and bar graphs as you would a single line or bar graph. Drew drove 60 mi after 1 h. Tip drove 30 mi after 1 h. Drew drove the farthest.

Exercises

1. Who drove the farthest after 2 h?

2. Who spent the least amount of time driving?

3. How many miles did each person drive?

4. How many more miles did Tip drive than Drew after 3 h?

5. How many miles did Tip drive between the second and third hours?

6. How many miles did Drew drive between the first and second hours?

Group Activity

7. In 10 s, write the digit 3 as many times as you can on a piece of paper. Rest for 5 s. Repeat the process for 1 min. Have a classmate time you. Now be the timer as your classmate writes. Make a double bar or line graph to compare your results with your classmate's. Who wrote the greater number of 3s in a 10-s period? What was the greatest difference in one period?

5-12 Computer: Graphs

Objective
Read computer graphs.

Elva Guerra keeps a record of her expenses on a computer. The chart shows her expenses for three months. She uses a **spreadsheet program** to show the data as a graph. Her computer can make different kinds of graphs for the same data.

Item	Amount Spent		
	May	June	July
Clothes	$145.23	$ 67.55	$253.56
Food	165.34	143.89	189.54
Transportation	63.50	73.45	89.60
Rent	300.00	300.00	300.00
Car payment	115.79	115.79	115.79
Entertainment	45.00	87.90	132.87

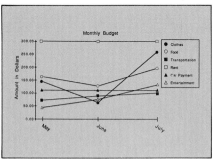
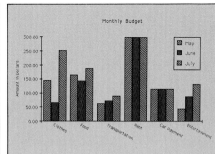
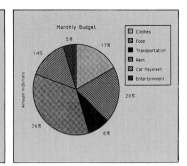

Exercises

1. What is the most that Elva spent on clothes in one month?

2. From which graph is it easiest to see what percent of her money Elva spent on rent?

3. Why do you think that Elva has the computer use different patterns on the circle graph?

4. Why do you think that Elva has the computer use different patterns on the bar graph?

5. What is the greatest amount shown on the vertical scale for the line graph? For the bar graph?

6. Is the horizontal scale the same for the bar and line graphs?

7. What is the total percent shown on the circle graph?

8. What items cost the same from month to month?

9. What was the total amount Elva spent for food in the three months?

10. From which graph is it easiest to compare the data? Explain why.

Suppose

11. Suppose Elva added another item, such as car insurance, to her record. In which graph would Elva have to tell the computer to add a new pattern?

TABLES, CHARTS, AND GRAPHS

CHAPTER 5 REVIEW

Identify each label.

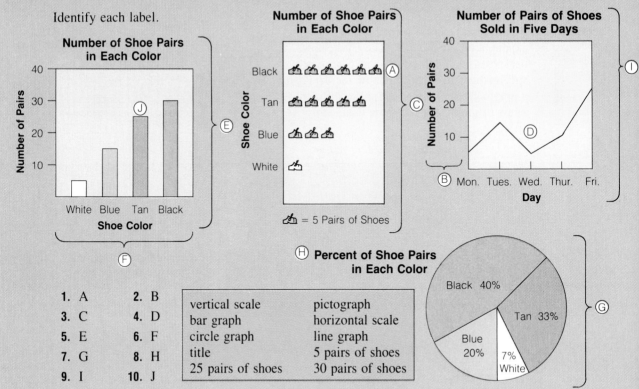

1. A
2. B
3. C
4. D
5. E
6. F
7. G
8. H
9. I
10. J

vertical scale pictograph
bar graph horizontal scale
circle graph line graph
title 5 pairs of shoes
25 pairs of shoes 30 pairs of shoes

Which type of graph would you draw for each set of data? Explain.

11. Tickets sold: Mon. 45, Tues. 53, Wed. 39, Thur. 62, Fri. 76
12. Percent of tickets sold: Play 22%, Concert 48%, Game 25%, Ballet 5%
13. Number of questions right: Amos 80, Misty 92, Tai 79, Ali 88

Choose a compatible number for adding to the given number.

14. 787
15. 821
16. 53
17. 642
18. 26.17
19. 174

Compatible Numbers

12	361	349
250	6.5	226
4.39	182	15

Which compensation method would you use to find each answer using mental math?

20. 9×43 A. $(9 \times 40) + 27$ B. $(9 \times 40) - 27$

21. $103 - 48$ A. $(103 - 50) + 2$ B. $(103 - 50) - 2$

22. $327 + 225$ A. $(325 + 225) + 2$ B. $(325 + 225) - 2$

23. Name three things you should evaluate when reading graphs.

CHAPTER 5 TEST

1. A radio station has a phone-in contest every 5 days and a giveaway every 3 days. Today the station had both events. How many days will it be before the station has both events on the same day again?
2. When did the class sell the most tickets? How many tickets did they sell?
3. Make a bar graph. Show the number of people who attended each night of the play: 182, 105, 93, 211, 226.
4. In which game did the quarterback complete the most passes? How many passes did he complete?
5. Make a pictograph. Show the number of students who attended each football game: 350, 411, 387, 503.
6. When was the stock price the highest? What was the price of the stock?
7. Make a line graph of the price of KJS stock on Monday through Friday: $36, $40, $38, $42, $46.

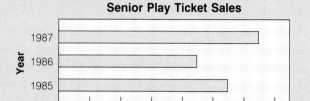

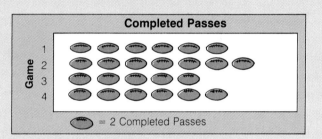

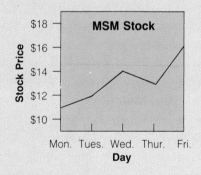

8. What percent more of the students like swimming better than like bowling? **17%**

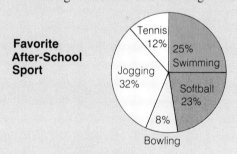

Estimate.   9. 48 + 176 + 253 + 325   10. 121 + 288 + 63 + 238

Find the answer. Use mental math.   11. 7 × $32   12. 476 − 248

13. Does the bar graph of *Senior Play Ticket Sales* clearly show the trend of sales for three years? Explain.
14. Read the question. Which question asks the same thing? *Question:* How many more students read magazines and books than read a newspaper?
    - **A.** How many students read magazines, books, and a newspaper?
    - **B.** How many fewer students read a newspaper than read magazines and books?

TABLES, CHARTS, AND GRAPHS

CHAPTER 5 CUMULATIVE REVIEW

1. Estimate.
   31×28
   A. 90
   B. 600
   C. 60
   D. 900

2. Subtract.
   $0.096 - 0.039$
   A. 0.043
   B. 0.057
   C. 0.67
   D. 0.067

3. Divide.
   $6\overline{)475}$
   A. 78 R5
   B. 7 R55
   C. 79 R1
   D. 79 R2

4. Estimate.
   $83 + 79 + 82 + 78$
   A. 320
   B. 240
   C. 280
   D. 360

5. Divide.
   $\$548.40 \div 8$
   A. $68.55
   B. $60.80
   C. $6.06
   D. $68.50

6. Multiply. Use mental math. 26×7
   A. 1442
   B. 56
   C. 144
   D. 182

7. Write as a standard numeral. 3^4
   A. 81
   B. 9
   C. 12
   D. 27

8. Find the difference between the number of votes for raspberries and for strawberries.
   A. 15 votes
   B. 35 votes
   C. 30 votes
   D. 5 votes

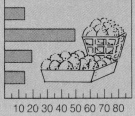

9. Add. Use mental math. $468 + 227$
   A. 685
   B. 241
   C. 695
   D. 6815

10. Round to the nearest ten. 4
    A. 5
    B. 40
    C. 10
    D. 0

11. Subtract. Use mental math. $\$56.75 - \$38$
    A. $18.75
    B. $28.25
    C. $28.75
    D. $16.75

12. Divide.
    $3300 \div 260$
    A. 12 R180
    B. 10 R70
    C. 13 R20
    D. 12 R18

13. Multiply.
    3812
    $\times 62$
    A. 195,244
    B. 236,344
    C. 235,344
    D. 236,244

14. Write the word name.
    1.05
    A. one and five tenths
    B. one hundred and five
    C. one and five hundredths
    D. ten and five tenths

15. How many runs did the Cowell baseball team score in May?
    A. 30 runs
    B. 25 runs
    C. 20 runs
    D. 10 runs

    Cowell Team Runs

    (graph: Number of Runs vs Month — Mar., Apr., May, June)

16. Which question asks the same thing?
    Question: How many more students liked salmon better than crab?
    A. How many students liked salmon and crab?
    B. What is the difference between the number of students who liked salmon and crab?

17. Divide.
    $0.45\overline{)9.765}$
    A. 21.7
    B. 2.17
    C. 0.217
    D. 217

18. Reuben had an account balance of $345.56. On May 19, he deposited $86.87. Find the new balance.
    A. $258.69
    B. $421.33
    C. $422.43
    D. $432.43

19. Arrange the decimals in order from least to greatest.
    4.06, 3.8, 4.26, 3.09
    A. 3.8, 3.09, 4.06, 4.26
    B. 3.09, 3.8, 4.06, 4.26
    C. 3.8, 3.09, 4.06, 4.26
    D. 3.8, 4.06, 4.26, 3.09

20. Compare. Use <, >, or =.
    3.08 ☐ 3.8
    A. <
    B. >
    C. =

21. Multiply.
    385 × 9
    A. 2765
    B. 2725
    C. 3465
    D. 3425

22. Divide.
    3.04 ÷ 1000
    A. 3040
    B. 0.304
    C. 0.0304
    D. 0.00304

23. Multiply.
    4.46 × 100
    A. 44.6
    B. 0.446
    C. 446
    D. 4460

24. Add. 4.05 + 37.8
    A. 41.85
    B. 78.3
    C. 31.85
    D. 33.75

25. Estimate. Round to the highest place.
    489 − 312
    A. 180
    B. 200
    C. 177
    D. 800

26. Estimate. Round to the highest place.
    3436 + 1582
    A. 5300
    B. 5000
    C. 4000
    D. 6000

27. Subtract.
    4002 − 987
    A. 3125
    B. 3025
    C. 3015
    D. 4015

28. Divide.
    2553 ÷ 23
    A. 111
    B. 108 R19
    C. 102 R7
    D. 101

Chapter 6 Overview

Key Ideas

- Use the problem-solving strategy *find a pattern*.
- Find customary and metric measures of length, capacity, weight, or mass.
- Read scales.
- Choose the likely temperature.
- Use formulas.
- Make group decisions.

Key Terms

- customary
- capacity
- temperature
- accurate
- weight
- formula
- measurement
- metric
- scale
- length
- mass

Key Skills

Multiply.

1. 8 × 3
2. 12 × 3
3. 36 × 4
4. 23 × 12
5. 12 × 7
6. 16 × 6
7. 125 × 3
8. 5280 × 3
9. 5280 × 25
10. 12 × 9
11. 36 × 35
12. 12 × 36
13. 34 × 10
14. 28 × 10
15. 75 × 100
16. 124 × 1000
17. 5.56 × 10
18. 0.44 × 100
19. 32.1 × 100
20. 0.3 × 1000
21. 0.024 × 10
22. 5.6 × 100

Divide.

23. 24 ÷ 3
24. 36 ÷ 3
25. 42 ÷ 4
26. 144 ÷ 12
27. 225 ÷ 12
28. 124 ÷ 36
29. 360 ÷ 36
30. 2356 ÷ 8
31. 3420 ÷ 12
32. 5660 ÷ 12
33. 23 ÷ 10
34. 35 ÷ 100
35. 124 ÷ 10
36. 345 ÷ 100
37. 12.4 ÷ 10
38. 0.34 ÷ 100
39. 9.033 ÷ 100
40. 234.5 ÷ 10
41. 456.3 ÷ 1000
42. 23 ÷ 1000

MEASUREMENT

6

6-1 Problem Solving: Learning to Use Strategies

Objective
Use the strategy find a pattern.

- SITUATION
- DATA
- **PLAN**
- ANSWER
- CHECK

Abigail is going to perform in a classical guitar concert. On Monday, she invited 2 people to the concert. On Tuesday, each of those people invited 2 people. On Wednesday, each of those people invited 2 people, and so on. If this rate continues, how many new people were invited on Saturday?

Some problems seem difficult until you **find a pattern.** Sometimes you need to draw a picture and/or make a table to find the pattern.

Example How many new people were invited on Saturday?

Draw a picture to help you see the pattern.

← Abigail
← invited Monday
← invited Tuesday
← invited Wednesday

Use the picture to help you make a table.

Day	New People Invited
Monday	2
Tuesday	4
Wednesday	8

Find the pattern. Complete the table to find the solution.

Day	New People Invited
Monday	2
Tuesday	4
Wednesday	8
Thursday	16
Friday	32
Saturday	64

The number of new people invited doubles every day.

There were 64 new people invited on Saturday.

Try This Solve.

On March 3, Arturo spent $2. On March 4, he spent $4. On March 5, he spent $6. If this rate continues, how much money will Arturo spend on March 10?

---

146 CHAPTER 6

Exercises

Solve. Use one or more of the problem-solving strategies.

Problem-Solving Strategies
- Guess and check.
- Choose an operation.
- Make an organized list.
- Act it out.
- **Find a pattern.**
- Write an equation.
- Use logical reasoning.
- Draw a picture or diagram.
- Make a table.
- Use objects.
- Work backward.
- Solve a simpler problem.

1. Norman made some designs using string. If he used 2 nails, he needed 1 string to connect the nails. If he used 3 nails, he needed 3 strings to connect each nail to the other nails. If he used 4 nails, he needed 6 strings, and so on. How many strings would he need to connect 8 nails?

2. Chang displayed some photographs in an art show. She had 32 more black-and-white photos than color photos. She had 88 photos in all. How many photos were in color?

3. Annette has a big part in the play *Cats*. She practices her lines every other night. On May 4, she learned 3 new lines. On May 6, she learned 5 new lines. On May 8, she learned 8 new lines. If this rate continues, how many new lines will she learn on May 30?

4. Winston had a square piece of cowhide that he wanted to cut into squares. If he made 2 cuts, he ended up with 4 square pieces. If he made 4 cuts, he ended up with 9 squares. If he made 6 cuts, he had 16 squares. How many squares would he get from 18 cuts?

5. The Kwans, Foltzes, and Machados are staying at a hotel near the Shakespeare Festival. The Machados must go up 9 floors to get to the Kwans' room. The Foltzes must go up 14 floors to get to the Kwans' room. The Machados are on the seventh floor. On what floor are the Foltzes?

6. LeRoy numbered the pages of his scrapbook. There were 50 pages. How many times did LeRoy write the digit 1?

7. Thelma had 7 goldfish and 2 large fishbowls. In how many different arrangements could she put the goldfish in the fishbowls?

8. Mr. Worthman is twice as old as Daniel. The sum of their ages is 48. How old are they?

Suppose

9. Suppose in the example on p. 146 that Abigail invited 3 people on Monday. On Tuesday, each of those people invited 3 people. On Wednesday, each of those people invited 3 people, and so on. How many people would be invited on Saturday?

Write Your Own Problem

10. Write a problem that you could solve using the strategy *find a pattern*. Use the information in the memo.

Memo

Ideas for Promoting School Play
1. Hand out flyers all over school.
2. Call 4 people. Ask each of those people to call 4 people.
3. Post 2 posters in each hall.
4. Run ads over the P.A. system.

MEASUREMENT

6-2 Customary Units of Length

Objective
Find likely customary units of length. Change customary units of length.

The **inch**, the **foot**, and the **yard** are the smaller units of length in the **customary system**. The largest unit is the **mile**.

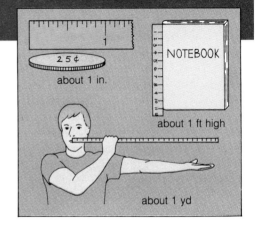

Customary Units of Length

Unit	Symbol	Relationship
inch	in.	
foot	ft	1 foot = 12 inches
yard	yd	1 yard = 3 feet = 36 inches
mile	mi	1 mile = 1760 yards = 5280 feet

Examples Give the likely measure: inches, feet, yards, or miles.

1. the distance from Brewster Green, MA, to San Antonio, TX — miles
2. the height of a basketball net — feet
3. the length of a punt in football — yards
4. the length of a piece of chalk — inches

Try This Give the likely measure: inches, feet, yards, or miles.

a. the length of this page
b. the length of your left leg
c. the length of a school hallway
d. the distance from New York to Rome

When changing customary units of length, multiply to change from a larger to a smaller unit. Divide to change from a smaller to a larger unit.

Examples Change the unit.

5. 13 ft = ___ in.   1 ft = 12 in. *Think: larger to smaller unit. Multiply.*
   13 × 12 = 156   13 ft times 12 in. in 1 ft
   13 ft = 156 in.

6. 7080 yd = ___ mi   1760 yd = 1 mi *Think: smaller to larger unit. Divide.*
   7080 ÷ 1760 = 4 R40 7080 yd divided by 1760 yd in 1 mi
   7080 yd = 4 mi 40 yd

Try This Change the unit.

e. 12 yd = ___ ft
f. 130 in. = ___ ft **10 in.**
g. 8800 yd = ___ mi

148 CHAPTER 6

Exercises

Give the likely measure: inches, feet, yards, or miles.
1. the distance to the sun
2. the width of a watch band
3. the width of your school desk
4. the height of your best friend
5. the distance you can kick a soccer ball
6. the distance you can run in one minute

Change the unit.
7. 5 ft = ___ in.
8. 24 ft = ___ yd
9. 4 mi = ___ ft
10. 15,840 ft = ___ mi
11. 276 in. = ___ ft
12. 7 yd = ___ ft
13. 19 ft = ___ yd
14. 59 ft = ___ yd
15. 29 in. = ___ ft
16. 7 mi = ___ ft
17. 35 in. = ___ ft
18. 324 in. = ___ yd
19. 579 in. = ___ ft 3 in.
20. 76 ft = ___ in.
21. 56 in. = ___ yd

Number Sense

Arrange the distances in order from longest to shortest.
22. 42 in., 1 yd, 16 ft, 5 yd, 0.5 mi, 1000 yd
23. 5 ft, 2 yd, 0.25 mi, 1500 ft, 144 in., 0.5 mi
24. 27 ft, 0.13 mi, 2000 yd, 2500 in., 3 mi, 0.9 mi

Data Hunt

25. Find the meanings of these units of length: **league, fathom, rod, cubit,** and **furlong.** Estimate your height in cubits.
26. Horses are measured in *hands*. Find the size of this unit. Give the height of a 15-hand horse in inches.

Problem Solving

27. Glenda had three boards already cut. Their lengths were 12 ft, 9 ft, and 7 ft. She needed to cut another board 10 ft long. Tell how she could measure 10 ft without going back for her tape measure. Tell what strategy you used.
28. Mr. Kowalski makes sales calls in several towns. Tuesday he drove 17 mi from his house to Westboro. He turned and came back along the same road 6 mi to Ridgefield. Later he drove in the same direction 24 mi to Logan. How far from home was Mr. Kowalski then?

Mixed Skills Review

Find the answer.
29. 12 × 6.4
30. 135 ÷ 36
31. 29.4 × 5.5
32. 297 ÷ 12

Divide by 100. Use mental math.
33. 479.6
34. 2000
35. 0.5
36. 64,231

MEASUREMENT

6-3 Accuracy of Measurement

Objective
Choose the degree of accuracy. Measure to the nearest quarter inch.

A carpenter is building a garden gate. The gardener is cutting large plant stakes in half to use with smaller plants. Both workers choose how accurate their measurements will be.

Example 1. Who needs the more accurate measurements: the carpenter or the gardener?

The carpenter must make many pieces of board fit together exactly. The gardener only needs a stake that is approximately the height of the plant. The carpenter needs the more accurate measurements.

Try This Who needs the more accurate measurements?

a. a gardener checking her cornstalks or a botanist measuring plant growth

b. Tammy cutting rope to tie a crate or Janet cutting ribbon for a hatband

To measure an object, line up the zero end of your ruler against one end of the object. Read the measure at the opposite end. Decide how accurate you want to be.

Example 2. Measure the pencil to the nearest $\frac{1}{4}$ in. Use a ruler.

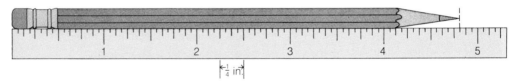

The tip of the pencil is closer to $4\frac{3}{4}$ in. than to 5 in. To the nearest $\frac{1}{4}$ inch, the pencil measures $4\frac{3}{4}$ in.

Try This Measure to the nearest $\frac{1}{4}$ inch. Use a ruler.

c.

d.

Exercises

Who needs the more accurate measurements?
1. a brick layer making a wall or a gym teacher marking a starting line
2. a tailor cutting out lining for a coat or Andy cutting gift wrap
3. a fisherman or a biologist describing the length of a fish

Measure to the nearest $\frac{1}{4}$ in.

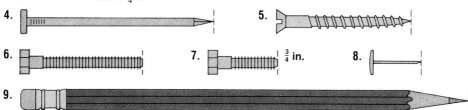

4.
5.
6.
7. $\frac{3}{4}$ in.
8.
9.

Use your ruler. Draw a line segment with the given length.
10. 5 in.   11. $3\frac{1}{2}$ in.   12. $7\frac{1}{4}$ in.   13. $4\frac{3}{4}$ in.   14. $\frac{1}{2}$ in.   15. $2\frac{1}{4}$ in.

Estimation   For Exercises 16–19, cut a piece of string to show your estimate of the length. Label the string with tape. Compare the estimate to the actual object.

16. the width of the doorway
17. the height of the chalkboard
18. your own height
19. the length of a yardstick
20. Measure the width of your thumb joint. Use your thumb joint to estimate the length of this book in inches.
21. Develop a plan to estimate distances using the length of your foot. Test your plan by measuring the width of the classroom.
22. Pirate maps in adventure stories often state distances in *paces*, or steps. Use the length of your step to estimate the measurements on this map.

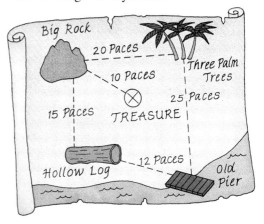

Group Activity   Do these activities with a partner or a team.
23. In 5 minutes, find as many objects as you can that measure an even number of inches.
24. Find something that measures $8\frac{1}{2}$ in. wide and 11 in. high.
25. Find 3 objects that together measure a total of 10 in.
26. Find an object that measures 1 in. in one dimension.

6-4 Customary Capacity

Objective
Find likely customary units of capacity. Change customary units of capacity.

The **gallon** is a customary unit of capacity. Other customary units of capacity are the **fluid ounce, cup, pint,** and **quart**.

Customary Units of Capacity

Unit	Symbol	Relationship
fluid ounce	fl oz	1 fl oz = 2 tablespoons
cup	c	1 cup = 8 fluid ounces
pint	pt	1 pint = 2 cups = 16 fluid ounces
quart	qt	1 quart = 2 pints = 4 cups
gallon	gal	1 gallon = 4 quarts = 8 pints

1 fl oz = 2 tablespoons 1 cup 1 qt of orange juice

1 gal of apple cider   1 pt = tall glass of water

Examples Give the likely measure: fluid ounces, cups, pints, quarts, or gallons.

1. the capacity of a bathtub — gallons
2. the capacity of a mug — fluid ounces
3. the capacity of a large pot — quarts

Try This Give the likely measure: fluid ounces, cups, pints, quarts, or gallons.

a. the capacity of a can of motor oil
b. the capacity of a juice glass
c. the capacity of a large milk carton
d. the capacity of a water heater

When changing units of capacity, multiply to change from a larger to a smaller unit. Divide to change from a smaller to a larger unit.

Examples Change the unit.

4. 19 pt = ___ qt   2 pt = 1 qt Think: smaller to larger unit. Divide.
   19 ÷ 2 = 9 R1   19 pt divided by 2 pt in 1 qt
   19 pt = 9 qt 1 pt

5. 6 gal = ___ qt   1 gal = 4 qt Think: larger to smaller unit. Multiply.
   6 × 4 = 24   6 gal times 4 qt in 1 gal
   6 gal = 24 qt

Try This Change the unit.

e. 23 pt = ___ qt   f. 20 fl oz = ___ c   g. 68 qt = ___ gal

Exercises

Give the likely measure: fluid ounces, cups, pints, quarts, or gallons.
1. the capacity of a water tank
2. the capacity of a shampoo bottle
3. the capacity of a large pitcher
4. the capacity of a soup spoon

Change the unit.
5. 8 qt = ___ pt
6. 7 c = ___ pt
7. 28 fl oz = ___ c
8. 56 pt = ___ qt
9. 9 gal = ___ qt
10. 64 qt = ___ gal
11. 48 fl oz = ___ c
12. 8 c = ___ pt
13. 43 pt = ___ qt
14. 17 qt = ___ gal
15. 16 pt = ___ gal
16. 6 gal = ___ pt
17. 24 fl oz = ___ pt
18. 40 pt = ___ gal
19. 12 c = ___ fl oz

Mixed Applications

20. Lela's car engine takes 1.5 gal of oil. There are 3 qt of oil in the engine. How much oil should Lela add?
21. The fuel tank of Lita's weed trimmer holds 2 pt. The fuel tank of her lawn mower holds 1.25 gal. How much more does the lawn mower tank hold than the weed trimmer tank?

Write Math

22. Write the steps that show how to find the number of fluid ounces in 1 gal.

Mental Math

Solve. Use mental math.
23. Lew plans to make punch for a party. He wants to make 4 times a 3-qt recipe. Lew will need a container that holds at least how many gallons?
24. A camper uses a 1-c dipper to fill a 2-qt canteen. How many dipperfuls does the camper need?

Calculator

In December 1986, Voyager flew around the world without refueling. The lightweight aircraft carried 1200 gal of fuel. Use your calculator to find how many times the tanks of each of these planes could be completely filled by the fuel in Voyager's tank.

Plane Tank Capacity	
25. Wright Brothers' Flyer	2 qt
26. Gas-Powered Model	2 fl oz
27. The Spirit of St. Louis	450 gal
28. Boeing 747-200B	52,400 gal

Group Activity

Do these activities with a partner or a team.
29. Challenge a partner to estimate the capacity of unmarked food cartons and cans. Use measuring cups to find who had the closest estimate.
30. With your team work out a plan for estimating the amount of water used to take a 4-minute shower. Calculate how much water the class uses in one day for showering.
31. Bring in containers to make a display of equivalent measures. Use the chart on p. 152 as a guide.

MEASUREMENT

6-5 Customary Weight

Objective
Find likely customary units of weight. Change customary units of weight.

Mass is the amount of matter in an object. Weight is the measurement of the quantity of mass in an object, and it depends on gravity. In the customary system, weight is measured in **ounces, pounds,** and **tons.**

Customary Units of Weight

Unit	Symbol	Relationship
ounce	oz	
pound	lb	1 pound = 16 ounces
ton	T	1 ton = 2000 pounds

Examples Give the likely measure: ounces, pounds, or tons.

1. the weight of an airplane — tons
2. the weight of a goldfish — ounces
3. the weight of a roasting chicken — pounds

Try This Give the likely measure: ounces, pounds, or tons.

a. the weight of a slice of bread
b. the weight of a truckload of wheat
c. the weight of a bag of potatoes
d. the weight of a bag of groceries

When changing customary units of weight, multiply to change from a larger to a smaller unit. Divide to change from a smaller to a larger unit.

Examples Change the unit.

4. 40 oz = ___ lb   16 oz = 1 lb *Think: smaller to larger unit. Divide.*
   40 ÷ 16 = 2 R8   40 oz divided by 16 oz in 1 lb
   40 oz = 2 lb 8 oz

5. 4.5 T = ___ lb   1 T = 2000 lb *Think: larger to smaller unit. Multiply.*
   4.5 × 2000 = 9000   4.5 T times 2000 lb in 1 T
   4.5 T = 9000 lb

Try This Change the unit.

e. 5000 lb = ___ T
f. 6.5 lb = ___ oz
g. 23 oz = ___ lb
h. 4 oz = ___ lb
i. 2.5 T = ___ lb
j. 2.25 lb = ___ oz

Exercises

Give the likely measure: ounces, pounds, or tons.
1. the weight of a tanker ship
2. the weight of a letter
3. the weight of a bag of fertilizer
4. the weight of an elephant
5. the weight of a bicycle
6. the weight of a bag of popcorn

Change the unit.
7. 9 lb = _____ oz
8. 3 T = _____ lb
9. 48 oz = _____ lb
10. 1.5 lb = _____ oz
11. 12 oz = _____ lb
12. 0.5 T = _____ lb
13. 56 oz = _____ lb
14. 3.5 lb = _____ oz
15. 80 oz = _____ lb
16. 3.5 T = _____ lb
17. 19 lb = _____ oz
18. 4.25 lb = _____ oz
19. 180 oz = _____ lb
20. 2 oz = _____ lb
21. 10 T = _____ lb
22. 0.25 T = _____ lb
23. 18 oz = _____ lb
24. 11,000 lb = _____ T

Mixed Applications

25. The shipping weight of a carton of books was 23 lb. How many ounces is this?
26. A country bridge has a weight limit of 9 T. Mr. Wettstein drives a truck whose weight was recorded at 19,345 lb. Is it safe for Mr. Wettstein to drive the truck over the bridge?

Estimation

Choose the likely weight.
27. the weight of an apple   A. 6 T   B. 6 oz   C. 6 lb
28. the weight of a puppy   A. 35 oz   B. 35 lb   C. 35 T
29. the weight of a softball   A. 7 oz   B. 7 T   C. 7 lb
30. the weight of a sumo wrestler   A. 290 oz   B. 290 T   C. 290 lb

Calculator

Use your calculator to check the answer. If the answer is wrong, give the correct answer.
31. 5.25 T = 166,000 oz
32. 8000 oz = 0.25 T
33. 8.8 T = 281,500 oz

Data Bank

34. Use the table on p. 490. Find how much you would weigh on each planet and on the moon.

Group Activity

Do these activities with a partner or a team.
35. Challenge a partner to estimate 1 lb. Find an object or group of objects that together weigh about 1 lb. Use a scale to find who had the closest estimate. You may want to choose some other weights to estimate.
36. Compare the weights of pairs of shoes by lifting them. Estimate who on your team has the lightest pair of shoes. Use a scale to find who had the closest estimate if the team members do not agree.
37. Hold a wristwatch in one hand. Find another object that seems to be about the same weight. Check the accuracy of your estimate with a scale.

MEASUREMENT

6-6 Consumer Math: Reading Customary Scales

Objective
Read customary scales.

- SITUATION
- DATA
- PLAN
- ANSWER
- CHECK

Holline visited Godfry's Market and weighed these potatoes. How much did they weigh?

Example

1. How much did the potatoes weigh?

Look at the scale. Find what number the needle points to on the scale. Read the weight from the scale.

The potatoes weighed 1 lb 8 oz.

Try This Read the weight on the scale.

a.  b.  c.

Example

2. Find how much vinegar is in the measuring cup.

The top of the liquid seems to curve. Read the mark at the bottom of the curve. The bottom is on the 5-fl-oz line.

The measuring cup contains 5 fl oz of vinegar.

Try This Find how much liquid is in the measuring cup.

d.  e.  f.

156 CHAPTER 6

Exercises

Read the weight on the scale.

1.
2.
3.
4.
5.
6.

Find how much liquid is in the measuring cup.

7.
8.
9.
10.
11.
12.

Number Sense

13. Arrange the pound weights in order from lightest to heaviest.

6-7 Metric Units of Length

Olympic Champions 1924
100-Meter Run: Harold Abrahams, G. Britain 10.6 seconds
400-Meter Run: Eric Liddell, G. Britain 47.6 seconds

Objective
Find likely metric units of length.

The movie *Chariots of Fire* tells the story of Harold Abrahams and Eric Liddell of Great Britain competing in the 1924 Olympics.

Understand the Situation
- Which man ran the longer race?
- What unit was used to measure the distances?

The **meter** is the base unit of the **metric system.** Each of the other units of length is based on the meter. Meters, centimeters, millimeters, and kilometers are the most commonly used metric units.

Metric Units of Length

Prefix	Meaning	Unit	Symbol	Size	Example
kilo	1000	kilometer	km	1000 m	length of 11 football fields
hecto	100	hectometer	hm	100 m	
deka	10	dekameter	dam	10 m	
—	base unit	meter	m	1 m	height of a doorknob from the floor
deci	0.1	decimeter	dm	0.1 m	
centi	0.01	centimeter	cm	0.01 m	width of a button on a standard phone
milli	0.001	millimeter	mm	0.001 m	thickness of the wire in a small paper clip

Examples Give the likely measure: millimeters, centimeters, meters, or kilometers.

1. the distance to the equator   kilometers
2. the length of your school desk   centimeters
3. the height of your school building   meters

Try This Give the likely measure: millimeters, centimeters, meters, or kilometers.

a. the thickness of an Olympic gold medal
b. the height of a hurdle
c. the distance from Baltimore to Chicago
d. the length of an Olympic swimming pool

158   CHAPTER 6

Exercises

Give the likely measure: millimeters, centimeters, meters, or kilometers.
1. the length of your bedroom
2. the distance to Mars
3. the thickness of a sunglass lens
4. the thickness of this book
5. the height of your teacher
6. the distance you can throw a ball
7. the height of your chair
8. the distance to the gym

Choose the likely length.
9. the height of a diving board: 10 mm, 10 m, 10 km, 10 dm
10. the distance from the earth to the moon: 340,000 m; 340,000 km; 340,000 mm; 340,000 cm
11. the thickness of a house key: 5 cm, 5 mm, 5 m, 5 km
12. the width of a large paper clip: 1 mm, 1 m, 1 cm, 1 km
13. the diameter of a large pizza: 50 mm, 50 cm, 50 m, 50 km

Group Activity

14. Stand against the chalkboard. Have your partner mark your height. Lift your arms out to your side, shoulder height. Have your partner mark where your fingertips end. Measure your height and your armspan in centimeters using a meter stick. How do the measures compare? How do your partner's measures compare?

Mixed Applications

Solve. Use the table.

15. How much longer did it take Johnny Weissmuller to swim the 100-m freestyle than it took Mark Spitz?

16. How much longer did it take Mark Spitz to swim 100 m than it took Evelyn Ashford to run the same distance?

US Olympic Heroes			
Athlete	Year	Event	Time
Johnny Weissmuller	1924	100-m freestyle	59 s
Mark Spitz	1972	100-m freestyle	51.22 s
Wilma Rudolph	1960	100-m run	11 s
Evelyn Ashford	1984	100-m run	10.97 s
Jackson Sholz	1924	200-m run	21.6 s
Carl Lewis	1984	200-m run	19.8 s

17. In 1984, Carl Lewis also won the 100-m run. Estimate his winning time.
18. At the 1924 Olympics, Paavo Nurmi won the 1500-m run, the 5000-m run, and the 10,000-m cross country run. Give his total winning distance in kilometers.

Practice Through Problem Solving

Write × or ÷ in each box to make a true equation.
19. 33.98 ☐ 1000 = 33,980
20. 0.57 ☐ 100 = 0.0057
21. 7.8 ☐ 10 = 0.78
22. 90 ☐ 1000 = 0.09
23. 42.9 ☐ 100 = 4290
24. 0.002 ☐ 10 = 0.02
25. 866 ☐ 100 = 8.66
26. 6 ☐ 100 = 0.06

MEASUREMENT

6-8 Changing Metric Units of Length

Objective
Change metric units of length.

Sarala measured the length of a lizard in biology class. The lizard was 80 mm long. How would Sarala state this measurement in centimeters?

Understand the Situation
- Are you changing from a smaller to a larger unit or from a larger to a smaller unit?
- How many millimeters are in one centimeter?
- After changing 80 mm to centimeters, will there be a greater number of units or a lesser number of units?

Think of place values when changing metric units. Each unit in the table is ten times as long as the unit to the right. When changing units, use mental math to multiply or divide.

Place Value	1000	100	10	1	0.1	0.01	0.001
Unit	km	hm	dam	m	dm	cm	mm

Examples
Change the unit.

1. 80 mm = ___ cm
8͜0 cm
80 mm = 8 cm

10 mm = 1 cm *Think: smaller to larger unit. Divide.* Move the decimal point one place to the left.
Sarala's lizard is 8 cm long.

2. 6 km = ___ cm
6.00000͜ cm
6 km = 600,000 cm

1 km = 100,000 cm *Think: larger to smaller unit. Multiply.* Move the decimal point five places to the right. Write zeros as needed to show place value.

3. 9.7 m = ___ cm
9.70͜ cm
9.7 m = 970 cm

1 m = 100 cm *Think: larger to smaller unit. Multiply.* Move the decimal point two places to the right. Write zeros as needed to show place value.

Try This
Change the unit.

a. 96 cm = ___ mm
b. 712 m = ___ km
c. 0.438 m = ___ cm
d. 19 mm = ___ cm
e. 2389 km = ___ m
f. 17.6 cm = ___ m

CHAPTER 6

Exercises

Change the unit.

1. 78 km = ____ cm
2. 23.78 hm = ____ m
3. 688 cm = ____ m
4. 3400 mm = ____ m
5. 5000 m = ____ km
6. 4986 m = ____ cm
7. 19,000 cm = ____ mm
8. 334 m = ____ km
9. 4.67 m = ____ cm
10. 89 mm = ____ cm
11. 9.6 m = ____ mm
12. 12.4 dm = ____ cm
13. 8.97 km = ____ m
14. 19.2 dm = ____ m
15. 13.8 mm = ____ cm
16. 3 km = ____ cm
17. 89 dm = ____ km
18. 0.09 m = ____ mm
19. 0.5 m = ____ cm
20. 0.83 m = ____ hm
21. 8 cm = ____ km
22. 7.9 cm = ____ m
23. 0.83 cm = ____ mm
24. 6 m = ____ mm

Mixed Applications

Use a metric ruler.

25. Measure the length of your notebook. State the length in centimeters and millimeters.

26. Measure the width of your desk. State the width in centimeters and meters.

27. The Komodo dragon is about 3 m long. The Gila monster is about 60 cm long. Which lizard is longer?

28. The six-lined race runner lizard can travel up to 29 km per hour. How many meters per hour is this?

Calculator

Write the calculator keystrokes you used to find each answer.

29. 46 mm = ____ km
    □□□□□□□□□

30. 12 cm = ____ m
    □□□□□□

31. 5.3 km = ____ mm
    □□□□□□□□□

32. 6.9 mm = ____ cm
    □□□□□□

Suppose

33. Suppose that each metric unit was five times as large as the next smaller unit. Explain how you could use a calculator to make changes in units.

34. Suppose that you had a meter stick marked only in centimeters. How could you measure 200 mm of ribbon?

35. Suppose that you had a stick 10 cm long. How could you measure 1.5 m of string?

Number Sense

Arrange the distances in order from longest to shortest.

36. 3 km, 300 cm, 45 m, 500 cm
37. 4500 mm, 8 m, 2 km, 350 cm
38. 0.009 km; 65 m; 675 cm; 109,800 mm
39. 5 m; 32,100 mm; 0.5 km; 890 cm

Problem Solving

40. In this table the lengths of the snakes are mixed up. Use the clues to find the correct lengths. Rewrite the table.

    The Eastern garter snake is 57 cm shorter than the corn snake. The black rat snake is 112 cm longer than the red milk snake.

Snake	Length
Eastern garter	1.83 m
Red milk	1.23 m
Black rat	0.66 m
Corn	0.71 m

MEASUREMENT

6-9 Metric Capacity

Objective
Change metric units of capacity.

Rudy works in a pharmacy part-time on weekends. Last weekend the pharmacist filled 250-milliliter bottles from 1-liter containers of rubbing alcohol. Rudy arranged the bottles on shelves.

Milliliters and **liters** are metric units of capacity. **Capacity** is the amount that a container will hold. In the metric system, the liter is the basic unit of capacity.

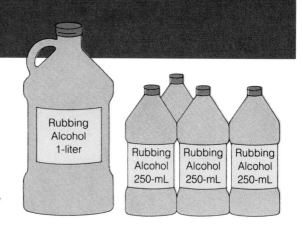

Metric Units of Capacity					
Prefix	Meaning	Unit	Symbol	Size	Example
kilo	1000	kiloliter	kL	1000 L	amount in four bathtubs
—	base unit	liter	L	1 L	amount in a small bucket
milli	0.001	milliliter	mL	0.001 L	amount in an eyedropper

Think of place values when changing metric units. Use mental math to multiply or divide by 1000.

Examples Change the unit.

1. 8 L = ___ mL
   8.000
   8 L = 8000 mL

   1 L = 1000 mL *Think: larger to smaller unit. Multiply.*
   Move the decimal point three places to the right. Write zeros as needed to show place value.

2. 28.9 mL = ___ L
   028.9
   28.9 mL = 0.0289 L

   1 mL = 0.001 L *Think: smaller to larger unit. Divide.*
   Move the decimal point three places to the left. Write zeros as needed to show place value.

3. 34,760 L = ___ kL
   34 760
   34,760 L = 34.760 kL

   0.001 L = 1 kL *Think: smaller to larger unit. Divide.*
   Move the decimal point three places to the left.

Try This Change the unit.

a. 54.7 mL = ___ L
b. 1.9 L = ___ mL
c. 46.2 L = ___ kL
d. 2384 L = ___ kL
e. 2 kL = ___ L
f. 0.09 L = ___ mL

Exercises

Change the unit.

1. 6 L = ___ kL
2. 33 L = ___ mL
3. 89 L = ___ mL
4. 78.9 L = ___ mL
5. 300 L = ___ mL
6. 0.56 L = ___ mL
7. 0.7 L = ___ mL
8. 32.6 L = ___ kL
9. 56.3 L = ___ mL
10. 2398 mL = ___ L
11. 123 L = ___ kL
12. 0.87 mL = ___ L
13. 1.2 L = ___ kL
14. 30 L = ___ mL
15. 30.8 mL = ___ L
16. 916.9 mL = ___ L
17. 302.8 mL = ___ L
18. 300 L = ___ kL
19. 1.22 kL = ___ L
20. 14,000 mL = ___ L
21. 99 L = ___ kL
22. 9.8 mL = ___ L
23. 16,000 L = ___ kL
24. 13.8 mL = ___ L

Choose the likely capacity.

25. the capacity of a drinking glass   A. 250 mL   B. 250 L   C. 250 kL
26. the capacity of a soup bowl   A. 400 kL   B. 400 L   C. 400 mL
27. the capacity of a garbage can   A. 80 kL   B. 80 L   C. 80 mL

Mixed Applications

28. The pharmacist has eight 1-L bottles of rubbing alcohol. How many 250-mL bottles can she fill?

29. The pharmacist has six 1-L bottles of mineral oil. She has two empty 150-mL bottles. How many more 150-mL bottles does she need to empty the 1-L bottles?

Number Sense

Which measurement is the same as the given measurement?

30. 67 L 23 mL   A. 90 mL   B. 67,023 mL   C. 67.23 mL
31. 12 kL 67 L   A. 12.67 kL   B. 79 L   C. 12,067 L
32. 2 kL 3L 24 mL   A. 29 mL   B. 2003.024 L   C. 23.24 kL

Mixed Skills Review

Solve. Use the pictograph for Exercise 33.

33. The shelf has how many bottles of rubbing alcohol? Of mineral oil? Of cough syrup?

34. The beginning balance in Rudy's checkbook was $19.42. Rudy entered these transactions: deposit $34.50, service charge $2.00, check #104 written for $14.95, #105 for $6.37, #106 for $10.13, deposit $10. Give the last balance in Rudy's checkbook.

Number of Bottles on Shelf

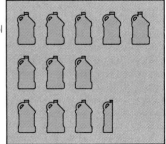

Rubbing Alcohol

Mineral Oil

Cough Syrup

 = four 250-mL bottles

6-10 Metric Mass

Objective
Find likely metric units of mass. Change metric units of mass.

Nutrients in food, such as protein and vitamins, help your body to grow and repair itself. The Food and Nutrition Board (FNB) recommends daily allowances of nutrients based on your age, weight, and height. The FNB states the mass of each nutrient in **grams** or **milligrams**. The gram is the basic metric unit of mass.

Metric Units of Mass

Prefix	Meaning	Unit	Symbol	Size	Example
		metric ton	t	1000 kg	a truckload of gravel
kilo	1000	kilogram	kg	1000 g	this book
—		base unit gram	g	1 g	a small paper clip
milli	0.001	milligram	mg	0.001 g	a fly's wing

Think of place values when changing metric units. Use mental math to multiply or divide by 1000.

Examples Change the unit.

1. 3286 mg = ___ g
   3.286
   3286 mg = 3.286 g

   1000 mg = 1 g *Think: smaller to larger unit. Divide.*
   Move the decimal point three places to the left.

2. 9.74 kg = ___ g
   9.740
   9.74 kg = 9740 g

   1 kg = 1000 g *Think: larger to smaller unit. Multiply.* Move the decimal point three places to the right. Write zeros as needed to show place value.

3. 2.5 t = ___ kg
   2.500
   2.5 = 2500 kg

   1 t = 1000 kg *Think: larger to smaller unit. Multiply.* Move the decimal point three places to the right. Write zeros as needed to show place value.

Try This Change the unit.

a. 8.23 kg = ___ g
b. 5 t = ___ kg
c. 650 g = ___ kg
d. 6.4 g = ___ mg
e. 986 mg = ___ g
f. 4000 kg = ___ t

Exercises

Change the unit.

1. 4379 mg = ___ g
2. 343 g = ___ kg
3. 231 mg = ___ g
4. 117 g = ___ mg
5. 435.9 g = ___ kg
6. 8 t = ___ kg
7. 12.8 t = ___ kg
8. 76 t = ___ kg
9. 0.06 mg = ___ g
10. 374 g = ___ kg
11. 89 mg = ___ g
12. 8.45 kg = ___ g
13. 6.7 kg = ___ g
14. 5400 kg = ___ t
15. 3.04 kg = ___ g
16. 4.37 mg = ___ g
17. 98.7 kg = ___ g
18. 6.89 mg = ___ g
19. 3298 g = ___ kg
20. 12,763 kg = ___ t
21. 19 kg = ___ t
22. 13.89 g = ___ mg
23. 2987 g = ___ kg
24. 0.5 t = ___ kg

Give the likely mass.

25. the mass of an apple — A. 175 mg B. 175 kg C. 175 g
26. the mass of a softball — A. 500 g B. 500 kg C. 500 mg
27. the mass of a pro-hockey player — A. 95 mg B. 95 kg C. 95 g
28. the mass of a Saturn V rocket — A. 2960 g B. 2960 kg C. 2960 t
29. the mass of vitamin B-6 in a tablet — A. 2 kg B. 2 g C. 2 mg
30. the mass of a hummingbird — A. 11 mg B. 11 g C. 11 kg

Problem Solving

31. A scale has 4 green chips and 3 pink chips on it. Another scale has 4 pink chips and 3 green chips on it. The first scale shows a mass of 10 g. The second scale shows a mass of 11 g. What is the total mass of 1 pink chip and 1 green chip?

32. Erin's pet guinea pig weighed 100 g at birth. It weighed 160 g at the end of the first week. It weighed 220 g at the end of the second week. It weighed 280 g at the end of the third week. At this rate, in how many weeks will the guinea pig reach its adult weight of 580 g?

Number Sense

Arrange the masses in order from heaviest to lightest.

33. 56 kg, 5000 mg, 100 g, 0.5 t
34. 0.9 t; 6000 g; 10,000 mg; 5 kg
35. 900 g, 50 mg, 4 kg, 0.0005 t
36. 0.5 g; 50,000 mg; 0.005 kg; 0.005 t

Mixed Applications

Solve. Use the table.

37. What is the total amount of protein in Samantha's dinner?
38. What is the total amount of vitamin C in grams?
39. How much iron would Samantha get from a dinner twice this size?

Samantha's Dinner	Protein	Calcium	Iron	Vitamin C
Lamb Chop	25 g	10 mg	1.5 mg	0
Broccoli	5 g	136 mg	1.2 mg	140 mg
Biscuit	2 g	34 mg	0.4 mg	trace
Orange	1 g	54 mg	0.5 mg	66 mg

6-11 Reading Metric Measures

Objective
Read metric measures of length, mass, and capacity.

Metric measures use decimal units. Standard metric rulers show centimeters and millimeters. Metric liquid measures show milliliters. A triple-beam balance scale shows total mass in 100-g, 10-g, and 1-g units.

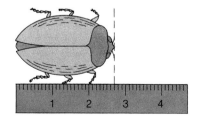

Examples

1. Read the ruler to the nearest centimeter.

   The measure is 2.7 cm. The nearest centimeter is 3.

2. Find how much liquid is in the beaker.

   The bottom of the curve is at 150 mL. The beaker holds 150 mL of liquid.

3. To measure mass on the triple-beam scale, slide the markers from zero to the right until the scale balances. Count the notches. Then add the measures to find the actual mass. Use mental math.

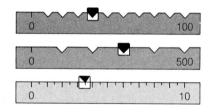

On the top beam, each notch is 10 g. Count 4 from zero. The top beam shows 40 g.

On the middle beam, each notch is 100 g. Count 3 from zero. The middle beam shows 300 g.

On the bottom beam, each long mark is 1 g. Count 3.5 from zero. The bottom beam shows 3.5 g.

Add the measures.   40 + 300 + 3.5 = 343.5 g

Try This
Find the metric measure. Use a metric ruler for **a**. Read to the nearest centimeter.

a.

b.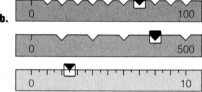

c.

166 CHAPTER 6

Exercises

Use a metric ruler. Measure the object to the nearest centimeter.

1.
2.
3.
4.
5.
6.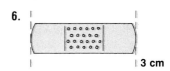

Find how much liquid is in the beaker.

7. 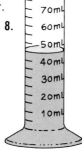 8. 9.

Find the mass.

10.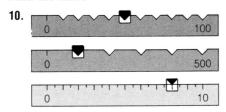
11.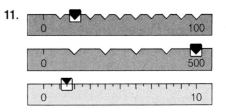

Use a metric ruler. Draw a line segment with the given length.

12. 4 cm   13. 15 cm   14. 5.7 cm   15. 18.2 cm   16. 6.5 cm

Estimation Without using a metric ruler, draw a line segment with the given length. Use this 1-cm line segment to help you estimate. ⊢——⊣ Use a metric ruler to check your estimate. How close did you come?

17. 7 cm   18. 9 cm   19. 14 cm   20. 21 cm   21. 18 cm

6-12 Temperature

Objective
Find the likely temperature in Celsius or Fahrenheit. Read Celsius and Fahrenheit thermometers.

The two scales most often used to measure temperature in the U.S. are the **Fahrenheit (F)** and the **Celsius (C)** scales. Temperature is measured in **degrees (°)**.

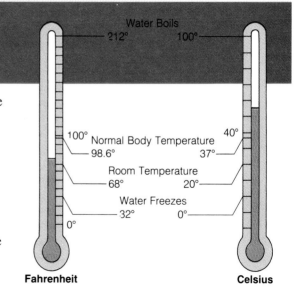

Understand the Situation
- At what temperature on each scale does water boil?
- What is a comfortable room temperature on each scale?

Example   Estimate the likely temperature.

1. a bowl of warm chicken soup   **A.** 40°F   **B.** 19°C   **C.** 104°F   **D.** 120°C

   40°F and 19°C are too cold. 120°C is too hot. The likely temperature is 104°F.

Try This   Estimate the likely temperature.

a. a day for wearing a swimsuit   **A.** 21°F   **B.** 36°C   **C.** 64°F   **D.** 145°C

Examples   Decide if the temperature is Fahrenheit or Celsius.

2. This thermometer shows a patient's body temperature.

   A temperature of 98.6° is normal body temperature on the Fahrenheit scale. The temperature is in Fahrenheit.

3. This oven thermometer is set for baking pizza.

   This temperature is high on the Celsius scale. It is too low for baking on the Fahrenheit scale. The temperature is in Celsius.

Try This   Decide if the temperature is Fahrenheit or Celsius.

b. indoors in the fall

c. outdoors in the summer

168 CHAPTER 6

Exercises

Estimate the likely temperature.

1. a blizzard   A. −4°C   B. 45°F   C. 35°F   D. 21°C
2. a summer day in south Texas   A. 100°C   B. 42°F   C. 43°C   D. 188°F
3. a snowball battle   A. 28°C   B. 26°F   C. −32°C   D. 65°F
4. riding a dirt bike   A. 69°F   B. 49°C   C. 12°F   D. 5°C
5. an ice cube   A. 62°C   B. 62°F   C. 26°C   D. 26°F

Decide if the temperature is Fahrenheit or Celsius.

6. The oven is set for baking bread.

7. This is an indoor thermometer on a winter day.

8. This meat thermometer shows the temperature for a medium-rare beef roast.

9. This thermometer shows the body temperature of a patient with a mild fever.

10. The Sunray High School Glee Club is planning a spring trip to Mexico. Mexico City's average temperature in April ranges from 11°C to 25°C. Should the club members pack for cold, mild, or hot weather?

Estimation   11. What temperatures on the Fahrenheit scale do you use as a guide to estimate weather conditions? On the Celsius scale?

Mixed Skills Review   Find the answer.

12. $108 \div 12$ 13. $52.11 \div 0.09$ 14. 10×460.9 15. 36×31

Evaluate the expression for $x = 5$.

16. $x \div 2$ 17. $7 + x$ 18. $12x$ 19. $9 - x$

6-13 Problem Solving: Using Formulas

Objective
Use formulas.

- **SITUATION**
- **DATA**
- **PLAN**
- **ANSWER**
- **CHECK**

Your plan for solving some problems can be to use a formula. A **formula** is a rule expressed in symbols. This table gives a few of the important formulas. A more complete list is in the Data Bank on p. 495.

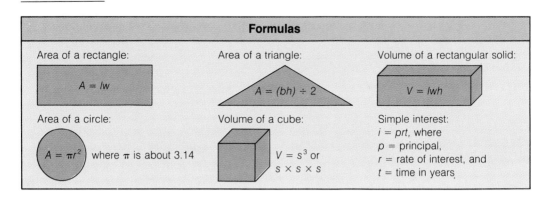

Example

Choose the correct formula. Solve.

Kwai took out a loan of $500. Find the interest on $500 at 8% annual interest, or 0.08, for 6 months.

Choose the correct formula. → Substitute values for the variables. → Solve. → State the answer in the correct units.

The correct formula is $i = prt$.

$p = \$500$
$r = 8\%$ or 0.08
$t = 6$ months or 0.5 year
$i = prt$
$= 500 \times 0.08 \times 0.5$

$i = 500 \times 0.08 \times 0.5$
$i = 40 \times 0.5$
$i = 20$

The interest on $500 at 8% for 6 months is $20.

Try This

Choose the correct formula.

a. the volume of a crate **b.** the area of a round rug **c.** interest earned on a savings account

Solve.

d. To buy a stereo, Ana will pay 10% annual interest, or 0.10, on a $335 loan for 2 years. How much interest will she pay in all?

Exercises

Choose the correct formula.

1. the volume of a rectangular jewelry box
2. the area of a round flower bed
3. the area of a three-cornered sail
4. the area of a page in this book
5. the volume of a box shaped like a cube
6. the annual interest on a savings account
7. the area of a triangular scarf
8. the interest charge on a credit card balance
9. the volume of a rectangular toolbox

10. Find the area of the piece of fabric.

11. Find the area of the coin.

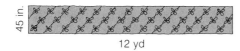

12. Find the volume of the container.

13. Find the area of the garden plot.

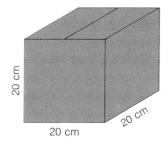

14. Find the area of a rectangular tabletop 4 ft wide and 3 ft long.
15. Find the volume of a toy block 2 cm high, 5 cm wide, and 7 cm long.
16. Find the interest on $200 at 9% interest, or 0.09, for one year.
17. Find the area of a circle with a radius of 2.2 cm.
18. Kwai earns interest of 5.5% per year, or 0.055, on his savings of $200. What will be the total amount in his account at the end of 2 years?

Write Math

19. Write a problem that you can solve using the formula for the area of a rectangle.
20. Write a problem that you can solve using the formula for the volume of a cube.
21. A square is a rectangle whose sides are all the same length. Write a formula to find the area of a square whose side is length s.

Evaluating Solutions

22. Here is how Susanna found the volume of a cube whose sides are 5 in. long: $V = s^3 = 5 \times 3 = 15$ in.^{3}. What is wrong with her solution?

MEASUREMENT

6-14 Problem Solving: Group Decisions

Objective
Make group decisions.

- **SITUATION**
- **DATA**
- **PLAN**
- **ANSWER**
- **CHECK**

Your group is on a 5-day river rafting trip. On the second day, rocks have damaged your supply raft. It cannot be repaired. You must decide which supplies you should take for the rest of the trip. The supplies must not weigh more than 30 lb.

Facts to Consider

- The temperature at night goes down to about 50°F.
- Here are the items in the supply raft.

12 dried soup packages: $\frac{1}{2}$ oz each
17 dried beef jerky packages: 1 oz each
3 boxes of matches: 1 oz each
3 knives: 4 oz each
2 hatchets: 1 lb each
4 life jackets: 1 lb each
4 sweaters: 32 oz
4 pairs of boots: 2 lb each
2 bottles salmon eggs for bait: 8 oz each
2 sauce pans: 2 lb each
1 can opener: 3 oz
150 ft of string: 1 oz
10 plastic bags: $\frac{1}{4}$ oz
1 camping stove: 10 lb

1 dried apple package: $\frac{1}{2}$ oz
15 powdered milk packages: 2 oz each
16 mixed nuts and fruits packages: $1\frac{1}{2}$ lb
1 package of cheese: 3 lb
12 hot cocoa packets: $\frac{1}{2}$ oz each
1 tent: 11 lb
4 sleeping bags: 3 lb each
20 juice cans: 3 oz each
4 fishing poles: 1 lb each
1 large frying pan: 2 lb
1 first-aid kit: 2 lb
10 cans stew: 1 lb each
10 cans soup: 1 lb each
2 lanterns: 5 lb each
1 can fuel: 3 lb

Plan and Make a Decision

- What are you being asked to do?
- What factors must you think about when choosing items to take?
- How might a list or chart help you reach your decision?
- How accurate must your calculations be? How might a calculator be of use?
- How will you reach a decision in your group?

Share Your Group's Decision

- Present your plan to the class. Explain how you made your decision.
- Compare your plan to other groups' plans.

Suppose   Suppose you could bring 50 lb from the supply raft. How would this change your decision?

6-15 Practice Through Problem Solving

Objective
Practice finding patterns.

Complete the sequence. Describe the pattern.

1. □ △ △ △ □ △ △ △
2. ○ △ □ ○ △ □ ○
3. △ □ △ △ □ △ △ △
4. ○ ○ △ △ △ ○ ○ △ △ △
5. △ ○ ○ ○ △ △ ○ ○ △ △ △
6. □ △ □ △ △ □ △ □ △ △ □

7. Find how many dots there are in a 6-dot by 6-dot square. Describe the pattern.

   1   4   9   ?

8. Find how many dots there are in a triangle with 6 dots along the base. Describe the pattern.

   1   3   6   ?

Complete the sequence. Describe the pattern.

9. 2, 22, 222, ___, ___, ___
10. 109876, 10987, 1098, ___, ___, ___
11. ab, abc, abcd, ___, ___, ___
12. aba, zyz, cdc, ___, ___, ___
13. 1 × 2, 2 × 3, 3 × 4, 4 × 5, 5 × 6, ___, ___, ___
14. (1 × 2) + 1, (2 × 2) + 2, (3 × 2) + 3, ___, ___, ___

Find how many numbers are in the sequence.

15. 1, 2, 3, 4, . . . , 25
16. 2, 4, 8, . . . , 32
17. 20, 17, 14, . . . , 2
18. 3, 6, 9, 12, . . . , 48
19. 5, 10, 15, . . . , 100
20. 39, 37, 35, . . . , 3

MEASUREMENT

6-16 Enrichment: Calculating Customary Measures

Objective
Use addition, subtraction, multiplication, and division with customary measures.

Vera poured 1.5 gal of water into her aquarium. She added 3 more quarts of water. How much water is in the aquarium now?

You can find the answer by adding the measures.

Examples Find the answer.

1. How many quarts of water are in the aquarium?

   1.5 gal + 3 qt =   Regroup 1.5 gal as 6 qt.
     6 qt + 3 qt = 9 qt   Add.
           = 2 gal 1 qt   Regroup to simplify.

   There are 2 gal 1 qt of water in the aquarium.

2.   1 ft 3 in.   1 ft 3 in.
   ×   5   ×   5   Multiply each unit.
               5 ft 15 in. = 6 ft 3 in.   Regroup to simplify.

3. 8 ft ÷ 6   8 ft ÷ 6   Regroup 8 ft as 96 in.
              96 in. ÷ 6 = 16 in.   Divide.
                     = 1 ft 4 in.   Regroup to simplify.

4.   5 ft 2 in.   5 ft 2 in. =   4 ft 14 in.   Regroup as 4 ft 14 in.
   −2 ft 8 in.   −2 ft 8 in. = −2 ft 8 in.   Subtract.
                                        2 ft 6 in.

Exercises

1. Find the total length of 3 pieces of chain. Each piece is 4 ft 8 in. long.

2. A rope 10 ft long is cut into 4 equal pieces. How long is each piece?

3. A bucket held 2 qt of water. Shelby added 2.25 more gallons of water. How much water is in the bucket now?

4. A piece of cord measures 3 ft long. A piece 1 ft 1 in. long is cut from it. How much cord is left?

5. How much wider is your desk than your math book?

6. What is the difference between your height and the height of the classroom door?

6-17 Calculator: Working with Measurement

Objective
Use a calculator to calculate with measures.

A dime is about 1 mm thick. Suppose you could stack dimes 1 km high. How much money would the stack be worth?

You can use your calculator to help you solve this problem.

Find the number of millimeters in one kilometer.

Think: 1 mm × 1000 = 1 m
1 m × 1000 = 1 km

[ON/C] 1 [×] 1000 [×] 1000 [=]

[1 0 0 0 0 0 0.]

Multiply your answer by $0.10.

[×] [.] 1 [=]

[1 0 0 0 0 0.]

The stack would be worth $100,000.

Exercises

Solve.

1. A quarter is about 1.5 mm thick. Would you rather have a stack of quarters 1 m high or a stack of dimes 4 m high?

2. Suppose someone gave you a stack of dimes equal to the measure of your height. How much money would you get?

3. A male African elephant weighs about 5.4 t. A house mouse weighs about 21 g. About how many house mice equal the weight of one male African elephant?

4. A bee hummingbird weighs about 3 g. A male African ostrich weighs about 140 kg. About how much more does the African ostrich weigh?

5. Suppose 0.2 oz of relish was put on each of 12 million hot dogs. How many 8-oz jars of relish would you need?

6. Suppose 0.5 fl oz of jam was put on each of 20 million pieces of toast. How many 1-pt jars of jam would you need?

7. The Seikan railway tunnel in Japan is 33.5 mi long. How many times longer is the length of the tunnel than the length of your pencil?

8. San Francisco's Golden Gate Bridge is 4200 ft long. Suppose you walked heel-to-toe across the bridge. About how many steps would you take?

9. Suppose the 1, 2, and [÷] keys on your calculator did not work. How could you find the number of feet in 300 in. on your calculator?

10. Suppose the 1, 6, and [×] keys on your calculator did not work. How could you find the number of ounces in 36 lb on your calculator?

MEASUREMENT

CHAPTER 6 REVIEW

Choose the one that does not belong in each group. Explain why not.

1. A. meter
   B. gram
   C. liter
   D. hectometer

2. A. foot
   B. yard
   C. ounce
   D. inch

3. A. degree
   B. ounce
   C. Fahrenheit
   D. Celsius

4. A. 3 ft = 1 yd
   B. 2 pt = 1 qt
   C. 2 lb = 32 oz
   D. 16 fl oz = 2 c

5. A. ruler
   B. thermometer
   C. customary
   D. beaker

6. A. 36 in.
   B. 5280 ft
   C. 1760 yd
   D. 1 mi

7. A. 3 km, 500 cm, 4 m
   B. 590 mg, 0.8 g, 2 kg
   C. 580 L, 0.1 kL, 99 mL
   D. 6.4 mm, 0.5 cm, 0.002 m

8. A. rate
   B. interest
   C. area
   D. principal

Find four different ways to complete each sentence to make it true.

9. To change from ___ to ___ move the decimal point three places to the right.

10. To change from ___ to ___ move the decimal point three places to the left.

11. Four statements have been scrambled. Unscramble each to state a true measurement relationship.
    3 yd = 2 T   4000 lb = 9 ft
    1 gal = 48 fl oz   6 c = 4 qt

Write the missing word.

12. Length is to width as base is to _____.

13. Gram is to kilogram as meter is to _____.

14. Degree is to thermometer as inch is to _____.

15. Fluid ounce is to pint as ounce is to _____.

16. 5280 ft is to 1 mi as 36 in. is to _____.

17. A gram is to mass as a _____ is to capacity.

Match.

18. area of a triangle
19. volume of a cube
20. area of a rectangle
21. area of a circle

A. lwh
B. πr^2
C. lw
D. prt
E. $(bh) \div 2$
F. s^3

22. Name two problem-solving strategies that can help you find a pattern.

23. Who needs the more accurate measurement: a teacher drawing a square on the chalkboard or an architect drawing a blueprint?

24. Who needs the less accurate measurement: a doctor measuring an amount of penicillin or a cook measuring an amount of olive oil?

CHAPTER 6 TEST

1. Rena called 2 club members. She asked each member to call 2 other members, and so on. It took 5 min to make the 2 calls. At this rate, how many members were called in 30 min?
2. Who needs the more accurate measurement: a person weighing himself or herself or a biologist weighing mice in an experiment?

Give the likely measure for items 3–6. Change the unit for items 7–18.
3. the length of a kite string: inches, feet, yards, or miles
4. the capacity of a teacup: fluid ounces, cups, pints, quarts, or gallons
5. the weight of a baby chick: ounces, pounds, or tons
6. the thickness of a contact lens: millimeters, centimeters, meters, or kilometers

7. 8 yd = _____ ft
8. 28 in. = _____ ft
9. 9 qt _____ gal
10. 17 c = _____ pt
11. 28 oz = _____ lb
12. 1.2 T = _____ lb
13. 5.6 cm = _____ mm
14. 3493 cm = _____ m
15. 453 L = _____ kL
16. 4.76 L = _____ mL
17. 56.4 g = _____ mg
18. 3.55 t = _____ kg

Find the customary measure. Measure to the nearest $\frac{1}{4}$ in. for item 21.

19.
20.
21.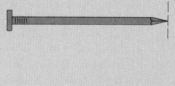

Find the metric measure.

22.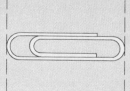
23.
24.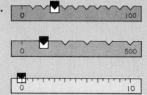

25. Estimate the likely temperature. a healthy human body
    A. 98.6°C   B. 37°C
    C. 37°F   D. 70°F

26. Decide if the temperature is Fahrenheit or Celsius. temperature for baking a casserole

MEASUREMENT 177

CHAPTER 6 CUMULATIVE REVIEW

1. Multiply.
   264 × 321
   A. 84,744
   B. 84,644
   C. 85,744
   D. 855,644

2. Tell the operation needed to solve this problem.
   Problem: The ice hockey team has a total of △ players. The soccer team has ☐ players. How many more players are on the ice hockey team?
   A. multiplication
   B. subtraction
   C. division
   D. addition

3. Inez's checking account balance is $244.50. On June 30, Inez deposited a $15 check. On July 4, Inez wrote a check for $28.79. What is her new balance?
   A. $230.71
   B. $288.29
   C. $258.29
   D. $229.50

4. Add. Use mental math.
   487 + 394
   A. 891
   B. 881
   C. 971
   D. 863

5. Divide.
   7)2109
   A. 30 R12
   B. 31 R2
   C. 300 R3
   D. 301 R2

6. Change the unit.
   630 L = ____ kL
   A. 0.63
   B. 6.3
   C. 0.063
   D. 63

7. Multiply.
   181 × 7
   A. 817
   B. 1267
   C. 1817
   D. 1256

8. Which question asks the same thing?
   Question: How many more students ride bikes than ride the bus?
   A. How many fewer students take the bus than ride bikes?
   B. How many students ride bikes or take the bus?

9. Change the unit.
   15 yd = ____ ft
   A. 26,400
   B. 45
   C. 540
   D. 180

10. Change the unit.
    17 qt = ____ gal
    A. 5 gal
    B. 4 gal 1 qt
    C. 5 gal 2 qt
    D. 4 gal 4 qt

11. Trevor bought 2 baskets of blueberries for $5. He paid for them with quarters and dimes only. What coins did he use?
    A. 10 quarters, 25 dimes
    B. 19 quarters, 1 dime
    C. 45 dimes, 5 quarters
    D. 5 quarters, 10 dimes

12. Subtract. Use mental math.
    $22.50 − $9.25
    A. $14.75
    B. $13.25
    C. $11.25
    D. $31.75

13. Measure to the nearest $\frac{1}{4}$ in.
    A. 1 in.
    B. $1\frac{1}{4}$ in.
    C. 2 in.
    D. $1\frac{1}{2}$ in.

14. Write as a standard numeral.
    four thousand, two hundred twenty-four
    A. 4224
    B. 40,224
    C. 42,240
    D. 4204

15. Divide.
    $3.25 \div 0.26$
    A. 13
    B. 12.5
    C. 0.125
    D. 1.25

16. What percent more of the students chose the natural history museum than chose the science museum?
    A. 15%
    B. 55%
    C. 30%
    D. 25%

Favorite Museum

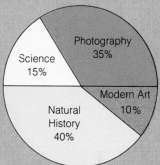

17. Subtract.
    $229 - 42$
    A. 187
    B. 227
    C. 177
    D. 287

18. Estimate.
    $29 + 96 + 98 + 33$
    A. 270
    B. 260
    C. 250
    D. 300

19. Divide.
    $8\overline{)22.4}$
    A. 2.8
    B. 28
    C. 0.28
    D. 2.08

20. Decide if an exact answer or an estimate is needed. The postal person needs to figure the cost of sending a 2-lb package from California to Texas.
    A. exact
    B. estimate

21. Multiply.
    $\$5.24 \times 25$
    A. $141
    B. $131.90
    C. $131
    D. $125.24

22. Estimate.
    $750 + 245$
    A. 1000
    B. 800
    C. 500
    D. 1400

23. Subtract.
    $\$8000 - \$1492$
    A. $6618
    B. $7508
    C. $7618
    D. $6508

24. Multiply.
    0.77×100
    A. 77
    B. 7.7
    C. 770
    D. 0.77

25. Divide.
    $248 \div 9$
    A. 20 R8
    B. 27
    C. 27 R5
    D. 26 R14

Chapter 7 Overview

Key Ideas

- Use the problem-solving strategies *use objects* and *act it out*.
- Find equivalent fractions.
- Write fractions in lowest terms.
- Write improper fractions as mixed numbers.
- Find the least common multiple.
- Find the least common denominator.
- Compare and order fractions and mixed numbers.
- Estimate gauge readings.
- Estimate using compatible numbers.
- Write fractions as decimals and decimals as fractions.

Key Terms

- fraction
- denominator
- mixed number
- repeating
- factor
- equivalent fractions
- lowest terms
- compatible numbers
- least common denominator
- greatest common factor
- numerator
- improper fraction
- terminating
- least common multiple

Key Skills

Multiply.

1. 4×5
2. 3×6
3. 8×6
4. 6×2
5. 9×9
6. 4×7
7. 6×7
8. 2×7
9. 4×3
10. 6×9
11. 8×5
12. 4×9
13. 3×8
14. 8×2
15. 6×6
16. 5×3
17. 7×7
18. 8×8

Divide.

19. $12 \div 4$
20. $32 \div 8$
21. $16 \div 4$
22. $28 \div 7$
23. $32 \div 4$
24. $36 \div 6$
25. $42 \div 7$
26. $25 \div 5$
27. $30 \div 6$
28. $48 \div 8$
29. $24 \div 4$
30. $28 \div 4$
31. $63 \div 7$
32. $81 \div 9$
33. $72 \div 9$
34. $56 \div 8$
35. $21 \div 7$
36. $30 \div 6$

Tell the place value of the underlined digit.

37. 0.<u>4</u>
38. 0.3<u>5</u>
39. 0.5<u>0</u>6
40. 0.0<u>3</u>
41. 0.00<u>2</u>
42. 0.<u>1</u>9
43. 0.29<u>7</u>
44. 0.33<u>2</u>
45. 0.<u>6</u>28
46. 0.<u>5</u>
47. 0.0<u>8</u>
48. 0.00<u>4</u>

FRACTIONS 7

7-1 Problem Solving: Learning to Use Strategies

Objective
Use the strategies use objects *and* act it out.

- SITUATION
- DATA
- **PLAN**
- ANSWER
- CHECK

Charmaine and a group of friends made sandwiches for the class picnic. They set up an assembly line. There were 3 steps: spreading mustard on the bread, putting on the lettuce and tomato, and putting on the meat. Each step took 5 seconds. How long would it take to make 3 sandwiches?

You can **use objects** or **act out** some problems that are difficult to solve.

Example Solve.

How long would it take to make 3 sandwiches?

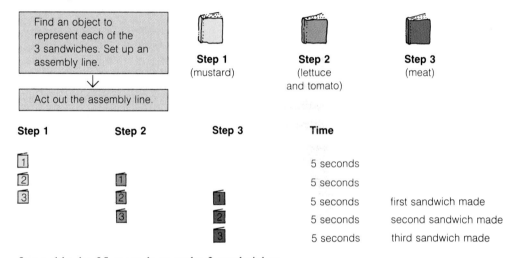

It would take 25 seconds to make 3 sandwiches.

Try This Solve.

a. Novian and his friends played 18 holes of miniature golf. They divided into 3 groups. Each group averaged 7 minutes per hole. About how long did it take everyone to play 5 holes?

b. The Bubble Shop car wash has 3 stations: wash, rinse, and dry. Each station takes about 5 minutes. About how long would it take 4 cars to be washed, rinsed, and dried?

Exercises

Use one or more of the problem-solving strategies.

Problem-Solving Strategies
- Guess and check.
- Choose an operation.
- Make an organized list.
- **Act it out.**
- Find a pattern.
- Write an equation.
- Use logical reasoning.
- Draw a picture or diagram.
- Make a table.
- **Use objects.**
- Work backward.
- Solve a simpler problem.

1. Janette set up the food for her buffet dinner. Each guest brought one of 4 types of dishes: salad, main dish, side dish, or dessert. Each person spent about 2 minutes at each of the types of dishes. About how long did it take 7 people to go through the line?

2. Che spent $24 without tax on napkins and paper plates. How many packages of each product did he buy?

3. At the spring dance, prizes were given to the fifth, eighth, and thirteenth persons who entered. If this pattern continued, who were the next 5 people who received prizes?

4. Every October Gina sets up a haunted house. She decorates 3 rooms in her house. It takes about 3 minutes for each group of visitors to go through each room. About how long would it take 5 groups to go through the haunted house?

5. The invitations for Wendell's barbecue asked guests to arrive at 3 p.m. By 3:05 p.m., 4 guests had arrived. By 3:10 p.m., 9 more guests had arrived. By 3:15 p.m., 16 more guests had arrived. At this rate, how many guests will arrive between 3:25 p.m. and 3:30 p.m.?

Suppose

6. Suppose in the example on p. 182 that there was a fourth step: putting the sandwich in a plastic bag. Suppose the fourth step took 5 seconds. How long would it take to make and wrap 3 sandwiches?

Write Your Own Problem

7. Write a problem that you could solve using the strategy *use objects* or *act it out*. Use the information in the note.

> Peter —
> The letters are done! Please have a group of people fold the letters and then stuff and seal the envelopes.
> Thanks,
> Greta

FRACTIONS

7-2 Using Fractions

Objective
Write fractions.

The Recreation Club is taking a travel survey. The survey page is divided into equal parts. What fraction of the survey page gives instructions?

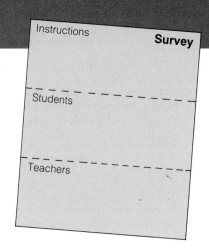

A **fraction** is a numeral used to name part of a set or whole.

Understand the Situation
- Into how many parts is the page divided?
- Are the parts the same size?
- How many of the parts give instructions?

Examples

1. Write the fraction for the part of the page that gives instructions.

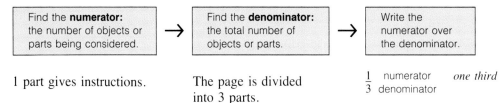

1 part gives instructions.   The page is divided into 3 parts. $\frac{1}{3}$ numerator / denominator   *one third*

One part out of 3 parts or $\frac{1}{3}$ of the page gives instructions.

2. For one class, the survey shows that 5 of the 19 students have traveled to Mexico. Write the fraction for students who have traveled to Mexico.

Find the numerator, the number of students being considered: 5

Find the denominator, the total number of students: 19

Five out of nineteen or $\frac{5}{19}$ of the students have traveled to Mexico.

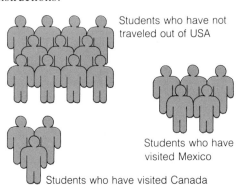

Try This

Write the fraction.
a. students who have visited Canada
b. students who have not visited Canada
c. the part of the page to be filled out by teachers or students

184   CHAPTER 7

Exercises

Write the fraction. Use the table.

1. the teachers who have visited two countries
2. the teachers who have visited three or more countries
3. the men who have visited four or more countries
4. the women who have never visited another country

**Travel by Teachers
Visits to Other Countries**

Number of Countries	Men	Women
4 or more	2	3
3	5	1
2	8	9
1	4	5
0	1	2

Write the fraction for the shaded part.

5.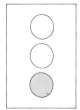
6.
7.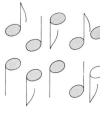

8.
9.
10.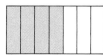

11.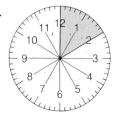
12.
13.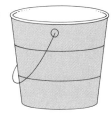

Write the word name for each fraction.

14. $\frac{5}{8}$ 15. $\frac{3}{4}$ 16. $\frac{2}{3}$ 17. $\frac{7}{6}$ 18. $\frac{4}{5}$ 19. $\frac{6}{7}$

Mixed Applications

20. Fernando needs $36 to buy a baseball glove. He has earned $20. What fraction of the total has he earned?

21. A soccer game is 60 min long. What part of the game has been played at the end of 48 min?

Show Math

22. Draw a rectangle. Shade $\frac{4}{5}$ of it.
23. Draw a circle. Shade $\frac{2}{3}$ of it.

Group Activity

24. Work with a group. Survey your classmates to find the number of countries each student has visited. Make a table of your data. Share the results with the class. Find what fraction of the class has visited at least one other country.

FRACTIONS

7-3 Equivalent Fractions

Objective
Write equivalent fractions.

A box of two-penny nails holds $\frac{5}{8}$ lb. A box of three-penny nails holds 10 oz, or $\frac{10}{16}$ lb. The weights of the two boxes are equal. $\frac{5}{8}$ and $\frac{10}{16}$ are **equivalent fractions**. They name the same number.

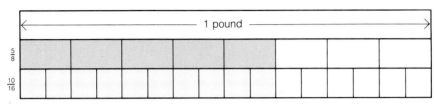

Example

1. Write an equivalent fraction for $\frac{5}{8}$.

Multiply the numerator and denominator of a fraction by the same nonzero number.

$$\frac{5}{8} = \frac{5 \times 2}{8 \times 2} = \frac{10}{16} \qquad \frac{5}{8} = \frac{5 \times 3}{8 \times 3} = \frac{15}{24} \qquad \frac{5}{8} = \frac{5 \times 4}{8 \times 4} = \frac{20}{32} \qquad \frac{5}{8} = \frac{10}{16} = \frac{15}{24} = \frac{20}{32}$$

Try This Write the equivalent fraction.

a. $\frac{3}{11} = \frac{3 \times 2}{11 \times 2} = \frac{\square}{\square}$ b. $\frac{1}{3} = \frac{1 \times 4}{3 \times 4} = \frac{\square}{\square}$ c. $\frac{5}{6} = \frac{5 \times 7}{6 \times 7} = \frac{\square}{\square}$

Examples Write the missing numeral.

2. $\frac{3}{4} = \frac{\square}{12}$ $\frac{3}{4} = \frac{\square}{12}$ (×3) $\frac{3}{4} \xrightarrow{\times 3} \frac{\square}{12}$ (×3) $\frac{3}{4} = \frac{9}{12}$ The missing numeral is 9.

Think: $4 \times 3 = 12$. Multiply: $3 \times 3 = 9$.

3. $\frac{6}{7} = \frac{24}{\square}$ $\frac{6}{7} \xrightarrow{\times 4} \frac{24}{\square}$ $\frac{6}{7} = \frac{24}{\square}$ (×4) $\frac{6}{7} = \frac{24}{28}$ The missing numeral is 28.

Think: $6 \times 4 = 24$. Multiply: $7 \times 4 = 28$.

Try This Write the missing numeral.

d. $\frac{7}{9} = \frac{\square}{27}$ e. $\frac{5}{4} = \frac{\square}{16}$ f. $\frac{3}{8} = \frac{30}{\square}$ g. $\frac{6}{7} = \frac{36}{\square}$ h. $\frac{16}{13} = \frac{\square}{52}$

Exercises

Write the equivalent fraction.

1. $\frac{4}{7} = \frac{4 \times 2}{7 \times 2} = \frac{\square}{\square}$
2. $\frac{3}{8} = \frac{3 \times 3}{8 \times 3} = \frac{\square}{\square}$
3. $\frac{5}{9} = \frac{5 \times 4}{9 \times 4} = \frac{\square}{\square}$
4. $\frac{7}{12} = \frac{7 \times 5}{12 \times 5} = \frac{\square}{\square}$
5. $\frac{2}{3} = \frac{2 \times 9}{3 \times 9} = \frac{\square}{\square}$
6. $\frac{3}{4} = \frac{3 \times 7}{4 \times 7} = \frac{\square}{\square}$

Write the missing numeral.

7. $\frac{1}{3} = \frac{\square}{9}$
8. $\frac{3}{7} = \frac{\square}{35}$
9. $\frac{8}{9} = \frac{32}{\square}$
10. $\frac{4}{3} = \frac{28}{\square}$
11. $\frac{3}{8} = \frac{18}{\square}$
12. $\frac{11}{15} = \frac{44}{\square}$
13. $\frac{9}{14} = \frac{\square}{42}$
14. $\frac{10}{9} = \frac{\square}{45}$
15. $\frac{5}{16} = \frac{\square}{48}$
16. $\frac{14}{8} = \frac{56}{\square}$

Write an equivalent fraction with a denominator of 36.

17. $\frac{1}{2}$
18. $\frac{2}{3}$
19. $\frac{3}{4}$
20. $\frac{4}{9}$
21. $\frac{7}{6}$
22. $\frac{17}{12}$

Number Sense

Write the missing equivalent fractions.

23. $\frac{2}{3} = \frac{4}{6} = \frac{6}{9} = \frac{\square}{\square} = \frac{\square}{\square} = \frac{\square}{\square}$
24. $\frac{7}{4} = \frac{14}{8} = \frac{21}{12} = \frac{\square}{\square} = \frac{\square}{\square} = \frac{\square}{\square}$
25. $\frac{5}{6} = \frac{\square}{\square} = \frac{15}{18} = \frac{\square}{\square} = \frac{\square}{\square} = \frac{30}{36}$
26. $\frac{9}{16} = \frac{\square}{\square} = \frac{27}{48} = \frac{\square}{\square} = \frac{\square}{\square} = \frac{54}{96}$

Write Math

27. Complete the following steps to find the cross products for each pair of fractions. Then write a rule about equivalent fractions.

| Multiply the numerator of the first fraction by the denominator of the second. | → | Multiply the numerator of the second fraction by the denominator of the first. |

a. $\frac{2}{3} = \frac{6}{9}$
b. $\frac{4}{5} = \frac{8}{10}$
c. $\frac{3}{4} = \frac{15}{20}$

Data Bank

Use the table of nail weights on p. 490. Find the difference in the number of nails in 1 lb. Use mental math.

28. 3d and 4d common
29. 2d and 3d casing
30. 2d and 3d finishing
31. 4d and 2d casing
32. 3d and 4d finishing
33. 2d and 4d common

Talk Math

Do the two boxes of nails weigh the same? Tell why or why not.

34.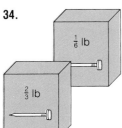
35.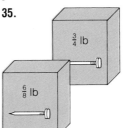
36.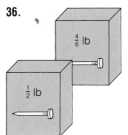

7-4 Writing Fractions in Lowest Terms

Objective
Find the greatest common factor. Simplify fractions using the GCF.

The 56 musicians in the Carlin Orchestra shown on p. 189 are seated by the type of instrument they play. You can write fractions for the sections of the orchestra. $\frac{12}{56}$ of the musicians play woodwind instruments. What is the greatest number that is a factor of both 12 and 56?

The **greatest common factor (GCF)** of two or more numbers is the greatest whole number that is a factor of each number.

Example 1. Find the greatest common factor of 12 and 56.

| Write the factors of each number. | → | Find the common factors. | → | Find the greatest common factor. |

12: 1, 2, 3, 4, 6, 12
56: 1, 2, 4, 7, 8, 14, 28, 56

12: **1, 2, 3, 4,** 6, 12
56: **1, 2, 4,** 7, 8, 14, 28, 56

12: 1, 2, 3, **4**, 6, 12
56: 1, 2, **4**, 7, 8, 14, 28, 56

4 is the greatest common factor of 12 and 56.

Try This Find the greatest common factor for each pair of numbers.
a. 14, 63   b. 28, 44   c. 24, 36   d. 60, 45   e. 15, 13

A fraction is in **lowest terms** when the greatest common factor of the numerator and denominator is 1.

Example 2. Write in lowest terms.

$\frac{18}{27}$

| Find the greatest common factor. | → | Divide the numerator and denominator by the GCF. |

18: 1, 2, 3, 6, **9**
27: 1, 3, **9**, 27

$\frac{18}{27} = \frac{18 \div 9}{27 \div 9} = \frac{2}{3}$

Try This Write in lowest terms.
f. $\frac{6}{9}$ g. $\frac{7}{28}$ h. $\frac{56}{32}$ i. $\frac{16}{45}$ j. $\frac{36}{84}$ k. $\frac{6}{8}$

CHAPTER 7

Exercises

Find the greatest common factor for each pair of numbers.

1. 6, 9
2. 8, 12
3. 10, 15
4. 18, 16
5. 28, 24
6. 11, 44
7. 18, 42
8. 22, 33
9. 6, 7
10. 15, 14
11. 25, 20
12. 12, 15
13. 17, 34
14. 21, 42
15. 81, 45
16. 12, 30
17. 54, 66
18. 70, 14
19. 60, 36
20. 27, 20

Write in lowest terms.

21. $\frac{24}{20}$
22. $\frac{20}{36}$
23. $\frac{27}{45}$
24. $\frac{16}{24}$
25. $\frac{25}{40}$
26. $\frac{14}{28}$
27. $\frac{18}{33}$
28. $\frac{24}{32}$
29. $\frac{28}{42}$
30. $\frac{35}{49}$
31. $\frac{28}{64}$
32. $\frac{38}{40}$
33. $\frac{27}{81}$
34. $\frac{40}{60}$
35. $\frac{36}{45}$
36. $\frac{56}{72}$
37. $\frac{48}{98}$
38. $\frac{28}{63}$
39. $\frac{25}{27}$
40. $\frac{12}{72}$
41. $\frac{24}{27}$
42. $\frac{35}{30}$
43. $\frac{13}{52}$
44. $\frac{25}{75}$

Write as a fraction in lowest terms.

45. What part of the musicians in the Carlin Orchestra play the tuba?
46. What part of the orchestra plays percussion instruments?
47. What part of the woodwind section plays the flute?
48. What part of the brass section plays the French horn?

Suppose Suppose the conductor adds two violas, a drum, and a French horn.

49. What part of the orchestra plays a string instrument?
50. What part of the orchestra plays a drum?
51. What part of the string section plays the viola?
52. What part of the brass section plays the French horn?

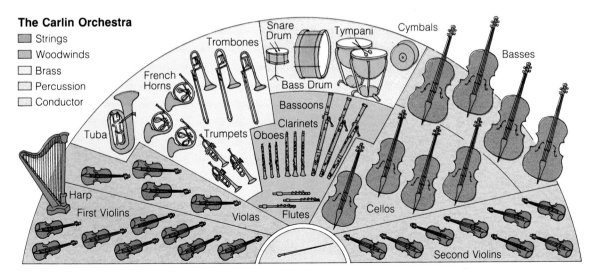

The Carlin Orchestra
- Strings
- Woodwinds
- Brass
- Percussion
- Conductor

7-5 Improper Fractions and Mixed Numbers

Objective
Write mixed numbers as improper fractions. Write improper fractions as mixed numbers or whole numbers.

The Oscar is an award for outstanding achievements in film. The Oscar weighs $8\frac{3}{4}$ lb and is $13\frac{1}{2}$ in. tall.

$8\frac{3}{4}$ and $13\frac{1}{2}$ are mixed numbers. A **mixed number** shows the sum of a whole number and a fraction. You can write a mixed number as an improper fraction. An **improper fraction** is a fraction with a numerator equal to or greater than the denominator.

©AMPAS

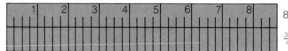

Example 1. Write $8\frac{3}{4}$ as an improper fraction.

| Multiply the whole number by the denominator. | → | Add the numerator. | → | Write the sum over the denominator. |

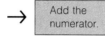 $4 \times 8 = 32$ $8\frac{3}{4}$ $32 + 3 = 35$ $8\frac{3}{4} = \frac{35}{4}$

Try This Write each mixed number as an improper fraction.

a. $5\frac{1}{4}$ b. $3\frac{2}{5}$ c. $1\frac{7}{12}$ d. $2\frac{5}{6}$ e. $11\frac{2}{3}$ f. $4\frac{3}{8}$

Example 2. Write the improper fraction $\frac{21}{9}$ as a mixed number or a whole number.

| Divide the numerator by the denominator. | → | Write the quotient as the whole number. | → | Write the remainder over the divisor as the fraction. |

$\overset{2}{}$ quotient
divisor $9\overline{)21}$
$\underline{18}$
3 remainder

$\frac{21}{9} = 2\frac{\square}{\square}$

$\frac{21}{9} = 2\frac{3}{9} = 2\frac{1}{3}$
Write all fractions in lowest terms.

Try This Write each improper fraction as a mixed number or a whole number.

g. $\frac{13}{7}$ h. $\frac{5}{4}$ i. $\frac{40}{8}$ j. $\frac{24}{9}$ k. $\frac{42}{18}$ l. $\frac{19}{19}$

Exercises

Write each mixed number as an improper fraction.

1. $2\frac{1}{3}$
2. $3\frac{1}{4}$
3. $1\frac{7}{8}$
4. $4\frac{2}{5}$
5. $5\frac{2}{7}$
6. $2\frac{7}{8}$
7. $1\frac{5}{12}$
8. $2\frac{9}{14}$
9. $7\frac{3}{4}$
10. $6\frac{2}{9}$
11. $2\frac{11}{15}$
12. $3\frac{7}{9}$
13. $8\frac{2}{3}$
14. $3\frac{5}{8}$
15. $5\frac{7}{10}$
16. $3\frac{15}{16}$
17. $4\frac{13}{24}$
18. $1\frac{7}{15}$
19. $18\frac{2}{3}$
20. $23\frac{1}{6}$
21. $11\frac{3}{5}$
22. $20\frac{4}{7}$
23. $15\frac{3}{8}$
24. $12\frac{5}{16}$

Write each improper fraction as a mixed number or a whole number.

25. $\frac{5}{3}$
26. $\frac{9}{4}$
27. $\frac{13}{8}$
28. $\frac{63}{7}$
29. $\frac{20}{9}$
30. $\frac{16}{6}$
31. $\frac{15}{12}$
32. $\frac{56}{14}$
33. $\frac{42}{6}$
34. $\frac{30}{9}$
35. $\frac{33}{15}$
36. $\frac{42}{20}$
37. $\frac{72}{12}$
38. $\frac{17}{1}$
39. $\frac{44}{10}$
40. $\frac{54}{16}$
41. $\frac{84}{24}$
42. $\frac{95}{6}$
43. $\frac{86}{5}$
44. $\frac{92}{6}$
45. $\frac{67}{9}$
46. $\frac{54}{3}$
47. $\frac{78}{4}$
48. $\frac{48}{3}$

Mixed Applications

49. Of the 270,000 people who work in the film industry, 4500 are eligible to vote for the Academy Awards. What fraction of the people who work in film can pick the winners?

50. The earliest movie theaters were the nickelodeons. A ticket cost 5¢. The average ticket price today is $3.50. What fraction of today's ticket price was the nickelodeon price?

Number Sense

51. Arrange each of the following on a number line. $1\frac{2}{3}$ $\frac{10}{4}$ $\frac{5}{10}$ $\frac{9}{6}$ $1\frac{6}{8}$ $\frac{5}{4}$

Mixed Skills Review

Complete each statement by writing each measure as a mixed number.

52. 20 in. = ___ ft
53. 3200 lb = ___ T
54. 9 pt = ___ qt
55. 45 in. = ___ yd
56. 44 oz = ___ lb
57. 11 ft = ___ yd
58. 10 c = ___ qt
59. 12 pt = ___ gal
60. 9 qt = ___ gal

Evaluate.

61. $x - 4$ for $x = 7$
62. $729 \div t$ for $t = 6$
63. $2y - 9$ for $y = 15$
64. $2s \div 4$ for $s = 12$
65. $33 + 2x$ for $x = 42$
66. $51 - 3p$ for $p = 13$
67. $5(f + 27)$ for $f = 3$
68. $57 \div 2r$ for $r = 6$
69. $(3x - 4) \div 4$ or $x = 6$

Give the formula you would use to solve each problem.

70. Find the area of a rectangular poster.
71. Find the area of a triangular sign.

7-6 Least Common Multiple

Objective
Find the least common multiple.

Susan designs table tops using tiles. Each tile is 4 in. wide and 5 in. long. What is the smallest square she can make using tiles?

Susan looks for the smallest number that is a multiple of both 4 and 5. This number is called the **least common multiple (LCM).**

Examples Find the least common multiple.

1. 4, 5

   Is the larger number a multiple of the smaller number?
   No, 5 is not a multiple of 4.
   List the multiples of both numbers.   Multiples of 5:   5, 10, 15, **20** . . .
   The LCM is 20.   Multiples of 4:   4, 8, 16, **20** . . .
   The smallest square that Susan can make will be 20 in. on each side.

2. 15, 5

   Is the larger number a multiple of the smaller number?
   Yes, 5 × 3 = 15. The LCM is 15.

Try This Find the least common multiple.

a. 4, 16   b. 6, 15   c. 27, 9   d. 14, 21   e. 9, 12   f. 8, 5

Exercises

Find the least common multiple.

1. 4, 12   2. 5, 30   3. 9, 6   4. 2, 3   5. 3, 8
6. 7, 5   7. 12, 6   8. 6, 16   9. 4, 18   10. 2, 5
11. 12, 18   12. 9, 15   13. 8, 14   14. 10, 8   15. 16, 20
16. 2, 3, 4   17. 3, 5, 6   18. 2, 4, 8   19. 4, 5, 10   20. 6, 8, 12

Number Sense Use the clues to find the missing number.

21. The LCM of two numbers is 42. One of the numbers is 14. The other number is larger than 20 but less than 40.

22. The LCM of two numbers is an odd number. The larger number is less than 40 and is a multiple of the smaller number. The smaller number is 9.

7-7 Least Common Denominator

Objective
Write equivalent fractions using the least common denominator.

Use the LCM to write fractions with the same denominator. The **least common denominator (LCD)** is the least common multiple of the denominators.

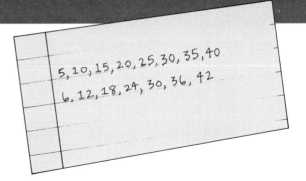

Examples Write equivalent fractions using the LCD.

1. $\frac{3}{5}, \frac{5}{6}$ The LCD is 30. $\frac{3}{5} = \frac{18}{30}$ $\frac{5}{6} = \frac{25}{30}$

2. $\frac{2}{9}, \frac{3}{4}$ The LCD is 36. $\frac{2}{9} = \frac{8}{36}$ $\frac{3}{4} = \frac{27}{36}$

3. $\frac{1}{2}, \frac{1}{3}, \frac{1}{4}$ The LCD is 12. $\frac{1}{2} = \frac{6}{12}$ $\frac{1}{3} = \frac{4}{12}$ $\frac{1}{4} = \frac{3}{12}$

Try This Write equivalent fractions using the LCD.

a. $\frac{1}{2}, \frac{1}{3}$ b. $\frac{3}{10}, \frac{2}{5}$ c. $\frac{7}{15}, \frac{4}{3}$ d. $\frac{3}{4}, \frac{1}{5}$ e. $\frac{2}{7}, \frac{1}{2}, \frac{5}{14}$

Exercises

Write equivalent fractions using the LCD.

1. $\frac{1}{2}, \frac{3}{4}$ 2. $\frac{4}{9}, \frac{7}{18}$ 3. $\frac{3}{50}, \frac{2}{25}$ 4. $\frac{15}{16}, \frac{5}{8}$ 5. $\frac{4}{5}, \frac{1}{3}$

6. $\frac{1}{6}, \frac{5}{9}$ 7. $\frac{8}{21}, \frac{3}{14}$ 8. $\frac{8}{9}, \frac{3}{7}$ 9. $\frac{5}{16}, \frac{1}{6}$ 10. $\frac{7}{10}, \frac{3}{4}$

11. $\frac{5}{9}, \frac{11}{12}$ 12. $\frac{7}{8}, \frac{5}{16}$ 13. $\frac{9}{20}, \frac{4}{15}$ 14. $\frac{3}{8}, \frac{13}{18}$ 15. $\frac{4}{3}, \frac{9}{6}$

16. $\frac{1}{2}, \frac{2}{3}, \frac{3}{4}$ 17. $\frac{1}{3}, \frac{5}{6}, \frac{2}{9}$ 18. $\frac{3}{8}, \frac{3}{4}, \frac{5}{16}$ 19. $\frac{9}{10}, \frac{4}{5}, \frac{1}{2}$

Problem Solving

20. Mrs. Leong visited 3 stores. She spent $\frac{1}{2}$ of her money at the first store. She spent $\frac{1}{2}$ of what she had left at the second store. Then she spent $\frac{1}{2}$ of what she had left at the last store. When she was through shopping Mrs. Leong had $15 left. How much money did she have when she began shopping?

FRACTIONS

7-8 Comparing Fractions and Mixed Numbers

Objective
Compare fractions and mixed numbers.

Jeanne makes quilts. She needs different kinds of fabrics and always looks for bargains. She finds a remnant that is $\frac{7}{8}$ yd long. Another remnant is 33 in. long. Which is the better buy?

Understand the Situation
- Is the price the same for all lengths?
- What fraction of a yard is 33 in.?

Examples Compare. Use <, >, or =.

1. $\frac{7}{8} \square \frac{11}{12}$

→ Write equivalent fractions using the LCD. → Compare the numerators. → Write the same symbol between the fractions.

$\frac{7}{8} = \frac{21}{24}$ and $\frac{11}{12} = \frac{22}{24}$ $21 < 22$ $\frac{7}{8} < \frac{11}{12}$ $\frac{11}{12}$ yd is the better buy.

2. $\frac{9}{6} \square \frac{6}{4}$ $\frac{9}{6} = \frac{18}{12}$ and $\frac{6}{4} = \frac{18}{12}$ $\frac{9}{6} = \frac{6}{4}$

Try This Compare. Use <, >, or =.

a. $\frac{3}{4} \square \frac{2}{3}$ b. $\frac{9}{14} \square \frac{6}{7}$ c. $\frac{7}{12} \square \frac{5}{8}$ d. $\frac{1}{5} \square \frac{2}{10}$ e. $\frac{8}{6} \square \frac{15}{12}$

Example 3. Compare $9\frac{2}{5}$ and $9\frac{1}{3}$. Use <, >, or =.

→ Compare the whole numbers. → If equal, compare the fractions. → Write the same symbol between the mixed numbers.

$9 = 9$ $\frac{2}{5} = \frac{6}{15}$ and $\frac{1}{3} = \frac{5}{15}$ $\frac{2}{5} > \frac{1}{3}$ $9\frac{2}{5} > 9\frac{1}{3}$

Try This Compare. Use <, >, or =.

f. $1\frac{2}{3} \square 2\frac{4}{5}$ g. $4\frac{1}{2} \square 4\frac{2}{4}$ h. $3\frac{4}{5} \square 3\frac{5}{8}$ i. $2\frac{3}{4} \square 2\frac{7}{16}$

Exercises

Compare. Use <, >, or =.

1. $\frac{1}{2} \square \frac{3}{4}$
2. $\frac{4}{9} \square \frac{7}{18}$
3. $\frac{4}{50} \square \frac{2}{25}$
4. $\frac{15}{16} \square \frac{5}{32}$
5. $\frac{4}{5} \square \frac{1}{3}$
6. $\frac{1}{6} \square \frac{5}{9}$
7. $\frac{8}{21} \square \frac{3}{14}$
8. $\frac{8}{9} \square \frac{3}{7}$
9. $\frac{5}{16} \square \frac{1}{6}$
10. $\frac{7}{10} \square \frac{3}{4}$
11. $\frac{5}{9} \square \frac{11}{12}$
12. $\frac{7}{8} \square \frac{5}{18}$
13. $\frac{9}{20} \square \frac{4}{15}$
14. $\frac{11}{12} \square \frac{13}{14}$
15. $\frac{20}{48} \square \frac{15}{36}$
16. $\frac{5}{12} \square \frac{11}{32}$
17. $\frac{13}{15} \square \frac{19}{30}$
18. $\frac{3}{7} \square \frac{3}{8}$
19. $\frac{40}{75} \square \frac{8}{15}$
20. $\frac{9}{16} \square \frac{3}{5}$
21. $1\frac{5}{6} \square 3\frac{2}{3}$
22. $4\frac{1}{4} \square 3\frac{3}{8}$
23. $1\frac{3}{7} \square 1\frac{19}{35}$
24. $2\frac{9}{56} \square 2\frac{11}{14}$
25. $6\frac{15}{80} \square 6\frac{3}{16}$
26. $2\frac{5}{9} \square 2\frac{6}{7}$
27. $1\frac{5}{8} \square 1\frac{3}{5}$
28. $7\frac{7}{11} \square 6\frac{2}{5}$
29. $8\frac{5}{9} \square 8\frac{15}{27}$
30. $2\frac{4}{5} \square 1\frac{24}{25}$
31. $1\frac{6}{12} \square 1\frac{1}{2}$
32. $3\frac{2}{3} \square 3\frac{3}{4}$

Arrange in order from least to greatest.

33. $\frac{1}{3}, \frac{1}{4}, \frac{1}{5}$
34. $\frac{3}{8}, \frac{1}{4}, \frac{2}{5}$
35. $\frac{1}{2}, \frac{2}{7}, \frac{1}{6}$
36. $\frac{1}{2}, \frac{2}{3}, \frac{3}{5}$
37. $1\frac{3}{4}, \frac{7}{9}, 1\frac{5}{36}$
38. $2\frac{1}{10}, 2\frac{2}{15}, 2\frac{3}{20}$
39. $6\frac{7}{45}, 6\frac{11}{15}, 6\frac{1}{9}$

Estimation Draw a number line from 0 to 2. Estimate and mark the location for each of the following.

40. $\frac{3}{4}$
41. $\frac{2}{3}$
42. $\frac{2}{5}$
43. $1\frac{1}{2}$
44. $1\frac{9}{16}$
45. $1\frac{1}{8}$

Mental Math Decide if each length of material is less than or greater than $\frac{1}{2}$ yd. Use mental math.

46. $\frac{3}{8}$ yd
47. $\frac{4}{9}$ yd
48. $\frac{3}{4}$ yd
49. $\frac{7}{12}$ yd
50. $\frac{1}{3}$ yd
51. $\frac{5}{6}$ yd

Write Math

52. Make a list of all lowest-terms fractions less than 1 that have denominators greater than 1 but less than 6. Arrange the fractions in order from least to greatest.

Mixed Skills Review Find the least common multiple.

53. 9, 4
54. 3, 21
55. 2, 11
56. 10, 4
57. 2, 5, 7
58. 3, 4, 6

Write as an improper fraction.

59. $8\frac{1}{4}$
60. $11\frac{5}{6}$
61. $3\frac{11}{15}$
62. $24\frac{2}{3}$
63. $4\frac{3}{10}$
64. $2\frac{5}{12}$

7-9 Consumer Math: Estimating Gauge Readings

Objective
Estimate gauge readings.

- SITUATION
- DATA
- PLAN
- ANSWER
- CHECK

A gasoline gauge shows what part of the tank is filled. You can use a fraction to estimate the amount of gasoline in the tank.

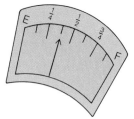

Example Read the gauge. Estimate what part of the tank is filled.

The marks divide the gauge into eight parts. The needle shows the tank is between $\frac{1}{4}$ and $\frac{1}{2}$ filled.
The tank is about $\frac{3}{8}$ filled.

Try This Read the gauge. Estimate what part of the tank is filled.

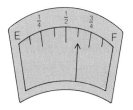 b. 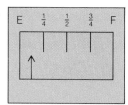 c.

Exercises

Read the gauge. Estimate what part of the tank is filled.

1. 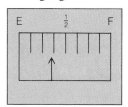   2.    3.

Show Math Draw a gauge for Exercises 4–5.

4. Show eight parts. Label the fourths. Show that the tank is about $\frac{7}{8}$ filled.

5. Show six parts. Label the thirds. Show that the tank is about $\frac{1}{6}$ filled.

Number Sense Decide which fraction on a gauge would show that a tank is about half filled.

6. $\frac{1}{3}$ 7. $\frac{4}{8}$ 8. $\frac{9}{16}$ 9. $\frac{5}{6}$ 10. $\frac{2}{4}$ 11. $\frac{2}{8}$

7-10 Estimating Fractions Using Compatible Numbers

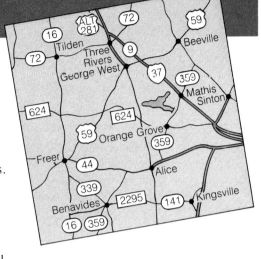

Objective
Estimate fractions using compatible numbers.

Tom records traffic patterns for the state highway department. He finds what fraction of all the cars on Route 16 went in each direction. He uses compatible numbers to give estimates of the fractions in his report.

Examples Find compatible numbers. Write in lowest terms.

1. In one hour, 91 cars traveled north on Route 16 to the intersection with Route 72. Of 91 cars, 33 cars turned east at the intersection.

   Turn east on Rte. 72 ⟶ $\frac{33}{91}$ is about $\frac{30}{90}$, or $\frac{1}{3}$.
   from north on Rte. 16. ⟶

	Direction	Fraction	Think	Lowest Terms
2.	Turn east on Rte. 72 from south on Rte. 16. ⟶	$\frac{24}{33}$	about $\frac{24}{32}$	$\frac{3}{4}$
3.	Turn north on Rte. 16 from west on Rte. 72. ⟶	$\frac{14}{41}$	about $\frac{15}{40}$	$\frac{3}{8}$

Try This Find compatible numbers. Write in lowest terms.

a. $\frac{16}{33}$ b. $\frac{21}{51}$ c. $\frac{45}{59}$ d. $\frac{69}{70}$ e. $\frac{34}{42}$

Exercises

Find compatible numbers. Write in lowest terms.

1. $\frac{6}{17}$ 2. $\frac{4}{25}$ 3. $\frac{11}{20}$ 4. $\frac{5}{16}$ 5. $\frac{17}{23}$ 6. $\frac{19}{27}$

7. $\frac{38}{49}$ 8. $\frac{26}{41}$ 9. $\frac{19}{23}$ 10. $\frac{39}{62}$ 11. $\frac{33}{45}$ 12. $\frac{29}{32}$

13. $\frac{14}{45}$ 14. $\frac{74}{99}$ 15. $\frac{29}{51}$ 16. $\frac{71}{83}$ 17. $\frac{13}{71}$ 18. $\frac{59}{98}$

Number Sense Which fractions are closest in value? Use compatible numbers to estimate.

19. $\frac{3}{7}, \frac{49}{101}, \frac{18}{23}$ 20. $\frac{3}{10}, \frac{4}{19}, \frac{11}{30}$ 21. $\frac{4}{9}, \frac{8}{21}, \frac{2}{5}$ 22. $\frac{24}{98}, \frac{6}{25}, \frac{7}{13}$

FRACTIONS

7-11 Writing Fractions as Decimals

Objective
Write fractions and mixed numbers as decimals.

Owls Win State Finals, 82–81

ATLANTA (AP) The Stone Mountain Owls defeated the Decatur Greyhounds to win the girls' state basketball title. The Owls made 36 field goals in 64 attempts.

Stone Mountain	FG	FT	Decatur	FG	FT
Moore	2/8	0/0	Chilton	5/12	0/2
Scannell	7/12	2/2	Reveille	9/15	3/3
Edwards	7/9	4/4	Davis	2/6	0/0
Tate	11/18	1/1	Gately	8/18	0/0
Hicks	2/7	2/2	Tucker	4/6	4/4
Anderson	7/10	1/3	Morris	5/10	0/0
			Lee	4/8	0/0
Totals	36/64	10/12	Totals	37/75	7/9

Understand the Situation

- What fraction of their field goal attempts did the Owls make?
- What fraction of their field goal attempts did the Greyhounds make?
- Are the fractions easy to compare?

You can write a fraction as a decimal. The fraction bar means *divided by*. The decimal equivalents of fractions are easy to compare.

Examples Write each fraction as a decimal. Use a calculator.

1. $\frac{36}{64}$ $\frac{36}{64} = 36 \div 64 = \boxed{0.5625}$

The answer 0.5625 is a **terminating decimal**.
The division ends with a remainder of 0.

$\frac{36}{64} = 0.5625$

2. $\frac{37}{75}$ $\frac{37}{75} = 37 \div 75 = \boxed{0.4933333}$

The division continues without a remainder of 0. Show the decimal repeats in the answer by placing a bar over the digit or group of digits that repeats. $0.49\overline{3}$ is a **repeating decimal**.

$\frac{37}{75} = 0.49\overline{3}$

You can easily compare decimal equivalents of fractions. $0.5625 > 0.49\overline{3}$

3. $6\frac{7}{11}$

$6\frac{7}{11} = 6 + (7 \div 11)$ $7 \div 11 = \boxed{0.6363636}$ $6\frac{7}{11} = 6.\overline{63}$

Try This Write each fraction as a decimal. Use a calculator.

a. $\frac{4}{5}$ b. $\frac{1}{3}$ c. $1\frac{1}{4}$ d. $2\frac{4}{9}$ e. $1\frac{1}{2}$ f. $\frac{1}{8}$

CHAPTER 7

Exercises

Write each fraction as a decimal. Use a calculator.

1. $\frac{1}{4}$ 2. $\frac{3}{5}$ 3. $\frac{7}{20}$ 4. $\frac{9}{25}$ 5. $\frac{5}{8}$ 6. $\frac{13}{16}$

7. $\frac{1}{2}$ 8. $\frac{17}{10}$ 9. $\frac{9}{5}$ 10. $\frac{2}{3}$ 11. $\frac{1}{6}$ 12. $\frac{7}{9}$

13. $\frac{11}{12}$ 14. $\frac{13}{32}$ 15. $\frac{9}{11}$ 16. $\frac{9}{4}$ 17. $\frac{5}{3}$ 18. $\frac{11}{8}$

19. $8\frac{31}{24}$ 20. $7\frac{3}{8}$ 21. $3\frac{5}{2}$ 22. $2\frac{17}{33}$ 23. $1\frac{50}{99}$ 24. $2\frac{1}{2}$

25. $3\frac{4}{5}$ 26. $1\frac{1}{12}$ 27. $2\frac{4}{9}$ 28. $1\frac{6}{9}$ 29. $5\frac{5}{11}$ 30. $8\frac{7}{9}$

Compare the fractions by comparing their decimal equivalents. Use <, >, or =.

31. $\frac{11}{12} \square \frac{9}{10}$ 32. $\frac{13}{18} \square \frac{17}{20}$ 33. $\frac{3}{8} \square \frac{10}{21}$ 34. $\frac{5}{11} \square \frac{7}{15}$

Number Sense

Look for a pattern. Then write repeating decimals.

35. $\frac{1}{9} = 0.\overline{1}, \frac{2}{9} = 0.\overline{2}, \frac{3}{9} = 0.\overline{3}, \frac{4}{9} = \square, \frac{5}{9} = \square, \frac{7}{9} = \square$

36. $\frac{1}{11} = 0.\overline{09}, \frac{2}{11} = 0.\overline{18}, \frac{3}{11} = 0.\overline{27}, \frac{4}{11} = \square, \frac{6}{11} = \square, \frac{10}{11} = \square$

37. $\frac{1}{99} = 0.\overline{01}, \frac{25}{99} = 0.\overline{25}, \frac{31}{99} = 0.\overline{31}, \frac{58}{99} = \square, \frac{74}{99} = \square, \frac{89}{99} = \square$

Find the missing numerals.

38. $\frac{3}{5} \rightarrow \square \div \square \rightarrow \square\overline{)\square}$ 39. $\frac{\square}{\square} \rightarrow 9 \div 4 \rightarrow \square\overline{)\square}$

40. $\frac{\square}{\square} \rightarrow 39 \div 15 \rightarrow \square\overline{)\square}$ 41. $\frac{\square}{\square} \rightarrow \square \div \square \rightarrow 12\overline{)7}$

Estimation

Estimate the first digit in the decimal equivalent for each fraction. Use a calculator to check your estimate.

42. $\frac{123}{999}$ 43. $\frac{217}{330}$ 44. $\frac{567}{789}$ 45. $\frac{468}{579}$ 46. $\frac{543}{654}$ 47. $\frac{303}{515}$

48. A scorekeeper writes the players' numbers on a floor plan to record field goal attempts. She circles the field goals scored. Write the fraction of shots made by each team. Write the decimal equivalents.

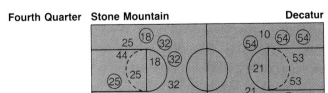

FRACTIONS **199**

7-12 Writing Decimals as Fractions

Objective
Write decimals as fractions and mixed numbers.

Rogers Hornsby won the National League Batting Championship six years in a row. A **batting average** is the number of hits divided by the number of times at bat. Hornsby earned his best average in 1924.

You can write 0.424 as a fraction.

Rogers Hornsby's Batting Averages

Year	Average
1920	.370
1921	.397
1922	.401
1923	.384
1924	.424
1925	.403
1928	.387

Examples Write each decimal as a fraction or as a mixed number in lowest terms.

1. 0.424

| Write the digits over the place value. | → | Write the fraction in lowest terms. |

$0.424 = \frac{424}{1000}$ $\frac{424 \div 8}{1000 \div 8} = \frac{53}{125}$

2. $0.6 = \frac{6}{10} = \frac{3}{5}$ **3.** $0.0625 = \frac{625}{10,000} = \frac{1}{16}$ **4.** $2.125 = 2 + 0.125$
$= 2 + \frac{125}{1000}$
$= 2\frac{1}{8}$

It is helpful to memorize the fraction equivalents for some repeating decimals.

Repeating Decimal	$0.08\overline{3}$	$0.1\overline{6}$	$0.\overline{3}$	$0.41\overline{6}$	$0.58\overline{3}$	$0.\overline{6}$	$0.8\overline{3}$	$0.91\overline{6}$
Fraction	$\frac{1}{12}$	$\frac{1}{6}$	$\frac{1}{3}$	$\frac{5}{12}$	$\frac{7}{12}$	$\frac{2}{3}$	$\frac{5}{6}$	$\frac{11}{12}$

5. $0.\overline{6} = \frac{2}{3}$

6. $3.8\overline{3} = 3 + 0.8\overline{3}$
$= 3 + \frac{5}{6}$
$= 3\frac{5}{6}$

Try This Write each decimal as a fraction or as a mixed number in lowest terms.
a. 0.84   **b.** 0.003   **c.** $0.1\overline{6}$ **d.** $10.\overline{3}$ **e.** 5.3

Exercises

Write each decimal as a fraction or as a mixed number in lowest terms.

1. 0.9
2. 0.4
3. 0.25
4. 0.36
5. 0.74
6. 4.3
7. 3.7
8. 4.13
9. 7.64
10. 5.75
11. 0.4375
12. 0.5625
13. 0.0006
14. 0.0875
15. 0.0044
16. 7.0625
17. 9.1875
18. 6.0125
19. 4.0075
20. 1.0052
21. $0.\overline{3}$
22. $0.\overline{3}$
23. $0.08\overline{3}$
24. $0.58\overline{3}$
25. $0.\overline{83}$
26. $2.\overline{6}$
27. $3.\overline{3}$
28. $1.1\overline{6}$
29. $5.41\overline{6}$
30. $2.08\overline{3}$

Arrange the numbers in order from least to greatest.

31. $0.4, \frac{3}{4}, 1, 1.2, \frac{7}{5}, 0.8$
32. $2, \frac{5}{2}, 2.4, 2\frac{1}{3}, 2.35, \frac{6}{2}$
33. $4\frac{3}{6}, 4.6, 5, \frac{24}{6}, 4.06, 4.1$
34. $\frac{36}{4}, 8\frac{7}{8}, 9.25, \frac{58}{7}, 8.75, 9.1$

Number Sense

35. If $0.\overline{3} = \frac{1}{3}$ and $0.0\overline{3} = \frac{1}{30}$, then $0.00\overline{3} = $ ____.
36. If $0.\overline{1} = \frac{1}{9}$ and $0.0\overline{1} = \frac{1}{90}$, then $0.00\overline{1} = $ ____.
37. If $0.75 = \frac{3}{4}$ and $0.075 = \frac{3}{40}$, then $0.0075 = $ ____.

Mental Math

Write each decimal as a fraction. Use mental math.

38. 0.025
39. $0.0\overline{6}$
40. $0.01\overline{6}$
41. $0.008\overline{3}$
42. $0.0041\overline{6}$

Data Hunt

43. Use an almanac to find the player who has the all-time best batting average for a single season in the American League. Write the decimal as a fraction in lowest terms.

Mixed Applications

Each team in the Connie Mack League has played 16 games. Use the decimal to find the number of wins and losses for each team in the league.
(Hint: Find equivalent fractions.
$0.750 = \frac{3}{4} = \frac{\square \text{ games won}}{16 \text{ games played}}$)

Connie Mack League Team Standings After 16 Games

	Team	W	L	Pct.
44.	Giants			.750
45.	Red Sox			.625
46.	Cardinals			.500
47.	Yankees			.125

Problem Solving

48. Thursday the Cubs' score was a multiple of 2 and one more run than they had scored on Wednesday. On Friday they scored half as many runs as Thursday. The total runs scored for the three days was 14. Give the daily scores.

49. Lilia Rivera has 9 coins that total $0.58. She does not have a 50¢ piece. What are the coins?

7-13 Problem Solving: Using Data from a Table

Objective
Use data from a table.

- **SITUATION**
- **DATA**
- **PLAN**
- **ANSWER**
- **CHECK**

In a stock table, stock prices appear as mixed numbers, whole numbers, or fractions. A stock price of $13\frac{5}{8}$ means that one share of stock sells for $13\frac{5}{8}$ dollars.

52 Week High	52 Week Low	Stock	Week High	Week Low	Current price
$27\frac{7}{8}$	$20\frac{5}{8}$	AT&T	$27\frac{3}{8}$	$25\frac{5}{8}$	$26\frac{1}{2}$
$36\frac{7}{8}$	$25\frac{3}{8}$	Avon	$30\frac{1}{4}$	$28\frac{3}{4}$	$29\frac{5}{8}$
$64\frac{7}{8}$	$45\frac{3}{4}$	Boeing	$53\frac{1}{8}$	$49\frac{7}{8}$	$51\frac{1}{4}$
57	$28\frac{5}{8}$	Disney	57	$31\frac{3}{4}$	57
101	$59\frac{1}{2}$	DuPont	101	$94\frac{1}{4}$	$95\frac{5}{8}$
$79\frac{1}{8}$	$46\frac{1}{2}$	EKodak	$79\frac{1}{8}$	$73\frac{7}{8}$	$75\frac{3}{4}$
$84\frac{5}{8}$	$48\frac{3}{4}$	Exxon	$79\frac{1}{4}$	$77\frac{5}{8}$	$79\frac{1}{4}$
$88\frac{1}{4}$	$65\frac{7}{8}$	GMot	$73\frac{5}{8}$	68	$70\frac{7}{8}$
$161\frac{7}{8}$	$115\frac{3}{4}$	IBM	$129\frac{7}{8}$	$126\frac{1}{4}$	129
$8\frac{1}{16}$	$3\frac{9}{16}$	NovaPh	$7\frac{7}{16}$	$3\frac{9}{16}$	$3\frac{9}{16}$
$50\frac{3}{8}$	$36\frac{1}{4}$	Sears	$47\frac{1}{8}$	$44\frac{5}{8}$	$47\frac{1}{8}$

Understand the Situation

- What is the current price of one share of Sears stock?
- What is the week's high for one share of Eastman Kodak stock?
- Which stock is currently selling at its 52-week high?

Example Solve. Round to the nearest cent.

Clare has $1200 to invest in stocks. Does she have enough money to buy 15 shares of General Motors stock?

Find the current price of one share of General Motors (GMot) stock.	Look at the table. The current price of one share is $70\frac{7}{8}$.
Change the price from a mixed number to a decimal.	$70\frac{7}{8}$ = $70.875
Multiply the price per share by the number of shares. Use a calculator.	15 × $70.875 = $1063.125, or $1063.13 to the nearest cent.
Compare the total cost to $1200.	$1063.13 < $1200
Check your answer. Use estimation.	$70 is a little less than $70\frac{7}{8}$. $70 × 15 = $1050. $1050 is a little less than $1063.13. The answer seems reasonable.

Clare has enough money to buy the stock.

Try This Solve. Round answers to the nearest cent.

a. What is the current cost of 10 shares of Nova Pharmaceutical stock?

b. An investor bought 20 shares of IBM at the 52-week low. The stock was sold at the current price. What was the amount of profit?

Exercises

Solve. Use the table on p. 202. Round answers to the nearest cent.

1. Find the current price of 5 shares of Sears stock.
2. Find the current price of 35 shares of Disney stock.
3. Darrell has $300 to invest. Does he have enough money to buy 20 shares of Avon stock?
4. What is the difference between the week's high and low for one share of Exxon stock?
5. What is the difference between the week's high and the current price for one share of Boeing stock?
6. What is the difference between the 52-week high and low for one share of Eastman Kodak stock?
7. What is the difference between the 52-week high and the current week's high for one share of AT&T stock?
8. An investor bought 40 shares of Exxon stock at the 52-week low. How much has the value of the investment increased?
9. Nora Ashley bought 50 shares of General Motors stock at $67\frac{3}{8}$. She sold the stock at the current price. Find the amount of profit.
10. Tim Mathis bought 16 shares of Nova Pharmaceutical at 7. He sold the stock at the current price. Find the amount of loss.

Mental Math

An investor has $500. Use mental math to find which of the following stocks the investor can buy.

11. 10 shares of Avon
12. 20 shares of Disney
13. 5 shares of DuPont
14. 7 shares of General Motors
15. 4 shares of IBM
16. 10 shares of Sears

Group Activity

17. Form partnerships of two or more students. Each partnership has $1000 with which to buy stock.
    a. Use the newspaper to decide which stocks to buy.
    b. Keep a record of the cost and the number of shares bought.
    c. Sell your shares at the market price in two weeks.
    d. Compare your results with those of other partnerships in your class.

Mixed Skills Review

Write each fraction in lowest terms.

18. $\frac{6}{8}$
19. $\frac{3}{9}$
20. $\frac{4}{6}$
21. $\frac{8}{10}$
22. $\frac{3}{12}$
23. $\frac{5}{20}$

24. $2\frac{3}{15}$
25. $3\frac{2}{8}$
26. $4\frac{4}{4}$
27. $2\frac{6}{9}$
28. $5\frac{3}{21}$
29. $7\frac{8}{24}$

Measure the line segment to the nearest $\frac{1}{4}$ in. Use a ruler.

30. ├──┤
31. ├─────┤
32. ├────┤
33. ├─┤

FRACTIONS

7-14 Problem Solving: Developing Thinking Skills

Objective
Use problem-solving strategies.

- SITUATION
- DATA
- PLAN
- ANSWER
- CHECK

Gary saved some money each week to buy tapes. When Music Inc. had a sale, he bought 9 tapes. He spent a total of $64 on tapes marked A and B. How many of each type of tape did he buy?

SALE!
All tapes marked:
A $8
B $6
C $5

Exercises

Follow the *Thinking Actions*. Answer each question.

Try these **before** starting to write.

Thinking Actions Before
- Read the problem.
- Ask yourself questions to understand it.
- Think of possible strategies.

1. What is the sale price of each type of tape: A, B, and C?
2. How much money did Gary spend?
3. What is the problem asking you to find?
4. What types of tapes did he buy?
5. What strategy, or strategies, might help you to solve this problem?
6. How much money would 2 A tapes cost?
7. How much money would 2 B tapes cost?
8. Can you make a table of the costs for different numbers of tapes bought?

Try these **during** your work.

Thinking Actions During
- Try your strategies.
- Stumped? Try answering these questions.
- Check your work.

9. Below are two correct solutions. Write the answer to the problem in a complete sentence. Name the strategies shown.

Solution 1

I will guess 3 A tapes and 6 B tapes.
(3 × $8) + (6 × $6) = $60 (too small)

I will guess 6 A tapes and 3 B tapes.
(6 × $8) + (3 × $6) = $66 (too large)

I will guess 5 A tapes and 4 B tapes.
(5 × $8) + (4 × $6) = $64 (It checks.)

Solution 2

Number: A	Cost	Number: B	Cost
1	$ 8	1	$ 6
2	$16	2	$12
3	$24	3	$18
4	$32	(4)	($24)
(5)	($40)	5	$30
6	$48	6	$36
7	$56	7	$42
8	$64	8	$48

Solve. Use one or more of the problem-solving strategies.

Problem-Solving Strategies
- Guess and check.
- Choose an operation.
- Make an organized list.
- Act it out.
- Find a pattern.
- Write an equation.
- Use logical reasoning.
- Draw a picture or diagram.
- Make a table.
- Use objects.
- Work backward.
- Solve a simpler problem.

10. Joel wants to buy a tape, a shirt, and a bike pump. He only earns enough money to buy one item per week. In how many different orders can he buy the items?

11. Cindy Ho made a long-distance call to her friend in Cleveland at 2 p.m. She talked for 5 minutes. Suppose she had called at 6 p.m. How much money would she have saved?

Period	First Minute	Each Additional Minute
8 a.m. to 5 p.m.	$0.40	$0.32
5 p.m. to 11 p.m.	$0.25	$0.20
11 p.m. to 8 a.m.	$0.19	$0.16

12. Jean, Danielle, Dao, and Mark applied for scholarships. Mark received his scholarship right after Jean did. Dao got his scholarship before one other person. Who got the first scholarship?

13. Theo bought some airline tickets at a special rate of $50 each. Later he bought some more tickets for $70 each. The total cost of 8 tickets was $440. How many of the tickets did Theo buy at the special rate?

14. Frank's family has a large vegetable garden. Frank made a graph showing how many zucchini the family picked each week. At this rate, how many zucchini will the family pick in the sixth week?

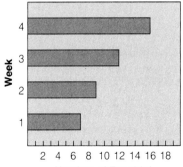
Number of Zucchini Picked

15. Three friends paid for a subscription to a weekly rock music magazine. Each keeps the magazine for one week, then passes it to the next person. How long does it take for the last person to get the fifth magazine?

Solve It Another Way

16. Gabriela solved this problem using the strategy *draw a picture*. Find another way to solve this problem.
    Problem: Ira checked out four books from the library: a western, a mystery, a book on photography, and a biography. In how many different orders can he read the books?

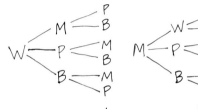

FRACTIONS

7-15 Enrichment: Elapsed Time

Objective
Find elapsed time.

Julio and Valene went hiking from 10:20 a.m. to 3:20 p.m. How long did they spend hiking?

Find the amount of time that **elapsed**, or passed.

Example

What is the elapsed time from 10:20 a.m. to 3:20 p.m.?

Find the amount of time before 12 p.m. and the amount of time after 12 p.m.

```
  12 h  0 min = 11 h 60 min
- 10 h 20 min = 10 h 20 min
                 1 h 40 min
```

Subtract to find the time before 12 p.m. Regroup. Subtract the same units.

time before 12 p.m.   1 h 40 min
time after 12 p.m.   +3 h 20 min
                       4 h 60 min = 5 h

Add the amount of time before 12 p.m. to the amount of time after 12 p.m. Add the same units. Regroup.

The elapsed time is 5 h.

Try This

Find the elapsed time.

a. 9:20 a.m. to 11:40 p.m.   **b.** 11:30 a.m. to 12:45 p.m.

Exercises

Find the elapsed time.

1. 8:10 a.m. to 10:20 a.m.   **2.** 2:40 p.m. to 6:50 p.m.   **3.** 5:45 a.m. to 9:32 a.m.

4. 3:38 p.m. to 11:29 p.m.   **5.** 4:40 a.m. to 1:20 p.m.   **6.** 6:47 a.m. to 2:18 p.m.

7. Giorgio rode his bike from 10:55 a.m. to 4:05 p.m. How long did Giorgio spend riding his bike?

8. Lana went horseback riding between 9:35 a.m. and 12:05 p.m. How long did Lana spend horseback riding?

Planning a Solution

9. Pick one or more of the statements that give the method that you could use to solve this problem.

Problem: Mei worked from 10:45 a.m. to 2:30 p.m. How long did she work?

A. Subtract 10 h 45 min from 2 h 30 min.

B. Subtract 10 h 45 min from 11 h 60 min. Add 2 h 30 min to the difference.

C. Add 2 h 30 min to 12 h. Subtract 10 h 45 min from the sum.

7-16 Computer: Telecommunications

Objective
Act out the telecommunications process. Use the baud rate.

All computer data is made up of electronic signals. These signals form a code that corresponds to 0s and 1s of the binary system. It is possible for one computer to send data to another computer over long distances.
Telecommunications is the sending and receiving of information.

A **modem** changes computer data to electrical pulses usually through wire, radio, or satellite. It can then send these pulses over telephone lines. The **baud rate** is the rate at which a modem transfers data. This rate is measured in **bits per second.**

Exercises

1. Act out the telecommunications process with three classmates.

   Student 1: Write a short question. Hand it to student 2.

   Student 2: Use the code A = 1, B = 2, C = 3, . . . , Z = 26 to change the question into numbers. Hand the coded question to student 3.

   Student 3: Change the numbers back to words. Use the same code. Hand the decoded question to student 4.

   Student 4: Answer the question. Hand the answer to student 3.

   Repeat the steps in the other direction until the answer gets back to student 1.

   Which students acted out the role of modems?

Suppose a modem transfers data at the rate of 1200 bits per second. Suppose it takes 8 bits to make each character (a letter, a punctuation mark, a space, and so on). Solve.

2. How many characters per second is this?

3. How long would it take the modem to change your first and last names?

4. Suppose the average word is 6 characters long. How many words per minute can this modem change data?

5. Anders wants to send a report of about 500 words. How long would it take the modem to change the report? (Use an average word length of 6 characters.)

FRACTIONS 207

CHAPTER 7 REVIEW

Choose the example of each.

1. fraction
   - A. 0.75
   - B. $\frac{1}{2}$
   - C. sixty-nine
   - D. 24

2. numerator of 8
   - A. $\frac{1}{8}$
   - B. $\frac{8}{9}$
   - C. 6×8
   - D. $6 + 8$

3. equivalent fractions
   - A. $\frac{1}{5}, \frac{4}{15}$
   - B. $\frac{2}{7}, \frac{4}{14}$
   - C. $\frac{1}{3}, 9$
   - D. $\frac{3}{4}, \frac{4}{12}$

4. repeating decimal
   - A. 3.75
   - B. 1.222
   - C. $6.\overline{6}$
   - D. 3.145

5. lowest terms
   - A. $\frac{3}{6}$
   - B. $\frac{4}{6}$
   - C. $\frac{1}{6}$
   - D. $\frac{6}{6}$

6. greatest common factor
   - A. 8, 6: 48
   - B. 16, 8: 2
   - C. 15, 13: 11
   - D. 24, 36: 12

Complete the sentence so that it is true.

7. A fraction is in lowest terms when the greatest common factor of the numerator and the denominator is _____ .
8. An improper fraction is a fraction with a numerator _____ than the denominator.
9. The least common multiple is the _____ .
10. The least common denominator is the _____ of the denominators.

Match one from each group.

	Group A	Group B	Group C
11. improper fraction	$\frac{2}{3}$	$\frac{3}{2}$	$\frac{6}{5}$
12. mixed number	$\frac{3}{8}$	$\frac{1}{3}$	$\frac{4}{7}$
13. denominator of 3	$\frac{7}{4}$	$\frac{4}{8}$	$4\frac{3}{8}$
14. numerator of 4	$\frac{4}{5}$	$3\frac{1}{4}$	$\frac{3}{4}$
15. less than $\frac{1}{2}$	$4\frac{3}{5}$	$\frac{5}{7}$	$\frac{1}{3}$

16. Write the fractions greater than $\frac{3}{4}$.
    $\frac{5}{8}, \frac{5}{6}, \frac{4}{5}, \frac{2}{3}, \frac{4}{7}, \frac{5}{9}$

17. Write the fractions less than $\frac{1}{4}$.
    $\frac{1}{2}, \frac{1}{3}, \frac{1}{8}, \frac{1}{5}, \frac{1}{16}, \frac{1}{9}$

Compare. Use <, >, or =.

18. $1\frac{5}{8} \square \frac{15}{8}$
19. $\frac{3}{4} \square 0.75$
20. $\frac{5}{9} \square \frac{5}{10}$
21. $\frac{1}{4} \square \frac{2}{7}$
22. $1\frac{2}{9} \square 1\frac{8}{36}$

CHAPTER 7 TEST

1. The science class is having a lab test. Experiments are set up at ten stations around the room. Group 1 begins at 10:00 a.m., followed by group 2, and then group 3. Each group averages 4 min at a station. At what time will group 3 be done?

Write the fraction for the shaded part.

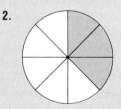

2. 3.

Write the missing numeral.   4. $\frac{2}{3} = \frac{\square}{12}$ 5. $\frac{5}{6} = \frac{10}{\square}$ 6. $\frac{3}{8} = \frac{15}{\square}$

Find the greatest common factor.   7. 9, 24   8. 5, 25   9. 5, 3   10. 36, 30

Write in lowest terms.   11. $\frac{11}{77}$ 12. $\frac{18}{36}$ 13. $\frac{10}{60}$ 14. $\frac{12}{16}$ 15. $\frac{3}{12}$

Write as an improper fraction.   16. $4\frac{5}{6}$ 17. $2\frac{3}{8}$ 18. $5\frac{2}{3}$

Write as a mixed number or a whole number.   19. $\frac{47}{7}$ 20. $\frac{5}{4}$ 21. $\frac{18}{6}$

Find the least common multiple.   22. 4, 5   23. 3, 5   24. 6, 2   25. 9, 4

Write equivalent fractions using the LCD.   26. $\frac{2}{3}, \frac{5}{8}$ 27. $\frac{5}{11}, \frac{7}{3}$ 28. $\frac{1}{5}, \frac{1}{10}$

Compare. Use <, >, or =.   29. $\frac{3}{5} \square \frac{2}{3}$ 30. $\frac{4}{16} \square \frac{2}{8}$ 31. $2\frac{6}{7} \square 3\frac{5}{6}$

32. Read the gauge. Estimate what part of the tank is filled.

Estimate. Find compatible numbers. Write in lowest terms.

33. $\frac{7}{13}$ 34. $\frac{9}{37}$ 35. $\frac{11}{30}$ 36. $\frac{11}{18}$ 37. $\frac{8}{23}$ 38. $\frac{21}{40}$

Write each fraction as a decimal.   39. $\frac{4}{16}$ 40. $\frac{3}{8}$ 41. $\frac{5}{9}$ 42. $1\frac{3}{4}$

Write each decimal as a fraction or as a mixed number in lowest terms.
43. 0.8   44. 0.375   45. 0.75   46. 1.6   47. 2.3   48. 3.5

49. The price for LSK stock is $7\frac{1}{4}$. Suza received $58 for her birthday. How many shares of LSK stock can she buy?

FRACTIONS

CHAPTER 7 CUMULATIVE REVIEW

1. Compare. Use <, >, or =.
   27.5 □ 26.9
   A. >
   B. <
   C. =

2. Choose the correct formula for the area of the garden plot.
   A. $A = lw$
   B. $i = prt$
   C. $V = lwh$
   D. $A = (bh) \div 2$

3. Choose the number that would be reasonable to calculate using mental math.
   197 + ___
   A. 13
   B. 95
   C. 42
   D. 16

4. Subtract.
   37.04 − 5.65
   A. 21.39
   B. 31.39
   C. 32.09
   D. 31.29

5. Choose the best pair of compatible numbers.
   223 + 625
   A. 223 + 630
   B. 250 + 650
   C. 200 + 600
   D. 225 + 625

6. Add.
   56.09 + 22.42
   A. 80.51
   B. 79.51
   C. 78.51
   D. 80.41

7. Multiply. Use mental math.
   8 × 42
   A. 320
   B. 336
   C. 332
   D. 316

8. Divide. Round to the nearest hundredth.
   6.40 ÷ 0.86
   A. 7.44
   B. 7.45
   C. 7.50
   D. 7.51

9. Estimate the likely temperature of a heated swimming pool.
   A. 50°F
   B. 92°F
   C. 110°C
   D. 26°C

10. Find the greatest common factor.
    12, 46
    A. 2
    B. 4
    C. 6
    D. 12

11. Change the unit.
    3.5 T = ___ lb
    A. 6500
    B. 5000
    C. 5500
    D. 7000

12. Write using exponents.
    16 squared
    A. 256^2
    B. 256
    C. 16^2
    D. 16^4

13. Compare. Use <, >, or =.
    $\frac{3}{8}$ □ $\frac{5}{6}$
    A. =
    B. <
    C. >

14. Multiply.
    229 × 63
    A. 14,490
    B. 14,467
    C. 14,497
    D. 14,427

15. Read the gauge. Estimate what part of the water tank is filled.
    A. $\frac{5}{6}$ B. $\frac{2}{4}$
    C. $\frac{1}{3}$ D. $\frac{6}{10}$

16. Write the fraction as a decimal.
    $3\frac{4}{5}$
    A. 4.5
    B. 3.8
    C. 4.8
    D. 3.5

17. Write the missing numeral.
    $\frac{5}{6} = \frac{\square}{24}$
    A. 25
    B. 20
    C. 30
    D. 10

18. Suppose you had $350. Which television could you buy?
    A. #2396
    B. #5604
    C. #1247
    D. #8930

TV Half-Price Sale	
Model #	Regular Price
2396	$ 895
5604	$1050
1247	$ 760
8930	$ 690

19. Round to the nearest dollar.
    $86.53
    A. $90
    B. $86.50
    C. $85
    D. $87

20. Maria bought 2 pairs of shoes for $23 each and a blouse for $29.95. Tax for these items was $8.60. How much did she pay all together?
    A. $83.99
    B. $61.55
    C. $84.55
    D. $63.89

21. Multiply.
    10.54 × 2.3
    A. 24.242
    B. 24.212
    C. 242.12
    D. 24.22

22. Use compatible numbers to estimate in lowest terms.
    $\frac{37}{49}$
    A. $\frac{9}{12}$
    B. $\frac{4}{5}$
    C. $\frac{7}{9}$
    D. $\frac{12}{25}$

23. Write as an improper fraction.
    $2\frac{3}{16}$
    A. $\frac{35}{16}$
    B. $\frac{23}{16}$
    C. $\frac{32}{16}$
    D. $\frac{5}{16}$

24. Change the unit.
    5.12 g = ___ mg
    A. 512
    B. 5120
    C. 0.0512
    D. 51.2

25. Divide.
    29.85 ÷ 15
    A. 0.19
    B. 109
    C. 1.99
    D. 19.9

26. Write as a fraction.
    0.215
    A. $\frac{215}{500}$ B. $\frac{215}{100}$
    C. $\frac{175}{500}$ D. $\frac{43}{200}$

27. Measure to the nearest centimeter.
    A. 29 cm
    B. 2.5 cm
    C. 3 cm
    D. 0.3 cm

FRACTIONS

Chapter 8 Overview

Key Ideas

- Use the problem-solving strategy *use logical reasoning*.
- Multiply and divide fractions and mixed numbers.
- Find the cost of buying in installments.
- Estimate using compatible numbers.
- Develop a plan to solve problems.
- Find overtime rates.

Key Terms

- logical reasoning
- reciprocal
- numerator
- compatible numbers
- denominator
- overtime
- installment

Key Skills

Multiply.

1. 6×5
2. 2×7
3. 8×6
4. 2×2
5. 9×9
6. 4×8
7. 6×7
8. 5×6
9. 3×4
10. 6×9
11. 8×7
12. 4×9
13. 8×3
14. 8×2
15. 6×6
16. 5×3
17. 8×9
18. 8×8

Divide.

19. $20 \div 4$
20. $24 \div 8$
21. $16 \div 4$
22. $35 \div 7$
23. $32 \div 4$
24. $48 \div 6$
25. $42 \div 7$
26. $25 \div 5$
27. $30 \div 6$
28. $40 \div 8$
29. $24 \div 4$
30. $28 \div 4$
31. $63 \div 7$
32. $81 \div 9$
33. $72 \div 9$
34. $56 \div 7$
35. $21 \div 7$
36. $36 \div 6$

Write each mixed number as an improper fraction.

37. $3\frac{1}{2}$
38. $2\frac{1}{3}$
39. $5\frac{1}{6}$
40. $3\frac{2}{5}$
41. $1\frac{2}{3}$
42. $4\frac{3}{4}$
43. $8\frac{1}{10}$
44. $2\frac{7}{12}$
45. $5\frac{4}{5}$
46. $1\frac{9}{10}$
47. $3\frac{1}{7}$
48. $9\frac{5}{8}$

MULTIPLICATION AND DIVISION OF FRACTIONS

8

8-1 Problem Solving: Learning to Use Strategies

Objective
Use the strategy use logical reasoning.

- **SITUATION**
- **DATA**
- **PLAN**
- **ANSWER**
- **CHECK**

Belinda, Rico, Glenn, and Darlene have four different summer jobs. No person's place of employment starts with the same letter as the person's name. Belinda gets a discount on clothes. Rico is the brother of the gas station attendant. Where does each person work?

department store

gas station

bookstore

restaurant

You cannot solve some problems using addition, subtraction, multiplication, or division. The strategy *use logical reasoning* can help you to solve some of these types of problems.

Example Solve.

Where does each person work?

Make a chart. Use the given data to fill in parts of the chart.

	Belinda	Rico	Glenn	Darlene
Department store	yes			no
Gas station		no	no	
Bookstore	no			
Restaurant		no		

Use logical reasoning to fill in more of the chart.

	Belinda	Rico	Glenn	Darlene
Department store	yes	no	no	no
Gas station	no	no	no	yes
Bookstore	no	yes		
Restaurant	no	no		

Complete the chart.

	Belinda	Rico	Glenn	Darlene
Department store	yes	no	no	no
Gas station	no	no	no	yes
Bookstore	no	yes	no	no
Restaurant	no	no	yes	no

Belinda works at a department store. Rico works at a bookstore. Glenn works at a restaurant. Darlene works at a gas station.

214 CHAPTER 8

Try This Solve.

Torr, Budd, Constance, and Dee each owns a different pet: a horse, a rabbit, a cat, and a dog. Budd does not feed his pet hay. Dee's pet catches mice. Constance feeds her pet lettuce. Which pet does each person own?

Exercises

Solve. Use one or more of the problem-solving strategies.

Problem-Solving Strategies
- Guess and check.
- Choose an operation.
- Make an organized list.
- Act it out.
- Find a pattern.
- Write an equation.
- **Use logical reasoning.**
- Draw a picture or diagram.
- Make a table.
- Use objects.
- Work backward.
- Solve a simpler problem.

1. Ramon has four coins in his pocket: a penny, a nickel, a dime, and a quarter. He pulls one coin out of his pocket at a time. There is only one coin larger in size than the third coin. The fourth coin is larger in size than only one coin. The first coin is five times the value of the third coin. In what order did Ramon pull the coins out of his pocket?

2. At the grocery store, Chad made a display of apples in a triangular pyramid. The first, or bottom, layer had 28 apples. The second layer had 21 apples. The third layer had 15 apples. If this pattern continued, how many apples would be in the fifth layer?

3. Natalie and three friends decided to make wristbands. They set up an assembly line with 4 stations. Each station took about 2 min. About how long did it take to make 6 wristbands?

4. Elisa delivers pizzas. The first hour she delivered one pizza. The second hour she delivered 2 pizzas. The third hour she delivered 4 pizzas. The fourth hour she delivered 8 pizzas. At this rate, how many pizzas would Elisa deliver the sixth hour?

5. Donna, Sara, Dayton, and Basil have part-time jobs. Each works on a different weekday, Monday through Thursday. Donna works before Sara. There are two full days between Sara's and Basil's workdays. Only one person works before Dayton. What day does each person work?

Suppose 6. Suppose in the example on p. 214 that Belinda received a discount on books. Darlene wore rubber gloves at work. Rico did not wear a tie to work. Where would each person work now?

Write Your Own Problem 7. Write a problem that you could solve using any of the problem-solving strategies. Use the information in this menu.

Brookside Cafe

Hamburger	$4.25
Hot dog	$3.10
Grilled Cheese	$2.95

MULTIPLICATION AND DIVISION OF FRACTIONS

8-2 Multiplying Fractions

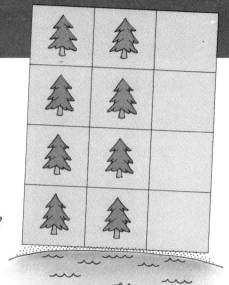

Objective
Multiply fractions.

A plan for a new resort shows that the resort is divided into 12 equal lots. Three of the lots are on the lakefront. Woods cover 8 lots. What fraction of the resort is wooded and on the lakefront?

Understand the Situation
- Into how many lots is the resort divided?
- What fraction of the resort is on the lakefront?
- What fraction of the resort is wooded?

To find $\frac{2}{3}$ of $\frac{1}{4}$, multiply.

Examples Multiply.

1. $\frac{2}{3} \times \frac{1}{4}$

Multiply the numerators. → Multiply the denominators. → Write in lowest terms.

$\frac{2}{3} \times \frac{1}{4} = \frac{2}{\square}$ $\qquad$ $\frac{2}{3} \times \frac{1}{4} = \frac{2}{12}$ $\qquad$ $\frac{2}{12} = \frac{2 \div 2}{12 \div 2} = \frac{1}{6}$

Two twelfths, or $\frac{1}{6}$, of the resort is wooded and on the lakefront.

2. $\frac{9}{10} \times \frac{3}{5} = \frac{27}{50}$ $\qquad$ **3.** $\frac{1}{5} \times 6 = \frac{1}{5} \times \frac{6}{1} = \frac{6}{5}$, or $1\frac{1}{5}$ $\qquad$ **4.** $\frac{1}{3} \times \frac{1}{2} \times \frac{1}{5} = \frac{1}{30}$

A short method helps to make some problems easier to solve. If you can, divide a numerator and a denominator by a common factor *before multiplying*.

5. $\frac{3}{8} \times \frac{5}{6}$ $\qquad\qquad$ **6.** $\frac{3}{5} \times 5$

$\frac{\overset{1}{\cancel{3}}}{8} \times \frac{5}{\underset{2}{\cancel{6}}} = \frac{5}{16}$ *Divide by 3.* $\qquad$ $\frac{3}{5} \times 5 = \frac{3}{\underset{1}{\cancel{5}}} \times \frac{\overset{1}{\cancel{5}}}{1} = 3$ *Divide by 5.*

Try This Multiply.

a. $\frac{1}{2} \times \frac{1}{3}$ $\qquad$ **b.** $\frac{5}{12} \times \frac{6}{5}$ $\qquad$ **c.** $\frac{3}{10} \times 20$ $\qquad$ **d.** $\frac{3}{4} \times \frac{4}{3}$ $\qquad$ **e.** $\frac{2}{3} \times \frac{3}{8} \times \frac{1}{4}$

Exercises

Multiply.

1. $\frac{1}{2} \times \frac{3}{4}$
2. $\frac{1}{5} \times \frac{2}{3}$
3. $\frac{7}{4} \times \frac{1}{7}$
4. $\frac{5}{4} \times \frac{1}{6}$
5. $\frac{4}{5} \times \frac{10}{11}$
6. $\frac{4}{9} \times \frac{3}{7}$
7. $\frac{3}{8} \times \frac{9}{10}$
8. $\frac{4}{3} \times \frac{9}{16}$
9. $\frac{3}{5} \times \frac{10}{9}$
10. $\frac{7}{9} \times \frac{27}{49}$
11. $\frac{5}{7} \times 1$
12. $\frac{3}{4} \times 0$
13. $\frac{1}{3} \times 6$
14. $8 \times \frac{1}{4}$
15. $\frac{7}{12} \times 18$
16. $\frac{8}{9} \times 9$
17. $4 \times \frac{7}{4}$
18. $24 \times \frac{9}{16}$
19. $\frac{5}{6} \times 8$
20. $6 \times \frac{5}{36}$
21. $\frac{3}{4} \times \frac{3}{4}$
22. $\frac{2}{3} \times \frac{2}{3}$
23. $\frac{7}{8} \times \frac{7}{8}$
24. $\frac{11}{12} \times \frac{11}{12}$
25. $\frac{3}{5} \times \frac{3}{5}$
26. $\frac{1}{2} \times \frac{1}{3} \times \frac{1}{6}$
27. $\frac{3}{4} \times \frac{1}{9} \times \frac{2}{3}$
28. $\frac{4}{5} \times \frac{2}{7} \times \frac{5}{3}$
29. $\frac{3}{5} \times \frac{1}{3} \times 4$
30. $\frac{8}{7} \times \frac{1}{4} \times 14$
31. $6 \times \frac{2}{5} \times \frac{10}{3}$
32. $\frac{7}{12} \times 8 \times \frac{3}{14}$
33. $\frac{1}{4} \times 9 \times \frac{1}{6}$
34. $\frac{3}{4}$ of $\frac{6}{7}$
35. $\frac{2}{3}$ of $\frac{9}{10}$
36. $\frac{1}{6}$ of $\frac{5}{6}$
37. $\frac{1}{5}$ of $\frac{3}{5}$
38. $\frac{2}{7}$ of $\frac{1}{4}$

Number Sense

 shows the number sentence $18 \times \frac{2}{3} \times \frac{1}{2} = 6$.

Write a number sentence for the diagram.

39.
40.
41.
42.

Mixed Applications

Solve.

43. A landscaping service planted 1200 trees in the new resort. How many trees of each kind did the landscaping service plant?

44. The Department of Forestry planted 450 dogwood trees around the lake. Two thirds of the trees were white. How many trees were not white?

Suppose

45. Suppose that the resort added 4 more lots. One lot was on the lakefront. None were wooded. Now what fraction of the resort is wooded and on the lakefront?

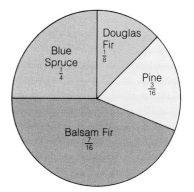

MULTIPLICATION AND DIVISION OF FRACTIONS

8-3 Multiplying with Mixed Numbers

Objective
Multiply with mixed numbers.

Jessica Liu is making curtains for her room. In her home economics class, Jessica learned to multiply the width of the window by 3 to allow for fullness in both sides of the curtain. What should the total width of both sides of the curtain be?

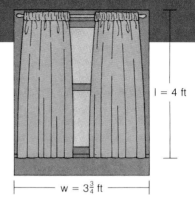

$l = 4$ ft

$w = 3\frac{3}{4}$ ft

Understand the Situation
- How wide is the window?
- By what number should Jessica multiply the width of the window?
- Is Jessica finding the fullness needed for both sides of the curtain?

To find the total width, multiply the width of the window by 3. You can estimate the product. $3\frac{3}{4} \times 3$ is about 4×3, or 12.

Examples Multiply.

1. $3\frac{3}{4} \times 3$

Write each number as a fraction. → Multiply the fractions.

$3\frac{3}{4} \times 3 = \frac{15}{4} \times \frac{3}{1}$ $\frac{15}{4} \times \frac{3}{1} = \frac{45}{4} = 11\frac{1}{4}$

Jessica should make each pair of curtains $11\frac{1}{4}$ feet wide.

The answer seems reasonable. $11\frac{1}{4}$ is close to the estimate, 12.

2. $\frac{1}{4} \times 1\frac{2}{3} = \frac{1}{4} \times \frac{5}{3} = \frac{5}{12}$

3. $9 \times 1\frac{5}{6} = \overset{3}{\cancel{\frac{9}{1}}} \times \frac{11}{\underset{2}{\cancel{6}}} = \frac{33}{2} = 16\frac{1}{2}$

4. $5\frac{1}{3} \times 1\frac{1}{8} = \overset{2}{\cancel{\underset{1}{\frac{16}{3}}}} \times \overset{3}{\cancel{\underset{1}{\frac{9}{8}}}} = 6$

5. $1\frac{3}{5} \times 1\frac{3}{5} = \frac{8}{5} \times \frac{8}{5} = \frac{64}{25} = 2\frac{14}{25}$

Try This Multiply.

a. $\frac{4}{5} \times 1\frac{1}{4}$ b. $2\frac{2}{3} \times \frac{11}{12}$ c. $1\frac{1}{6} \times 18$ d. $4\frac{2}{7} \times 1\frac{1}{5}$ e. $2\frac{5}{8} \times \frac{3}{7} \times \frac{1}{3}$

CHAPTER 8

Exercises

Multiply.

1. $1\frac{1}{8} \times \frac{3}{5}$
2. $\frac{7}{18} \times 2\frac{1}{2}$
3. $3\frac{3}{4} \times \frac{7}{8}$
4. $\frac{5}{9} \times 1\frac{4}{7}$
5. $2\frac{1}{3} \times \frac{9}{10}$
6. $4 \times 3\frac{1}{2}$
7. $2\frac{2}{7} \times 14$
8. $3 \times 2\frac{4}{5}$
9. $1\frac{1}{3} \times 9$
10. $4 \times 1\frac{3}{8}$
11. $2\frac{1}{2} \times 1\frac{3}{4}$
12. $3\frac{1}{3} \times 4\frac{1}{3}$
13. $6\frac{5}{12} \times 1\frac{1}{2}$
14. $4\frac{1}{6} \times 2\frac{2}{5}$
15. $3\frac{1}{8} \times 2\frac{2}{5}$
16. $1\frac{1}{4} \times 3\frac{3}{8}$
17. $2\frac{2}{3} \times 1\frac{3}{16}$
18. $2\frac{5}{6} \times 2\frac{3}{5}$
19. $3\frac{3}{5} \times 1\frac{1}{9}$
20. $1\frac{5}{6} \times 1\frac{1}{11}$
21. $2\frac{1}{9} \times \frac{3}{14}$
22. $\frac{4}{7} \times 1\frac{3}{4}$
23. $\frac{24}{13} \times 2\frac{1}{6}$
24. $1\frac{7}{8} \times \frac{18}{11}$
25. $\frac{5}{7} \times 1\frac{3}{10}$
26. $2\frac{2}{3} \times \frac{3}{10} \times 5$
27. $3\frac{1}{4} \times 9 \times \frac{1}{6}$
28. $3\frac{3}{5} \times 3\frac{1}{3} \times 4$
29. $\frac{4}{3} \times 2\frac{3}{4} \times 3$
30. $7\frac{1}{2} \times 10 \times 1\frac{2}{3}$
31. $4\frac{9}{13} \times 8\frac{1}{2} \times 0$
32. $3\frac{2}{5} \times 1 \times 4\frac{1}{17}$
33. $\frac{5}{4} \times 2 \times 3\frac{1}{2}$
34. $\frac{3}{4}$ of $1\frac{1}{2}$
35. $\frac{1}{5}$ of $2\frac{1}{3}$
36. $\frac{3}{5}$ of $1\frac{1}{4}$
37. $\frac{5}{6}$ of $1\frac{2}{3}$
38. $\frac{5}{3}$ of $2\frac{2}{5}$

Show Math

Use a ruler for Exercises 39–41.

Line segment AB is $2\frac{1}{2}$ in. long.   A •————————————————————• B

39. Draw a line segment $\frac{1}{2}$ as long as AB. Label the segment EF.
40. Draw a line segment $1\frac{1}{2}$ as long as AB. Label the segment GH.
41. Draw a line segment $\frac{3}{4}$ as long as AB. Label the segment IJ.

Mixed Applications

Jessica decided to make a dust ruffle to cover three sides of her bed. The clerk at the fabric store told her how to find the amount of fabric needed for fullness. Add the lengths of the sides to be covered. Multiply the sum by $2\frac{1}{2}$.

42. What measurement should Jessica use for a dust ruffle to cover both lengths and one width of her bed?
43. Find the measurement of a dust ruffle to cover both lengths and one width of a bed that is 7 ft by 78 in.

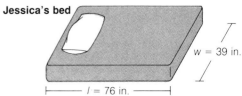

Jessica's bed
$w = 39$ in.
$l = 76$ in.

Mixed Skills Review

Estimate.

44. 223×61
45. $482 + 356$
46. $5802 - 3293$
47. $5422 \div 89$
48. $3533 \div 69$
49. $6993 - 3241$
50. $623 + 481$
51. 414×53

Change the unit.

52. 63 m = _____ cm
53. 4 kg = _____ g
54. 8 mL = _____ L
55. 9.2 cm = _____ m

8-4 Consumer Math: Installment Buying

Objective
Find the cost of installment buying.

- SITUATION
- DATA
- PLAN
- ANSWER
- CHECK

On an **installment plan**, a customer pays part of the cost when buying the item and a certain amount of the remaining cost plus interest each month. The **down payment** is the amount paid when buying the item.

Sid's Cycles Installment Plan				
Motorcycle	List Price	Down Payment	Monthly Payment	Terms in Months
Quad ATV	$1600	$\frac{1}{4}$	$72.09	18
Maxim 700	$2400	$\frac{1}{3}$	$140.66	12
Sportster	$2800	$\frac{1}{5}$	$105.70	24

The cost of an item on an installment plan is the sum of the down payment and the monthly payments. A finance charge is included in the monthly fee.

Example Find the cost of the Maxim 700 on the installment plan.

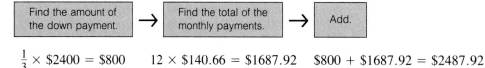

$\frac{1}{3} \times \$2400 = \$800$ $12 \times \$140.66 = \$1687.92$ $\$800 + \$1687.92 = \$2487.92$

The cost of the motorcycle on the installment plan is $2487.92.

Try This Find the cost of each motorcycle on the installment plan.
a. Sportster   **b.** Quad ATV

Exercises

Find the cost of each motorcycle on the installment plan.

	Motorcycle	List Price	Down Payment	Monthly Payment	Terms in Months
1.	650 Tempter	$2100	$\frac{1}{4}$	$94.59	18
2.	Chopper	$1800	$\frac{1}{5}$	$126.72	12
3.	700 Intruder	$3600	$\frac{1}{3}$	$111.58	24
4.	Sportster	$2800	$\frac{1}{4}$	$69.50	36

8-5 Reciprocals

Objective
Find the reciprocal.

Two numbers whose product is 1 are **reciprocals** of each other.

$\frac{3}{4} \times \frac{4}{3} = 1$ $\frac{3}{4}$ and $\frac{4}{3}$ are reciprocals.

$1\frac{2}{3} \times \frac{3}{5} = \frac{5}{3} \times \frac{3}{5} = 1$ $1\frac{2}{3}$ and $\frac{3}{5}$ are reciprocals.

$1 \times 1 = 1$ The number 1 is its own reciprocal.

To find the reciprocal of a fraction, invert, or reverse, the positions of the numerator and the denominator. To find the reciprocal of a mixed number, first change it to an improper fraction.

Examples Find the reciprocal.

1. $\frac{7}{10}$ $\frac{7}{10} \bowtie \frac{10}{7}$ 2. 5 $5 = \frac{5}{1} \bowtie \frac{1}{5}$ 3. $1\frac{3}{4}$ $1\frac{3}{4} = \frac{7}{4} \bowtie \frac{4}{7}$

Try This Find the reciprocal.

a. $\frac{2}{3}$ b. 21 c. $\frac{5}{8}$ d. $3\frac{1}{2}$ e. 1 f. $5\frac{3}{4}$

Exercises

Find the reciprocal.

1. $\frac{9}{10}$ 2. $\frac{4}{5}$ 3. $\frac{8}{9}$ 4. $\frac{7}{4}$ 5. 3 6. 17

7. $\frac{11}{16}$ 8. $\frac{13}{24}$ 9. $1\frac{3}{5}$ 10. $3\frac{2}{3}$ 11. $2\frac{5}{6}$ 12. $4\frac{11}{20}$

Show Math 13. Copy the number line. Mark and label the reciprocal of each point.

Number Sense

14. Is the reciprocal of a whole number always less than the whole number?
15. When is the reciprocal of a number greater than that number?
16. How would you find the reciprocal of 0.5?
17. Is there a reciprocal of 0? Explain why or why not.

8-6 Dividing Fractions

Objective
Divide fractions.

Dwight has 2 small pizzas. He serves $\frac{1}{4}$ of a pizza to each friend. How many pieces of pizza are there all together?

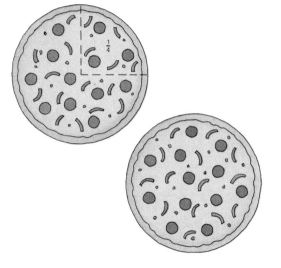

Understand the Situation
- How many pizzas are there?
- What fraction of a pizza does Dwight serve to each friend?

You can divide to find the number of pieces.

Examples Divide.

1. $2 \div \frac{1}{4}$ *Think: 2 pizzas divided into $\frac{1}{4}$s. How many $\frac{1}{4}$s are in 2?*

Find the reciprocal of the divisor. → Multiply by the reciprocal of the divisor.

$\frac{1}{4}$ is the divisor. $2 \div \frac{1}{4} = 2 \times \frac{4}{1} = \frac{2}{1} \times \frac{4}{1} = 8$
The reciprocal of $\frac{1}{4}$ is $\frac{4}{1}$.
There are 8 pieces of pizza.

2. $\frac{3}{4} \div \frac{7}{8} = \frac{3}{4} \times \frac{\overset{2}{8}}{7} = \frac{6}{7}$ **3.** $\frac{5}{6} \div 3 = \frac{5}{6} \times \frac{1}{3} = \frac{5}{18}$ **4.** $2 \div \frac{1}{3} = 2 \times 3 = 6$

5. $9 \div \frac{3}{2} = \frac{\overset{3}{9}}{1} \times \frac{2}{3} = 6$ **6.** $\frac{4}{3} \div \frac{4}{3} = \frac{\overset{1}{4}}{3} \times \frac{\overset{1}{3}}{4} = 1$ **7.** $0 \div \frac{71}{100} = 0$

Try This   Divide.

a. $\frac{7}{8} \div \frac{4}{5}$ **b.** $\frac{3}{4} \div \frac{1}{2}$ **c.** $\frac{5}{12} \div 10$ **d.** $3 \div \frac{1}{4}$ **e.** $\frac{3}{2} \div \frac{2}{3}$

f. $\frac{1}{6} \div 12$ **g.** $\frac{5}{4} \div \frac{5}{4}$ **h.** $\frac{1}{4} \div \frac{1}{16}$ **i.** $14 \div \frac{7}{12}$ **j.** $100 \div \frac{1}{50}$

Exercises

Divide.

1. $\frac{1}{3} \div \frac{1}{2}$
2. $\frac{1}{6} \div \frac{1}{4}$
3. $\frac{2}{3} \div \frac{1}{3}$
4. $\frac{1}{5} \div \frac{2}{5}$
5. $\frac{1}{4} \div \frac{1}{8}$
6. $\frac{2}{7} \div \frac{3}{7}$
7. $\frac{3}{7} \div \frac{2}{7}$
8. $\frac{4}{9} \div \frac{5}{9}$
9. $\frac{5}{9} \div \frac{4}{9}$
10. $\frac{6}{11} \div \frac{6}{11}$
11. $\frac{5}{12} \div \frac{5}{9}$
12. $\frac{1}{6} \div \frac{3}{4}$
13. $\frac{5}{8} \div \frac{7}{8}$
14. $\frac{1}{3} \div \frac{5}{9}$
15. $\frac{3}{4} \div \frac{5}{3}$
16. $\frac{3}{4} \div \frac{7}{12}$
17. $\frac{11}{15} \div \frac{11}{15}$
18. $\frac{7}{10} \div \frac{2}{5}$
19. $\frac{5}{6} \div \frac{10}{12}$
20. $\frac{12}{15} \div \frac{5}{8}$
21. $2 \div \frac{1}{4}$
22. $7 \div \frac{14}{9}$
23. $3 \div \frac{6}{7}$
24. $10 \div \frac{5}{2}$
25. $6 \div \frac{9}{7}$
26. $\frac{3}{10} \div 6$
27. $\frac{5}{12} \div 8$
28. $\frac{4}{5} \div 7$
29. $\frac{12}{11} \div 5$
30. $\frac{16}{21} \div 24$
31. $\frac{11}{36} \div \frac{2}{3}$
32. $\frac{3}{8} \div \frac{9}{10}$
33. $\frac{21}{40} \div \frac{7}{8}$
34. $\frac{8}{9} \div \frac{12}{11}$
35. $\frac{14}{15} \div 21$
36. $12 \div \frac{5}{7}$
37. $\frac{21}{25} \div \frac{13}{20}$
38. $\frac{35}{40} \div \frac{7}{8}$
39. $\frac{17}{36} \div \frac{15}{36}$
40. $\frac{7}{10} \div \frac{7}{100}$

Show Math Draw a picture for the problem. Then solve the problem.

41. How many $\frac{1}{3}$s are in 2?
42. How many $\frac{1}{2}$s are in 3?
43. How many $\frac{1}{8}$s are in $\frac{3}{4}$?
44. How many $\frac{1}{6}$s are in $\frac{4}{3}$?

Mixed Applications

45. A bead is $\frac{1}{2}$ in. long. How many beads do you need to make a bracelet 6 in. long? a necklace 18 in. long?
46. Each loop in a spring takes $\frac{3}{8}$ in. of wire. How many loops can you make from 60 in. of wire?

Problem Solving

47. A highway crew placed reflector lights at the beginning of the highway and every $\frac{2}{3}$ mi after that. How many lights are needed for a strip 8 mi long?
48. Lucia can program her CD player to play songs in any order. In how many ways can she program the CD player to play 4 songs?

Mixed Skills Review

Find the answer. Use mental math.

49. $50 - 3$
50. 33×8
51. $248 - 50$
52. $327 + 125$
53. $425 + 440$
54. $\$13.25 - \$10.75$
55. 9×62
56. 202×6

Compare. Write <, >, or =.

57. $0.45 \square 0.405$
58. $\frac{1}{2} \square \frac{2}{5}$
59. $7.843 \square 78.43$
60. $1\frac{2}{3} \square 1\frac{4}{6}$
61. $1284 \square 1428$
62. $3\frac{5}{10} \square 2\frac{5}{10}$

MULTIPLICATION AND DIVISION OF FRACTIONS

8-7 Dividing with Mixed Numbers

Objective
Divide with mixed numbers.

Cindy Shock owns a small business that makes sportswear for men, women, and children. Each week she plans the schedule for her four employees based on the number of orders she has received.

Cindy's Closet, Inc. Weekly Orders

Item	Sewing Time	Number of Orders
Skirt	$\frac{3}{4}$ h	40
Shirt	$1\frac{1}{2}$ h	32
Slacks	$1\frac{2}{3}$ h	36
Jacket	$2\frac{1}{4}$ h	18

Exercises

1. How much time is needed for an employee to make one skirt?

2. How many skirts can one employee make in 3 hours?

Hours	1	2	3	4	5	6	7	8
Shirts		1		2		3		4

3. Cindy uses a chart to show the schedule. How many shirts can an employee make in 6 hours?

4. Cindy can divide 6 by $1\frac{1}{2}$ to find the exact number. Her answer is 4. Is this answer correct?

$$6 \div 1\frac{1}{2} = \frac{6}{1} \div \frac{3}{2}$$
$$= \frac{6}{1} \times \frac{2}{3} = \frac{12}{3} = 4$$

5. One working day is 8 hours. Estimate how many shirts one employee can make in one working day.

6. Jim said one employee could make 5 shirts in 8 hours. Eileen said the answer was $5\frac{1}{3}$. Find the exact answer. Who was correct?

7. Draw a chart to show the number of jackets that one employee could make in 8 hours.

8. Find the exact number of jackets that one employee could make in 8 hours.

9. There are $12\frac{3}{4}$ working hours left in the week. Is it enough time for one employee to make 6 jackets?

10. One employee went home sick and missed $6\frac{1}{2}$ hours of work. How many pairs of slacks could the employee have made in this amount of time?

11. How many hours will it take to make all of the slacks orders?

12. How many hours will it take to make all the jacket orders?

13. Suppose Cindy received 8 more orders for shirts. How many hours will it take to make all the shirt orders?

Decide if the quotient is correct. If it is not correct, explain why not.

14. $6 \div 1\frac{1}{2} = 4$
15. $2\frac{2}{3} \div \frac{5}{8} = 1\frac{2}{3}$
16. $1\frac{3}{10} \div 1\frac{4}{5} = \frac{13}{18}$

$\frac{6}{1} \div \frac{3}{2} = \frac{\overset{2}{\cancel{6}}}{1} \times \frac{2}{\underset{1}{\cancel{3}}} = 4$

$\frac{8}{3} \div \frac{5}{8} = \frac{8}{3} \times \frac{5}{8} = \frac{5}{3} = 1\frac{2}{3}$

$\frac{13}{10} \div \frac{9}{5} = \frac{13}{\underset{2}{\cancel{10}}} \times \frac{\overset{1}{\cancel{5}}}{9} = \frac{13}{18}$

Divide.

17. $5 \div 2\frac{1}{2}$
18. $3 \div 1\frac{3}{4}$
19. $4\frac{1}{2} \div 2$
20. $3 \div 1\frac{2}{3}$

21. $6\frac{1}{3} \div 4$
22. $3\frac{5}{6} \div 1\frac{2}{3}$
23. $3\frac{3}{8} \div 1\frac{3}{4}$
24. $2\frac{1}{4} \div 2\frac{2}{3}$

25. $2 \div 1\frac{2}{3}$
26. $6 \div 2\frac{1}{3}$
27. $1\frac{3}{4} \div \frac{3}{5}$
28. $2\frac{1}{4} \div \frac{5}{8}$

29. $\frac{2}{3} \div 1\frac{3}{4}$
30. $\frac{7}{8} \div 2\frac{1}{2}$
31. $1\frac{7}{10} \div 5$
32. $2\frac{5}{8} \div 7$

33. $1\frac{5}{6} \div 6$
34. $4\frac{2}{5} \div 11$
35. $3\frac{3}{4} \div 1\frac{4}{5}$
36. $4\frac{1}{6} \div 1\frac{9}{10}$

Show Math Draw a picture for the problem.

37. $3 \div 1\frac{1}{2}$
38. $6 \div 2\frac{1}{4}$
39. $4\frac{1}{3} \div 1\frac{2}{3}$
40. $5\frac{3}{4} \div 2$

Mixed Applications Solve. Use the chart on p. 224.

41. There are $18\frac{1}{2}$ working hours left in the week. Is this enough time for one of Cindy's employees to make 7 jackets?

42. One of the employees agreed to work $6\frac{3}{4}$ hours of overtime to finish a rush order for 3 shirts. Is this enough time?

43. Decide if Cindy's four employees can finish all the orders for the week. Make a table to show your results. (Assume a 40-hour week.)

Write Math 44. Make a flowchart of the steps used when dividing with mixed numbers.

Working with Variables Solve and check.

$\frac{x}{10} = 2$ You can undo the action of dividing by multiplying. To get the variable alone on one side, multiply by 10.

$\frac{x}{10} \times 10 = 2 \times 10$ Remember to use the same action on both sides of an equation.

$x = 20$ $\frac{20}{10} = 2$ Check your answer by substituting it for the variable. The solution is 20.

Solve and check.

45. $\frac{p}{6} = 3$
46. $\frac{y}{9} = 4$
47. $\frac{x}{25} = \frac{1}{5}$

48. $\frac{a}{14} = \frac{2}{7}$
49. $\frac{1}{12} = \frac{k}{4}$
50. $\frac{3}{16} = \frac{t}{8}$

MULTIPLICATION AND DIVISION OF FRACTIONS

8-8 Estimating Products and Quotients Using Compatible Numbers

Objective
Estimate products and quotients using compatible numbers.

The maintenance department is putting new storage racks in the equipment room at Towson High School. About how long a rack will hold 16 basketballs?

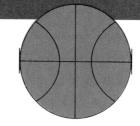

Basketball diameter = $9\frac{4}{17}$ in.

Baseball diameter = $2\frac{7}{23}$ in.

Understand the Situation
- What is the diameter of a basketball?
- What operation will you use to answer the question?

Examples

1. Estimate the length of a rack that will hold the basketballs.
   Estimate the product of 16 and $9\frac{4}{17}$. Use compatible numbers.

   $9\frac{4}{17}$ is about 10.   *Think of a number that is close in value and that is easier to use.*

   $16 \times 10 = 160$ in.   A rack about 160 in. long will hold all of the basketballs.

2. Estimate the number of baseballs that will fit on a rack that is 32 in. long.
   Estimate $32 \div 2\frac{7}{23}$.

   $2\frac{7}{23}$ is about 2.   *Think of a compatible number.*

   $32 \div 2 = 16$ The rack will hold about 16 baseballs.

Try This Estimate. Use compatible numbers.

a. $\frac{10}{21} \times 32$ b. $1\frac{30}{32} \times 18$ c. $27 \div \frac{11}{24}$ d. $36 \div 1\frac{37}{40}$ e. $24 \times \frac{5}{11}$

Exercises

Estimate. Use compatible numbers.

1. $\frac{3}{7} \times 16$ 2. $\frac{18}{19} \times 24$ 3. $9 \div \frac{15}{32}$ 4. $32 \div \frac{7}{16}$ 5. $6\frac{30}{39} \times 22$

Data Bank Use the Data Bank on p. 490. Solve.

6. Estimate the number of tennis balls that fit on a rack 23 in. long.
7. Estimate the length of a rack needed for 15 softballs.
8. Estimate the length of a rack needed for 20 soccer balls.

8-9 Problem Solving: Developing a Plan

Objective
Decide if an estimate or an exact answer is needed. Choose a calculation method.

- SITUATION
- DATA
- **PLAN**
- ANSWER
- CHECK

OVER _____ ATTEND FIRST FOUR GAMES

What number might a newspaper reporter place in the headline?

Example Tell if the problem needs an exact answer or an estimate. Choose a calculation method. Solve the problem.

Exact answer or estimate: *A newspaper headline usually reports the approximate numbers of people that attend games. You could estimate the total. Round each number to the nearest thousand and add.*
23,000 + 28,000 + 34,000 + 27,000 = ?

Calculation method: *The rounded numbers are still large. You could find the answer using a calculator.* 23,000 + 28,000 + 34,000 + 27,000 =

> 112000.

A newspaper reporter might write 112,000.

Exercises

Tell if the problem needs an exact answer or an estimate. Choose a calculation method. Solve the problem.

1. Darian bought a jacket for $54. She gave the clerk a $100 bill. How much change should Darian receive?

2. A pair of shorts cost $8.95. Jolene wants to buy 5 pairs. She has $44. Does she have enough money?

3. Jeremiah bought 2 ties and 4 shirts. He wrote a check for the total amount. For how much money did he write the check?

4. You earn $4.95 an hour. How long would it take you to make enough money to buy 3 sweaters?

Clothing Sale
Pants	$24.95
Shirts	$15
Sweaters	$35
Ties	$4
Shoes	$22–45

Price with tax

Write Your Own Problem

5. Write a problem that calls for an estimate and that you might solve using mental math.

6. Write a problem that calls for an exact answer and that you might solve using a calculator.

MULTIPLICATION AND DIVISION OF FRACTIONS

8-10 Problem Solving: Using Data from a Chart

Objective
Solve problems using data from a chart.

- SITUATION
- **DATA**
- PLAN
- ANSWER
- CHECK

To solve some problems you need to use data from a chart. You can use fractions to change the recipes to serve any number of people.

Meat Loaf
Serves 8.
$1\frac{1}{2}$ lb ground beef
1 c tomato juice
$\frac{3}{4}$ c oats, uncooked
1 egg, beaten
$\frac{1}{4}$ c chopped onion
1 tsp salt
$\frac{1}{4}$ tsp pepper

Oatmeal Cookies
Makes 5 dozen.
$\frac{3}{4}$ c vegetable shortening
1 c brown sugar, packed
$\frac{1}{2}$ c granulated sugar
1 tsp vanilla
3 c oats, uncooked
1 c flour
$\frac{1}{2}$ tsp soda

Oatmeal

Servings	1	2	6
water	$\frac{3}{4}$ c	$1\frac{1}{2}$ c	4 c
salt	$\frac{1}{8}$ tsp	$\frac{1}{4}$ tsp	$\frac{3}{4}$ tsp
oats	$\frac{1}{3}$ c	$\frac{2}{3}$ c	2 c

Suppose you wanted to make only 6 servings of the meat loaf. Find what fraction of the ingredients you would use.

You plan to serve 6 people. $\frac{6}{8} = \frac{3}{4}$ You would use $\frac{3}{4}$ of each ingredient.
The recipe serves 8 people.

Find the amounts of ground beef, tomato juice, and oats you would use for 6 servings.

$1\frac{1}{2}$ lb beef $\times \frac{3}{4} = \frac{3}{2} \times \frac{3}{4} = \frac{9}{8} = 1\frac{1}{8}$ lb   tomato juice 1 c $\times \frac{3}{4} = \frac{3}{4}$ c

oats $\frac{3}{4}$ c $\times \frac{3}{4} = \frac{9}{16}$ c   Use compatible numbers to estimate this amount.

$\frac{9}{16}$ is close to $\frac{8}{16}$. Measure about $\frac{1}{2}$ c of oats.

Exercises

Write a recipe for each of the following. Tell which amounts you would likely estimate.

1. oatmeal for 4 people
2. twelve dozen oatmeal cookies
3. three dozen oatmeal cookies
4. oatmeal for 9 people
5. meat loaf for 14 people
6. meat loaf for 20 people

8-11 Mental Math: Break Apart Numbers

Objective
Use the mental math strategy break apart numbers.

The Bentley Chamber of Commerce keeps records about different jobs in the city. People who work at each of the jobs listed are paid $1\frac{1}{2}$ times the hourly rate when they work overtime.

You can solve some fraction problems by using the mental math strategy *break apart numbers*.

Employment Data: Bentley Township

Job	Regular Hours per Week	Hourly Rate
Construction worker	36	$10.40
Telephone operator	40	$ 8.00
Hospital aide	$37\frac{1}{2}$	$ 6.88
Supermarket cashier	36	$ 7.34
Assembly line worker	35	$ 9.50

Examples Find the overtime rate for the hourly rate. Use mental math.

1. a telephone operator

   To find the overtime rate, multiply the hourly rate by $1\frac{1}{2}$.

   $1 \times \$8 = \$8$ and $\frac{1}{2} \times \$8 = \$4$ *Think: Break apart $1\frac{1}{2}$ into 1 and $\frac{1}{2}$.*

   $1\frac{1}{2} \times \$8 = \$8 + \$4 = \$12$

2. a construction worker

   $1\frac{1}{2} \times \$10.40 = \$10.40 + \$5.20$
   $= \$10.40 + \$5.20 = \$15.60$

Try This Find the overtime rate for the hourly rate. Use mental math.
   a. $10   b. $8.40   c. $9.50   d. $6.70   e. $3.40   f. $6.60

Exercises

Find the overtime rate for the hourly rate. Use mental math.

1. $4   2. $18   3. $7   4. $13   5. $3.60   6. $7.40
7. $13.70   8. $11.40   9. $5.25   10. $6.88   11. $12.20   12. $7.34

Write Math 13. A telephone operator works overtime on a Sunday. The operator is paid $2\frac{1}{2}$ times the regular hourly rate. Write the steps for finding this overtime rate using mental math.

8-12 Problem Solving: Developing Thinking Skills

Objective
Use problem-solving strategies.

- SITUATION
- DATA
- PLAN
- ANSWER
- CHECK

At a Chinese restaurant, John, Sue, Lois, Amnon, and Drew were seated at a round table with a "lazy Susan" in the middle. The waiter placed a dish in front of each person except Drew. It took about one minute for each person to take food from one dish. Everyone took some food from each dish. About how long did it take for everyone to be served?

Exercises

Follow the *Thinking Actions*. Answer each question.

Try these **before** starting to write.

Thinking Actions Before
- Read the problem.
- Ask yourself questions to understand it.
- Think of possible strategies.

1. How many people were there?
2. How many dishes were there?
3. How long did it take each person to take food from one dish?
4. What strategy, or strategies, might help you to solve this problem?

Try these **during** your work.

Thinking Actions During
- Try your strategies.
- Stumped? Try answering these questions.
- Check your work.

5. How could you act out the situation to help solve the problem?
6. Could a table help you to solve the problem?
7. How long would it take everyone to take food from at least one dish?

8. Here are two correct solutions. Write the answer to the problem in a complete sentence. Name the strategies shown.

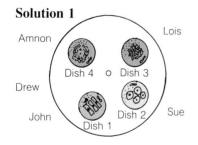

Solution 1

Amnon, Lois, Drew, John, Sue around a circle with Dish 4, Dish 3, Dish 2, Dish 1

Solution 2

	1 min	2 min	3 min	4 min	5 min
John	Dish 1	—	Dish 4	Dish 3	Dish 2
Sue	Dish 2	Dish 1	—	Dish 4	Dish 3
Lois	Dish 3	Dish 2	Dish 1	—	Dish 4
Amnon	Dish 4	Dish 3	Dish 2	Dish 1	—
Drew	—	Dish 4	Dish 3	Dish 2	Dish 1

CHAPTER 8

Solve. Use one or more of the problem-solving strategies.

Problem-Solving Strategies
- Guess and check.
- Choose an operation.
- Make an organized list.
- Act it out.
- Find a pattern.
- Write an equation.
- Use logical reasoning.
- Draw a picture or diagram.
- Make a table.
- Use objects.
- Work backward.
- Solve a simpler problem

9. To go from The Corner Deli to Paula's house, you go 3 blocks east, 2 blocks south, and 1 block east. To go from Paula's house to Bert's house, you go 1 block south, 5 blocks west, 2 blocks south, and 1 block west. Who lives closer to The Corner Deli? How many blocks closer?

10. For a club meeting, Marta bought 10 pizzas with one topping. She spent a total of $78. How many large and medium pizzas did she buy?

PIZZA—one topping
Medium $7
Large $9

11. Cleve works at Southside Sandwich Shop. When he makes sandwiches, he puts on meat, lettuce, cheese, and sprouts. In how many different orders can he put the meat, lettuce, cheese, and sprouts on the sandwiches?

12. At a school dinner, Oscar and his friends were in charge of making blended fruit drinks. They made the drinks in an assembly line. There were 4 different jobs. It took 30 s for each person's job. How long did it take to make 5 drinks?

13. Hester, Ken, Colleen, and Josh want to go out to dinner. Each wants to go to a different type of restaurant: Japanese, Mexican, Chinese, and Italian. Hester likes tempura. Colleen and Ken dislike pasta. Ken's sister in the group likes tacos. Which restaurant does each person like?

Suppose

14. Suppose in the example on p. 230 that Lydia joined her friends for dinner. They ordered one more dish. About how long would it take for everyone to be served?

Evaluating Solutions

15. Dana solved this problem. Is his solution correct? Explain why or why not.
    Problem: Sari, Warren, Lea, and Benito each ordered different vegetables: carrots, corn on the cob, peas, and green beans. Sari dislikes peas. Benito's vegetable was not green. Warren did not need his fork to eat his vegetable. What vegetable did each person order?

	Sari	Warren	Lea	Benito
Carrots	no	yes	no	no
Corn	no	no	no	yes
Peas	no	no	yes	no
Green beans	yes	no	no	no

MULTIPLICATION AND DIVISION OF FRACTIONS

8-13 Enrichment: Federal Income Tax Form 1040EZ

Objective
Complete a federal income tax form.

Terry worked part-time at Ryan's Auto Shop in 1987. In January of 1988, his employer gave him a W-2 form to use when filing his taxes.

Each taxpayer in the U.S. must complete a tax form for the past year's wages. The easiest form to complete is the 1040EZ. You must be single with no dependents in order to file this form.

Example
Terry earned $15 in interest. He is single with no dependents.

Follow the directions on each line of the form. To find the amount of tax on the taxable income, read the tax table.

Exercises

Complete a tax form for each person. Each person is single with no dependents.

	Name	Social Security Number	Wages	Federal Tax Withheld	Interest Earned	Contributions to Charities
1.	Sharon Crosby	000-37-9742	$7533.00	$329.00	$55.20	$102.00
2.	Matthew Chin	000-71-3318	$7313.37	$374.78	$10.71	$ 52.00

232 CHAPTER 8

8-14 Calculator: Multiplying and Dividing Fractions

Objective
Use a calculator to multiply and divide fractions.

The following calculator codes show three ways to multiply fractions. Some of the keys are not named. Fill in the missing operation signs to multiply the fractions using each method.

$\frac{2}{5} \times \frac{3}{10} = 0.12$

Round these answers to the nearest hundredth.

A. [MRC] [ON/C] [2] [ ] [5] [ ] [3] [ ] [10] [=]

1. Use method A to multiply.
$\frac{32}{45} \times \frac{17}{48} = \underline{\qquad}$

B. [MRC] [ON/C] [5] [ ] [10] [M+] [2] [ ] [3] [ ] [MRC] [=]

2. Use Method B to multiply.
$\frac{16}{55} \times \frac{28}{47} = \underline{\qquad}$

C. [MRC] [ON/C] [2] [ ] [3] [ ] [5] [ ] [10] [=]

3. Use Method C to multiply.
$\frac{72}{105} \times \frac{84}{95} = \underline{\qquad}$

Use any method to find each product. Round to the nearest hundredth.

4. $\frac{29}{30} \times \frac{18}{35}$ 5. $\frac{201}{220} \times \frac{118}{205}$ 6. $\frac{14}{135} \times \frac{18}{135}$ 7. $\frac{160}{259} \times \frac{284}{295}$ 8. $\frac{156}{233} \times \frac{110}{129}$

These calculator codes show three ways to divide fractions. Fill in the missing operation signs to divide the fractions using each method.

$\frac{3}{4} \div \frac{2}{5} = 1.875$

Round these answers to the nearest hundredth.

D. [MRC] [ON/C] [3] [ ] [4] [ ] [5] [ ] [2] [=]

9. Use Method D to divide.
$\frac{17}{34} \div \frac{15}{26} = \underline{\qquad}$

E. [MRC] [ON/C] [2] [ ] [5] [M+] [3] [ ] [4] [ ] [MRC] [=]

10. Use Method E to divide.
$\frac{37}{50} \div \frac{18}{37} = \underline{\qquad}$

F. [MRC] [ON/C] [4] [ ] [5] [M+] [3] [ ] [2] [ ] [MRC] [=]

11. Use Method F to divide.
$\frac{23}{40} \div \frac{17}{45} = \underline{\qquad}$

Use any method to find each quotient. Round to the nearest hundredth.

12. $\frac{7}{15} \div \frac{8}{35}$ 13. $\frac{3}{100} \div \frac{9}{8}$ 14. $\frac{18}{25} \div \frac{22}{25}$ 15. $\frac{71}{240} \div \frac{18}{265}$ 16. $\frac{69}{150} \div \frac{24}{175}$

MULTIPLICATION AND DIVISION OF FRACTIONS

CHAPTER 8 REVIEW

Fill in the blank.

1. $\frac{3}{4}$ is to $\frac{4}{3}$ as

   $\frac{7}{10}$ is to _____.

2. $\frac{10}{15}$ is to $\frac{2}{3}$ as

   $\frac{8}{24}$ is to _____.

3. $1\frac{1}{6}$ is to $\frac{7}{6}$ as

   $2\frac{2}{3}$ is to _____.

4. $1\frac{3}{5} \times 1\frac{3}{5}$ is to

   $\frac{8}{5} \times \frac{8}{5}$ as $2\frac{1}{2} \times 2\frac{1}{2}$

   is to _____.

5. $\frac{2}{6} \div \frac{4}{3}$ is to

   $\frac{2}{6} \times \frac{3}{4}$ as $\frac{1}{5} \div \frac{10}{2}$

   is to _____.

6. $1\frac{4}{12}$ is to $1\frac{1}{3}$ as

   $2\frac{5}{10}$ is to _____.

Arrange the steps in the proper order.

7. Multiply.

   $2\frac{1}{4} \times 2$

   Step x Multiply the fractions.
   Step y Write in lowest terms.
   Step z Write each mixed number as a fraction.

8. Divide.

   $\frac{2}{3} \div \frac{1}{3}$

   Step a Write in lowest terms.
   Step b Find the reciprocal of the divisor.
   Step c Multiply the fractions.

Match.

9. installment plan
10. reciprocals
11. divisor
12. quotient
13. compatible number

A. two numbers whose product is 1
B. the answer in division
C. a number that is close in value and that is easier to use
D. a number by which the dividend is divided
E. down payment plus monthly payment with interest
F. list price

14. Which installment plan is the best?

    A. list price $500
       down payment $\frac{1}{4}$
       monthly payment $35
       terms in months 12

    B. list price $500
       down payment $\frac{1}{5}$
       monthly payment $26
       terms in months 18

    C. list price $500
       down payment $\frac{1}{8}$
       monthly payment $75
       terms in months 6

15. Describe a situation where you would need an exact answer.

16. Describe a situation where you would need an estimate.

17. Describe a situation where you could use mental math.

18. Describe a situation where you could use a calculator.

CHAPTER 8 TEST

1. Five students work in the town library after school. Each student works on a different day of the week. Ben cannot work on Monday, Wednesday, or Friday. Adam cannot work on Tuesday or Thursday. Chris works the day before Ben. Melissa works 2 days after Chris, but not on Wednesday. Betsy works the day after Adam. What day does each student work?

Multiply.

2. $\frac{3}{4} \times \frac{5}{6}$ 3. $\frac{8}{9} \times 6$ 4. $7 \times \frac{3}{28}$ 5. $\frac{10}{11} \times \frac{10}{11}$ 6. $\frac{4}{5} \times \frac{7}{9} \times \frac{3}{4}$

7. $2\frac{2}{3} \times 3\frac{3}{4}$ 8. $5 \times 6\frac{4}{5}$ 9. $3\frac{1}{3} \times 9$ 10. $6\frac{2}{9} \times 3\frac{3}{14}$ 11. $5\frac{5}{11} \times 33 \times 3\frac{7}{12}$

Find the installment cost for the motorcycle.

	Motorcycle	List Price	Down Payment	Monthly Payment	Terms in Months
12.	Chopper	$2200	$\frac{1}{5}$	$160.60	12
13.	Sportster	$3300	$\frac{1}{3}$	$ 67.83	36

Find the reciprocal.   14. $\frac{1}{7}$ 15. $\frac{9}{5}$ 16. 8 17. $\frac{2}{5}$ 18. $3\frac{2}{7}$

Divide.

19. $\frac{3}{5} \div \frac{3}{4}$ 20. $\frac{4}{7} \div \frac{2}{7}$ 21. $\frac{14}{16} \div \frac{7}{8}$ 22. $8 \div \frac{4}{9}$ 23. $\frac{24}{25} \div 8$

24. $30 \div 3\frac{3}{4}$ 25. $5\frac{3}{5} \div 7$ 26. $\frac{2}{3} \div 3\frac{5}{9}$ 27. $6\frac{8}{11} \div \frac{37}{44}$ 28. $4\frac{1}{6} \div 1\frac{1}{10}$

Estimate. Use compatible numbers.   29. $\frac{4}{9} \times 28$ 30. $\frac{3}{7} \times 34$ 31. $25 \div 2\frac{19}{41}$

32. Decide if the problem needs an estimate or an exact answer. Choose a calculation method. Solve the problem.

    Problem: Isabel took Luther to lunch for his birthday. The bill was $18.56. Isabel paid with a $20 bill. How much change should she receive?

33. Rewrite this recipe for 12 people. Tell which amounts you would likely estimate.

Meat Loaf	
Serves 8.	
$1\frac{1}{2}$ lb	ground beef
1 c	tomato juice
$\frac{3}{4}$ c	oats, uncooked
1	egg, beaten
$\frac{1}{4}$ c	chopped onion
1 tsp	salt
$\frac{1}{4}$ tsp	pepper

Find the overtime rate (time and a half) for the hourly rate. Use mental math.

34. $6   35. $9   36. $6.20

MULTIPLICATION AND DIVISION OF FRACTIONS

CHAPTER 8 CUMULATIVE REVIEW

1. Divide.
   24.5 ÷ 10
   A. 2.45
   B. 0.245
   C. 245.0
   D. 240.5

2. Multiply.
   $\frac{3}{6} \times \frac{4}{8}$
   A. $\frac{18}{24}$
   B. $\frac{1}{4}$
   C. 1
   D. $\frac{12}{24}$

3. Estimate. Round to the highest place.
   108 + 296
   A. 400
   B. 300
   C. 290
   D. 350

4. Lisa put $\frac{1}{3}$ down payment on a bicycle that cost $225. She has to make 12 monthly payments of $15.58. What is the installment cost of the bicycle?
   A. $411
   B. $186
   C. $261.96
   D. $209.50

5. In how many ways can you pick 3 out of 6 posters to be framed? Name the strategy that Siobahn used to solve this problem.
   A. Make an organized list.
   B. Draw a picture or diagram.
   C. Use objects.
   D. Guess and check.

ABC   ACD   ADE   AEF   10
ABD   ACE   ADF
ABE   ACF
ABF

BCD   BDE   BEF   6
BCE   BDF
BCF

CDE   CEF   3
CDF

DEF   1
      20

6. Find the reciprocal.
   $3\frac{6}{8}$
   A. $\frac{30}{8}$
   B. $\frac{8}{30}$
   C. $\frac{17}{8}$
   D. $\frac{8}{17}$

7. Estimate the amount of sales for July and August.
   A. $60,000
   B. $68,000
   C. $58,000
   D. $70,000

Sales for April–September

8. Multiply.
   492 × 230
   A. 114,160
   B. 113,260
   C. 113,160
   D. 114,260

9. People who work overtime are usually paid $1\frac{1}{2}$ times their hourly rate. Find the overtime rate for $10.40. Use mental math.
   A. $20.80
   B. $5.20
   C. $10.80
   D. $15.60

10. Write the decimal as a fraction in lowest terms.
    0.204
    A. $\frac{204}{1000}$
    B. $\frac{51}{250}$
    C. $\frac{1}{5}$
    D. $2\frac{4}{100}$

11. Saul saved $2 the first month. He saved $4 the second month. He saved $6 the third month. At this rate, how much did he save the sixth month?
   A. $26
   B. $10
   C. $12
   D. $18

12. How many people said that they used peaches for baking?
   A. 9 people
   B. 12 people
   C. 40 people
   D. 45 people

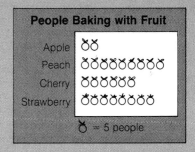

13. Divide.
   $6 \div \frac{2}{9}$
   A. $\frac{1}{27}$
   B. $\frac{12}{9}$
   C. 27
   D. 56

14. Estimate.
   $42.09 + $19.35 + $7.89
   A. $69
   B. $24
   C. $100
   D. $50

15. Find the difference. Use mental math.
   $182.50 − $35
   A. $148.50
   B. $149
   C. $149.50
   D. $147.50

16. Estimate.
   129×42
   A. 5200
   B. 7000
   C. 1294
   D. 3200

17. Subtract.
   300.2 − 49.53
   A. 195.1
   B. 350.73
   C. 350.77
   D. 250.67

18. What was the total amount deducted for this pay period? Use the pay stub shown.
   A. $131.97
   B. $530
   C. $387.04
   D. $142.96

SOC. SEC. NO. 952-00-4950 ROMERO, GERALD			PERIOD ENDING 01-31-88		**ACE GRAPHICS**	CHECK NO. 546663	
RATE	HOURS	GROSS	TYPE OF DEDUCTION	AMOUNT	YEAR TO DATE	AMOUNT	
13.25	40	530.00	FICA	36.57	Total Gross	530.00	
			Federal Tax	95.40	FICA	36.57	
			State Tax	10.99	Fed. Tax	95.40	
					State Tax	10.99	
CURRENT GROSS	530.00		CURRENT DEDUCTIONS		CURRENT NET	387.04	

19. Divide.
   $12,443 \div 23$
   A. 541
   B. 584 R11
   C. 500 R9
   D. 54

20. Round to the nearest tenth.
   9.016
   A. 9.01
   B. 9.02
   C. 10
   D. 9.0

21. Multiply.
   5.4×0.36
   A. 1.944
   B. 19.44
   C. 0.1944
   D. 194.4

Chapter 9 Overview

Key Ideas
- Use the problem-solving strategy *work backward*.
- Add and subtract fractions and mixed numbers with like denominators.
- Add and subtract fractions and mixed numbers with unlike denominators.
- Use the mental math strategy *compatible numbers*.
- Estimate using rounding.
- Find dimensions for picture frames.
- Check your understanding of a problem situation.
- Make group decisions.

Key Terms
- numerator
- lowest terms
- compatible numbers
- denominator
- sum
- least common multiple
- mixed number
- difference

Key Skills

Find the least common multiple.

1. 6, 12
2. 4, 16
3. 2, 3
4. 3, 4
5. 5, 6
6. 3, 5
7. 3, 8
8. 4, 10
9. 6, 8
10. 5, 8
11. 5, 16
12. 5, 12

Find the greatest common factor.

13. 6, 12
14. 5, 15
15. 12, 18
16. 15, 30
17. 18, 20
18. 24, 36
19. 35, 41
20. 42, 60
21. 38, 64
22. 50, 125
23. 25, 60
24. 36, 48

Write in lowest terms.

25. $\frac{5}{3}$
26. $\frac{6}{2}$
27. $\frac{8}{3}$
28. $\frac{9}{4}$
29. $\frac{7}{5}$
30. $3\frac{8}{6}$
31. $2\frac{5}{4}$
32. $3\frac{8}{5}$
33. $7\frac{3}{2}$
34. $5\frac{5}{3}$
35. $12\frac{3}{2}$
36. $23\frac{6}{4}$
37. $13\frac{5}{4}$
38. $30\frac{15}{10}$
39. $4\frac{18}{16}$
40. $4\frac{9}{8}$
41. $8\frac{4}{3}$
42. $6\frac{11}{9}$

ADDITION AND SUBTRACTION OF FRACTIONS

9

9-1 Problem Solving: Learning to Use Strategies

Objective
Use the strategy work backward.

- SITUATION
- DATA
- **PLAN**
- ANSWER
- CHECK

You can solve some problems by **working backward** to find the solution. It helps to make a flowchart using the data in the problem. Then you can work backward to solve the problem.

Example Solve.

Jana wanted to buy a used car. The car she wanted was too expensive in January. In March, the owner lowered the price by $100. In May, the owner cut the March price in half. Jana bought the car for $600. What was the price of the car in January?

Make a flowchart. Use the data from the problem. Start with the unknown price. End with the final known price.

$$\boxed{?} \to \boxed{-100} \to \boxed{\div 2} \to \boxed{\$600}$$

Undo each operation in the first flowchart. Multiplication can undo division. Addition can undo subtraction. These are **inverse operations**.

$$\boxed{?} \leftarrow \boxed{+100} \leftarrow \boxed{\times 2} \leftarrow \boxed{\$600}$$

Work backward to solve.

$$\boxed{\$1300} \leftarrow \boxed{+100} \leftarrow \boxed{\times 2} \leftarrow \boxed{\$600}$$

The price of the car in January was $1300.

Try This Solve.

Driver Education is an elective at Paul's school. There were 10 more students enrolled in winter than in fall. There were twice as many students enrolled in spring as in winter. There were 50 students signed up in the spring. How many students were in the course in the fall?

Exercises

Solve. Use one or more of the problem-solving strategies.

1. Stan won free tickets to an antique car show. He gave 2 tickets to his brother. He divided the rest equally among himself and 3 friends. He kept 2 tickets. How many tickets did he win in all?

Problem-Solving Strategies
- Guess and check.
- Choose an operation.
- Make an organized list.
- Act it out.
- Find a pattern.
- Write an equation.
- Use logical reasoning.
- Draw a picture or diagram.
- Make a table.
- Use objects.
- **Work backward.**
- Solve a simpler problem.

2. At the Thrifty Car Wash, each car goes through 3 cycles. Each cycle takes 30 s. How long does it take for 5 cars to go through the car wash?

3. Lydell bought a seatcover for his car for $10.49 and some parts for his engine for three times as much. He had $8.75 left over. How much money did he have to start with?

4. Barb, Jose, Juanita, and Jack get to school in different ways. One drives a car. One rides a bike. One rides the train. One rides a unicycle. Barb needs to buy gasoline about twice a month. Jack loves to be different. Jose hates to exercise. How does each person get to school?

5. Since Joel bought a used car, his repair bills have been $7 the first month, $21 the second month, and $35 the third month. If this rate continues, how much will he pay for repairs the eighth month?

6. After lying on the beach, a crab was trying to swim back into the ocean. It took the crab 1 min to travel 12 m. Then a wave would wash the crab back 4 m. The crab would rest 1 min before moving again. If it continued in this way, how long would it take the crab to move 44 m from the beach?

7. Mrs. Carlos has 4 family photos. Each photo is in a frame. She wants to hang the photos on a wall. In how many different orders can she hang the photos in a row?

Suppose

8. Suppose in the example on page 240 that Jana bought the car for $850. What was the price of the car in January?

Write Your Own Problem

9. Write a problem that you can solve working backward. Use the data in the table.

10. Write a problem that you can solve using the *guess and check* strategy. Use the data in the table.

Car Wash	Number of Customers
9 a.m.–10 a.m.	⊮ I
10 a.m.–11 a.m.	⊮ ⊮ ⊮ ⊮ IIII
11 a.m.–12 noon	⊮ ⊮ ⊮ ⊮

9-2 Adding and Subtracting Fractions with Like Denominators

Objective
Add and subtract fractions with like denominators.

The city library is open from 9 a.m. to 9 p.m. The children's librarian works from 9 a.m. to 3 p.m. The reference librarian works from 3 p.m. to 7 p.m. What fraction of the library's open hours is at least one of these librarians on duty?

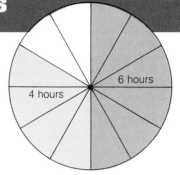

Number of Open Hours in One Day

Understand the Situation
- How many hours is the library open?
- What fraction of the library's open hours does the reference librarian work?

Examples

1. Add. $\frac{6}{12} + \frac{4}{12}$

   | Add the numerators. Write the sum over the denominator. | → | Write the sum in lowest terms. |

   $$\begin{array}{r}\frac{6}{12} \\ +\frac{4}{12} \\ \hline \frac{10}{12}\end{array} \quad \text{like denominators}$$

   $$\frac{10}{12} = \frac{5}{6}$$

   One of the two librarians is on duty $\frac{5}{6}$ of the library's open hours.

2. Subtract. $\frac{11}{32} - \frac{3}{32}$

   | Subtract the numerators. Write the difference over the denominator. | → | Write the difference in lowest terms. |

   $$\begin{array}{r}\frac{11}{32} \\ -\frac{3}{32} \\ \hline \frac{8}{32}\end{array}$$

   $$\frac{8}{32} = \frac{1}{4}$$

Try This Add or subtract.

a. $\frac{2}{9} + \frac{5}{9}$ b. $\frac{11}{12} - \frac{7}{12}$ c. $\frac{7}{8} - \frac{3}{8}$ d. $\frac{3}{10} + \frac{3}{10}$ e. $\frac{25}{27} - \frac{7}{27}$

Exercises

Add or subtract.

1. $\frac{3}{7}$
   $+\frac{2}{7}$

2. $\frac{3}{8}$
   $+\frac{3}{8}$

3. $\frac{5}{11}$
   $-\frac{4}{11}$

4. $\frac{9}{14}$
   $-\frac{3}{14}$

5. $\frac{1}{4} + \frac{1}{4}$
6. $\frac{9}{14} + \frac{3}{14}$
7. $\frac{5}{12} + \frac{1}{12}$
8. $\frac{7}{16} + \frac{5}{16}$
9. $\frac{11}{15} - \frac{7}{15}$
10. $\frac{5}{8} - \frac{3}{8}$
11. $\frac{5}{9} - \frac{2}{9}$
12. $\frac{21}{30} - \frac{12}{30}$
13. $\frac{11}{28} + \frac{5}{28}$
14. $\frac{7}{15} + \frac{2}{15}$
15. $\frac{17}{16} - \frac{5}{16}$
16. $\frac{5}{6} - \frac{1}{6}$
17. $\frac{12}{13} - \frac{6}{13}$
18. $\frac{19}{36} - \frac{11}{36}$
19. $\frac{1}{10} + \frac{7}{10}$
20. $\frac{17}{8} - \frac{13}{8}$
21. $\frac{2}{9} + \frac{1}{9} + \frac{3}{9}$
22. $\frac{1}{6} + \frac{2}{6} + \frac{1}{6}$
23. $\frac{1}{15} + \frac{7}{15} + \frac{2}{15}$

The head librarian posted this schedule of working hours from 9 a.m. to 9 p.m. Give a fraction for each part.

24. the part of the time exactly 2 people are working
25. the part of the time 3 people are working
26. the part of the time exactly 1 person is working
27. the part of the time the head librarian works more than the assistant librarian
28. the part of the time the children's librarian works more than the reference librarian

Schedule of Working Hours

	9	10	11	12	1	2	3	4	5	6	7	8	9
			Children's Librarian										
								Reference Librarian					
		Head Librarian											
							Assistant Librarian						

Suppose

29. Suppose the library hired another assistant to work from 9 a.m. to 3 p.m. What part of the day would 3 people be on duty?

30. Suppose the children's librarian worked 3 hours less. What fraction of the working hours would this librarian work?

Mental Math

Find the total weight for each set of keys. Use mental math.

31. Skeleton key $\frac{4}{12}$ oz, File cabinet $\frac{2}{12}$ oz, Library office $\frac{5}{12}$ oz

32. Key ring $\frac{3}{8}$ oz, Safe deposit box $\frac{3}{8}$ oz, Post office box $\frac{1}{8}$ oz

33. Rental car, trunk $\frac{3}{24}$ oz, Key ring $\frac{9}{24}$ oz, Hotel room $\frac{5}{24}$ oz, Rental car, ignition $\frac{3}{24}$ oz

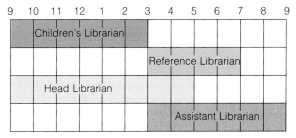

9-3 Adding and Subtracting Mixed Numbers with Like Denominators

Objective
Add and subtract mixed numbers with like denominators.

The Sports Club is selling strips of tickets to Field Day. Each strip is divided into six tickets. Ray has sold 4 strips and $\frac{5}{6}$ of another strip. Lorna has sold $3\frac{1}{6}$ strips. How many more strips has Ray sold?

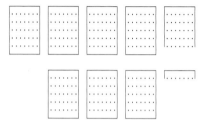

Understand the Situation
- Do the fractions of ticket strips have the same denominators?
- What operation would you use to compare how many strips of tickets each student sold?

Examples
Add or subtract.

1. $4\frac{5}{6} - 3\frac{1}{6}$

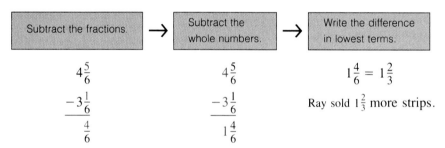

2. $3\frac{5}{7} + 1\frac{4}{7}$

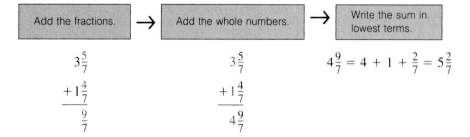

Try This
Add or subtract.

a. $5\frac{13}{16} - 4\frac{9}{16}$ b. $5\frac{11}{12} + \frac{7}{12}$ c. $6\frac{1}{2} - 5\frac{1}{2}$ d. $3\frac{3}{10} + 1\frac{9}{10}$

Exercises

Add or subtract.

1. $9\frac{5}{11}$
   $-4\frac{4}{11}$

2. $6\frac{11}{15}$
   $+\frac{7}{15}$

3. $24\frac{5}{8}$
   $+22\frac{3}{8}$

4. $7\frac{5}{9}$
   $-7\frac{2}{9}$

5. $1\frac{3}{7} + \frac{5}{7}$
6. $\frac{4}{5} + 2\frac{3}{5}$
7. $\frac{5}{8} + 1\frac{5}{8}$
8. $4\frac{1}{6} + \frac{5}{6}$
9. $3\frac{5}{9} - \frac{2}{9}$
10. $4\frac{7}{10} - \frac{3}{10}$
11. $6\frac{1}{3} - 4$
12. $5\frac{7}{8} - 5$
13. $3\frac{1}{4} + 2\frac{3}{4}$
14. $4\frac{5}{9} + 5\frac{7}{9}$
15. $1\frac{7}{12} + 1\frac{7}{12}$
16. $6\frac{15}{16} + 2\frac{11}{16}$
17. $4\frac{13}{24} - 2\frac{3}{24}$
18. $7\frac{7}{18} - 6\frac{5}{18}$
19. $7\frac{1}{2} - 6\frac{1}{2}$
20. $9\frac{2}{3} - 1\frac{2}{3}$
21. $2\frac{1}{2} + 1\frac{1}{2}$
22. $3\frac{2}{3} + 1\frac{2}{3}$
23. $4\frac{7}{8} + 2\frac{3}{8}$
24. $4\frac{11}{20} + \frac{9}{20}$
25. $6\frac{7}{8} - 3\frac{3}{8}$
26. $8\frac{7}{24} - 8\frac{1}{24}$
27. $7\frac{3}{8} - 2\frac{3}{8}$
28. $12\frac{4}{5} - 12$

Write the mixed number in lowest terms.

29. $1\frac{9}{7}$
30. $4\frac{5}{3}$
31. $2\frac{10}{8}$
32. $3\frac{16}{12}$
33. $6\frac{14}{7}$
34. $3\frac{7}{2}$
35. $3\frac{20}{15}$
36. $1\frac{9}{6}$
37. $5\frac{4}{4}$
38. $2\frac{13}{5}$
39. $1\frac{20}{12}$
40. $1\frac{12}{6}$

Number Sense

Continue the pattern.

41. $\frac{2}{5}, \frac{4}{5}, 1\frac{1}{5},$ ___, ___, ___
42. $\frac{1}{4}, 1, 1\frac{3}{4},$ ___, ___, ___
43. $\frac{3}{10}, \frac{7}{10}, 1\frac{1}{10},$ ___, ___, ___
44. $2\frac{1}{8}, 3\frac{3}{8}, 4\frac{5}{8},$ ___, ___, ___

Problem Solving

45. The Sports Club Boosters sold $37\frac{5}{6}$ strips of Field Day tickets last week and $29\frac{5}{6}$ strips this week. Suppose they sell 5 more strips. They will have sold twice as many ticket strips as they sold last year. How many ticket strips did the Sports Club Boosters sell last year?

Mixed Skills Review

46. 1 ft 7 in.
    $+1$ ft 9 in.

47. 3 yr 9 mo
    $+2$ yr 6 mo

48. 5 gal 3 qt
    $+1$ gal 2 qt

49. 2 h 50 min
    $+7$ h 25 min

Find equivalent fractions.

50. $\frac{4}{5} = \frac{\square}{10} = \frac{\square}{15} = \frac{\square}{20}$
51. $\frac{5}{9} = \frac{\square}{18} = \frac{\square}{27} = \frac{\square}{36}$
52. $\frac{2}{3} = \frac{\square}{6} = \frac{\square}{9} = \frac{\square}{12}$
53. $\frac{3}{7} = \frac{\square}{14} = \frac{\square}{21} = \frac{\square}{28}$

ADDITION AND SUBTRACTION OF FRACTIONS

9-4 Adding and Subtracting Fractions with Unlike Denominators

Objective
Add and subtract fractions with unlike denominators.

What fraction of the students chose either the beach or the mountains?

Understand the Situation

- Do the fractions for beach and mountains have the same denominator?
- What operation would you use to find the fraction of students who chose either the beach or the mountains?

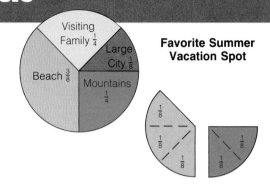

Favorite Summer Vacation Spot

You can add and subtract fractions with different denominators. Write equivalent fractions with common denominators. Use the LCD.

Examples Add or subtract.

1. $\frac{3}{8} + \frac{1}{4}$

Are the denominators the same? → If not, find the LCD. → Find the equivalent fractions with the LCD. → Add or subtract.

$\frac{3}{8}$
$+\frac{1}{4}$ No

The LCD is 8.

$\frac{3}{8} = \frac{3}{8}$
$+\frac{1}{4} = \frac{2}{8}$

$\frac{3}{8} = \frac{3}{8}$
$+\frac{1}{4} = \frac{2}{8}$
$\phantom{+\frac{1}{4}} = \frac{5}{8}$

Of the students, $\frac{5}{8}$ chose either the beach or the mountains for a vacation.

2. $\frac{3}{4} - \frac{2}{3}$

$\frac{3}{4} = \frac{9}{12}$
$-\frac{2}{3} = \frac{8}{12}$
$\phantom{-\frac{2}{3}} = \frac{1}{12}$

The LCD is 12.

The difference is $\frac{1}{12}$.

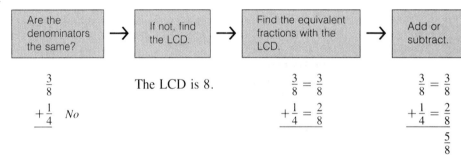

Try This

a. $\frac{2}{3} + \frac{1}{5}$ b. $\frac{1}{8} + \frac{7}{12}$ c. $\frac{6}{7} - \frac{5}{14}$ d. $\frac{4}{5} - \frac{3}{4}$

Exercises

Add or subtract.

1. $\frac{7}{9} + \frac{2}{3}$
2. $\frac{3}{7} + \frac{9}{14}$
3. $\frac{1}{2} + \frac{4}{5}$
4. $\frac{5}{6} + \frac{1}{9}$
5. $\frac{3}{4} - \frac{1}{2}$
6. $\frac{2}{3} - \frac{4}{9}$
7. $\frac{3}{4} - \frac{7}{16}$
8. $\frac{5}{8} - \frac{1}{3}$
9. $\frac{4}{8} + \frac{3}{8}$
10. $\frac{9}{14} + \frac{3}{4}$
11. $\frac{3}{4} + \frac{1}{3}$
12. $\frac{1}{6} + \frac{5}{8}$
13. $\frac{7}{8} - \frac{1}{3}$
14. $\frac{3}{4} - \frac{2}{9}$
15. $\frac{3}{2} - \frac{3}{4}$
16. $\frac{7}{6} - \frac{2}{6}$
17. $\frac{3}{5} + \frac{5}{6}$
18. $\frac{9}{16} + \frac{1}{2}$
19. $\frac{7}{18} + \frac{12}{18}$
20. $\frac{5}{7} + \frac{5}{8}$
21. $\frac{21}{32} - \frac{8}{32}$
22. $\frac{9}{8} - \frac{4}{5}$
23. $\frac{13}{24} - \frac{1}{6}$
24. $\frac{1}{4} - \frac{1}{9}$
25. $\frac{1}{2} + \frac{2}{3} + \frac{5}{6}$
26. $\frac{1}{6} + \frac{3}{4} + \frac{1}{9}$
27. $\frac{4}{9} + \frac{11}{45} + \frac{3}{5}$
28. $\frac{2}{7} + \frac{1}{3} + \frac{8}{21}$

29. What fraction of the people surveyed chose either visiting family or visiting a large city?
30. How much larger was the fraction of the people who chose the beach than the fraction of those who chose the mountains?

Number Sense Use Egyptian fractions to write the following sums. Use only fractions that have 1 as the numerator.

> **Egyptian Fractions**
> $\frac{1}{2} + \frac{1}{3} = \frac{5}{6}$
> $\frac{1}{3} + \frac{1}{4} = \frac{7}{12}$
> $\frac{1}{2} + \frac{1}{3} + \frac{1}{5} = \frac{14}{15}$

31. $\frac{3}{4}$
32. $\frac{7}{10}$
33. $\frac{5}{8}$
34. $\frac{4}{9}$
35. $\frac{13}{16}$
36. $\frac{11}{18}$

Problem Solving The winner in a TV game show spins for a trip with all expenses paid. Give the chance of winning each trip as a fraction.

37. cruise
38. European tour
39. rafting
40. World Series or the Super Bowl
41. a water vacation
42. European tour or World Series

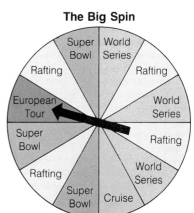

The Big Spin

Use Objects 43. Use fraction strips like those in Example 2 on p. 246. Show $\frac{1}{2} + \frac{1}{3}$ using the fraction strips.

9-5 Adding and Subtracting Mixed Numbers with Unlike Denominators

Objective
Add and subtract mixed numbers with unlike denominators.

The chart shows the amount of snowfall during the first three months of one year. Compare the February and March snowfall in Chicago.

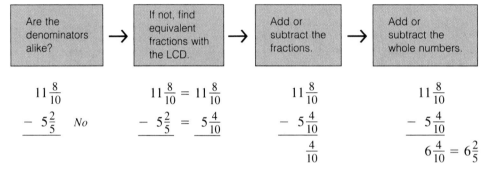

Snowfall in inches

	January	February	March
Boston	$4\frac{1}{4}$	$3\frac{5}{8}$	$1\frac{1}{2}$
Chicago	$17\frac{7}{10}$	$11\frac{8}{10}$	$5\frac{2}{5}$
Denver	$3\frac{3}{4}$	$4\frac{1}{8}$	$1\frac{1}{2}$
Duluth	$14\frac{1}{5}$	$16\frac{1}{2}$	$4\frac{3}{4}$

Examples Subtract.

1. $11\frac{8}{10} - 5\frac{2}{5}$

Are the denominators alike?	→	If not, find equivalent fractions with the LCD.	→	Add or subtract the fractions.	→	Add or subtract the whole numbers.

$$\begin{array}{r} 11\frac{8}{10} \\ -\ 5\frac{2}{5} \\ \hline \end{array} \text{ No} \qquad \begin{array}{r} 11\frac{8}{10} = 11\frac{8}{10} \\ -\ 5\frac{2}{5} = 5\frac{4}{10} \\ \hline \end{array} \qquad \begin{array}{r} 11\frac{8}{10} \\ -\ 5\frac{4}{10} \\ \hline \frac{4}{10} \end{array} \qquad \begin{array}{r} 11\frac{8}{10} \\ -\ 5\frac{4}{10} \\ \hline 6\frac{4}{10} = 6\frac{2}{5} \end{array}$$

In Chicago $6\frac{2}{5}$ in. less snow fell in March than in February.

2. Find the total amount of snowfall in Duluth for January, February, and March.

$$14\frac{1}{5} + 16\frac{1}{2} + 4\frac{3}{4}$$

$$\begin{array}{r} 14\frac{1}{5} = 14\frac{4}{20} \\ 16\frac{1}{2} = 16\frac{10}{20} \\ +\ 4\frac{3}{4} = 4\frac{15}{20} \\ \hline 34\frac{29}{20} = 35\frac{9}{20} \end{array}$$

The total amount of snowfall was $35\frac{9}{20}$ in.

Try This Add or subtract.

a. $8\frac{5}{6} - 7\frac{5}{8}$ **b.** $4\frac{1}{7} + 2\frac{3}{4}$ **c.** $117\frac{5}{9} + 84\frac{2}{3}$ **d.** $23\frac{9}{10} - 18\frac{1}{3}$

Exercises

Add or subtract.

1. $\frac{3}{10} + 1\frac{4}{5}$
2. $2\frac{3}{8} + \frac{1}{4}$
3. $4\frac{5}{9} - \frac{1}{6}$
4. $8\frac{7}{16} - \frac{4}{16}$
5. $5\frac{2}{3} + 4\frac{5}{6}$
6. $1\frac{2}{4} + 5\frac{3}{4}$
7. $5\frac{3}{4} - 4\frac{1}{5}$
8. $7\frac{3}{8} - 1\frac{1}{4}$
9. $9\frac{5}{8} - 2\frac{7}{12}$
10. $6\frac{4}{9} - 5\frac{1}{6}$
11. $3\frac{5}{16} + 1\frac{1}{6}$
12. $6\frac{1}{4} + 5\frac{3}{7}$
13. $5\frac{3}{4} - 3\frac{2}{5}$
14. $4\frac{5}{6} - 1\frac{4}{9}$
15. $6\frac{3}{7} + 2\frac{1}{6}$
16. $7\frac{3}{8} + 3\frac{4}{7}$
17. $6\frac{3}{6} + 5\frac{4}{6} + 8\frac{5}{6}$
18. $4\frac{1}{6} + 9\frac{3}{4} + 7\frac{1}{9}$
19. $10\frac{1}{4} + \frac{5}{12} + 6\frac{3}{8}$

Solve. Use the chart on page 248.

20. Find the difference in snowfall for the months of January and February in Duluth.
21. Find the total amount of January snowfall for all four cities.
22. Compare the total snowfall in Chicago and Duluth during the months of January and February.
23. Find the total amount of March snowfall for all four cities. Compare this total with the January total for all four cities.

Show Math Draw a picture to show the answer to each problem.

24. $2\frac{3}{5} + 3\frac{1}{5}$
25. $1\frac{3}{4} + 2\frac{3}{4}$
26. $4\frac{1}{2} + \frac{1}{2}$
27. $\frac{4}{3} + \frac{5}{4}$
28. $\frac{1}{4} + \frac{4}{4}$

Working with Variables

Solve and check.

$10x = 20$ You can undo the action of multiplying by dividing. To get the variable alone on one side, divide by 10.

$\frac{10x}{10} = \frac{20}{10}$ Remember to use the same action on both sides of an equation to keep the sides equal.

$x = 2$ $10 \times 2 = 20$ Check your answer by substituting it for the variable. The solution is 2.

Solve and check.

29. $8y = 40$
30. $96 = 16r$
31. $\frac{2}{3}y = 12$
32. $5t = \frac{2}{5}$
33. $13b = 156$
34. $900 = 36c$
35. $\frac{1}{2}n = 15$
36. $36 = \frac{3}{4}v$

Practice Through Problem Solving

37. Arrange the digits 4, 5, 6, 8 into two fractions that will give the least sum. Use each digit once.
38. Arrange the digits 4, 5, 6, 8 into two fractions that will give the greatest sum. Use each digit once.

9-6 Mental Math: Compatible Numbers

Objective
Use the mental math strategy compatible numbers.

Ken stocks shelves at an auto parts store after school. How many hours did Ken work last week?

Fractions that can be added using mental math to give a sum of 1 are **compatible**. Use compatible numbers to solve this problem.

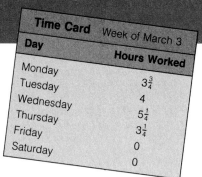

Examples Add. Use mental math.

1. $3\frac{3}{4} + 4 + 5\frac{1}{4} + 3\frac{1}{4}$ Think: $3\frac{3}{4} + 5\frac{1}{4} = 8 + 1 = 9$ $9 + 4 + 3\frac{1}{4} = 16\frac{1}{4}$

Ken worked $16\frac{1}{4}$ hours last week.

2. $2\frac{3}{5} + 3 + 1\frac{1}{5}$ Think: $2\frac{3}{5} + 1\frac{1}{5} = 3\frac{4}{5}$ $3\frac{4}{5} + 3 = 6\frac{4}{5}$

3. $1\frac{1}{8} + 4\frac{1}{2} + 3\frac{7}{8}$ Think: $1\frac{1}{8} + 3\frac{7}{8} = 5$ $5 + 4\frac{1}{2} = 9\frac{1}{2}$

Try This Add. Use mental math.

a. $3\frac{1}{6} + 2 + 1\frac{5}{6}$ b. $2\frac{1}{2} + 4\frac{1}{2} + 1\frac{1}{2}$ c. $5\frac{4}{7} + 3\frac{1}{4} + \frac{3}{7}$ d. $\frac{11}{16} + 1\frac{5}{16} + 1\frac{3}{8}$

Exercises

Add. Use mental math.

1. $4\frac{3}{4} + 2 + 3\frac{1}{4}$ 2. $1\frac{3}{5} + 5 + 6\frac{2}{5}$ 3. $7\frac{1}{3} + 3\frac{1}{3} + 2\frac{1}{3}$ 4. $3\frac{5}{6} + 1\frac{1}{6} + 5$

5. $2\frac{7}{12} + 9 + 3\frac{2}{12}$ 6. $5\frac{3}{8} + 7 + 3\frac{3}{8}$ 7. $\frac{5}{11} + 2 + 4\frac{6}{11}$ 8. $7\frac{1}{4} + 6\frac{3}{4} + \frac{1}{4}$

Find the missing number. Use mental math.

9. $3\frac{2}{3} + \square = 4$ 10. $\square + 2\frac{3}{8} = 4$ 11. $1\frac{1}{4} + \square + \frac{1}{4} = 4$

Mixed Applications

Ken boxes special auto parts to ship to customers. List the weights of 2 parts that will pack in each box to total the exact weight given.

12. 4 lb   13. 5 lb
14. 6 lb   15. 7 lb

Weights of Parts in Pounds		
$1\frac{1}{8}$	$1\frac{1}{2}$	$2\frac{7}{8}$
$2\frac{1}{2}$	$3\frac{3}{8}$	$4\frac{1}{4}$
$2\frac{3}{4}$	$3\frac{1}{2}$	$1\frac{5}{8}$

9-7 Regrouping Mixed Numbers

Objective
Regroup mixed numbers.

You sometimes need to regroup mixed numbers in order to subtract.

The drawing shows $2\frac{1}{3}$ regrouped as $1\frac{4}{3}$.

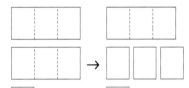

Examples Regroup.

1. $9\frac{1}{4} = 8\frac{\square}{\square}$

| Write the mixed number as a sum. | → | Regroup the whole number. | → | Write the 1 as a fraction with the same denominator. | → | Add the fractions. |

$9\frac{1}{4} = 9 + \frac{1}{4}$ $8 + 1 + \frac{1}{4}$ $8 + \frac{4}{4} + \frac{1}{4}$ $9\frac{1}{4} = 8\frac{5}{4}$

2. $4\frac{5}{8} = 3\frac{\square}{\square} = 3\frac{13}{8}$ 3. $1\frac{1}{6} = \frac{\square}{\square} = \frac{7}{6}$ 4. Regroup 10 as a mixed number using thirds. $10 = 9\frac{3}{3}$

Try This Regroup.

a. $2\frac{1}{2} = 1\frac{\square}{\square}$ b. $4\frac{2}{3} = 3\frac{\square}{\square}$ c. $8\frac{4}{9} = 7\frac{\square}{\square}$ d. $12\frac{4}{5} = 11\frac{\square}{\square}$ e. $1\frac{5}{7} = \frac{\square}{\square}$

Exercises

Regroup.

1. $2\frac{1}{4} = 1\frac{\square}{\square}$ 2. $3\frac{1}{5} = 2\frac{\square}{\square}$ 3. $5\frac{3}{8} = 4\frac{\square}{\square}$ 4. $7\frac{2}{3} = 6\frac{\square}{\square}$

5. $2\frac{5}{6} = 1\frac{\square}{\square}$ 6. $3\frac{3}{7} = 2\frac{\square}{\square}$ 7. $8\frac{11}{12} = 7\frac{\square}{\square}$ 8. $9\frac{13}{16} = 8\frac{\square}{\square}$

9. $5\frac{17}{20} = 4\frac{\square}{\square}$ 10. $2\frac{9}{24} = 1\frac{\square}{\square}$ 11. $1\frac{7}{8} = \frac{\square}{\square}$ 12. $1\frac{4}{9} = \frac{\square}{\square}$

Regroup as a mixed number using eighths.   13. 4   14. 18   15. 30

Regroup as a mixed number using fifths.   16. 11   17. 20   18. 9

Show Math
19. Draw a picture to show why $4\frac{6}{4}$ is the same as $5\frac{1}{2}$.
20. Draw a picture to show why $3\frac{1}{5}$ is the same as $2\frac{6}{5}$.

9-8 Subtracting Mixed Numbers with Regrouping

Objective
Subtract mixed numbers with regrouping.

Three American women have won the Olympic gold medal for the high jump. In 1948 Alice Coachman broke the Olympic record set by Jean Shiley. By how much did Coachman break the old record?

Understand the Situation
- What operation will give the answer?
- Is the difference in feet or inches?
- Are the denominators of the fraction alike?

Jean Shiley 1932	5 ft $5\frac{1}{4}$ in.
Alice Coachman 1948	5 ft $6\frac{1}{8}$ in.
Mildred McDaniel 1956	5 ft $9\frac{1}{4}$ in.

Examples
Subtract.

1. $6\frac{1}{8} - 5\frac{1}{4}$

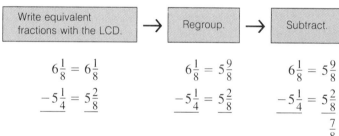

Alice Coachman broke the old Olympic record by $\frac{7}{8}$ in.

2. $4\frac{2}{3} - \frac{4}{5}$

$$4\frac{2}{3} = 4\frac{10}{15} = 3\frac{25}{15}$$
$$-\frac{4}{5} = \frac{12}{15} = \frac{12}{15}$$
$$\phantom{-\frac{4}{5} = \frac{12}{15} = } 3\frac{13}{15}$$

3. $5 - 2\frac{3}{4}$

$$5 = 4\frac{4}{4}$$
$$-2\frac{3}{4} = 2\frac{3}{4}$$
$$\phantom{-2\frac{3}{4} = } 2\frac{1}{4}$$

Try This
Subtract.

a. $5\frac{1}{6} - 4\frac{5}{6}$ **b.** $4\frac{2}{9} - \frac{1}{2}$ **c.** $3 - 1\frac{1}{2}$ **d.** $5\frac{1}{4} - 3\frac{5}{8}$

Exercises

Subtract.

1. $2\frac{1}{3} - \frac{1}{2}$
2. $1\frac{3}{8} - \frac{3}{4}$
3. $4\frac{1}{6} - \frac{4}{9}$
4. $3\frac{1}{2} - \frac{7}{8}$
5. $3\frac{2}{5} - 1\frac{3}{4}$
6. $5\frac{2}{3} - 2\frac{5}{9}$
7. $7\frac{1}{2} - 3\frac{5}{7}$
8. $9\frac{2}{3} - 3\frac{5}{6}$
9. $6\frac{2}{3} - 5\frac{4}{5}$
10. $8\frac{1}{9} - 1\frac{3}{4}$
11. $4\frac{3}{14} - 2\frac{1}{2}$
12. $3\frac{5}{6} - 1\frac{5}{8}$
13. $9\frac{1}{4} - 3\frac{5}{16}$
14. $7\frac{4}{5} - 5\frac{2}{3}$
15. $2\frac{1}{8} - 1\frac{2}{3}$
16. $6\frac{1}{6} - 2\frac{1}{2}$
17. $10\frac{5}{8} - 2\frac{3}{4}$
18. $13\frac{1}{2} - 9\frac{5}{6}$
19. $4\frac{5}{18} - 1\frac{8}{9}$
20. $7\frac{3}{8} - 5\frac{6}{7}$
21. $28\frac{1}{3} - 9\frac{3}{4}$
22. $16\frac{3}{7} - 8\frac{2}{3}$
23. $6\frac{3}{4} - 3\frac{9}{14}$
24. $9\frac{3}{10} - 4\frac{4}{5}$

25. By how many inches did Mildred McDaniel break the record Jean Shiley set for the high jump?

26. At the 1984 Olympics a record of 6 ft $7\frac{1}{2}$ in. for the women's high jump was set. Find the difference between this jump and Coachman's Olympic record.

Mental Math

Is the difference more or less than the given answer? Use < or >.

27. $5\frac{1}{2} - 3\frac{3}{4} \square 2$
28. $7\frac{5}{6} - 3\frac{2}{3} \square 4$
29. $12\frac{3}{8} - 8\frac{9}{16} \square 3$
30. $8\frac{2}{7} - 1\frac{3}{5} \square 7$

Data Bank

Use the information about the Olympic men's high jump in the Data Bank on page 491.

31. In 1936 Cornelius Johnson set an Olympic record in the high jump. By how much did he break the old record?

32. Which gold medalist broke the Olympic record by the greatest number of inches?

33. Dick Fosbury won the gold medal in 1968. How many more inches would he need to tie the 1984 Olympic record?

34. Australian John Winter won the men's high jump at the 1948 Olympics. How many more inches did he need to tie the existing record?

Mixed Skills Review

Solve and check.

35. $3\frac{2}{3} + x = 7$
36. $1\frac{2}{5} + y = 5\frac{3}{10}$
37. $t + 1\frac{3}{4} = 2\frac{1}{8}$
38. $f + 4\frac{5}{6} = 8\frac{1}{2}$
39. $9\frac{1}{3} - r = 6\frac{1}{4}$
40. $s - 5\frac{7}{8} = 2\frac{1}{6}$

Find the missing number.

41. $\frac{2}{3}$ yd = $\square$ ft
42. 8 in. = $\square$ ft
43. 4 ft = $\square$ yd

ADDITION AND SUBTRACTION OF FRACTIONS

9-9 Estimating Sums and Differences Using Rounding

Objective
Estimate sums and differences using rounding.

A cross-country ski trail begins at the Chalet. The Ski Club plans to ski about 6 miles to a lodge, and then eat lunch. Which lodge would be best?

Understand the Situation

- Do the skiers need an estimate or exact answer?
- Is there more than one way to get to each lodge?

To estimate with fractions and mixed numbers, round to the nearest whole number or to the nearest $\frac{1}{2}$.

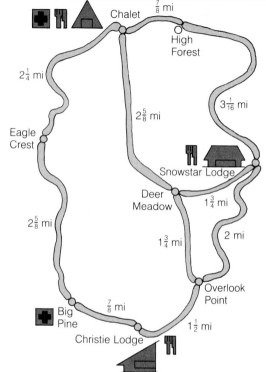

Examples Estimate.

1. $\frac{7}{8} + 3\frac{1}{16}$

   $\frac{7}{8}$ mi is about 1 mi.

   $3\frac{1}{16}$ mi is about 3 mi.

   $1 + 3 = 4$ The distance is about 4 mi to Snowstar Lodge.

2. $2\frac{1}{4} + 2\frac{5}{8} + \frac{7}{8}$

   $2\frac{1}{4}$ is about $2\frac{1}{2}$ mi.
   $2\frac{5}{8}$ is about $2\frac{1}{2}$ mi.
   $\frac{7}{8}$ is about 1 mi.

   $2\frac{1}{2} + 2\frac{1}{2} + 1 = 6$

   The distance is about 6 mi to Christie Lodge.

3. $2\frac{5}{8} + 1\frac{3}{4} + 1\frac{1}{2}$

   $2\frac{5}{8}$ is about $2\frac{1}{2}$ mi.
   $1\frac{3}{4}$ is about 2 mi.
   $1\frac{1}{2}$ does not need to be rounded.

   $2\frac{1}{2} + 2 + 1\frac{1}{2} = 6$

   Using a different route, it is about 6 mi to Christie Lodge.

Try This Estimate.

a. $2\frac{1}{8} + 6\frac{4}{5}$ b. $3\frac{1}{2} - 1\frac{1}{3}$ c. $4\frac{3}{16} + \frac{5}{9}$ d. $7\frac{1}{3} - 4\frac{3}{10}$ e. $5\frac{2}{3} + 6\frac{2}{7}$

Exercises

Estimate.

1. $3\frac{3}{4} + 1\frac{2}{5}$
2. $6\frac{1}{2} + 9\frac{5}{6}$
3. $5\frac{2}{3} - 2\frac{4}{9}$
4. $7\frac{1}{8} - 4\frac{5}{7}$
5. $4\frac{4}{5} - 3\frac{1}{10}$
6. $2\frac{1}{1} - 1\frac{5}{8}$
7. $4\frac{5}{12} - 1\frac{9}{16}$
8. $8\frac{5}{9} - 2\frac{1}{2}$
9. $2\frac{5}{11} + 4\frac{11}{12}$
10. $4\frac{5}{32} - 1\frac{1}{4}$
11. $12\frac{13}{15} - 10\frac{2}{3}$
12. $2\frac{1}{8} + 6\frac{4}{5}$
13. $1\frac{3}{4} + 1\frac{1}{8} + \frac{5}{8}$
14. $4\frac{1}{4} - 2\frac{3}{8}$
15. $3\frac{1}{6} - 1\frac{3}{5}$
16. $4\frac{6}{7} + 6\frac{2}{3} + \frac{3}{7}$

17. A skier fell on the hill at Overlook Point. Where is the closest first aid station? About how far is it?

18. Which trail from Snowstar Lodge to the Chalet is shorter? Estimate the difference.

19. A skier at Eagle Crest wants to return to Snowstar Lodge. Is it shorter to go through High Forest or Deer Meadow? Estimate the difference.

20. A snow-making machine is halfway between the Chalet and Big Pine. About how many miles from the Chalet is the machine located?

Show Math

21. Triangle Ski Resort advertises that each loop trail is about 6 miles long. Design a layout for the resort using the following distances: $1\frac{3}{4}$, $1\frac{1}{2}$, $3\frac{3}{10}$, $\frac{5}{8}$, $4\frac{2}{5}$, $1\frac{1}{8}$, $2\frac{3}{5}$, $\frac{3}{4}$, $2\frac{2}{3}$.

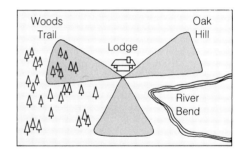

Estimation

22. Alana is packing bundles of clothes. She needs to make each bundle weigh about 12 lb. Use estimation to choose the bags Alana should pack together.

Practice Through Problem Solving

Suppose you were given the following number cards. Make a list of all the ways you could make a true sentence using the number cards.

| 1 | 2 | 3 | 4 | 5 | 6 |

23. + < 1

24. − > 2

9-10 Consumer Math: Framing Pictures

Objective
Use fractions in consumer problems.

- SITUATION
- DATA
- PLAN
- ANSWER
- CHECK

Jeanne and Tom Sakai bought a painting and a frame at an auction. Tom plans to cut a colored mat to border the painting. How wide should he make the colored mat to fit between the painting and the frame?

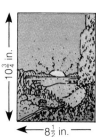

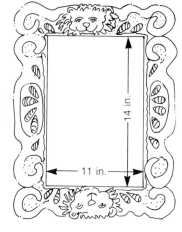

Drawing a picture will help you understand the problem. Find the difference between the width of the frame and the width of the painting. Divide this amount by 2.

Example

The inside of the frame is 11 in. wide. The painting is $8\frac{1}{2}$ in. wide.

$11 - 8\frac{1}{2} = 2\frac{1}{2}$ The difference is $2\frac{1}{2}$ in.

Divide the difference by 2 to get the size of each border.

$2\frac{1}{2} \div 2 = \frac{5}{2} \times \frac{1}{2} = \frac{5}{4} = 1\frac{1}{4}$

The mat border will be $1\frac{1}{4}$ in. on each side.

Try This Use the measurements in the drawing.

Find how wide the mat border will be at the top and bottom of the Sakais' painting.

Exercises

Find the size of the mat border for both dimensions of each project.

1. Inside of frame:
   8 in. wide by 10 in. high
   Certificate:
   $5\frac{1}{4}$ in. wide by $6\frac{1}{2}$ in. high

2. Inside of frame:
   5 in. wide by 7 in. high
   Picture:
   $3\frac{3}{8}$ in. wide by $4\frac{5}{8}$ in. high

3. Inside of frame:
   4 in. wide by 6 in. high
   Embroidery:
   $2\frac{3}{4}$ in. wide by $3\frac{7}{16}$ in. high

4. Inside of frame:
   20 in. wide by 16 in. high
   Painting:
   $15\frac{7}{8}$ in. wide by $11\frac{3}{16}$ in. high

5. Inside of frame:
   $6\frac{1}{2}$ in. wide by 5 in. high
   Autograph:
   $4\frac{1}{2}$ in. wide by $3\frac{1}{4}$ in. high

6. Inside of frame:
   $12\frac{3}{4}$ in. wide by $19\frac{5}{8}$ in. high
   Newspaper clipping:
   $9\frac{1}{8}$ in. wide by $15\frac{1}{2}$ in. high

7. The Sakais are making photo panels. What is the total width of the panel in the drawing?

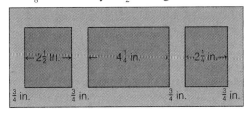

8. A family photograph measured 10 in. high by 12 in. wide. The mat added $1\frac{7}{8}$ in. on each side and $2\frac{1}{2}$ in. at the top and the bottom. What size frame should Jeanne buy?

Problem Solving

9. Jeanne's mother wants to use a wide border around an old sampler. The sampler is $11\frac{1}{4}$ in. wide. She will add $3\frac{5}{8}$ in. of border on each side. Each frame strip is $1\frac{1}{4}$ in. wide. What will be the finished width of frame, border, and sampler?

10. Tom had a piece of mat board 36 in. wide. He cut 2 large mats of equal size and a small mat $5\frac{3}{4}$ in. wide from the board. He had a piece $7\frac{1}{2}$ in. wide left over. How wide was each of the large mats?

11. Jeanne is mounting 3 horse show ribbons inside a frame 14 in. wide. The top of each ribbon is 3 in. in diameter. Jeanne spaces the ribbons evenly inside the frame. How much space does she leave between each ribbon and between the ribbons and the frame?

Estimation

Estimate. Use < or >.

12. $3\frac{9}{10} + 4\frac{7}{8} \square 8\frac{1}{2}$
13. $8 - 2\frac{4}{5} \square 6$
14. $2\frac{7}{8} + 3\frac{7}{8} + 4\frac{7}{8} \square 12$
15. $5\frac{2}{3} - 1\frac{1}{5} \square 4$
16. $\frac{3}{4} + 4\frac{3}{8} \square 5$
17. $1\frac{1}{2} - \frac{7}{8} \square 1$
18. $6\frac{1}{12} + 4\frac{2}{5} + \frac{3}{5} \square 11$
19. $\frac{1}{3} + 1\frac{9}{10} \square 2$
20. $3\frac{1}{8} + 2\frac{1}{32} \square 5\frac{1}{2}$

Data Hunt

21. Call an art supply store. Find what standard sizes of mat board they carry. What size board would you need to make 3 mats $11\frac{1}{4}$ in. high by $14\frac{3}{4}$ in. wide? Draw a picture to show how you would cut the board.

Mixed Skills Review

Change the unit.

22. 3 c = ___ fl oz
23. 10 ft = ___ yd **1 ft**
24. 247 g = ___ kg

Write using standard numerals.

25. $(2.05)^2$
26. 15 squared
27. 100^3

ADDITION AND SUBTRACTION OF FRACTIONS   **257**

9-11 Problem Solving: Understanding the Situation

Objective
State the question in the problem.

- **SITUATION**
- **DATA**
- **PLAN**
- **ANSWER**
- **CHECK**

Use the weights in the chart. Read each question carefully. Then decide which of the questions that follow asks the same thing.

Weights in Ounces
Gerbil	$4\frac{2}{5}$
Giant bat	$1\frac{9}{10}$
Golden hamster	$4\frac{1}{5}$
Hummingbird	$\frac{2}{5}$
Mole	$3\frac{1}{4}$
Mouse	$\frac{4}{5}$
Shrew	3

1. *Question:* Two of which animal will weigh the same as one mouse?
   A. Which animal weighs half as much as a mouse?
   B. Does the mouse weigh twice as much as 2 of these animals?

2. *Question:* Two gerbils weigh the same as 3 of what other animal?
   A. The total weight of 2 gerbils equals $\frac{1}{3}$ the weight of what other animal?
   B. The total weight of 2 gerbils is the same as 3 times the weight of what other animal?

3. *Question:* A shrew is $1\frac{1}{10}$ oz heavier than a giant bat. Does the shrew weigh more or less than a mole?
   A. Is $1\frac{1}{10}$ more or less than $3\frac{1}{4}$?
   B. Is the weight of a mole more or less than $1\frac{1}{10}$ oz + $1\frac{9}{10}$ oz?

4. *Question:* What is the difference between the weight of a mole and the combined weight of a mouse and a giant bat?
   A. $\frac{4}{5}$ plus $1\frac{9}{10}$ equals how much?
   B. The weight of the mole less the weights of the mouse and the bat equals how many ounces?

5. Write a question about adding some of these animal weights. Then solve the problem.

6. Write a question about the differences in the weights of the animals. Then solve the problem.

Strategy Practice

7. In a biology laboratory Tim had set up cages that looked like this. He wanted to remove exactly 6 dividers to make 2 square cages. Which dividers must he remove?

1	2	3
4	5	6

8. At a math contest, 25 problems were given. The scoring gave 4 points for each correct answer and deducted 2 points for each incorrect answer. Fran's score was 64. How many correct answers did she have?

9. On a difficult hike, Lou, Sirina, and Sanh took turns leading. Lou led for $9\frac{1}{2}$ mi. This was 2 more miles than Sirina led. Sirina led twice as far as Sanh did. Find the total distance of the hike.

9-12 Problem Solving: Group Decisions

Objective
Make group decisions.

- SITUATION
- DATA
- PLAN
- ANSWER
- CHECK

To promote school spirit, your class has decided to make and sell banners at school football games. You must make a complete plan for the project. The plan should include design, costs, schedule, price, and a production plan.

Facts to Consider
- The banners must use two school colors.
- Felt costs $6.83 per yard. It is 60 in. wide.
- Sticks to hold the banners come in 4-ft lengths at 15¢ per foot.
- Glue is $1.50 per 8-oz bottle.
- The sports department is willing to lend the class money for supplies. The class must pay back the loan.
- The activities committee has information on attendance at the games.

Plan and Make a Decision
- Tell what your group is being asked to do.
- How will your group choose the design of the banner?
- How will you organize the class to make the banners?
- What factors will determine the amount and cost of supplies? the selling price?
- Would a chart or graph help you organize and analyze the data you collect?
- How will you organize the class to sell the banners? How might a diagram or chart help you make a decision?

Share Your Group's Decision
- Make a presentation of your completed plan to the class.
- Compare your plan with plans of other groups. Point out the best features of each.

Suppose
- Suppose felt was on sale at $4 per yard. How would this affect your decisions?

9-13 Enrichment: Time Zones

Objective
Calculate time in different time zones.

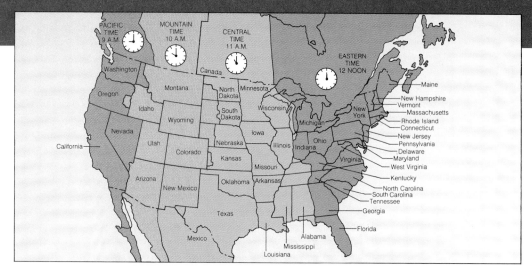

The map shows where the world time-zone lines cross the continental United States. As you travel west, the time is 1 hour earlier for each time-zone line you cross.

Example Jan lives in Albany, New York. She calls her cousin in Denver, Colorado, at 4:15 p.m. (EST). What time is it in Denver?

There are 2 time-zone lines between New York and Colorado. Colorado is west of New York. The time in Colorado will be 2 hours earlier than in New York.

It is 2:15 p.m. (MST) in Denver, Colorado.

Try This
a. When it is 4 p.m. in Salt Lake City, Utah, what time is it in St. Louis, Missouri?
b. When it is 1 p.m. in Miami, Florida, what time is it in Billings, Montana?

Exercises

Many states have cities with the same names as famous cities in Europe. Use the map to find what time it is in each city when it is 12:00 noon in Paris, Texas.

1. Rome, Pennsylvania
2. Dublin, California
3. Florence, Arizona
4. Berlin, Wisconsin
5. Aberdeen, Washington
6. Glasgow, Montana
7. Naples, Florida
8. Milan, Missouri
9. Toledo, Ohio
10. Odessa, Texas

9-14 Computer: Loops

Objective
Use loops.

A computer follows a series of single steps in order. You can write a computer program that sends the computer from one line to another line that is not in sequence. The **GOTO** command followed by a line number tells a computer to go to that line number. This can create a loop in the program. **Loops** are parts of a program that repeat an action.

Example

Identify the loop in this program.
Decide what the program will print.

```
NEW
10 PRINT ''XXX''
20 GOTO 10
RUN
```

Line 10   tells the computer to print three Xs.
Line 20   tells the computer to go back to line 10 and do the same thing all over again. The computer could display Xs forever. You can stop the loop by pressing CTRL C.

Try This

Identify the loop in this program. Decide what the program will print.

```
NEW
10 PRINT ''THE QUICK BROWN FOX''
20 PRINT ''JUMPS OVER THE LAZY DOG''
30 GOTO 10
RUN
```

Exercises

Identify the loop in the program. Decide what the program will print.

1.
```
NEW
10 PRINT ''THE QUICK BROWN FOX''
20 PRINT ''JUMPS OVER THE LAZY DOG''
30 GOTO 20
RUN
```

2.
```
NEW
10 PRINT ''THE QUICK BROWN FOX''
20 GOTO 10
30 PRINT ''JUMPS OVER THE LAZY DOG''
RUN
```

3. Change line 10 in the example program to a message you want to repeat.

4. Make a flowchart of the example program.

CHAPTER 9 REVIEW

Continue the pattern.

1. $\frac{1}{3}$, 1, $1\frac{2}{3}$, ___, ___, ___
2. $\frac{3}{5}$, $1\frac{2}{5}$, $2\frac{1}{5}$, ___, ___, ___
3. $\frac{1}{8}$, 1, $1\frac{7}{8}$, ___, ___, ___
4. $1\frac{4}{9}$, $2\frac{5}{9}$, $3\frac{2}{3}$, ___, ___, ___
5. $2\frac{1}{4}$, $3\frac{1}{2}$, $4\frac{3}{4}$, ___, ___, ___
6. $\frac{1}{2}$, 3, $5\frac{1}{2}$, ___, ___, ___

Write true or false.

7. $\frac{3}{10} + 1\frac{2}{5} = 1\frac{7}{10}$
8. $4\frac{7}{8} - 3\frac{3}{4} = 1\frac{1}{8}$
9. $2\frac{3}{5} + 3\frac{1}{4} = 5\frac{13}{20}$
10. $4\frac{2}{3} - 1\frac{3}{4} = 3\frac{11}{12}$
11. $5\frac{1}{2} - 2\frac{7}{9} = 2\frac{13}{18}$
12. $3\frac{1}{5} - 2\frac{1}{3} = \frac{13}{15}$
13. $1\frac{1}{2} + 2\frac{1}{4} + 3\frac{1}{4} = 7$
14. $1\frac{1}{6} + 1\frac{1}{3} + 2\frac{1}{2} = 5\frac{1}{3}$

Use mental math to complete each statement with <, >, or =.

15. $2\frac{1}{12} + 1\frac{5}{12}$ ___ $3\frac{1}{2}$
16. $3\frac{2}{3} + 4\frac{1}{2}$ ___ 8
17. $2\frac{1}{3} + 9\frac{1}{3}$ ___ $11\frac{1}{2}$
18. $1\frac{3}{7} + 1\frac{4}{9}$ ___ 3
19. $7\frac{1}{3} - 5\frac{1}{2}$ ___ 2
20. $8\frac{5}{6} - 7\frac{1}{2}$ ___ 1
21. $12\frac{7}{8} - 6\frac{3}{8}$ ___ $6\frac{3}{4}$
22. $9\frac{5}{8} - 2\frac{3}{8}$ ___ $7\frac{1}{4}$
23. $1\frac{4}{9} - \frac{7}{9}$ ___ $\frac{2}{3}$
24. $15\frac{1}{5} - 3\frac{1}{4}$ ___ 12

Choose a number from the box to complete each sentence.

25. $2\frac{1}{8}$ + ___ is about 7.

26. $3\frac{1}{7}$ + ___ is about 7.

27. $12\frac{4}{5}$ − ___ is about 6.

28. $4\frac{8}{9}$ + ___ is about 13.

29. $6\frac{1}{7}$ − ___ is about 3.

30. $9\frac{12}{25}$ − ___ is about 4.

$3\frac{1}{4}$	$9\frac{1}{3}$	$5\frac{1}{2}$
$4\frac{5}{7}$	$3\frac{7}{9}$	$2\frac{1}{2}$
$6\frac{2}{3}$	$8\frac{1}{3}$	$1\frac{4}{5}$

CHAPTER 9 TEST

1. Three high school clubs sold tickets to a dance to raise money. They paid expenses of $77.34 from the total ticket sales. Then each club got $89.22. How much money did the dance raise?

Add or subtract.

2. $\frac{3}{7} + \frac{2}{7}$
3. $\frac{8}{11} - \frac{3}{11}$
4. $\frac{7}{12} + \frac{1}{12}$
5. $\frac{17}{18} - \frac{11}{18}$
6. $2\frac{1}{5} + \frac{4}{5}$
7. $4\frac{7}{9} - \frac{4}{9}$
8. $2\frac{3}{5} + 1\frac{4}{5}$
9. $2\frac{8}{15} - 2\frac{2}{15}$
10. $\frac{1}{2} + \frac{2}{3}$
11. $\frac{3}{5} - \frac{1}{10}$
12. $\frac{5}{8} + \frac{5}{12}$
13. $\frac{8}{15} - \frac{2}{5}$
14. $\frac{3}{4} + 1\frac{1}{2}$
15. $2\frac{4}{5} - \frac{1}{3}$
16. $3\frac{5}{12} + 1\frac{5}{9}$
17. $5\frac{7}{10} - \frac{3}{5}$

Add. Use mental math.

18. $1\frac{2}{3} + 4 + 2\frac{1}{3}$
19. $3\frac{3}{5} + 2 + 5\frac{2}{5}$
20. $4\frac{5}{12} + 3 + 1\frac{1}{12}$

Regroup.

21. $3\frac{4}{5} = 2\frac{\square}{\square}$
22. $5\frac{7}{9} = 4\frac{\square}{\square}$
23. $6\frac{11}{15} = 5\frac{\square}{\square}$

Subtract.

24. $3\frac{1}{3} - \frac{2}{3}$
25. $5\frac{3}{8} - 3\frac{3}{4}$
26. $4\frac{5}{12} - \frac{5}{6}$
27. $7\frac{5}{9} - 6\frac{2}{3}$

Estimate.

28. $2\frac{5}{6} + 5\frac{2}{9}$
29. $12\frac{4}{7} - 4\frac{5}{8}$
30. $2\frac{1}{3} + 3\frac{2}{5} + \frac{1}{2}$

Find the size of the mat border for both dimensions of each project.

31. inside of frame: 9 in. wide by 12 in. high
    picture: $7\frac{1}{4}$ in. wide by $10\frac{1}{2}$ in. high

32. inside of frame: 24 in. wide by 40 in. high
    painting: $20\frac{3}{4}$ in. wide by $36\frac{3}{8}$ in. high

Use the weights in the chart. Read the question carefully. Then decide which of the questions that follow asks the same thing.

33. *Question:* Five mice weigh $\frac{1}{5}$ oz less than which animal?

    The weight of 5 mice plus $\frac{1}{5}$ oz equals the weight of which other animal?

    B. One-fifth of the weight of which animal is the same as the weight of a mouse?

Weights in Ounces	
Gerbil	$4\frac{2}{5}$
Giant bat	$1\frac{9}{10}$
Golden hamster	$4\frac{1}{5}$
Hummingbird	$\frac{2}{5}$
Mole	$3\frac{1}{4}$
Mouse	$\frac{4}{5}$
Shrew	3

ADDITION AND SUBTRACTION OF FRACTIONS

CHAPTER 9 CUMULATIVE REVIEW

1. Add.
   $\frac{5}{6} + \frac{3}{4}$
   A. $\frac{8}{10}$
   B. $1\frac{7}{12}$
   C. $1\frac{2}{3}$
   D. $\frac{15}{24}$

2. Divide.
   $6\frac{2}{3} \div 3$
   A. $2\frac{2}{9}$
   B. 20
   C. $1\frac{2}{9}$
   D. 11

3. Find the size of the mat border for both dimensions of this project.
   inside of frame: 18 in. wide by 14 in. high
   photograph: $13\frac{3}{4}$ in. wide by $9\frac{3}{16}$ in. high
   A. $2\frac{5}{8}$ in., $2\frac{13}{32}$ in.
   B. 4 in., $3\frac{3}{16}$ in.
   C. 2 in., $2\frac{9}{32}$ in.
   D. $2\frac{1}{8}$ in., $2\frac{13}{32}$ in.

4. Change the unit.
   5 gal = ___ qt
   A. 18
   B. 22
   C. 20
   D. 19

5. Subtract.
   $\frac{17}{36} - \frac{11}{36}$
   A. $\frac{1}{6}$
   B. $\frac{4}{9}$
   C. $\frac{7}{9}$
   D. $\frac{16}{36}$

6. Find the least common multiple of 8 and 28.
   A. 2
   B. 56
   C. 14
   D. 4

7. Write using exponents.
   8 to the sixth power.
   A. 48
   B. 8^6
   C. 262,144
   D. 8_6

8. Add. Use mental math.
   $5\frac{3}{8} + 4\frac{2}{3} + \frac{5}{8}$
   A. $9\frac{8}{8}$
   B. $9\frac{10}{19}$
   C. $10\frac{1}{6}$
   D. $10\frac{2}{3}$

9. Multiply.
   $1\frac{3}{8} \times 2\frac{4}{5}$
   A. $2\frac{17}{20}$
   B. $2\frac{33}{40}$
   C. $3\frac{17}{20}$
   D. $3\frac{4}{5}$

10. Write the fraction as a decimal.
    $\frac{3}{8}$
    A. 0.375
    B. 0.0375
    C. 3.75
    D. 0.475

11. Subtract.
    $7\frac{2}{3} - 5\frac{7}{8}$
    A. $2\frac{19}{24}$
    B. $2\frac{5}{24}$
    C. $1\frac{3}{4}$
    D. $1\frac{19}{24}$

12. Divide.
    $87 \div 1000$
    A. 8.7
    B. 870
    C. 0.087
    D. 0.870

13. Estimate. Round to the highest place.
    78 + 32
    A. 130
    B. 110
    C. 100
    D. 90

14. Change the unit.
    120 in. = ____ ft
    A. 15 ft
    B. 11 ft, 10 in.
    C. 10 ft
    D. 10 ft, 4 in.

15. Write as a standard numeral. six and four hundred twelve thousandths
    A. 0.6412
    B. 6.4012
    C. 6.412
    D. 604.012

16. Add.
    329
    +347
    A. 676
    B. 666
    C. 675
    D. 686

17. Multiply.
    46.836 × 100
    A. 468.36
    B. 5793.6
    C. 0.46836
    D. 4683.6

18. Estimate.
    $\frac{10}{22} \times 24$
    A. 14
    B. 12
    C. 9
    D. 8

19. Regroup.
    $3\frac{2}{3} = 2\frac{\square}{\square}$
    A. $\frac{5}{3}$
    B. $\frac{1}{3}$
    C. $1\frac{5}{3}$
    D. $\frac{4}{3}$

20. Divide.
    24)40.8
    A. 1.9
    B. 1.68
    C. 2.3
    D. 1.7

21. Change the unit.
    19,350 L = ____ kL
    A. 193,500
    B. 1.9350
    C. 19.350
    D. 193.50

22. Find the reciprocal.
    $5\frac{3}{8}$
    A. $\frac{8}{16}$
    B. $\frac{8}{43}$
    C. $5\frac{8}{3}$
    D. $4\frac{11}{8}$

23. Write the missing numeral.
    $\frac{8}{11} = \frac{\square}{55}$
    A. 48
    B. 21
    C. 40
    D. 5

24. Subtract.
    $122\frac{4}{5} - 85\frac{2}{3}$
    A. $47\frac{2}{15}$
    B. $37\frac{2}{15}$
    C. $37\frac{1}{3}$
    D. $36\frac{8}{15}$

25. Multiply. Use mental math.
    $7.75 × 8
    A. $60.75
    B. $62
    C. $56.75
    D. $61.50

26. How much lemon juice is in the measuring cup?
    A. 4 fl oz
    B. 8 fl oz
    C. 10 fl oz
    D. 6 fl oz

ADDITION AND SUBTRACTION OF FRACTIONS

Chapter 10 Overview

Key Ideas
- Use the problem-solving strategy *guess and check*.
- Find rates.
- Find the better buy.
- Solve proportions.
- Use similar figures and scale drawings.
- Evaluate the reasonableness of answers.
- Write ratios as percents.
- Write percents as decimals and fractions.
- Write decimals and fractions as percents.
- Collect and analyze data.

Key Terms
- ratio
- proportion
- rate
- percent
- scale drawing
- similar figures
- cross products
- corresponding
- unit rate
- unit price

Key Skills

Write in lowest terms.

1. $\frac{4}{8}$
2. $\frac{3}{9}$
3. $\frac{3}{12}$
4. $\frac{5}{25}$
5. $\frac{6}{8}$
6. $\frac{4}{24}$
7. $\frac{12}{36}$
8. $\frac{10}{50}$
9. $\frac{18}{40}$
10. $\frac{6}{32}$
11. $\frac{8}{36}$
12. $\frac{20}{100}$
13. $\frac{25}{80}$
14. $\frac{32}{60}$
15. $\frac{28}{51}$
16. $\frac{34}{60}$
17. $\frac{25}{100}$
18. $\frac{50}{75}$

Write two fractions that are equivalent to the one given.

19. $\frac{1}{3}$
20. $\frac{1}{5}$
21. $\frac{2}{5}$
22. $\frac{3}{8}$
23. $\frac{1}{6}$
24. $\frac{5}{8}$
25. $\frac{3}{10}$
26. $\frac{3}{4}$
27. $\frac{5}{12}$
28. $\frac{21}{100}$
29. $\frac{1}{200}$
30. $\frac{1}{1000}$

Tell which digit is in the hundredths place.

31. 0.45
32. 2.04
33. 0.738
34. 0.5
35. 0.97
36. 0.1032

Arrange the numbers in order from least to greatest.

37. 0.56, 0.65, 0.08, 0.5, 0.068
38. 0.4, 0.004, 0.040, 4.00, 0.44

RATIO, PROPORTION, AND PERCENT 10

10-1 Problem Solving: Learning to Use Strategies

Objective
Use the strategy guess and check.

- SITUATION
- DATA
- PLAN
- ANSWER
- CHECK

Ekwa collects jazz and classical cassette tapes. She has a total of 115 tapes. There are 65 more jazz tapes than classical tapes. How many classical tapes does she have?

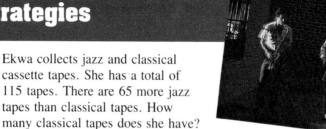

The best way to solve some problems is by trial and error. This strategy is called **guess and check**. You can guess the answer and then check to see if your answer is correct.

Example Solve.

How many classical tapes does Ekwa have?

| Guess the number of classical tapes. Check your guess. | Guess = 15 | Check. Number of classical tapes: 15 Number of jazz tapes: $15 + 65 = 80$ Total number of tapes: $15 + 80 = 95$ $95 < 115$ The sum is less than the given total. |

↓

| Guess another number. Check your guess. | Guess = 30 | Check. Number of classical tapes: 30 Number of jazz tapes: $30 + 65 = 95$ Total number of tapes: $30 + 95 = 125$ $125 > 115$ The sum is greater than the given total. |

↓

| Guess another number. Check your guess. | → Guess = 25 | Check. Number of classical tapes: 25 Number of jazz tapes: $25 + 65 = 90$ Total number of tapes: $25 + 90 = 115$ $115 = 115$ It checks. |

Try This

a. For the past two weeks, Barry's band practiced a total of 65 h. They practiced 25 more hours the second week than the first week. How many hours did they practice the first week?

b. Tuwanna has a total of 56 albums and CDs. She has 32 fewer CDs than albums. How many albums does she have?

Exercises

Solve. Use one or more of the problem-solving strategies.

1. In a survey of 78 freshmen, 20 more students liked playing sports after school than liked watching TV after school. How many students liked playing sports after school?

Problem-Solving Strategies
- Guess and check.
- Choose an operation.
- Make an organized list.
- Act it out.
- Find a pattern.
- Write an equation.
- Use logical reasoning.
- Draw a picture or diagram.
- Make a table.
- Use objects.
- Work backward.
- Solve a simpler problem.

2. Maria bought a whole block of tickets to a rock concert. After selling 10 tickets, she divided the rest of the tickets equally among herself and 4 friends. She ended up with 3 tickets. How many tickets did she buy?

3. The local radio station played 220 songs in 18 h. Three times as many songs were rock as were other types of music. How many songs were not rock music?

4. Jim, Kameel, Robyn, and Tae all like different kinds of music. One evening, they got their tapes mixed up. Each had someone else's tapes. Kameel did not have Jim's tapes. Jim and Tae did not have each other's tapes. Jim found Robyn's name on the tapes he had. Who had whose tapes?

	Jim	Kameel	Robyn	Tae
Jim's tapes				
Kameel's tapes				
Robyn's tapes				
Tae's tapes				

5. Alison earned enough money to buy a set of drums. She went to three music stores in her area. MusicTime is 15 mi west of Alison's house. Copp's Music is 22 mi east of MusicTime. Guitar 'n' Things is 5 mi east of Copp's. How far is each store from Alison's house?

6. Mr. Desai was building a square pen. He used 8 vertical posts on each side of the pen. How many posts did he use?

7. Mrs. Iwata planted 60 vegetable seeds in a 5-day period. Each day she planted 3 more seeds than the day before. How many seeds did Mrs. Iwata plant each day?

Suppose

8. Suppose in the example on p. 268 that Ekwa has 43 more jazz tapes than classical tapes. How many jazz tapes does she have?

Write Your Own Problem

9. Write a problem that you could solve using the *guess and check* strategy. Use the school dance survey results.

Survey Results: School Dance

Live Band/Higher Admission Price—83 students
Taped Music/Lower Admission Price—64 students

RATIO, PROPORTION, AND PERCENT

10-2 Ratios

Objective
Write ratios.

A **ratio** is a comparison of two numbers. Marta wanted to know the ratio of wins to losses for the Beacon Blast.

County Soccer League

Team	Won	Lost
Beacon Blast	34	14
Pinole Knights	32	16
Richmond Rangers	31	17
Los Trancos Stallions	20	28
Harbor City Pirates	18	30
Mill Valley Stars	15	33

Understand the Situation

- How many games did the Blast win?
- How many games did the Blast lose?

You can write the ratio of wins to losses for the Beacon Blast in three ways.

34 to 14   34:14 $\frac{34}{14}$

You read each ratio as *34 to 14*.

You can write a ratio as a fraction in lowest terms.

Examples Write the ratio as a fraction in lowest terms.

1. Beacon Blast
wins to losses

   wins → $\frac{34}{14} = \frac{17}{7}$ Write wins over losses. Write the ratio
   losses →   as a fraction in lowest terms.

   The ratio of wins to losses for the Beacon Blast is $\frac{17}{7}$.

2. Los Trancos Stallions
wins to total games played

   wins → $\frac{20}{20 + 28} = \frac{20}{48} = \frac{5}{12}$
   total games →

3. 84 students for
14 teachers

   students → $\frac{84}{14} = \frac{6}{1}$ *Leave the*
   teachers →   *denominator as 1.*

4. 80 heartbeats per
60 seconds

   heartbeats → $\frac{80}{60} = \frac{4}{3}$
   seconds →

5. 9 days absent out of
180 school days

   days absent → $\frac{9}{180} = \frac{1}{20}$
   school days →

Try This Write the ratio as a fraction in lowest terms.

a. Pinole Knights wins to losses   **b.** Harbor City Pirates losses to wins

c. 72 heartbeats per 60 seconds   **d.** 6 tickets for $42

e. 4 in. of rain for 28 days   **f.** 216 students for 6 buses

Exercises

Write the ratio as a fraction in lowest terms.

NHL Campbell Conference Norris Division

	Wins	Losses	Ties
Chicago	26	21	7
St. Louis	23	22	7
Minnesota	20	24	8
Toronto	15	32	5
Detroit	12	37	5

Chicago
1. wins to losses
2. losses to wins
3. wins to total games played

St. Louis
4. wins to losses
5. losses to wins
6. losses to total games played

Minnesota
7. wins to losses
8. ties to wins
9. ties to total games played

Toronto
10. wins to ties
11. ties to total games
12. total games played to losses

13. 20 km in 7 h
14. $16.75 interest for 5 months
15. 2 lb hamburger for 8 servings
16. 92 passengers on 2 buses
17. 7 apples for 89¢
18. 8 lemons for 96¢
19. 8 in. of snowfall for 30 days
20. 3 parts hydrogen for 2 parts oxygen

Problem Solving

21. Willa scored a total of 50 points in 5 basketball games. In each game, she scored 3 more points than the game before. How many points did Willa score in each game?

22. In a tennis match, Boris Becker defeated John McEnroe. Becker had a ratio of 39 games won to 33 games lost. Fill in the missing set scores. (Hint: The first player to win 6 games and to win by at least 2 games wins a set.)

Set	1	2	3	4	5
Score	4–☐	☐–13	☐–☐	☐–2	☐–2

Show Math

Draw a picture to represent each ratio.

23. tennis balls to basketballs 5:3
24. soccer balls to hockey pucks 2:3

Mixed Skills Review

Write each fraction in lowest terms.

25. $\frac{30}{15}$ 26. $\frac{11}{55}$ 27. $\frac{4}{18}$ 28. $\frac{20}{25}$ 29. $\frac{6}{16}$ 30. $\frac{18}{27}$

Tell if you need an estimate or an exact answer.

31. You have $5. You want to know if you can buy 3 magazines for $1.25 each.
32. You sold your old bike for $45. You had to make change for three $20 bills.

RATIO, PROPORTION, AND PERCENT

10-3 Proportions

Objective
State if two ratios are equal.

Elston caught 6 fish in 2 h. Dina caught 9 fish in 3 h. Are the ratios of the number of fish caught to the time spent fishing equal for both Elston and Dina?

Understand the Situation
- What is Elston's ratio of the number of fish caught to the time spent fishing?
- What is Dina's ratio of the number of fish caught to the time spent fishing?

A **proportion** is a sentence that states that two ratios are equal. Two ratios are equal if their **cross products** are equal.

Examples State if the ratios are equal. Use = or ≠.

1. $\frac{6}{2} \square \frac{9}{3}$ Read as *6 is to 2 as 9 is to 3*.

   Find the cross products. → Are the cross products equal? If yes, the ratios are equal.

   $\frac{6}{2} \times \frac{9}{3}$
   $6 \times 3 = 18$
   $2 \times 9 = 18$

   $18 = 18$

   Yes, the ratios of the number of fish caught to the time spent fishing are equal for Elston and Dina.

2. $\frac{1.5}{5} \square \frac{0.6}{3}$ $1.5 \times 3 = 4.5$
   $5 \times 0.6 = 3$
   $4.5 \neq 3$, so $\frac{1.5}{5} \neq \frac{0.6}{3}$

3. $\frac{\frac{2}{3}}{\frac{3}{4}} \square \frac{8}{9}$ $\frac{2}{3} \times 9 = 6$
   $\frac{3}{4} \times 8 = 6$
   $6 = 6$, so $\frac{\frac{2}{3}}{\frac{3}{4}} = \frac{8}{9}$

Try This State if the ratios are equal. Use = or ≠.

a. $\frac{9}{21} \square \frac{6}{14}$ b. $\frac{24}{15} \square \frac{32}{28}$ c. $\frac{1}{1.2} \square \frac{2.5}{3}$ d. $\frac{\frac{3}{5}}{2} \square \frac{5}{\frac{5}{2}}$

CHAPTER 10

Exercises

State if the ratios are equal. Use = or ≠.

1. $\frac{2}{3} \square \frac{6}{9}$ 2. $\frac{3}{7} \square \frac{8}{21}$ 3. $\frac{3}{4} \square \frac{6}{8}$ 4. $\frac{2}{5} \square \frac{4}{9}$ 5. $\frac{7}{8} \square \frac{4}{5}$

6. $\frac{10}{16} \square \frac{5}{8}$ 7. $\frac{6}{4} \square \frac{3}{2}$ 8. $\frac{16}{20} \square \frac{2}{5}$ 9. $\frac{5}{6} \square \frac{4}{5}$ 10. $\frac{7}{10} \square \frac{5}{7}$

11. $\frac{9}{15} \square \frac{3}{5}$ 12. $\frac{6}{21} \square \frac{4}{14}$ 13. $\frac{6}{20} \square \frac{9}{27}$ 14. $\frac{33}{44} \square \frac{36}{48}$ 15. $\frac{4}{3} \square \frac{32}{24}$

16. $\frac{13}{9} \square \frac{39}{28}$ 17. $\frac{40}{35} \square \frac{24}{21}$ 18. $\frac{48}{60} \square \frac{75}{100}$ 19. $\frac{32}{52} \square \frac{24}{39}$ 20. $\frac{30}{27} \square \frac{20}{18}$

21. $\frac{3.2}{3.6} \square \frac{4}{3}$ 22. $\frac{6.3}{7} \square \frac{7.2}{8}$ 23. $\frac{\frac{3}{5}}{\frac{4}{5}} \square \frac{3}{4}$ 24. $\frac{\frac{2}{3}}{\frac{3}{3}} \square \frac{\frac{5}{6}}{\frac{4}{4}}$ 25. $\frac{\frac{1}{2}}{\frac{4}{4}} \square \frac{\frac{1}{4}}{\frac{2}{2}}$

Show Math Finish each picture to make equal ratios of white to black objects.

26.    27.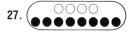

28.    29.

Number Sense Use the clues to write a proportion.

30. Each ratio is equal to $\frac{2}{3}$. The numerator of one ratio is 6. The denominator of the other ratio is 21.

31. Each ratio is equal to $\frac{3}{4}$. The numerator of one ratio is 12. The cross product of the other ratio is equal to 240.

32. Find three proportions. Use the ratio of the number of fish caught to the time spent fishing.

Name	Number of Fish Caught	Time Spent Fishing
Tanya	6	2 h
Brent	4	3 h
Tonio	2	4 h
Lin Su	12	4 h
Heather	4	8 h
Cleo	8	6 h

Mixed Skills Review Change the unit.

33. 1 qt = ___ fl oz   34. 4 lb = ___ oz   35. $1\frac{1}{2}$ qt = ___ pt
36. 5 ft 2 in. = ___ in.   37. 11 yd = ___ ft   38. 2 mi = ___ yd

Estimate.

39. 643 ÷ 82   40. 408 × 12   41. 2077 + 3690   42. 555 − 348

RATIO, PROPORTION, AND PERCENT

10-4 Solving Proportions

Objective
Solve proportions.

A state police survey reported that an average of 4.5 people in 7 cars used seat belts. The police surveyed 140 cars. How many people were wearing seat belts?

BUCKLE UP!

Understand the Situation

- What is the ratio of the average number of people wearing seat belts to the number of cars?
- How many cars did the police survey?

You can use a proportion to solve this problem.

Examples

1. How many people were wearing seat belts?

   Write a proportion. → Solve. Use cross products.

   people → $\frac{4.5}{7} = \frac{p}{140}$ ← people
   cars → ← cars

   $\frac{4.5}{7} = \frac{p}{140}$

   $4.5 \times 140 = 7 \times p$
   $630 = 7 \times p$
   $630 \div 7 = (7 \div 7) \times p$
   $90 = p$

   Ninety people were wearing seat belts.

2. One gallon of paint will cover 450 sq ft. How many gallons will you need to paint 1500 sq ft?

   gallons → $\frac{1}{450} = \frac{g}{1500}$ ← gallons
   sq ft → ← sq ft

   $\frac{1}{450} = \frac{g}{1500}$

   $1 \times 1500 = 450 \times g$
   $1500 = 450 \times g$
   $1500 \div 450 = (450 \div 450) \times g$
   $3\frac{1}{3} = g$

   You will need $3\frac{1}{3}$ gal of paint to cover 1500 sq ft.

Try This   Solve.

a. Two polo shirts cost $13. How much do three shirts cost?

b. Elka Morgan earns $112 for 8 h of work. She earned $147. How many hours did she work?

Exercises

Solve.

1. $\frac{3}{5} = \frac{n}{15}$
2. $\frac{6}{p} = \frac{2}{3}$
3. $\frac{9}{12} = \frac{12}{m}$
4. $\frac{h}{21} = \frac{21}{49}$
5. $\frac{w}{30} = \frac{2}{5}$
6. $\frac{16}{b} = \frac{24}{15}$
7. $\frac{28}{40} = \frac{35}{d}$
8. $\frac{7}{8} = \frac{k}{4}$
9. $\frac{r}{9} = \frac{4}{27}$
10. $\frac{3}{y} = \frac{7}{14}$
11. $\frac{5}{6} = \frac{3}{s}$
12. $\frac{2}{c} = \frac{2}{8}$
13. $\frac{3}{7} = \frac{7}{t}$
14. $\frac{11}{v} = \frac{5}{11}$
15. $\frac{5}{2} = \frac{x}{11}$

16. Workers used 25 kg of cement to make 125 kg of concrete. How many kilograms of cement will they need for 75 kg of concrete?

17. Fifteen pounds of moon rocks weigh 90 lb on earth. Suppose an astronaut weighs 102 lb on earth. How much will she weigh on the moon?

18. Tyrell Justin mixes $\frac{1}{2}$ c of powdered paint with 2 c of water to make paint for his art class. How much powder should he mix with 6 c of water?

19. Norma Spinelli answered 3 out of 4 questions correctly on her driver's license exam. How many questions did she get right out of 32 questions?

20. Two pounds of chicken serves 4 people. How much chicken should serve 14 people?

21. Lee Mi Chao bought 2 colors of lip gloss for $10. How many colors could she buy for $35?

Estimation Tell if each estimate for *n* is reasonable.

22. $\frac{15}{20} = \frac{n}{4}$
Estimate: 3

23. $\frac{n}{30} = \frac{36}{72}$
Estimate: 10

24. $\frac{90}{60} = \frac{n}{45}$
Estimate: 30

25. $\frac{700}{n} = \frac{300}{250}$
Estimate: 700

Talk Math Name the missing item.

26. A dime is to a dollar as a penny is to a _____.

27. An egg is to a dozen eggs as an _____ is to a foot.

28. A quart is to a _____ as a quarter is to a dollar.

29. A _____ is to a deck as a week is to a year.

Evaluating Solutions Tell if the solution is correct. If it is not correct, explain why not.

30. $\frac{5}{8} = \frac{n}{24}$
$5 \times 24 = 8 \times n$
$120 = 8 \times n$
$120 \times 8 = n$
$960 = n$

31. $\frac{10}{n} = \frac{3}{16}$
$10 \times 16 = n \times 3$
$160 = n \times 3$
$160 \div 3 = n$
$53\frac{1}{3} = n$

32. $\frac{6}{n} = \frac{12}{5}$
$6 \times 12 = n \times 5$
$72 = n \times 5$
$72 \div 5 = n$
$14\frac{2}{5} = n$

Problem Solving

33. Mr. Muñoz bought some fruit. He bought twice as many peaches as plums and 2 more bananas than peaches. Ten pieces were either bananas or apples. There were 4 apples. How many plums did Mr. Muñoz buy?

RATIO, PROPORTION, AND PERCENT

10-5 Rates

Objective
Find the unit rate.

The *Godspeed* is a copy of the seventeenth-century bark. It followed the 1607 route from England to Jamestown, Virginia. The ship sailed about 14 nautical miles in 4 h. About how many nautical miles per hour did the *Godspeed* sail?

Understand the Situation

- How far did the *Godspeed* sail in 4 h?
- How far do you think the *Godspeed* could have sailed in 2 h?
- What is the ratio of the number of nautical miles the *Godspeed* sailed to the number of hours it sailed?

A **rate** is a ratio that you use to compare two measures. A **unit rate** is the rate for one unit of measure.

Examples

1. About how many nautical miles per hour did the *Godspeed* sail?

   | Write the rate as a ratio. | → | Write a proportion. | → | Solve. |

   $$\text{miles} \rightarrow \frac{14}{4} \qquad \text{miles} \rightarrow \frac{14}{4} = \frac{m}{1} \leftarrow \text{miles} \qquad \frac{m}{1} = m, \text{ so}$$
   $$\text{h} \rightarrow \qquad\qquad \text{h} \rightarrow \qquad\qquad \leftarrow \text{h}$$
   $$\frac{14}{4} = m$$
   $$3\frac{1}{2} = m$$

   The *Godspeed* sailed about $3\frac{1}{2}$ nautical miles per hour.

2. Faith Hermann packs oysters in a canning factory. She packed 39 jars of oysters in 3 min. How many jars per min can Faith pack?

   $$\text{jars} \rightarrow \frac{39}{3} = \frac{j}{1} \leftarrow \text{jars} \qquad \frac{j}{1} = j$$
   $$\text{min} \rightarrow \qquad\qquad \leftarrow \text{min} \qquad \frac{39}{3} = j$$
   $$13 = j$$

   Faith can pack 13 jars of oysters per min.

Try This

a. An economy car travels 115.5 mi on 3 gal of gasoline. How many miles per gallon can the car travel?

b. Kip Meyer serves 6 kg of chicken to 16 people. How many kilograms per person is this? Round to the nearest tenth.

Exercises

Solve.

1. On a summer vacation, the Charlton family traveled 1929 km in 6 days. How many kilometers per day is this?

2. Mark did 98 sit-ups in 2 min. How many sit-ups per minute did Mark do?

3. An average family uses 1252 gal of water in 4 days. How many gallons of water per day does the family use?

4. Alexa swims 50 m in 38 s. How many meters per second does Alexa swim? Round to the nearest tenth.

5. A car travels 1064 mi on 56 gal of gasoline. How many miles per gallon does the car travel?

6. A ski trail guard earns $33.75 for working 9 h. What is the hourly wage?

7. A radio disk jockey played 15 records in 1 h. At this rate, how long will it take to play the Top 40?

8. A long-distance call from Atlanta to Savannah costs 65¢ for 3 min. At this rate, how long can you talk for $3?

9. Two cassette tapes cost $11.50. At this rate, how many tapes can you buy for $28.75?

10. Patty Boyette's weekly salary is $250. Her employer takes out $72.50 for taxes each week. At this rate, how much will her employer take out from a salary of $625?

Mental Math

Write the unit rate.

11. $\dfrac{\$180}{3 \text{ months}}$ 12. $\dfrac{240 \text{ words}}{4 \text{ min}}$ 13. $\dfrac{75 \text{ car washes}}{5 \text{ h}}$ 14. $\dfrac{324 \text{ letters}}{4 \text{ sentences}}$

Problem Solving

Solve.

15. Dwaine bought 2 items. He gave the clerk $30. The clerk gave him $5.75 in change. Which 2 items did Dwaine buy?

Item	Price
Book	$4.75
Socks	$6.75
Plant	$5.25
Shirt	$17.00
Shorts	$19.50

Mixed Skills Review

Estimate.

16. $48 + 52 + 51 + 47 + 100$ 17. $224 + 126 + 123 + 225$

Find the answer.

18. 3.6×1000 19. $27{,}492 + 3{,}629$ 20. $\$858 \div 33$ 21. $4\tfrac{1}{2} - 2\tfrac{2}{3}$

Write the standard numeral.

22. 5^2 23. 3^3 24. 2^4 25. 6^2 26. 4^3 27. 1^5

RATIO, PROPORTION, AND PERCENT

10-6 Consumer Math: Better Buy

Objective
Find the better buy.

The **unit price** is the cost of one unit of an item, such as 1 fl oz of orange juice. Unit prices are usually calculated to the nearest tenth of one cent. Which orange juice is the better buy?

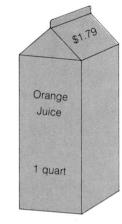

Understand the Situation
- How much does the carton of orange juice cost?
- How many fluid ounces of orange juice are in the carton?
- What operation do you use to find the unit price?

When comparing products of equal quality, the item with the lower unit price is the better buy.

Example Find the better buy of orange juice.

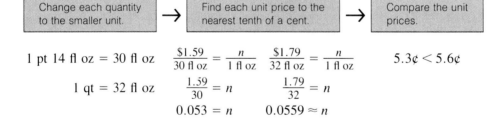

1 pt 14 fl oz = 30 fl oz $\dfrac{\$1.59}{30 \text{ fl oz}} = \dfrac{n}{1 \text{ fl oz}}$ $\dfrac{\$1.79}{32 \text{ fl oz}} = \dfrac{n}{1 \text{ fl oz}}$ 5.3¢ < 5.6¢

1 qt = 32 fl oz $\dfrac{1.59}{30} = n$ $\dfrac{1.79}{32} = n$

                     $0.053 = n$ $0.0559 \approx n$

                     $5.3¢ = n$ $5.6¢ \approx n$

The can of orange juice is the better buy.

Try This Find the better buy.

a. detergent
   64 oz for $2.89
   40 oz for $1.79

b. paper towels
   88 sq ft for 59¢
   125 sq ft for 79¢

c. pet food
   5 cans for $1.55
   3 cans for $0.99

Exercises

Find the better buy.

1. cereals
   10 oz for $1.79
   13 oz for $2.13

2. potatoes
   2 lb 5 oz for $1.29
   5 lb   for $2.50

3. gift wrap (same width)
   7 ft 9 in. for $2.39
   4 ft 6 in. for $1.09

4. peanut butter
   1 lb 2 oz for $1.86
   1 lb 12 oz for $2.89

5. lemonade
   $1\frac{1}{2}$ qt for $0.89
   2 qt 1 pt for $1.50

6. twine
   16 yd for $1.25
   50 ft for $1.29

7. mineral water
   28 fl oz for $0.98
   9.6 fl oz for $0.49

8. vinegar
   1 pt for $0.49
   1 qt for $1.15

9. nutmeg
   60 g for $2.59
   22.6 g for $0.69

Calculator Find the missing numbers. Round to the nearest hundredth.

10. Roasted Peanuts
Net Wt	Unit Price	Total Price
0.84 lb	$1.32 per lb	

11. Spaghetti
Net Wt	Unit Price	Total Price
1.74 lb		$0.68

12. Raisins
Net Wt	Unit Price	Total Price
	$0.98 per lb	$1.90

Mental Math Find the unit price to the nearest tenth of a cent. Use mental math.

13. liquid detergent
    50 fl oz for $3.90
    coupon 40¢

14. toothpaste
    5 oz for $1.45
    coupon 35¢

15. napkins
    60 napkins for 55¢
    coupon 25¢

Data Hunt

16. Find the cost of three different sizes of milk. Find the unit price of each item. Which size has the lowest unit price?

Mixed Applications

Solve. Use the information on the rice packages.

17. How much more does the 10-lb bag of rice cost than the 2-lb bag?

18. What is the difference between the unit prices for the two bags?

19. Suppose that there is a 5-lb bag of rice on sale for $1.70. Which bag of rice is the best buy?

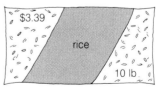

RATIO, PROPORTION, AND PERCENT

10-7 Similar Figures and Proportions

Objective
Find missing lengths using similar figures.

Fran is making a quilt. The quilt pattern uses two triangles that are **similar** (~), or have the same shape. To cut out material, Fran needs to know the length of side *DE*.

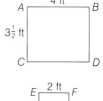

Understand the Situation

- What is the length of side *AC*?
- What is the length of side *DF*?
- What side in the smaller triangle matches side *AC*?

In similar figures, the lengths of **corresponding** sides, or matching sides, are proportional. The corresponding angles are **congruent** (≅), or are the same size.

Examples

Find the missing length.

1. Find the length of side *DE*.

| Write a proportion. | → | Solve. |

$$AC \rightarrow \frac{6 \text{ in.}}{3 \text{ in.}} = \frac{10 \text{ in.}}{l \text{ in.}} \leftarrow AB$$
$$DF \rightarrow \qquad \qquad \leftarrow DE$$

$$\frac{6}{3} = \frac{10}{l}$$

$6 \times l = 3 \times 10$
$6 \times l = 30$
$(6 \div 6) \times l = 30 \div 6$
$l = 5$

The length of side *DE* is 5 in.

2. Find the length of *EG*.

$$AC \rightarrow \frac{3\frac{1}{2} \text{ ft}}{x \text{ ft}} = \frac{4 \text{ ft}}{2 \text{ ft}} \leftarrow AB$$
$$EG \rightarrow \qquad \qquad \leftarrow EF$$

$$\frac{3\frac{1}{2}}{x} = \frac{4}{2}$$

$3\frac{1}{2} \times 2 = x \times 4$
$7 = x \times 4$
$7 \div 4 = x \times (4 \div 4)$
$1\frac{3}{4} = x$

The length of side *EG* is $1\frac{3}{4}$ ft.

Try This

Find the missing length.

a.

b.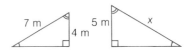

280 CHAPTER 10

Exercises

Find the missing length.

1.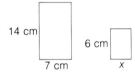
2.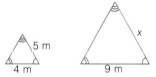
3.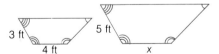
4.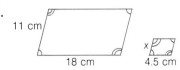
5.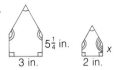
6.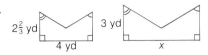
7.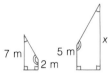
8.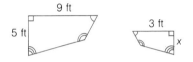

Mixed Applications

Solve.

9. Zuni drew a design $2\frac{1}{8}$ in. by $3\frac{1}{4}$ in. for a cloth 10 in. wide. She wants to cross-stitch it on a cloth 15 in. wide. What are the measurements of the new design?

10. Ricky made a design $2\frac{1}{2}$ in. by $3\frac{1}{4}$ in. for a cloth 12 in. wide. He wants to print it on a cloth 15 in. wide. What are the measurements of the new design?

Group Activity

11. Find the height of the flagpole at school. Measure the length of the flagpole's shadow in centimeters. Stand next to the flagpole. Have someone measure the length of your shadow in centimeters. Have someone measure your height in centimeters. Use similar figures to find the height of the flagpole.

Problem Solving

12. Each week Roland added pennies to a jar. The first week he added 1¢. The second week he added 2¢, the third week 4¢, and the fourth week 8¢. If this rate continued, how many pennies did Roland add the tenth week?

13. Mrs. Santos spent $58 on bath and hand towels. Bath towels cost $7. Hand towels cost $5. How many towels did Mrs. Santos buy?

RATIO, PROPORTION, AND PERCENT

10-8 Scale Drawings

Objective
Use scales to find actual measurements. Find scale measurements.

Dora Anderson collects model trains. Each car is a smaller copy of an actual car. What are the actual measurements of the caboose?

Understand the Situation
- What is the length of the model?
- What is the height of the model?
- What is the scale?

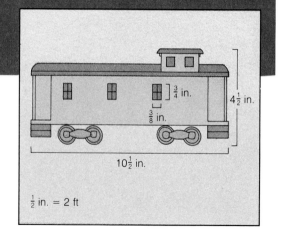

You can use proportions to find the actual measurements from a scale drawing or model.

Example 1. Find the actual length of the caboose.

Write a proportion. Use the scale as one ratio. → Solve.

scale → $\dfrac{\frac{1}{2}\text{ in.}}{2\text{ ft}} = \dfrac{10\frac{1}{2}\text{ in.}}{n\text{ ft}}$ ← length of model
← length of caboose

$\dfrac{\frac{1}{2}}{2} = \dfrac{10\frac{1}{2}}{n}$

$\frac{1}{2} \times n = 2 \times 10\frac{1}{2}$

$\frac{1}{2} \times n = 21$

$\left(\frac{1}{2} \div \frac{1}{2}\right) \times n = 21 \div \frac{1}{2}$

$n = 42$

The actual length of the caboose is 42 ft.

Try This a. Find the actual height of the caboose.

You can use proportions to find the scale measurements of an actual object.

Example 2. Find the scale measurement for a length of 10 ft. Use a scale of $\frac{1}{2}$ in. = 8 ft.

scale → $\dfrac{\frac{1}{2}\text{ in.}}{8\text{ ft}} = \dfrac{n\text{ in.}}{10\text{ ft}}$ ← model
← actual

$\dfrac{\frac{1}{2}}{8} = \dfrac{n}{10}$

$\frac{1}{2} \times 10 = 8 \times n$

$5 = 8 \times n$

$5 \div 8 = (8 \div 8) \times n$

$\frac{5}{8} = n$

The scale measurement for 10 ft is $\frac{5}{8}$ in.

Try This Find the scale measurement. Use a scale of $\frac{1}{2}$ in. = 10 ft.

b. 12 ft **c.** 14 ft **d.** 50 ft **e.** 20 ft **f.** $10\frac{1}{2}$ ft **g.** $8\frac{1}{2}$ ft

Exercises

Use a ruler and the map. Find the distances.
1. McComb to Brookhaven
2. Forest to Meridian
3. Batesville to Durant
4. Greenville to Greenwood
5. Yazoo City to Gulfport
6. Poplarville to Lucedale

Find the scale measurement.
Use a scale of $\frac{1}{2}$ in. = 10 ft.

7. 15 ft **8.** 25 ft **9.** 32 ft
10. 24 ft **11.** 5 ft **12.** 6 ft
13. 8 ft **14.** 40 ft **15.** 9 ft
16. $3\frac{1}{3}$ ft **17.** $4\frac{1}{2}$ ft **18.** $6\frac{3}{4}$ ft

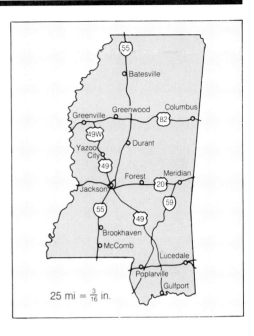

25 mi = $\frac{3}{16}$ in.

Show Math 19. Make a scale drawing of a room in your house. Follow these steps:
   A. Measure the length and width of the room.
   B. Decide upon the scale.
   C. Draw the room on paper. Label the measurements using the scale.
   D. Measure the length and width of each piece of furniture in the room.
   E. Draw the furniture on cardboard. Label the measurements using the scale.
   F. Cut out the furniture. Arrange the pieces on your floor plan.
   G. Now arrange the furniture a different way.

Mixed Skills Review Find the answer.

20. $32 × 89 **21.** $8\frac{1}{2} × 2$ **22.** $5\frac{1}{4} + 2\frac{2}{3}$ **23.** 59.4 ÷ 54

24. Make a bar graph of the numbers of students who attended the first four swim meets: 24, 36, 28, 32.

Estimate.
25. 28 × 11 **26.** 361 + 239 **27.** 1422 − 325 **28.** 2379 ÷ 61

RATIO, PROPORTION, AND PERCENT

10-9 Percents

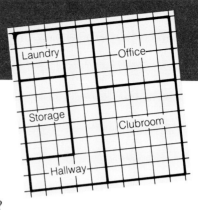

Objective
Write a ratio as a percent. Add and subtract percents.

Jobe and Helen Hundt are planning the layout of their new basement. They used a 10 by 10 square grid to represent the basement. What percent of the space is the clubroom?

Understand the Situation
- How many small squares are in the 10 by 10 grid?
- How many small squares are in the clubroom?

The ratio of a number to 100 is a **percent**. Percent means *per hundred*.

Example Solve.

1. What percent of the space is the clubroom?

$$\frac{\text{number of small squares in clubroom}}{\text{number of small squares in grid}} \rightarrow \frac{30}{100}, \text{ or } 30\% \text{ of the space is the clubroom.}$$

Try This Solve.

a. What percent of the space is the laundry room?

b. What percent of the space is the office?

The ratio $\frac{100}{100}$, or 100%, means 1 or all of something. Notice that the sum of all the percents in the basement layout is 100%.

Example Solve.

2. A floor of an office building uses 70% of the space for offices and 21% of the space for hallways. What percent of the space is left for washrooms?

| Add the known percents. | → | Subtract the sum from 100%. |

$70\% + 21\% = 91\%$ $100\% - 91\% = 9\%$

9% of the space is left for washrooms.

Try This Solve.

c. A floor of a condominium building uses 68% of the space for apartments and 19% for hallways. What percent of the space is left for storage?

Exercises

This 10 by 10 grid shows the layout of an apartment. Solve.

1. What percent of the space is the kitchen?
2. What percent of the space is the living room?
3. What percent of the space is the dining room?
4. What percent of the space is the bathroom and the bedroom?

Find what percent of the space is left.

5. recreation building

Room	Percent of Space
Gym	60%
Weight room	10%
Storage	8%
Racquetball and lockers	?%

6. laboratory

Room	Percent of Space
Storage	12%
Lab	55%
Office	?%

Show Math

7. Draw a 10 by 10 grid to represent a basement. Divide up the space as shown in the table. Find the percent of the space for the clubroom.

Room	Percent of Space
Laundry	18%
Workshop	25%
Storage	20%
Clubroom	?%

Number Sense

Copy the number line. Mark and label the percent or fraction on the number line.

8. 45%   9. 70%   10. 5%   11. $\frac{60}{100}$ 12. 95.3%   13. $\frac{2}{100}$

14. 80%   15. 50%   16. $\frac{80}{100}$ 17. $33\frac{1}{3}\%$ 18. 15%   19. 25%

Estimation

Estimate the percent for each letter.

20.
   0%   x   y 100%

21.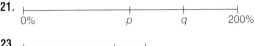
   0%   p   q   200%

22.
   0%   a   b   100%

23.
   0%   m   100%

Estimate. Label the 100% point.

24. |—————+—————|
    0%   50%

25. |———————————————|
    0%   300%

10-10 Percents and Decimals

Objective
Write percents as decimals.
Write decimals as percents.

By 1987, there were 16 James Bond films. Sean Connery played Agent 007 in 44% of the films. What is this percent written as a decimal?

To write a percent as a decimal, remove the percent sign. Multiply the number by 0.01. This moves the decimal point two places to the left.

James Bond
Sean Connery 44%
Roger Moore 44%
George Lazenby 6%
Timothy Dalton 6%

Examples Write as a decimal.

1. 44%

| Remove the % sign. | → | Multiply by 0.01. |

$44\% \rightarrow 44$ $44 \times 0.01 = 0.44$ *Move the decimal two places to the left.*

Forty-four percent is equal to 0.44.

2. 6%
$6\% \rightarrow 6$
$6 \times 0.01 = 0.06$

3. 113%
$113\% \rightarrow 113$
$113 \times 0.01 = 1.13$

Try This Write as a decimal.

a. 63%   **b.** 1.9%   **c.** 8%   **d.** 25%   **e.** 146.8%

To write a decimal as a percent, divide by 0.01. This moves the decimal point two places to the right. Write a % sign.

Examples Write as a percent.

4. 0.5
$0.5 = 50\%$

5. 1.26
$1.26 = 126\%$

6. $0.\overline{6}$
$0.\overline{6} = 66\frac{2}{3}\%$ or $\approx 66.7\%$

Try This Write as a percent.

f. 0.96   **g.** 0.436   **h.** 2.46   **i.** 1   **j.** $0.\overline{3}$

286 CHAPTER 10

Exercises

Write as a decimal.

1. 46% 2. 19% 3. 65% 4. 124% 5. 3.5%
6. 9% 7. 6.5% 8. 1.2% 9. 0.03% 10. 0.15%
11. 0.24% 12. 0.09% 13. $12\frac{1}{2}\%$ 14. 232% 15. 114%

Write as a percent.

16. 0.83 17. 0.07 18. 5 19. 0.14 20. 0.17
21. 1.4 22. 2.93 23. 1.08 24. 7.568 25. 0.004
26. 0.9 27. 0.018 28. 0.709 29. 1.508 30. 1.2

Solve.

31. Ernst Blofeld was the villain in 19% of the James Bond films. Write 19% as a decimal.

32. Sean Connery as James Bond drove an Aston Martin DB-5 in 0.125 of the films. Write 0.125 as a percent.

Number Sense

Copy the number line. Label each of the following on your number line.

|———————————————————————|
0%   100%

33. 0.25 34. 30% 35. 0.1 36. 75% 37. 0.8 38. $66\frac{2}{3}\%$

State if the circle graph is reasonable. If it is not reasonable, explain why not.

39. 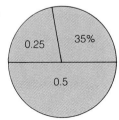 40. 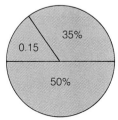 41.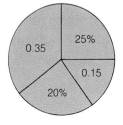

Data Bank

42. Use the data in the circle graph on p. 491. Write each percent as a decimal.

Calculator

43. Choose the calculator keys that you would press to find 3.5 as a percent. Arrange the keys in the proper order.

10-11 Percents and Fractions

Objective
Write percents as fractions.
Write fractions as percents.

What percent of the employees brought their lunch?

Lunch Survey of Wilton Factory Employees	
Brought lunch	23
Bought lunch	19
Ate no lunch	8

Understand the Situation
- How many employees were surveyed?
- How many employees brought their lunch?
- What fraction of the sample brought their lunch?

Examples Write as a percent.

1. $\frac{23}{50}$

$\frac{23}{50} = \frac{x}{100}$ $23 \times 100 = 50 \times x$ *Use proportions to change*
                                    $2300 = 50 \times x$ *fractions to percents.*
                     $2300 \div 50 = (50 \div 50) \times x$
                                    $46 = x$

$\frac{23}{50} = \frac{46}{100} = 46\%$ 46% of the employees brought their lunch.

2. $\frac{46}{80}$

$\frac{46}{80} \rightarrow 80\overline{)46.000}^{\,0.575}$ *Change the fraction to a decimal.*
                                                              Then write the decimal as a percent.

$\frac{46}{80} = 0.575 = 57.5\%$ or $57\frac{1}{2}\%$

Try This Write as a percent.

a. employees who bought lunch   b. $\frac{5}{8}$ c. $\frac{2}{3}$ d. $\frac{1}{5}$ e. $\frac{1}{2}$

To change a percent to a fraction, first change the percent to a decimal.

Examples Write as a fraction.

3. 68%

$68\% = 0.68 = \frac{68}{100} = \frac{17}{25}$ *Write in lowest terms.*

4. 6.25%

$6.25\% = 0.0625 = \frac{625}{10000} = \frac{1}{16}$

Try This Write as a fraction.

f. 29%   g. 32%   h. 2.5%   i. 1.25%   j. 6%   k. 88.1%

Exercises

Write as a percent.

1. $\frac{3}{5}$ 2. $\frac{1}{8}$ 3. $\frac{3}{4}$ 4. $\frac{3}{8}$ 5. $\frac{3}{16}$ 6. $\frac{1}{20}$

7. $\frac{4}{5}$ 8. $\frac{3}{10}$ 9. $\frac{5}{6}$ 10. $\frac{1}{3}$ 11. $\frac{1}{12}$ 12. $\frac{4}{25}$

13. $\frac{89}{100}$ 14. $\frac{7}{20}$ 15. $\frac{9}{10}$ 16. $\frac{11}{15}$ 17. $1\frac{1}{4}$ 18. $2\frac{2}{5}$

Write as a fraction.

19. 64%   20. 17%   21. 29%   22. 87%   23. 3.5%   24. 14.6%

25. 3.25%   26. 35.18%   27. 112%   28. 100%   29. 0.8%   30. 250%

31. 2%   32. 380%   33. 9.7%   34. 0.4%   35. 3%   36. 0.82%

Solve. Use the table on p. 288.

37. Find the percent of employees who ate no lunch.

38. Find the percent of employees who bought a lunch and who brought a lunch.

Calculator

Solve. Round to the nearest tenth of a percent.

39. In the lunch survey, 5 of the people who ate no lunch were on a diet. What percent of this group was on a diet?

40. In the lunch survey, 9 of the people who bought lunch went to the factory cafeteria. What percent of this group ate at the cafeteria?

Evaluating Solutions

41. Art changed $\frac{4}{5}$ to a percent using the method shown. Explain Art's method. Use his method to change $\frac{18}{25}$ to a percent.

$$\frac{4}{5} \times \frac{20}{20} = \frac{80}{100} = 80\%$$

Talk Math

42. Explain why the percents in the graph total more than 100.

Data Hunt

43. Survey students who bring their lunch to school. Find the reasons why they bring their lunch. How do your results compare with the results in the graph?

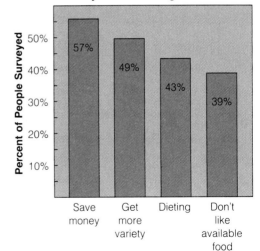

10-12 Comparing Percents, Decimals, and Fractions

Objective
Compare percents, decimals, and fractions.

Juan finds the same jam box on sale at two different stores. Where will he get the better buy?

Bargain Barn
30% Discounts on all Stereo Items

Soundwaves
$\frac{1}{3}$ off Regular Price

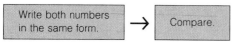

Understand the Situation

- What is a discount?
- Are the discounts given in the same form?
- Can you compare these discounts?

Examples

1. Compare 30% and $\frac{1}{3}$. Use percent. Use <, >, or =.

   Write both numbers in the same form. → Compare.

   Write $\frac{1}{3}$ as a percent.

   $\frac{1}{3} = 0.\overline{3} = 33\frac{1}{3}\%$ $30\% < 33\frac{1}{3}\%$

   One third off the regular price is the better buy.

2. Compare 30% and $\frac{1}{3}$. Use fractions. Use <, >, or =.

   Write 30% as a fraction. $30\% = 0.3 = \frac{3}{10}$ $\frac{3}{10} \square \frac{1}{3}$

   Compare using the LCD. $\frac{9}{30} < \frac{10}{30}$.   so $30\% < \frac{1}{3}$

3. Compare 0.435 and $43\frac{1}{2}\%$. Use decimals. Use <, > or =.

   Write $43\frac{1}{2}\%$ as a decimal. $43\frac{1}{2}\% = 0.435$

   $0.435 = 0.435$ so $0.435 = 43\frac{1}{2}\%$

4. Compare 0.296 and 225%. Use percent. Use <, >, or =.
   Write 0.296 as a percent. $0.296 = 29.6\%$
   $29.6\% < 225\%$ so $0.296 < 225\%$

Try This Compare. Use <, >, or =.

a. $\frac{3}{5} \square 55\%$ b. $0.7 \square 71\%$ c. $\frac{12}{7} \square 127\%$ d. $0.49 \square 49\%$

Exercises

Compare. Use <, >, or =.

1. $\frac{2}{3}$ ☐ 75%
2. 0.23 ☐ 24%
3. $\frac{4}{5}$ ☐ 80%
4. 0.95 ☐ 95%
5. 6.25% ☐ 0.62
6. 22% ☐ $\frac{2}{9}$
7. 4.3% ☐ 0.34
8. 16% ☐ $\frac{1}{6}$
9. $\frac{1}{4}$ ☐ 25%
10. 0.575 ☐ 57%
11. $\frac{1}{5}$ ☐ 22%
12. 0.638 ☐ 64%
13. 3.2% ☐ 0.32
14. 33% ☐ $\frac{1}{3}$
15. 8.1% ☐ 0.8
16. 44% ☐ $\frac{4}{11}$
17. $\frac{4}{7}$ ☐ 60%
18. 0.043 ☐ 4%
19. $\frac{6}{25}$ ☐ 15%
20. 0.099 ☐ 1%
21. 31% ☐ 0.30
22. 100% ☐ $\frac{8}{8}$
23. $\frac{11}{16}$ ☐ 70%
24. 7.9% ☐ 7.9

Number Sense

Arrange in order from least to greatest.

25. $\frac{3}{4}$, 67%, 0.87, 1.2, 67.9%, 132%
26. $\frac{2}{5}$, $\frac{1}{7}$, 0.76, 91%, 1.9%, $\frac{1}{3}$
27. 1.78, $\frac{2}{3}$, 16%, 2.4, 23.5%, $\frac{3}{8}$
28. 0.09, $\frac{9}{8}$, 0.8, 89%, $\frac{8}{9}$, 88%
29. 1.8, $\frac{13}{12}$, 81%, 18, $\frac{1}{8}$, 108%
30. 100%, $\frac{1}{4}$, $\frac{79}{64}$, 64%, $\frac{5}{6}$, 0.79

Write Math

Tell which form you would choose to compare each set of numbers: percents, decimals, or fractions. Tell why.

31. 33%, 0.39, $\frac{3}{10}$
32. $81\frac{1}{2}$%, 0.72, $\frac{64}{75}$
33. 25%, 0.5, $\frac{3}{8}$

Mixed Applications

Solve.

34. Which store has the greater discount on tapes?
35. On which items does the Bargain Barn have the lesser discount?
36. How does Soundwaves' discount on CDs compare to Bargain Barn's discount on tapes?
37. How much would a $6 album cost at the Bargain Barn with the discount?
38. How much would a $12 CD cost at Soundwaves with the discount?

Discount	Bargain Barn	Soundwaves
CDs	15%	$\frac{1}{8}$
LPs	$\frac{1}{4}$	30%
Tapes	$\frac{1}{3}$	35%

Problem Solving

39. Juan bought some blank audiotapes and videotapes at the Bargain Barn. The audiotapes were packaged in boxes of 4 each, and the videotapes were packaged in boxes of 3 each. Juan bought 30 tapes all together. How many boxes of tapes did he buy?

RATIO, PROPORTION, AND PERCENT

10-13 Problem Solving: Determining Reasonable Answers

Objective
Decide if an answer is reasonable.

- **SITUATION**
- **DATA**
- **PLAN**
- **ANSWER**
- **CHECK**

What fraction of the people surveyed visited bookstores? Marianne thought that about $\frac{1}{3}$ of the people surveyed visited bookstores. Is her answer reasonable?

Stores You Visit at the Mall

Major chain stores	81%
Shoe stores	66%
Local department stores	61%
Women's clothing stores	53%
Bookstores	49%
Men's clothing stores	48%

You can decide if an answer is reasonable without solving the problem. Use estimation to find an answer. Decide if the given answer is reasonably close to your estimate.

Examples

Decide if each answer is reasonable without solving the problem. If it is not reasonable, tell why not.

1. Is Marianne's answer reasonable?

49% of the people surveyed visited bookstores. 49% is about 50%, or $\frac{1}{2}$. Marianne's answer is not reasonable.

2. Problem: What percent more of the people surveyed visited local department stores than visited women's clothing stores?

Answer: 8% more of the people surveyed visited local department stores than visited women's clothing stores.

About 60% of the people surveyed visited local department stores. About 50% of the people surveyed visited women's clothing stores. 60% − 50% is 10%. Because 10% is about 8%, the answer seems reasonable.

Try This

Decide if each answer is reasonable without solving the problem. If it is not reasonable, tell why not.

a. Problem: What percent more of the people surveyed visited women's clothing stores than visited bookstores?

Answer: 10% more of the people surveyed visited women's clothing stores than visited bookstores.

b. Problem: What percent more of the people surveyed visited major chain stores than visited men's clothing stores?

Answer: 33% more of the people surveyed visited major chain stores than visited men's clothing stores.

Exercises

Decide if each answer is reasonable without solving the problem. If it is not reasonable, tell why not.

1. *Problem:* What percent more of the people surveyed worried about making the right decision than worried about getting along with their in-laws?

   Answer: 30% more of the people surveyed worried about making the right decision.

2. *Problem:* What fraction of the people surveyed worried about messing up during the ceremony?

   Answer: About $\frac{2}{5}$ of the people surveyed worried about messing up during the ceremony.

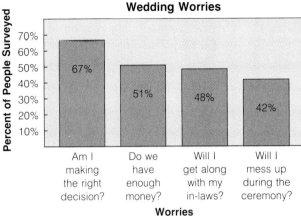

3. *Problem:* Write as a decimal the percent of people who worried about having enough money.

   Answer: 51% is equal to 5.1.

4. *Problem:* How many more newspapers print *Garfield* than print *Blondie*?

   Answer: Eighty-nine more newspapers print *Garfield*.

5. *Problem:* What is the ratio of the number of newspapers that print *Beetle Bailey* to the number of newspapers that print *Blondie*?

   Answer: The ratio is about 9:1.

Comic Strips Printed in Greatest Number of Newspapers	
Comic	Number of Newspapers
Peanuts	1201
Garfield	1085
Blondie	991
Beetle Bailey	964
Frank and Ernest	744

Suppose

6. *Problem:* Suppose that *Frank and Ernest* appeared in 277 more newspapers. How many more newspapers would print *Peanuts* than would print *Frank and Ernest*?

   Answer: Two hundred eighty more newspapers would print *Peanuts*.

Write Your Own Problem

7. Write a problem. Use the data in the table. Write an answer that is reasonable or that is not reasonable. Trade problems with a classmate and solve.

Amount of Money Spent on TV Ads*	
Cars	$641
Food	$638
Medicine	$304
Health/Beauty aids	$298
Restaurants	$263

*approximate amounts in millions of dollars

10-14 Problem Solving: Data Collection and Analysis

Objective
Collect and analyze data.

- **SITUATION**
- **DATA**
- **PLAN**
- **ANSWER**
- **CHECK**

Stores often send out advertisements to promote sales and to acquaint customers with their products and prices. This is an advertisement sent out by a lumber store.

AT J&J's FRIENDLY LUMBER YOU GET THE LUMBER AND PLYWOOD YOU NEED AT THE PRICES YOU CAN AFFORD!

PINE BOARDS

Longer lengths also in stock	6 foot	8 foot	10 foot	12 foot
1" × 3"	1.29	1.79	—	—
1" × 4"	1.99	2.49	3.49	3.99
1" × 6"	2.49	3.99	4.99	5.99
1" × 8"	3.99	4.99	6.99	7.99
1" × 10"	4.99	6.49	8.49	9.99
1" × 12"	5.99	7.99	10.99	12.99

4' × 8' BIRCH & OAK PLYWOOD

3/4" BIRCH $30.00 SHEET

Situation

Work in groups.

Your school is changing several rooms into a recreation/lounge area for students. Your group has been asked to design a wood unit to hold stereo equipment, records, tapes, and any other equipment you wish. You will need to make a scale drawing of your cabinet. You should decide how much wood you will need. You should determine the cost of the wood.

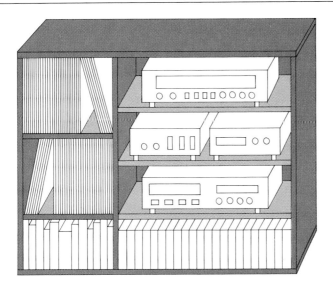

Data Collection

Find the exact measurements of the equipment you will be placing in the cabinet. Make sure the measurements of your cabinet are appropriate.

Facts to Consider

- You can use plywood, boards, or a combination of both.
- In the advertisements, the measurements for boards show the thickness, width, and length. Suppose the table in the ad shows that a board is $1'' \times 4''$ and 10 foot. This means the board is $\frac{3}{4}$ in. thick, $3\frac{1}{2}$ in. wide, and 10 ft long. Check with a hardware store for true measurements.

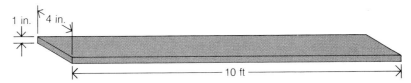

- The ad for the plywood gives the size and the thickness of the sheet of plywood.
- Plywood must be bought in whole sheets.
- For your scale drawing, use a scale of 1 in. = 1 ft.

Share Your Group's Plan

- Display your scale drawing. Explain your plan for the wood unit. State the total cost of the wood.
- Compare your design to other groups' designs. Discuss the advantages of each idea.

Suppose

- Suppose you find out that your budget is cut by 20%. How would you change your design to fit the new budget?

10-15 Enrichment: Telephone Costs

Objective
Calculate telephone costs.

Fran received this information about phone service in her area.

Type of Service	Basic Cost per Month	Additional Charge
Unlimited	$16.89	none—unlimited number of outgoing local calls
Limited (includes 65 local calls)	$ 9.60	$0.09 charge for each outgoing local call over the 65 call limit
Economy	$ 5.80	$0.09 charge for each outgoing local call

There are three rates for long distance calls.

Long Distance Calls Rate Table							
	Mon.	Tues.	Wed.	Thurs.	Fri.	Sat.	Sun.
8 a.m.–5 p.m.	W	W	W	W	W	N	N
5 p.m.–11 p.m.	E	E	E	E	E	N	E
11 p.m.–8 a.m.	N	N	N	N	N	N	N

W = weekday, full rate
E = evening, 40% less than W
N = night, 60% less than W

Examples

1. Find the cost for each type of service for 70 local calls in a month.

   Unlimited
   $16.89

   Limited
   $ 0.09
   ×   5 calls more than 65
   $ 0.45
   +9.60 basic cost
   $10.05 cost of 70 calls

   Economy
   $ 0.09
   ×   70 number of calls
   $ 6.30
   +5.80 basic cost
   $12.10 cost of 70 calls

2. Fran called her cousin on Monday at 10 a.m. The long distance call cost $3.50. What would the call cost at 6 p.m. on Monday?

   $ 3.50 weekday rate
   ×0.40 evening rate discount
   $ 1.40 evening discount

   $ 3.50 weekday rate
   −1.40 less evening discount
   $ 2.10 cost of call

Exercises

Find the cost of each type of service for these local calls in one month.

1. 25 calls   2. 65 calls   3. 130 calls
4. Tyrone called his aunt on Monday at 10 a.m. The long distance call cost $2.50. What would the call cost if he had waited until 6 p.m. to call? Until 7 a.m. on Tuesday?

10-16 Calculator: Scale Drawings

Objective
Use a calculator to find actual measurements on a scale drawing.

Marshall School Floor Plan

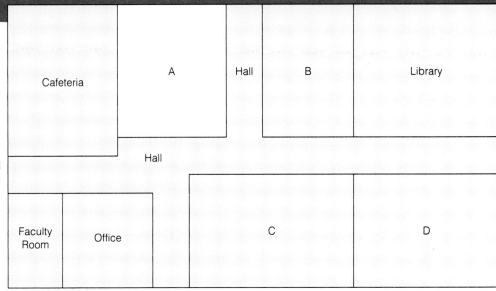

Scale: 1 cm = 3.2 m

Example Find the actual dimensions of room A on the floor plan of Marshall School.

Measure the length and width of room A on the floor plan.
$l = 3.5$ cm, $w = 3$ cm

Use the multiplying constant on a calculator to find the actual dimensions.
Scale: 1 cm = 3.2 m. Enter the constant, 3.2, and the operation sign, multiplication. [3][.][2][×] Enter the length. [3][.][5] Press [=] | 11.2 |

Without clearing the calculator, enter the width. [3] Press [=] | 9.6 |

Room A is actually 11.2 m long and 9.6 m wide.

Exercises

Find the actual dimensions of each room on the floor plan.

1. room B
2. room C
3. room D
4. cafeteria
5. office
6. library
7. faculty room

8. Give the actual dimensions of the library if the scale of the floor plan is 1 cm = 2.8 m.

9. The perimeters of which two rooms are equal?

RATIO, PROPORTION, AND PERCENT

CHAPTER 10 REVIEW

Match the appropriate ratio to the statement. The Rangers played 20 games, of which they won 6 and tied 2.

1. wins to total games played
2. losses to wins
3. losses to total games played A
4. wins to losses
5. ties to wins

A. $\dfrac{12}{20} = \dfrac{3}{5}$ B. $\dfrac{6}{12} = \dfrac{1}{2}$

C. $\dfrac{6}{20} = \dfrac{3}{10}$ D. $\dfrac{2}{6} = \dfrac{1}{3}$

E. $\dfrac{12}{6} = \dfrac{2}{1}$

Write true or false.

6. $\dfrac{2}{3} = \dfrac{10}{15}$ _____
7. $\dfrac{11}{33} \neq \dfrac{22}{66}$ _____
8. $\dfrac{5}{7} = \dfrac{7}{14}$ _____
9. $\dfrac{\frac{3}{8}}{\frac{1}{8}} = \dfrac{3}{1}$ _____

Solve each proportion. After solving for the variable, write the matching letter. Unscramble the letters to spell a word.

10. $\dfrac{2}{5} = \dfrac{n}{25}$
11. $\dfrac{8}{n} = \dfrac{1}{3}$
12. $\dfrac{12}{8} = \dfrac{24}{n}$
13. $\dfrac{n}{6} = \dfrac{5}{30}$
14. $\dfrac{6}{4} = \dfrac{n}{16}$
15. $\dfrac{8}{n} = \dfrac{24}{21}$
16. $\dfrac{n}{50} = \dfrac{100}{500}$

A	B	C	D	E	H	J	L	M	O	P	R	S	T	U
8	20	10	5	16	3	9	22	4	1	15	24	6	7	11

Write the missing number.

17. One cup of flour is to 2 c of flour as 3 c of flour is to _____ c of flour.
18. One-third cup of milk is to 4 c of milk as 1 c of milk is to _____ c of milk.
19. Find the better buy.
    pet food: 3 cans for $1.74 or 5 cans for $2.85

Write the percent as a decimal.

20. 18%   A. 0.18   B. 1.8   C. 18.0   D. 180
21. 8.23% A. 823   B. 82.3 C. 0.0823 D. 8230

Compare. Use <, >, or =.

22. $\dfrac{2}{5}$ ☐ 20%
23. 0.15 ☐ $1\dfrac{1}{2}$
24. $33\dfrac{1}{3}$ ☐ $\dfrac{3}{10}$

CHAPTER 10 TEST

1. Beth Lum made $147 in two weeks. She made $53 more the first week than the second week. How much money did she make the first week?

Write the ratio.   **2.** 5 tickets for $40   **3.** 400 mi in 8 h

State if the ratios are equal. Use = or ≠.   **4.** $\frac{4}{5} \square \frac{12}{15}$ **5.** $\frac{7}{11} \square \frac{21}{35}$ **6.** $\frac{2.5}{5} \square \frac{4}{8}$

Solve.   **7.** $\frac{n}{4} = \frac{6}{8}$ **8.** $\frac{8}{18} = \frac{12}{y}$ **9.** $\frac{4}{7} = \frac{v}{28}$ **10.** $\frac{15}{s} = \frac{45}{12}$ **11.** $\frac{3}{2} = \frac{x}{5}$

12. Clearview's airport is open 15 h per day. Every day 180 planes land. About how many planes land every hour?

13. Find the better buy.   yarn: 24 yd for $7.36 or 90 ft for $10.30

14. Find the missing length.

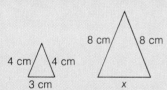

15. On a map of Pennsylvania, 1 in. represents 25 mi. The map shows Pittsburgh to be 14 in. from Philadelphia. What is the actual distance?

Find the scale measurement. Use a scale of $\frac{1}{2}$ in. = 25 ft.   **16.** 5 ft   **17.** 50 ft

18. A studio has a closet, a work area, and a bathroom. The closet takes up 12% of the space. The bathroom takes up 15%. What percent is the work area?

Write as a decimal.   **19.** 45.8%   **20.** 305%   **21.** $11\frac{1}{2}$%   **22.** 7%

Write as a percent.   **23.** 8   **24.** 0.04   **25.** $\frac{2}{3}$ **26.** $3\frac{3}{8}$

Write as a fraction.   **27.** 37.5%   **28.** 5%   **29.** 300%   **30.** 0.4%

Compare. Use <, >, or =.   **31.** 125% $\square$ $1\frac{1}{4}$ **32.** 0.33 $\square$ $33\frac{1}{3}$%

Decide if the answer is reasonable. If it is not reasonable, tell why not.

33. *Problem:* Of 200 people surveyed, 154 people said *yes* and 36 people said *no*. The rest had no opinion. What percent had no opinion?
Answer: Of the people surveyed, 10% had no opinion.

RATIO, PROPORTION, AND PERCENT

CHAPTER 10 CUMULATIVE REVIEW

1. Harlan drove 3042 mi across the country in 12 days. How many miles per day is this?
   A. 253.5 mi per day
   B. 365 mi per day
   C. 221.6 mi per day
   D. 263.5 mi per day

2. Multiply.
   4192×64
   A. 258,288
   B. 268,788
   C. 245,088
   D. 268,288

3. The overtime rate is $1\frac{1}{2}$ times the regular rate. Find the overtime rate for an hourly rate of $10.40. Use mental math.
   A. $20.80
   B. $5.20
   C. $13.00
   D. $15.60

4. How much liquid is in the beaker?
   A. 50 mL
   B. 75 mL
   C. 25 mL
   D. 60 mL

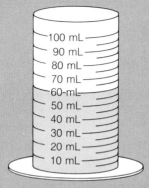

5. Arrange the fractions in order from least to greatest.
   $\frac{1}{4}, \frac{4}{9}, \frac{2}{3}, \frac{3}{8}$
   A. $\frac{2}{3}, \frac{4}{9}, \frac{3}{8}, \frac{1}{4}$
   B. $\frac{3}{8}, \frac{1}{4}, \frac{4}{9}, \frac{2}{3}$
   C. $\frac{4}{9}, \frac{3}{8}, \frac{2}{3}, \frac{1}{4}$
   D. $\frac{1}{4}, \frac{3}{8}, \frac{4}{9}, \frac{2}{3}$

6. Write as a percent.
   $\frac{3}{8}$
   A. 37.5%
   B. 3.75%
   C. 375%
   D. 0.375%

7. Write the ratio.
   8 pears for $1.84
   A. $\frac{1}{0.23}$
   B. 0.23
   C. 0.1472
   D. $\frac{1}{24}$

8. Add.
   $\frac{4}{5} + \frac{5}{14}$
   A. $\frac{19}{30}$
   B. $\frac{9}{11}$
   C. $1\frac{49}{30}$
   D. $1\frac{11}{70}$

9. Subtract.
   $3\frac{1}{2} - 1\frac{3}{8}$
   A. $1\frac{1}{8}$
   B. $2\frac{1}{4}$
   C. $1\frac{3}{4}$
   D. $2\frac{1}{8}$

10. Divide.
    $6 \div 2\frac{1}{3}$
    A. 14
    B. $2\frac{4}{7}$
    C. $\frac{42}{3}$
    D. $2\frac{1}{2}$

11. Estimate the likely temperature of a package of frozen chicken.
    A. 60°F
    B. 0°F
    C. 43°C
    D. 10°C

12. Find the missing length of the triangle.
    A. 8 ft
    B. 12 ft
    C. 10 ft
    D. 11 ft

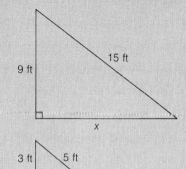

13. Jen's starting account balance was $2386.97. On August 4, Jen wrote check #101 for $35.43. On August 6, Jen deposited $152.83. On August 12, Jen wrote check #102 for $322.71. What is Jen's new balance?
    A. $2897.94
    B. $2181.66
    C. $2521.42
    D. $2081.66

14. Compare. Use <, >, or =.
    6.732 □ 6.76
    A. <
    B. >
    C. =

15. Regroup.
    $4\frac{5}{6} = 3\frac{\square}{\square}$
    A. $3\frac{9}{6}$
    B. $3\frac{1}{6}$
    C. $3\frac{11}{6}$
    D. $\frac{1}{6}$

16. Multiply.
    $\frac{2}{3} \times 6\frac{3}{4} \times \frac{1}{7}$
    A. $\frac{9}{14}$
    B. $\frac{13}{42}$
    C. $6\frac{9}{14}$
    D. $6\frac{1}{14}$

17. Write as a decimal.
    282.6%
    A. 28.26
    B. 282.6
    C. 2.826
    D. 0.2826

18. Divide.
    6)4208
    A. 702
    B. 701 R2
    C. 701
    D. 700 R5

19. Write in lowest terms.
    $\frac{36}{40}$
    A. $\frac{9}{10}$
    B. $\frac{2}{5}$
    C. $\frac{3}{5}$
    D. $\frac{18}{20}$

20. Find the better buy of tomato sauce.
    an 8 oz can for $0.72 or a 16 oz can for $1.56
    A. 8 oz can
    B. 16 oz can

21. Change the unit.
    7040 yd = _____ mi
    A. 3 mi, 1060 yd
    B. 4 mi
    C. 3 mi, 960 yd
    D. 3 mi

22. Compare. Use <, >, or =.
    $\frac{3}{5}$ □ 60%
    A. <
    B. >
    C. =

RATIO, PROPORTION, AND PERCENT

Chapter 11 Overview

Key Ideas
- Use the problem-solving strategy *solve a simpler problem*.
- Solve percent problems.
- Solve percent problems using mental math.
- Estimate with percent.
- Find simple interest.
- Find percent of increase and decrease.

Key Terms
- percent
- principal
- discount
- rate of interest
- compatible numbers
- proportion
- simple interest

Key Skills

Write each percent as a decimal.

1. 24%
2. 32%
3. 8%
4. 2%
5. 125%
6. 200%
7. 24.5%
8. 36.8%
9. $10\frac{1}{2}\%$
10. $4\frac{3}{4}\%$
11. 100%
12. 0.4%
13. 0.01%
14. 89.9%
15. 101%
16. 220%
17. $100\frac{1}{2}\%$
18. $33\frac{1}{3}\%$

Write each percent as a fraction.

19. 45%
20. 20%
21. 18%
22. 50%
23. 8%
24. 75%
25. 48%
26. 4.5%
27. 8.2%
28. $10\frac{1}{2}\%$
29. $4\frac{3}{4}\%$
30. 120%
31. 225%
32. 0.5%
33. 0.07%
34. $33\frac{1}{3}\%$
35. $50\frac{1}{2}\%$
36. 0.12%

Write each fraction as a percent. Round to the nearest tenth.

37. $\frac{2}{5}$
38. $\frac{3}{4}$
39. $\frac{3}{8}$
40. $\frac{1}{10}$
41. $\frac{2}{3}$
42. $\frac{4}{9}$
43. $\frac{12}{20}$
44. $\frac{24}{36}$
45. $\frac{8}{3}$
46. $\frac{6}{5}$
47. $\frac{4}{3}$
48. $\frac{11}{10}$

USING PERCENT

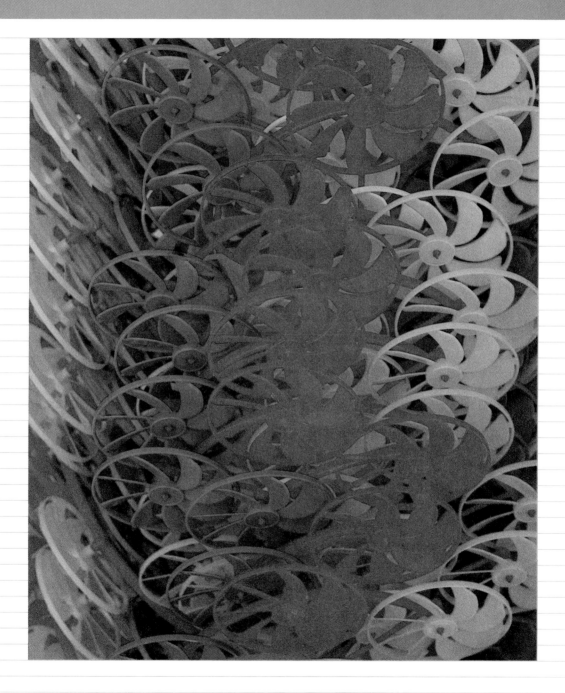

11-1 Problem Solving: Learning to Use Strategies

Objective
Use the strategy solve a simpler problem.

- SITUATION
- DATA
- PLAN
- ANSWER
- CHECK

Problems with large numbers are sometimes difficult to solve. You can use the strategy **solve a simpler problem**. Solve the same problem but use smaller numbers to help you develop a plan for solving the original problem. Then go back and solve the problem again using the larger numbers.

Example Solve.

For every ton of paper people recycle, they save 17 trees. At the end of one year, Rockdale recycled enough paper to save 8789 trees. Rockdale made $24,816 from recycling paper. How much money did Rockdale make for each ton of paper?

| Restate the problem using smaller numbers. |
| Solve the problem. |
| Solve the original problem the same way using the larger numbers. |

Suppose for every ton of paper, people save 2 trees. Rockdale saved 20 trees. They made $50.

$20 \div 2 = 10$ ← tons of paper recycled
$\$50 \div 10 = \$5$ ← price for one ton of paper

$8789 \div 17 = 517$ ← tons of paper recycled
$\$24,816 \div 517 = \$48$ ← price for one ton of paper

Rockdale made $48 for each ton of paper.

Try This Solve.

a. The Alpine High School freshman class planned a school breakfast for 1780 students. They spent $373.80 on eggs. The eggs cost $0.84 per dozen. About how many eggs will they use for each person?

b. Students sold calendars to 958 people. The price per calendar was $8. They made a total of $15,328. Suppose each person bought about the same number of calendars. About how many calendars did each person buy?

c. Students could buy a class picture for $12.85. The photographer made $3 on each sale. The remaining money went to supplies. The photographer made $1305. How much money went to supplies?

d. Students sold school buttons to 1548 people. The price per button was $0.70. They made a total of $2167.20. Suppose each person bought about the same number of buttons. About how many buttons did each person buy?

Exercises

Solve. Use one or more of the problem-solving strategies.

Problem-Solving Strategies
- Guess and check.
- Choose an operation.
- Make an organized list.
- Act it out.
- Find a pattern.
- Write an equation.
- Use logical reasoning.
- Draw a picture or diagram.
- Make a table.
- Use objects.
- Work backward.
- **Solve a simpler problem.**

1. Anna worked part-time at the office of a local politician. She sent out letters to voters. There are about 7300 voters in the area. Anna spent $365 photocopying letters. Each copy cost 2.5¢. How many pages was each letter?

2. The school formed a student task force to improve the cafeteria lunch lines. David timed students going through the line. It took each student an average of 2 minutes at each stop: salad, main dish, drinks, cash register. How long did it take 6 people to go through the line?

3. It takes about 245 h for a rock group to record 35 min of useable music. How much does it cost to produce one minute of useable music?

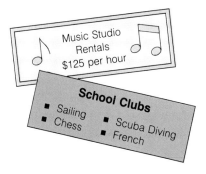

4. Jan, Barb, Travis, and Nate were all members of different school clubs. Barb does poorly in foreign languages. Jan needs to buy flippers. Travis and Barb dislike the water. Who belongs to each club?

5. The Town Council wants to put a new stop sign at Brady St. and Winters St. They timed cars driving by the streets. Out of 155 cars, 4 times as many were going the speed limit as were speeding. How many cars were speeding?

Suppose

6. Suppose in the example on p. 304 that Rockdale recycled enough paper to save only 8483 trees. How much money did Rockdale make recycling paper?

Write Your Own Problem

7. Write a problem that you could solve by solving a simpler problem. Use the data at right.

11-2 Finding a Percent of a Number

Objective
Find a percent of a number.

Nan plans to buy a new helmet to wear when riding her dirt bike. How much is the discount?

Spring Sale
Helmets—20% off
Regular price $75

Understand the Situation
- What is the regular price of a new helmet?
- What is the rate of discount?
- What does Nan need to find?

To find a percent of a number, write and solve an equation. Change the percent to a decimal or fraction and multiply.

Examples

1. Find the discount on the helmet.

   | Write as an equation. Write the percent as a decimal or a fraction. | → | Solve. |

   20% of $75 is what number? $0.2 \times \$75 = n$
   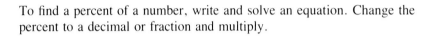
   $0.2 \times \$75 = n$ $\$15 = n$

   The discount on the helmet is $15.

2. Bike Club dues this year are 125% of last year's dues. Last year's dues were $12. What are club dues this year?

   What number is 125% of $12? $n = 1\frac{1}{4} \times \$12$
   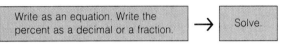
   $n = 1\frac{1}{4} \times \$12$ $n = \frac{5}{4} \times \$12$
                                  $n = \$15$

   The Bike Club dues are $15 this year.

Try This Solve.

a. There are 32 students in Rita's class. 25% of the students wear glasses. How many students wear glasses?

b. Richard weighs 140 lb. 67% of his weight is water. How much of Richard's weight is water?

c. There are 35 members in the Bike Club. Of those members, 20% ride mountain bikes. How many members ride mountain bikes?

Exercises

Solve.

1. What number is 40% of 20?
2. 60% of 50 is what number?
3. 70% of $200 is what amount?
4. What number is 23% of 300?
5. 85% of $41 is what amount?
6. What amount is 65% of $34?
7. What number is 17.9% of 0.52?
8. 32.2% of 14.5 is what number?
9. 180% of 60 is what number?
10. What number is 133% of 40?
11. $12\frac{1}{2}$% of 72 is what number?
12. What number is $66\frac{2}{3}$% of $\frac{1}{2}$?
13. Victor's lunch bill was $15. He left a 20% tip. How much money did he leave for a tip?
14. Roberta bought a backpack for $40. She had to pay 5.6% tax. How much did she pay for tax?
15. Stella Ranch has 56 acres of crops. Farm workers planted 20% of the acres with corn. How many acres did they plant with corn?
16. The cost of a new fender was $29 last year. The cost this year is 120% of last year's price. What is the price of a new fender this year?

Calculator Write the calculator keystrokes that you would need to solve each problem. Show two ways to solve each problem.

17. The bike shop had 15 jackets on sale. 20% of the jackets were blue. How many of the jackets were blue?
    A. ☐☐☐☐☐☐
    B. ☐☐☐☐☐
18. There were 6 pairs of gloves in stock. The bike shop ordered 350% more. How many pairs did they order?
    A. ☐☐☐☐☐☐
    B. ☐☐☐☐☐

Talk Math

19. In which problem would you change the percent to a fraction? Tell why.
    A. $33\frac{1}{3}$% of the 75 bikes in stock were black. How many bikes were black?
    B. Betty drew a bike for a newspaper ad. The bike was $1\frac{1}{2}$ in. long. She decided to enlarge it 300%. How long was the bike then?

Mixed Applications Find the regular price, the amount of discount, and the sale price for each. Use the ad.

20. one pair of goggles
21. two tires
22. one set of handle bars and one fender
23. one pair of goggles and one tire
24. two fenders and one set of handle bars
25. one set of handle bars, one tire, and one helmet

Spring Sale
20% off everything
Helmets: reg. $75
Tires: reg. $55
Fenders: reg. $21
Goggles: reg. $25
Handle bars: reg. $35

USING PERCENT

11-3 Finding What Percent One Number Is of Another

Objective
Find what percent one number is of another.

Tracey worked at a toll booth checking traffic patterns. After a one-hour period during rush hour, she had counted 70 cars. What percent of these cars held three or more people?

Traffic Tally

1 person	|||| |||| |||| |||| || |||| |||| |||| ||||
2 people	|||| |||| ||||
3 people or more	|||| |||| ||||

Understand the Situation

- How many cars held three or more people?
- How many cars did Tracey count in one hour?

To find what percent one number is of another, write and solve an equation.

Examples

1. What percent of the cars held three or more people?

Write as an equation. → Solve.

$$14 \text{ is what percent of } 70?$$
$$\downarrow \quad \downarrow \quad \quad \downarrow \quad \quad \downarrow \quad \downarrow$$
$$14 \;=\; \quad n \quad \times \; 70$$

$14 = n \times 70$

$\dfrac{14}{70} = n \times \dfrac{70}{70}$ *Divide both sides by 70.*

$\dfrac{14}{70} = n$

$0.2 = n$ *Find the percent.*

$20\% = n$

20% of the cars held three or more people.

Another way to find what percent one number is of another is to solve a proportion.

2. What percent of the cars held one person?

$\dfrac{n}{100} = \dfrac{42}{70}$ ← cars with one person
 ← total cars   *Write a proportion.*

$70n = 42 \times 100$ *Solve.*

$70n = 4200$

$\dfrac{70n}{70} = \dfrac{4200}{70}$

$n = 60$ 60% of the cars held one person.

Try This

a. What percent of the cars held two or more people?

b. What percent of the cars held one or two people?

Exercises

Solve.

1. 30 is what percent of 60?
2. What percent of 50 is 20?
3. What percent of 140 is 35?
4. 50 is what percent of 250?
5. 19 is what percent of 400?
6. What percent of 300 is 35?
7. 7.2 is what percent of 8?
8. What percent of 24 is 1.4?
9. What percent of 3.5 is 7?
10. 7.2 is what percent of 9?
11. $\frac{3}{8}$ is what percent of $1\frac{1}{2}$?
12. What percent of $3\frac{1}{3}$ is $\frac{5}{6}$?
13. On the history quiz, Marsha got 16 questions right out of 25. What percent of the questions did Marsha get right?
14. Luke made 30 out of 36 first serves in a tennis set. What percent of Luke's first serves were good?
15. There are 50 parking spots in one of the school's parking lots. There were cars in 30 of the spots by 9 a.m. There were cars in what percent of the parking spots by 9 a.m.?
16. Last June, 21 out of the 30 days were sunny. What percent of the days were sunny?

Suppose Suppose Tracey only counted 60 cars in a one-hour period. Solve Exercises 17–18.

17. Tracey counted 36 cars that held one person. What percent of the cars held one person?
18. Tracey counted 12 cars that held three or more people. What percent of the cars held three or more people?

Mixed Applications In another survey, Tracey counted the types of cars that went through the toll booth. In one period she counted 103 cars. She made a bar graph of her findings.

19. Estimate the number of each type of car.
20. Find the percent of each type of car.
21. Estimate the difference between the number of compact cars and the number of sports cars.

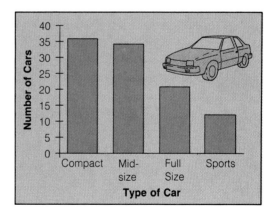

22. Estimate the total number of compact and mid-size cars.
23. Suppose there were 20 more sports cars. Estimate the percent of sports cars.

Problem Solving

24. Mr. McClintock owns a car lot. Of the cars on his lot, $\frac{1}{2}$ are sports cars, $\frac{1}{4}$ are compact cars, and the rest are mid-size cars. There are 8 mid-size cars on the lot. How many cars are on the lot?

11-4 Finding a Number Given a Percent

Objective
Find a number given a percent.

A ranger counted the number of seals in different herds. The count for Herd A was only 80% of last year's count. How many seals were there in Herd A last year?

Understand the Situation

- How many seals did the ranger count in Herd A this year?
- What percent of last year's count was Herd A?

To find a number given a percent, write and solve an equation.

Examples

1. How many seals were there in Herd A last year?

   Write as an equation. Write the percent as a decimal or a fraction. → Solve.

   36 is 80% of what number?
   ↓ ↓ ↓ ↓ ↓
   $36 = 0.8 \times n$

   $36 = 0.8 \times n$
   $\frac{36}{0.8} = \frac{0.8}{0.8} \times n$ Divide both sides by 0.8.
   $\frac{36}{0.8} = n$
   $45 = n$

   There were 45 seals in Herd A last year.

Another way to find a number given a percent is to solve a proportion.

2. 84 is 200% of what number?

   $\frac{84}{n} = \frac{200}{100}$ Write a proportion.
   $84 \times 100 = n \times 200$ Solve.
   $8400 = 200n$
   $\frac{8400}{200} = \frac{200n}{200}$
   $42 = n$

Try This Solve.

a. Herd B was 90% of last year's count. How many seals were there in Herd B last year?

b. Herd C was 114% of last year's count. How many seals were there in Herd C last year?

Exercises

Solve.

1. 56 is 80% of what number?
2. 24% of what number is 24?
3. 290 is 20% of what number?
4. 3% of what number is 216?
5. 18 is 6% of what number?
6. 9% of what number is 27?
7. 7% of what number is 6.37?
8. 12.15 is 15% of what number?
9. 40% of what number is $\frac{1}{5}$?
10. $\frac{1}{6}$ is 10% of what number?
11. 150% of what number is 66?
12. 24 is 300% of what number?
13. Duarte's Department Store pays their sales clerks 20% of their total sales. Manny made $1360. What were his total sales?
14. There are 42 children in the village of Hamlen. This is 30% of the total population. How many people live in Hamlen?
15. Jill has 18 buffalo nickels in her coin collection. This number is 0.2% of her collection. How many coins does Jill have?
16. Mr. Santos left a $6 tip for his dinner. This amount was 15% of the bill. What was the bill for dinner?

Evaluating Solutions

17. Yoshi solved the problem at the right. Was the solution correct?

21 is 300% of what number?
$\frac{1}{3}$ of 21 = 7
$n = 7$

Problem Solving

Solve. Use the table shown.

18. The count for Herd X this year is 60% of last year's count for Herd Y. The count for Herd Z this year is 150% of the count for Herd X last year. The total count last year for all these herds was 84. Find last year's count for each herd.

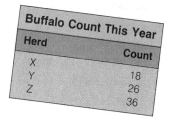

Buffalo Count This Year

Herd	Count
X	18
Y	26
Z	36

19. The number of rangers this year is 20% more than last year. The number of park guides this year is 35% more than the number of rangers this year. There were 10 rangers last year. How many rangers and guides are there this year?

Mixed Skills Review

Find the answer.

20. $\frac{1}{2} \times 3\frac{1}{4}$
21. $7000 - 560.3$
22. 100×8.74
23. $0.62 \div 100$

Solve.

24. $\frac{16}{2} = \frac{n}{8}$
25. $\frac{3}{4} = \frac{12}{n}$
26. $\frac{n}{3} = \frac{9}{10}$
27. $\frac{5}{n} = \frac{8}{11}$
28. $\frac{5}{6} = \frac{n}{0.3}$

11-5 Mental Math: Percents

Objective
Use mental math with percents.

When Sal goes shopping, he uses mental math to find the amount of discount. How much is the discount on the jeans?

Examples Solve. Use mental math.

1. How much is the discount on the jeans?
   25% of $28

   $\frac{1}{4}$ of $28   Think: $25\% = \frac{1}{4}$

   $\frac{1}{4} \times 28 = 7$ Think: $28 \div 4 = 7$

   The discount is $7.

2. 50% of 16

   $\frac{1}{2}$ of 16   Think: $50\% = \frac{1}{2}$

   $\frac{1}{2} \times 16 = 8$ Think: $16 \div 2 = 8$

3. 10% of 120

   0.10 of 120   Think: $10\% = 0.10$

   $0.10 \times 120 = 12$ Think: Move the decimal point one place to the left.

Try This Solve. Use mental math.

a. 50% of 24   b. 10% of 300   c. 25% of 160   d. 10% of 280

Exercises

Solve. Use mental math.

1. 10% of $125
2. 25% of 36 **9**
3. 50% of $180
4. 10% of 540
5. 50% of 680
6. 10% of 2400 **240**
7. 25% of 200
8. 25% of 1600

Number Sense

Fill in each blank.

9. 20% of 110 = ____   Think: 10% of 110 = ____ → 11 × 2 = ____

   30% of 120 = ____   Think: 10% of 120 = ____ → ____ × 3 = ____

   40% of 160 = ____   Think: 10% of 160 = ____ → ____ × ____ = ____

   60% of 200 = ____   Think: 10% of 200 = ____ → ____ × ____ = ____

   70% of 500 = ____   Think: 10% of 500 = ____ → ____ × ____ = ____

Mixed Practice

Find the answer.

1. $2\frac{3}{4} + 3\frac{1}{12}$
2. $3602 - 1987$
3. $\$308 \times 52$
4. $4\frac{1}{6} \div 1\frac{2}{5}$
5. $3276 \div 52$
6. $1.026 - 0.09$
7. $8\frac{1}{3} - 5\frac{3}{5}$
8. $12{,}892 + 1{,}436$
9. $2\frac{2}{9} \times 1\frac{1}{5}$
10. $45.3 + 25.62$
11. $\$62.40 \div 13$
12. 4.12×8.5

13. Make a bar graph of these data.

Your Favorite Snack	
Snack	Votes
Popcorn	27
Fruit	25
Nuts	16

14. Make a pictograph of these data.

Your Favorite Drink with Dinner	
Drink	Votes
Milk	22
Water	32
Juice	16

Estimate.

15. 312×52
16. $3598 \div 61$
17. $349 + 148$
18. $48 + 49 + 52$
19. $4475 \div 92$
20. $2833 - 1792$
21. 619×22
22. $32 + 29 + 31$

Evaluate.

23. $36t$ for $t = 24$
24. $1966 + m$ for $m = 2707$
25. $28.62 - s$ for $s = 0.753$
26. $3.829 \div n$ for $n = 100$

Solve.

27. What number is 60% of 48?
28. 13 is what percent of 39?
29. 22 is $33\frac{1}{3}\%$ of what number?
30. 36% of 25 is what number?
31. What percent of 48 is 12?
32. 9 is 12.5% of what number?

Arrange in order from least to greatest.

33. 6.08, 6.5, 6.015, 6.8, 6.51
34. $2\frac{3}{4}, 2\frac{7}{8}, 2\frac{5}{12}, 2\frac{5}{6}, 2\frac{1}{2}$
35. 50%, $\frac{2}{5}$, 0.6, 61%, $\frac{5}{8}$, 0.63
36. 22%, 0.23, $\frac{1}{5}$, 19%, 0.18, $\frac{2}{9}$

Round to the given place.

37. 38.74
   tens
38. $12.95
   dollar
39. 0.034
   tenths
40. 0.0549
   hundredths
41. $3.62
   dime

Write the correct formula.

42. for the area of a rectangle
43. for the area of a triangle
44. for the volume of a cube
45. for the area of a circle

USING PERCENT

11-6 Estimating Percents Using Compatible Numbers

Objective
Estimate percents using compatible numbers.

A recent poll surveyed 1980 adults to find what electronic equipment they use in their homes. About what percent of the adults use a home computer?

Electronic Equipment Owners

Equipment	Number
Color TV	1801
Push-Button Phone	1049
Garage Door Opener	277
Home Computer	217
CD Player	79

Understand the Situation

- How many people did the poll survey?
- How many people use a home computer?

To estimate percents, find pairs of compatible numbers.

Examples

1. Estimate the percent of adults using a home computer.

   217 is what percent of 1980?   Rewrite using compatible numbers.
   200 is what percent of 2000?   217 is about 200. 1980 is about 2000.
   $200 = n \times 2000$

   $\frac{200}{2000} = n$

   $10\% = n$

   About 10% of the adults use a home computer.

2. The survey showed that 45% of the homes have microwave ovens. Estimate the number of homes with microwave ovens.

   45% of 1980 = n   Rewrite using compatible numbers.
   50% of 2000 = n   45% is about 50%. 1980 is about 2000.

   $\frac{1}{2} \times 2000 = n$

   1000 = n   About 1000 homes have microwave ovens.

3. Chun-Mei left a 15% tip for a $20.13 bill. About how much tip did he leave?

   15% of 20.13 = n   Rewrite using compatible numbers.
   15% of 20 = n   20.13 is about 20.
   (10% of 20) + (5% of 20) = n   Break 15% into 10% and 5%.
   10% of 20 = 2
   5% of 20 = 1   5% is half of 10%.
   2 + 1 = 3   Add both answers.

   He left a tip of about $3.

Try This

Estimate.

a. 12 is what percent of 37?   **b.** What is 34% of 312?

Exercises

Estimate.
1. 16 is what percent of 38?
2. What percent of 39 is 11?
3. What percent of 53 is 9?
4. 13 is what percent of 23?
5. 31% of 313 is what number?
6. What number is 47% of 210?
7. What number is 31% of 61?
8. 9.5% of 1200 is what number?
9. 9 is 9% of what number?
10. 24% of what number is 31?
11. 149% of what number is 8?
12. 19.6 is 200% of what number?

Estimate the amount of tip for each bill. Leave a 15% tip.
13. bill: $59.37
14. bill: $11.09
15. bill: $19.14
16. bill: $28.53
17. bill: $32.03
18. bill: $12.08
19. bill: $102.56
20. bill: $79.24

Choose the best estimate. State if the estimate is high or low.
21. 25% of 78   A. 10   B. 20   C. 30   high   low
22. 52% of 64   A. 25   B. 32   C. 40   high   low
23. 9% of 60   A. 6   B. 4   C. 8   high   low
24. 97% of 36   A. 30   B. 27   C. 36   high   low

Estimate the percent of each shaded area.

25.
26.
27.

28.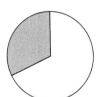
29.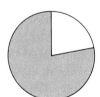
30.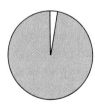

Mixed Skills Review

Write each fraction as a decimal.
31. $\frac{3}{4}$
32. $\frac{5}{8}$
33. $\frac{1}{2}$
34. $\frac{2}{5}$
35. $\frac{11}{10}$
36. $2\frac{4}{5}$

Write the ratio as a fraction in lowest terms.
37. 100 miles in 2 hours
38. 6 wins to 10 losses

USING PERCENT

11-7 Consumer Math: Simple Interest

Objective
Find simple interest.

- SITUATION
- DATA
- PLAN
- ANSWER
- CHECK

Devon saved $500 to buy her first car. She found a good used car that cost $1200. Her father helped her by borrowing $700 from his credit union. Devon agreed to pay the loan plus interest at the rate of 9% per year for two years.

Definitions

Principal (p)—the amount of money borrowed or loaned

Interest (i)—the amount of money paid for the use of money

Time (t)—the length of time the money is borrowed or loaned

Rate (r)—the percent of the principal charged as interest

Understand the Situation

- What is the principal of the loan?
- What rate of interest will Devon pay?
- How long does Devon have to repay the loan?

The amount of interest paid on a loan depends on the **principal**, the **rate** of interest, and the **time**. Use the formula to find simple interest. Be sure to state the rate and the length of time in the same units of time.

Interest equals principal times rate times length of time. $i = prt$

Examples

1. Find the total interest that Devon must pay on her loan.

   Substitute values for the variables. → Compute.

   $p = \$700$
   $r = 9\%$ per year
   $t = 2$ years

   $i = \$700 \times 0.09 \times 2$
   $i = \$126$

   Devon must pay $126 in interest.

   Find the total interest.

2. $p = \$2000$
   $r = 1.5\%$ per month
   $t = 5$ months
   $i = \$2000 \times 0.015 \times 5 = \$150$

3. $p = \$2500$
   $r = 10.5\%$ per year
   $t = 6$ months   *Think: 0.5 years*
   $i = \$2500 \times 0.105 \times 0.5 = \$131.25$

Try This Find the total interest.

a. $p = \$650$
   $r = 1.8\%$ per month
   $t = 8$ months

b. $p = \$1250$
   $r = 11\%$ per year
   $t = 18$ months

c. $p = \$16,500$
   $r = 15\%$ per year
   $t = 25$ years

Exercises

Find the total interest.

1. $p = \$250$
   $r = 1.8\%$ per month
   $t = 8$ months

2. $p = \$1800$
   $r = 11\%$ per year
   $t = 18$ months

3. $p = \$15{,}500$
   $r = 15\%$ per year
   $t = 25$ years

4. $p = \$1950$
   $r = 11\%$ yearly
   $t = 6$ months

5. $p = \$3500$
   $r = 11.5\%$ per year
   $t = 6$ years

6. $p - \$800$
   $r = 1.8\%$ monthly
   $t = 1$ year

Solve.

7. Mrs. Tanouye borrowed $500 at 12% interest per year. She paid back the loan in 3 years. What is the total amount she paid back?

8. Mr. Stetson deposited $400 in a savings account. The money earned 5% simple interest per year. How much would Mr. Stetson have in his savings account after four years?

9. Jason Masimore borrowed $900 for 5 years from a savings and loan association to buy a new car. The savings and loan charged an interest rate of 10.5% per year. What is the total amount Jason must repay?

10. Milly Ng bought a $500 certificate for 3 months. The certificate earns interest at the rate of 9.5% per year. What is the total value of the certificate in 3 months?

11. Terri Mercer borrowed $3000 for one year. She repaid $3360 at the end of one year. What rate of interest did she pay?

12. Suppose that you want to earn $1000 in interest in one year. How much money would you have to invest at 12% yearly interest in order to earn the $1000?

Write Math

13. Write a formula for finding the total amount *(A)* to be repaid on a principal *(p)* borrowed at a rate of *r*% for a period of time *(t)*.

Data Hunt

Call a local bank, savings and loan, or credit union to answer Exercises 14–15.

14. Find how much interest $350 in a savings account would earn in one year. Is the interest rate the same for $3500?

15. Find the current rate of interest on a new-car loan. Find the total cost of a loan of $10,700 for 4 years. Is the interest rate the same for 3 years?

Problem Solving

16. Suppose you deposited $100 in a savings account on January 1. On the last day of January the bank paid you 0.5% interest on the total amount you had on deposit. You deposited the interest and began February with $100 plus January's interest in your account. On the last day of February the bank paid you interest on the new total. Continue the pattern to find the total amount you would have on the next New Year's Day. Round to the nearest cent.

11-8 Percent of Increase or Decrease

Objective
Find the percent of increase or decrease.

Frank Torockio collects baseball cards. He bought a Don Mattingly card for $6. Two years later Frank sold the card at an auction for $15. What was the percent of increase or decrease in the value of the card?

Understand the Situation

- What did Frank pay for the Don Mattingly card?
- For how much did Frank sell the card?
- Did the value of the card increase or decrease?

To find a percent of change, find the amount of increase or decrease. Then find what percent that change is of the first amount.

Examples

1. Find the percent of increase or decrease in the value of the Don Mattingly card.

   | Find the amount of increase or decrease. | → | Find what percent the amount of increase or decrease is of the first amount. |

   $6 to $15 $15 - 6 = 9$
   The value increased $9.

   9 is what percent of 6?
   $p = \frac{9}{6} = 1.5 = 150\%$

   The percent of increase was 150%.

2. Darla Herbold bought a 1949 Franklin half dollar for $200. Later she had to sell the coin for $175. Find the percent of increase or decrease in the coin's value.

   $200 to $175 $200 - 175 = 25$
   The value decreased $25.

   What percent of $200 is $25?
   $p = \frac{25}{200} = 0.125 = 12.5\%$

   The percent of decrease was 12.5%.

Try This Find the percent of increase or decrease.

a. Chris Bess bought an 1893 Columbian one-cent stamp for $20. Later she sold it for $28. What was the change in the value of the stamp?

b. Rafael Perez bought a 1907 Jamestown two-cent stamp for $32. One month later he sold the stamp for $28. What was the change in the value of the stamp?

Exercises

Find the percent of increase or decrease.

1. 10 to 12
2. $30 to $25
3. 1000 to 1050
4. 12 to 6
5. 225 to 300
6. 36 to 16
7. 45 to 80
8. $50 to $30
9. 8 to 15
10. $175 to $35

11. Patti Gorsuch bought a Mike Scott baseball card for $5. She later sold the card for $9. What was the percent of change in the value of the card?

12. At an auction, Ken Flickinger bought an 1889CC Morgan silver dollar for $350. He later sold the coin for $325. What was the percent of change in the value of the coin?

13. Dale Koo collects Hummel figures. She bought a figure for $1200. She later sold it for $1550. What was the percent of change in the value of the figure?

14. Bill Fritz collects mint sheets of history stamps. He bought a sheet of the 1956 Alamo nine-cent stamp for $25. He later sold the sheet for $30. What was the percent of change in the value of the sheet?

15. Janine Kelly bought 10 Tony Gwynn baseball cards for $0.75 each. She later sold all of the cards for a total of $25. What was the percent of change in value of the cards?

16. Emma Avila bought a 1927 ten-cent Lindbergh airmail stamp for $14.50. Later she sold the stamp at an auction for $19. She paid the auctioneer 15% of the sale price. How much profit did she make on the stamp? What was the percent of profit?

Mixed Skills Review

Solve. Use the table for Exercises 17–20.

17. Draw a line graph to show the change in value from 1978 to 1988.
18. Give the year and percent of greatest change.
19. Find the difference between the greatest and least values for the card.
20. How many times greater was the value in 1984 than in 1979?

Find each answer.

21. $1\frac{2}{3} \times 4\frac{1}{2}$
22. $612 \div 12$
23. $4\frac{1}{5} - 2\frac{2}{3}$
24. $78 + 54 + 22$
25. 2.34×40.7
26. $9845 \div 100$

Eddie Murray Card

Year	Value
1978	$0.75
1979	$1.25
1980	$1.75
1981	$2.25
1982	$3.00
1983	$5.25
1984	$10.00
1985	$17.00
1986	$20.00
1987	$25.00
1988	$28.00

USING PERCENT

11-9 Problem Solving: Using Data from a Table

Objective
Find sales tax using data from a table.

- SITUATION
- **DATA**
- PLAN
- ANSWER
- CHECK

To solve some problems, you need to use data from a table. Many states charge sales tax on the sale of certain items. Each state sets its own rate. Sales clerks can use a table to find the amount of sales tax.

Here is part of the table for a 6% sales tax. The complete table is in the Data Bank on p. 492.

6% Sales Tax Reimbursement Schedule

Transaction	Tax	Transaction	Tax	Transaction	Tax
.01– .10	.00	8.42– 8.58	.51	16.92–17.08	1.02
.11– .22	.01	8.59– 8.74	.52	17.09–17.24	1.03
.23– .39	.02	8.75– 8.91	.53	17.25–17.41	1.04
.40– .56	.03	8.92– 9.08	.54	17.42–17.58	1.05
.57– .73	.04	9.09– 9.24	.55	17.59–17.74	1.06
.74– .90	.05	9.25– 9.41	.56	17.75–17.91	1.07
.91–1.08	.06	9.42– 9.58	.57	17.92–18.08	1.08
1.09–1.24	.07	9.59– 9.74	.58	18.09–18.24	1.09
1.25–1.41	.08	9.75– 9.91	.59	18.25–18.41	1.10
1.42–1.58	.09	9.92–10.08	.60	18.42–18.58	1.11

Examples

1. Find the sales tax on a sale of $18.20. Use the table.

> Look in the *Transaction* column. Find the line that has the cost of the item. → Read the tax for that line.

$18.20 is between $18.09 and $18.24.   The sales tax is $1.09.

2. Find the sales tax on $35.50.

If the amount is greater than the table shows, break the amount apart.
Read the tax for each part.
Add the tax.

$35.50 = $18 + $17.50
Tax on $18   = $1.08
Tax on $17.50 = $1.05
Total tax due   = $2.13

Try This

Find the sales tax. Use the table.

a. $0.64   **b.** $9.87 + $6.43 + $1.16   **c.** $24.02

Exercises

Find 6% sales tax. Use the table in the Data Bank on page 492.

1. $21.35
2. $16.71
3. $7.43 + $4.46
4. $5 + $8 + $12
5. $0.87
6. $0.91 + $1.45
7. $2.04
8. $9.56
9. $19.78 + $3.30
10. $22.38
11. $16.50
12. $8.99
13. $32
14. $43.98
15. $78.08
16. $25.73
17. $54.13
18. $29.05
19. $87 + $25 + $17
20. $121.10

Calculator Find the total cost with 6% sales tax for Exercises 21–22. Use a calculator.

21. Jay bought 4 T-shirts at $12.99 each, 6 pairs of socks at $3.49 each, and a pair of athletic shoes for $27.50.

22. Lucy bought 6 plastic drop cloths at $1.25 each, 3 quarts of paint at $12.95 each, and 2 paint brushes at $2.49 each.

Many states do not charge sales tax on food or prescription medicines. Find the total of the list of mixed items. Taxable items are marked T. The sales tax is 5.5%. Use a calculator. Round to the nearest cent.

23. $1.43
    3.10 T
    0.97 T
    4.38
    0.36
    0.52

24. $7.32
    0.58
    2.60 T
    0.98
    3.85 T
    0.46 T

25. $13.45 T
    8.00 T
    0.85
    1.26
    20.00 T
    7.03

Estimation

26. You can estimate sales tax when shopping. With your group, write a plan for estimating your sales tax. Tell what strategy you used.

Estimate the tax in your state for these amounts.

27. $17
28. $14.98
29. $24.52
30. $36.23

Problem Solving

Items you buy from out-of-state businesses are not subject to sales tax. The cost of shipping and handling is usually added to the cost of the item. Decide if you would buy the following items from a local store or through the mail.

31. SLR camera body
32. fl.8 lens
33. zoom flash

Item	Local Price	Sales Tax	Mail Order Price	Handling Charges	Shipping/Postage
SLR Camera Body	$239.50	6%	$218.45	$5.00	$5.25
50 mm fl.8 lens	$64.95	7%	$59.79	$3.50	$3.78
Zoom flash	$59.99	5%	$48.30	$4.25	$4.62

USING PERCENT

11-10 Problem Solving: Developing Thinking Skills

Objective
Use problem-solving strategies.

- **SITUATION**
- **DATA**
- **PLAN**
- **ANSWER**
- **CHECK**

Rashan went canoeing. In 1 h, he traveled 2 mi upstream. After paddling for 1 h, he needed to rest for 1 h. While he rested, the canoe floated back downstream 1 mi. At this rate, how long did it take Rashan to go 4 mi upstream?

Exercises

Follow the *Thinking Actions*. Answer each question.

Try these **before** starting to write.

Thinking Actions Before
- Read the problem.
- Ask yourself questions to understand it.
- Think of possible strategies.

Try these **during** your work.

Thinking Actions During
- Try your strategies.
- Stumped? Try answering these questions.
- Check your work.

1. How far upstream did Rashan paddle in 1 h?
2. How far back did the canoe float while Rashan rested?
3. What strategy, or strategies, might help you to solve this problem?
4. Can you draw a picture showing how far Rashan canoed?
5. Can you make a table showing the distances Rashan traveled?
6. How far upstream was Rashan after 2 h?
7. How far upstream was Rashan after 3 h?

8. Write the answer in a complete sentence. Name the strategies shown.

Solution 1

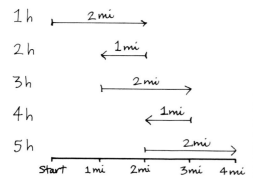

Solution 2

Hour	Distance Traveled
1 h	2 mi
2 h	2 mi − 1 mi = 1 mi
3 h	1 mi + 2 mi = 3 mi
4 h	3 mi − 1 mi = 2 mi
5 h	2 mi + 2 mi = 4 mi

Solve. Use one or more of the
problem-solving strategies.

Problem-Solving Strategies
- Guess and check.
- Choose an operation.
- Make an organized list.
- Act it out.
- Find a pattern.
- Write an equation.
- Use logical reasoning.
- Draw a picture or diagram.
- Make a table.
- Use objects.
- Work backward.
- Solve a simpler problem.

9. Orin, Mori, Sachiko, and Corinne hiked to the top of Half Dome in Yosemite National Park. Orin was not first. Sachiko was not last. Corinne had one person behind her. Mori was right behind Sachiko. In what order did they reach the top?

10. Midge knew some Italian before she went to Italy. She learned many new words once she arrived. She learned 35 new words the first week, 32 words the second week, and 29 words the third week. At this rate, how many new words did she learn the ninth week?

11. Norma's choir group went to London. One afternoon some of the students went to a play. Orchestra seats were $12. Balcony seats were $8. They spent a total of $76 on 8 tickets. How many orchestra and balcony seats did they buy?

12. Marco packed 3 shirts (green, blue, and yellow), 3 pants (tan, grey, and blue), and 2 coats (navy and white). How many different outfits could he wear?

13. Kwang and his father went on a tour of a castle in England. They followed their guide through 7 rooms. They stayed in each room about 5 min. Whenever they left a room, the next tour group would enter. About how long would it take 4 tour groups to complete their tours?

14. Roberto's biking club took a tour of Canada. The second day, they biked half as far as they did the first day. The third day, they went 12 mi less than on the second day. They rode 20 mi the third day. How many miles did they ride the first day?

15. On a trip to Spain a group of 63 students had their choice of going to a bullfight or seeing some ancient ruins. Thirty-five more students wanted to see the bullfight. How many students wanted to see the ruins?

Finishing a Solution

16. Finish the solution to this problem.

    Problem: Jamie, Ruth, Frank, and Sarah each speak a different foreign language. No one speaks a language that starts with the same letter as his or her first name. Frank can read characters. Ruth never did well in French. Sarah lived in Paris for ten years. What language can each person speak?

	Jamie	Ruth	Frank	Sarah
Japanese	no		yes	
Russian		no		
French			no	
Spanish				no

USING PERCENT

11-11 Enrichment: Commission

Objective
Calculate earnings on commission.

Jasper is a sales clerk in a men's store. He earns a salary of $4.50 per hour plus 5% commission on sales above $500 per week. What are Jasper's earnings for the week?

Wilson, Jasper — May 5–11

	Mon.	Tue.	Wed.	Thurs.	Fri.	Sat.	Sun.	Total
Hours worked	2.0	2.0	4.0	—	—	6.0	6.0	20.0
Sales	$80	$95	$110	—	—	$240	$225	$750

Pay per hour: $4.50
Commission: 5% above $500 per week

Understand the Situation
- How much money does Jasper earn per hour?
- What was Jasper's sales total for the week?
- On how much money in sales will Jasper earn commission?

Example Find Jasper's earnings for the week.

$4.50 × 20 h = $100.00   Basic salary
($750 − $500) × 0.05 = $12.50   Commission
$112.50   Total earnings

Jasper's earnings for the week were $112.50.

Try This Find the earnings for the week.

a. Pearl is paid 6% commission on all sales plus $75 per week. She sold $1900 in goods last week.

b. Toshio earns a straight commission of 9% on sales. He sold $1950 in goods last week.

Exercises

Find the earnings for the week.

1. Naveen works on commission at a shoe store. He is paid $100 plus 3% on sales above $1100. His sales totaled $2600 in one week.

2. Ella is a sales clerk at Platter World. She earns $75 plus 9% on sales above $650. Her sales for the week totaled $982.

3. In the summer Ernie drives a lunch truck. He is paid a commission of 19% on sales above $500. In one week he sold a total of $865 in food.

4. Sylvia works at a clothing store. She earns a commission of 11% on all sales above $250. She sold $1080 in clothing last week.

11-12 Computer: Percents

Objective
Use a computer to calculate percents.

You are shopping for a skateboard. You find one that costs $69.95 plus tax. Tax is 6.5% of the price. How much money do you need to buy the skateboard?

You can program your computer to find the answer.

Type the following program. Do not forget to type the punctuation exactly as shown. RUN your program to find the total cost of the skateboard.

```
10 INPUT "Item Name (press 'RETURN' if no more items) ";ITEM$
20 IF ITEM$ = "" GOTO 70
30 INPUT "Price "; PRICE: INPUT "% tax rate "; RTE
40 TAX = PRICE * RTE / 100: PRINT"Tax = ";TAX
50 TTAL = PRICE + TAX: PRINT "Total = "; TTAL
60 GOTO 10
70 END
```

Line 10 Asks you for an item name. The computer stores the name in a variable called ITEM$. $ at the end of a variable name means that the variable is words and not numbers.

Line 20 If there are no more items, the line will be blank. This is shown by "". The computer will go to line 70.

Line 70 Exits the program.

Line 30 Asks you to enter a price and a tax rate. The computer stores the price under the variable name PRICE and the tax rate under the variable name RTE.

Line 40 Computes the tax. Multiplies the price by the percent tax rate divided by 100 (the decimal notation of the tax rate).

Line 50 Computes the total amount of money spent. Adds the price to the tax.

Line 60 Loops the program. You can keep entering items, prices, and tax rates. If you do not, the program exits in line 70.

Exercises

1. Make a flowchart of the program.
2. What does line 40 print?
3. What does line 50 print?
4. Which line sends the computer back to the start of the program?
5. How does the computer know when to exit the loop?
6. Which line computes the amount of tax?

USING PERCENT

CHAPTER 11 REVIEW

Decide if you need to find a percent of a number, what percent one number is of another, or a number given a percent.

1. There are 28 students in Niles' class. Of the students, 12% have blue eyes. How many students have blue eyes?
2. Emilio left a $5 tip for his dinner. This amount was 15% of the bill. What was the bill for his dinner?
3. Annamaria bought a denim skirt for $48 plus tax. Tax on goods is 7%. How much tax did she pay?
4. Si Lu got 18 questions right out of 20. What percent of the questions did she get right?

Compare. Use <, >, or =.

5. 40% of 20 ☐ 7
6. 125% of 62 ☐ 77.5
7. 3.5% of 36 ☐ 126
8. 22% of $\frac{2}{5}$ ☐ 0.88

State if each statement is true or false. Use mental math.

9. 10% of 180 is 1.8.
10. 25% of 400 is 100.
11. 50% of 45 is 90.
12. 25% of 28 is 6.

Choose the best estimate.

13. 15 is what percent of 29? A. 30% B. 40% C. 50% D. 60%
14. What number is 25% of 41? A. 5 B. 10 C. 15 D. 20
15. 11 is 11% of what number? A. 80 B. 90 C. 100 D. 200
16. 98% of 492 is what number? A. 492 B. 500 C. 4920 D. 5000

Decide if the total interest is greater or less than the amount shown. Use < or >.

17. $p = \$2000$
    $r = 11\%$ per year
    $t = 5$ years
    ☐ $1000

18. $p = \$15{,}500$
    $r = 9\%$ per year
    $t = 4$ years
    ☐ $5600

19. $p = \$3500$
    $r = 12\%$ per year
    $t = 10$ months
    ☐ $420

Write the amount of tax for each purchase. Use the tax table.

20. three pairs of socks for $16.95
21. a blouse for $17.47
22. a pair of shorts for $17.72

6% Sales Tax					
Transaction	Tax	Transaction	Tax	Transaction	Tax
.01–.10	.00	8.42–8.58	.51	16.92–17.08	1.02
.11–.22	.01	8.59–8.74	.52	17.09–17.24	1.03
.23–.39	.02	8.75–8.91	.53	17.25–17.41	1.04
.40–.56	.03	8.92–9.08	.54	17.42–17.58	1.05
.57–.73	.04	9.09–9.24	.55	17.59–17.74	1.06

Find the percent of increase or decrease for each item.

23. item A
    bought for $10
    sold for $15
24. item B
    bought for $25
    sold for $50
25. item C
    bought for $4000
    sold for $2800

CHAPTER 11 TEST

1. Pasco's Sport Shop bought 150 baseball caps for $520.50. The shop sold each cap for $5.25. How much more was the selling price than the wholesale price of each cap?

Solve.

2. What number is 35% of 50?
3. 40% of $80 is what amount?
4. 16.5% of 120 is what number?
5. 250% of what number is 325?
6. 13 is what percent of 39?
7. What percent of 50 is 125?
8. $1\frac{3}{4}$ is what percent of 17.5?
9. What percent of 1.68 is 1.4?
10. 5% of what number is 15?
11. 32.5 is 250% of what number?
12. 135 is 65% of what number?
13. 25% of what number is $\frac{1}{2}$?

Solve. Use mental math.

14. 25% of 240
15. 10% of 6300
16. 50% of $480
17. 25% of $320

Estimate.

18. 9 is what percent of 31?
19. 18 is 24% of what number?
20. What number is 53% of 221?
21. What percent of 83 is 43?
22. 341 is what percent of 34?
23. 62% of what number is 29?

Find the total interest.

24. $p = \$600$
    $r = 9.6\%$ per year
    $t = 2$ years

25. $p = \$4500$
    $r = 1.05\%$ per month
    $t = 30$ months

26. $p = \$1800$
    $r = 11\%$ per year
    $t = 18$ months

Find the percent of increase or decrease.

27. Lee Sung bought a stereo set for $450. Two years later he sold it for $360. What was the percent of change in the value of the stereo?
28. Anita Ray collects Wedgwood pottery. She bought a vase for $65. Four years later she sold it for $95. What was the percent of change in the value of the vase?

Find the sales tax. Use the 6% sales tax table.

29. $0.49
30. $13.02 + $4.03
31. $6.34 + $0.79 + $1.68

6% Sales Tax					
Transaction	Tax	Transaction	Tax	Transaction	Tax
.01–.10	.00	8.42–8.58	.51	16.92–17.08	1.02
.11–.22	.01	8.59–8.74	.52	17.09–17.24	1.03
.23–.39	.02	8.75–8.91	.53	17.25–17.41	1.04
.40–.56	.03	8.92–9.08	.54	17.42–17.58	1.05
.57–.73	.04	9.09–9.24	.55	17.59–17.74	1.06

USING PERCENT

CHAPTER 11 CUMULATIVE REVIEW

1. Multiply.
   $1\frac{3}{9} \times 2\frac{2}{5}$
   A. $2\frac{2}{15}$
   B. $3\frac{1}{5}$
   C. $1\frac{1}{9}$
   D. $2\frac{3}{5}$

2. Find the least common multiple.
   8, 22
   A. 44
   B. 48
   C. 64
   D. 88

3. Solve. Use mental math.
   50% of 48
   A. 25
   B. 96
   C. 12
   D. 24

4. State if the ratios are equal. Use = or ≠.
   $\frac{3.4}{3} \square \frac{2.1}{2}$
   A. =
   B. ≠

5. Find the reciprocal.
   $4\frac{2}{3}$
   A. $\frac{14}{3}$
   B. $\frac{3}{14}$
   C. 3
   D. $\frac{10}{3}$

6. Shigeru won free tickets to a circus. He gave 3 tickets to his sister. He divided the rest of the tickets equally among himself and 5 friends. He kept 2 tickets. How many tickets did he win in all?
   A. 20 tickets
   B. 16 tickets
   C. 10 tickets
   D. 15 tickets

7. Choose the correct formula for the volume of a rectangular solid.
   A. $V = lwh$
   B. $V = s^3$
   C. $A = lw$
   D. $i = prt$

8. The crafts fair had 125 different booths selling goods. 12% of the booths sold pottery. How many booths sold pottery?
   A. 35 booths
   B. 18 booths
   C. 15 booths
   D. 10 booths

9. Divide.
   $\frac{3}{10} \div 2\frac{3}{4}$
   A. $\frac{33}{40}$
   B. $2\frac{2}{5}$
   C. $\frac{6}{55}$
   D. $2\frac{9}{40}$

10. What fraction of the circle is shaded?
    A. $\frac{5}{8}$
    B. $\frac{3}{8}$
    C. $\frac{3}{5}$
    D. $\frac{3}{6}$

11. Two pairs of running shoes cost $68. How much money do 3 pairs of shoes cost?
    A. $102
    B. $136
    C. $108
    D. $96

12. Add.
$22\frac{4}{5} + 19\frac{5}{8}$
A. $41\frac{9}{13}$
B. $42\frac{17}{40}$
C. $42\frac{7}{40}$
D. $41\frac{17}{40}$

13. Estimate.
$\frac{11}{21} \times 28$
A. 14
B. 12
C. 17
D. 11

14. Change the unit.
704 mg = ___ g
A. 7.04
B. 0.704
C. 7040
D. 70.4

15. Estimate.
$3\frac{4}{8} + 6\frac{2}{3} + 8\frac{7}{16}$
A. 24
B. 21
C. 17
D. 19

16. Write as a mixed number.
$\frac{62}{18}$
A. $3\frac{1}{2}$
B. $3\frac{4}{9}$
C. 4
D. $2\frac{13}{9}$

17. Write the ratio.
8 crates of tangerines for $48
A. 8
B. $\frac{1}{8}$
C. $\frac{1}{6}$
D. 0.16

18. Which graph has the greater growth in car sales from November to February?
A. Graph I
B. Graph II
C. Both Graph I and Graph II are the same.
D. Neither Graph I nor Graph II has an increase in sales.

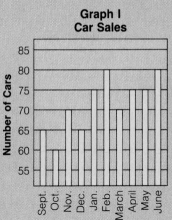

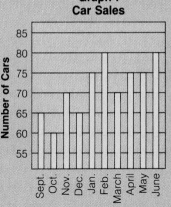

19. Find the amount of liquid in the measuring cup.
A. 14 fl oz
B. 12 fl oz
C. 8 fl oz
D. 9 fl oz

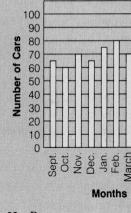

20. Regroup.
$3\frac{4}{5} = 2\frac{\square}{\square}$
A. $\frac{9}{5}$
B. $1\frac{4}{5}$
C. $\frac{8}{5}$
D. $\frac{5}{5}$

USING PERCENT

Chapter 12 Overview

Key Ideas

- Use several problem-solving strategies.
- Name points, lines, and line segments.
- Draw and name angles.
- Name parallel, skew, and perpendicular lines.
- Name triangles, quadrilaterals, and other polygons.
- Find perimeter and circumference.
- Make circle graphs.
- Make group decisions.

Key Terms

- point
- angle
- congruent
- triangle
- line
- right angle
- perpendicular
- quadrilateral
- segment
- acute angle
- parallel
- perimeter
- ray
- obtuse angle
- polygon
- circumference
- vertex
- skew
- plane
- regular polygon
- equilateral triangle
- isosceles triangle
- scalene triangle
- radius
- diameter
- parallelogram
- rectangle
- rhombus
- square
- trapezoid

Key Skills

Write each percent as a decimal.

1. 20%
2. 30%
3. 6%
4. 12%
5. 85%
6. 90%
7. 34.5%
8. 96.8%
9. $20\frac{1}{2}\%$
10. $7\frac{3}{4}\%$
11. 100%
12. 0.7%

Multiply.

13. 3.14×5
14. 3.14×7
15. 3.14×12
16. 3.14×2.5
17. $2 \times 3.14 \times 4$
18. $2 \times 3.14 \times 10$
19. $2 \times 3.14 \times 2.5$
20. $2 \times 3.14 \times 5.5$

Evaluate.

21. $2t + 2w$ for $t = 3$ and $w = 5$
22. $4s$ for $s = 5.75$
23. $3.14d$ for $d = 0.25$
24. $2l + 2y$ for $l = 0.8$ and $y = 1.2$

GEOMETRY 12

12-1 Problem Solving: Learning to Use Strategies

Objective
Use problem-solving strategies.

- SITUATION
- DATA
- **PLAN**
- ANSWER
- CHECK

Joan felt that she could win the badminton contest by drawing a picture to find the answer.

Badminton Contest

Win a free racquet!

Find the number of matches that need to be played to find the winner in this badminton tournament:
- Singles tournament
- Single elimination
- 16 players

Number: _____ matches

Name: _____

Understand the Situation

- What did Joan need to find?
- How many players were in the tournament?
- If Player A faces Player B, is that one match or two?
- If Player A faces Player B, will they play each other again?

Example Find the number of matches that need to be played to finish a tournament with 16 players.

Solve a simpler problem first. **Draw a picture** to help. Draw pictures for tournaments of 2, 3, 4, and 5 players. The vertical line segment that connects two players represents one match.

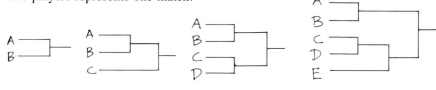

Make a table of the number of players and matches.

Players   2   3   4   5
matches   1   2   3   4
           −1   −1   −1   −1

Find a pattern. The number of matches that need to be played is one less than the number of players. There are 16 players in the tournament. There would be 15 matches.

Try This

a. Find the number of matches that need to be played to finish a single-elimination volleyball tournament with 52 teams.

b. Find the number of oranges in this stack.

Exercises

Problem-Solving Strategies
- Guess and check.
- Choose an operation.
- Make an organized list.
- Act it out.
- Find a pattern.
- Write an equation.
- Use logical reasoning.
- Draw a picture or diagram.
- Make a table.
- Use objects.
- Work backward.
- Solve a simpler problem.

1. How many blocks are in this stack?

   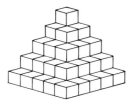

2. How many squares can you find in this four-by-four square?

3. How many matchsticks would you need to make a pattern like this that is 20 squares long?

4. How many matchsticks would you need to make a pattern like this that is 15 triangles long?

5. Delia started to save money for a new saddle. On the first day, she deposited $1 in the bank. On the third day, she deposited $3. On the fifth day, she deposited $5. Suppose she continues to save at this rate. How much money will she have in the bank after the ninety-ninth day?

Use Objects

6. Fold a square piece of paper as shown. Punch a hole where shown. Guess how many holes you made in the sheet. Take another square piece of paper and draw where you think the holes would be.

      punch

Talk Math

7. Jason painted the outside of this stack of white cubes blue. Suppose you took the cubes apart. How many of the cubes do you think will have only one side painted blue? Two sides painted blue? Three sides painted blue? Four sides painted blue? Explain your answers.

   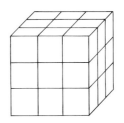

Write Your Own Problem

8. Write a problem for this picture.

   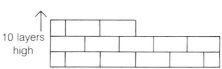

   10 layers high

GEOMETRY **333**

12-2 Points, Lines, and Segments

Objective
Name and draw points, lines, and segments.

Carlos uses plans called **blueprints** to map out his work. These blueprints are for placing a pipe under a road.

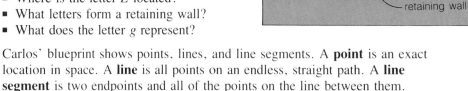

Understand the Situation

- Where is the letter *E* located?
- What letters form a retaining wall?
- What does the letter *g* represent?

Carlos' blueprint shows points, lines, and line segments. A **point** is an exact location in space. A **line** is all points on an endless, straight path. A **line segment** is two endpoints and all of the points on the line between them.

Examples
Use Carlos' blueprint to name the following.

1. two points

Points *E* and *F* are locations on the edge of the road. Points *A*, *B*, *C*, and *D* are locations on the retaining wall.

2. two lines

$\overleftrightarrow{EF}$ is a line that represents one edge of the road. $\overleftrightarrow{g}$ is a line that represents the other edge of the road.

3. two line segments

$\overline{EF}$ is a segment of the roadside. $\overline{AB}$, $\overline{BC}$, and $\overline{CD}$ are segments that show the location of the retaining wall.

Try This
Use this drawing to name each of the following.

a. two points   **b.** two lines   **c.** two line segments

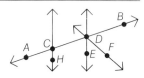

Examples
Draw and label each of the following.

4. Draw two points, *S* and *T*. Draw a line segment between the points.

5. Draw two points, *M* and *N*. Draw a line through the points. Put arrows on the ends.

6. Draw a point, *Z*. Draw a line through *Z*. Put arrows on the ends. Label the line *w*.

Try This
Draw and label each of the following.

d. $\overline{JK}$ **e.** $\overleftrightarrow{DE}$ **f.** point *Q* on $\overleftrightarrow{r}$ **g.** $\overline{OP}$

Exercises

Use this blueprint to name each of the following.

1. three points
2. four line segments
3. two lines

Use this drawing to name each of the following.

4. five points
5. six line segments
6. five lines

Draw and label each of the following.

7. $\overline{BC}$
8. point P on $\overleftrightarrow{ON}$
9. $\overleftrightarrow{s}$
10. point D on $\overleftrightarrow{c}$
11. $\overleftrightarrow{KL}$
12. point R on $\overline{MY}$
13. $\overline{UV}$
14. point F on $\overleftrightarrow{g}$

Draw and label each of the following.

15. $\overline{MN}$ with point P halfway between points M and N
16. a pair of lines, named s and t, that meet at point J
17. $\overleftrightarrow{AB}$, $\overleftrightarrow{AC}$, and $\overleftrightarrow{BD}$ so that C and D are on opposite sides of $\overleftrightarrow{AB}$
18. $\overleftrightarrow{RS}$ and $\overleftrightarrow{TU}$ so that they will never meet

Group Activity

19. List uses of the words *point*, *line*, and *segment* in our everyday talk. How do these uses compare to the geometric meaning of the words?
20. List examples of points, lines, and segments in the classroom. Give examples of segments and lines that meet and that do not meet.

Mixed Skills Review

Find each answer.

21. $2\frac{3}{4} + 3\frac{1}{8}$
22. $7245 \div 69$
23. $93.45 + $67
24. $8.06 - 7.7$
25. 7.89×45.2
26. $4\frac{1}{6} \times 2\frac{2}{5}$
27. $5\frac{2}{3} - 1\frac{3}{4}$
28. $64.03 \div 100$

Change these units.

29. 34 in. = ___ ft **10 in.**
30. 8 yd = ___ ft
31. 6 mi = ___ yd

Solve.

32. Ray has 28 pens. 25% of them are red. How many pens are red?
33. Lola bought a skirt for $40. She paid 6.5% tax. How much tax did she pay?

GEOMETRY 335

12-3 Rays and Angles

Objective
Name and measure rays and angles.

Victoria drew a plan for a light show. She planned to put two mirrors in a corner and to shine three colored light beams on them.

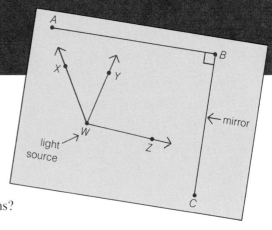

Understand the Situation

- What points are shown in the drawing?
- What segments represent mirrors?
- How did Victoria represent the light beams?

A **ray** has one endpoint and goes off endlessly in one direction along a straight path. Two rays that have a common endpoint, or **vertex,** form an **angle.**

Examples Use the light show plan to name each of the following.

1. three rays
   $\overrightarrow{WX}$, $\overrightarrow{WY}$, and $\overrightarrow{WZ}$ are rays.

2. four angles
   $\angle ABC = \angle CBA = \angle B$; $\angle XWY = \angle YWX$
   $\angle XWZ = \angle ZWX$; $\angle YWZ = \angle ZWY$

Try This Use the drawing to name each of the following.

a. six rays   b. six angles

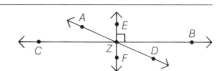

Example

3. Measure $\angle BOA$.

   A. Place the center of the protractor on the vertex.
   B. Place the 0 degree mark on one side of the angle.
   C. Read the measure of the angle at the other side.

   $\angle BOA$ has a measure of 50 degrees. Write $m\angle BOA = 50°$.

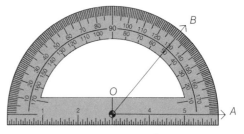

Try This Measure each angle.

c.    d.    e.

336 CHAPTER 12

Exercises

Use these drawings to name the following.
1. six rays
2. three angles

Use this drawing to name the following.
3. six rays
4. four angles

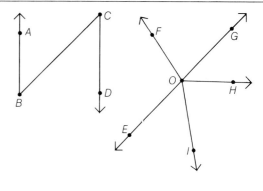

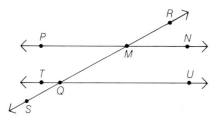

Measure these angles.
5. ∠MOP
6. ∠HOP
7. ∠ROB
8. ∠TOB
9. ∠BOP
10. ∠SOP
11. ∠TOP
12. ∠MOH

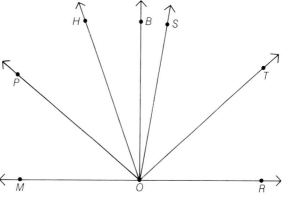

Estimation   Estimate the measure of each angle.

13.
14.
15.

16.
17.
18. (figure with O, K, L)

Data Bank   19. Use the blueprint in the Data Bank on p. 491. Find the angle measures.

Mixed Skills Review   Find each answer.

20. 3.14×5
21. $3\frac{1}{3} + 2\frac{2}{9}$
22. $540 \div 5$
23. $2 + 9 + 20 + 17$

24. $30 + 12 + 7 + 5$
25. $4\frac{2}{7} \div 1\frac{1}{14}$
26. 8×3.14
27. $720 \div 6$

GEOMETRY

12-4 Drawing and Naming Angles

Objective
Draw angles. Name acute, obtuse, and right angles.

Hector is designing a poster. He uses a protractor for his drawing. He wants to divide the rising sun into angles measuring 30°, 40°, 50°, and 60°.

Understand the Situation
- Why is Hector using a protractor?
- What are the measures of the angles he is drawing?
- What shape will the angles form together?

Example

1. Draw $\angle RST$ with a measure of 40°.

   Draw $\vec{SR}$ to form one side of the angle.

   Place the center of the protractor on point S. Mark point T at 40°.

   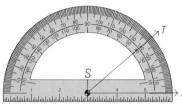

   Draw $\vec{ST}$ to finish the 40° angle.

   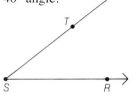

Try This

Draw each angle.

a. m$\angle ACT = 45°$ b. m$\angle T = 90°$ c. m$\angle TAX = 167°$ d. m$\angle C = 180°$

You can name special types of angles. A **right angle** has a measure of 90°. An **acute angle** has a measure less than 90°. An **obtuse angle** has a measure greater than 90° and less than 180°.

Examples Name each angle as right, acute, or obtuse.

2.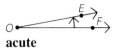
   acute
   The angle has a measure less than 90°.

3.
   obtuse
   The angle has a measure greater than 90°.

4. 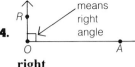   means right angle
   right
   The angle has a measure equal to 90°.

Try This

Name each angle as right, acute, or obtuse.

e. $\angle AXB$ f. $\angle CXD$ g. $\angle DXB$ h. $\angle BXC$

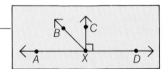

338   CHAPTER 12

Exercises

Draw each angle.

1. m∠APE = 40°
2. m∠Y = 85°
3. m∠SON = 35°
4. m∠T = 160°
5. m∠RAN = 30°
6. m∠GAB = 115°
7. m∠X = 37°
8. m∠MET = 175°

Name each angle as right, acute, or obtuse.

9.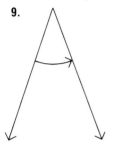
10.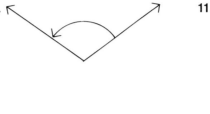
11.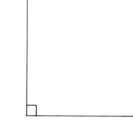

12.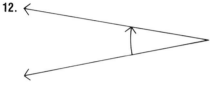

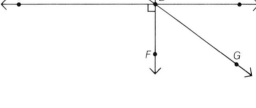

13. ∠BDE
14. ∠CDA
15. ∠FDE
16. ∠GDF
17. ∠CDB
18. ∠GDB

Mental Math

19. m∠A = 30°. Bob drew an angle 5 times this measure. What is the measure of Bob's angle?

20. m∠CAT = 68°. What is the measure of an angle half its measure?

21. m∠A = 20° and m∠B = 50°. Put the two angles together. What measure does the new angle have?

22. m∠Z = 140° and m∠S = 95°. How much greater is the measure of ∠Z?

Talk Math

23. Add the measures of two acute angles. When will the sum be greater than 90°? Why? Draw angles to help you explain.

24. Subtract the measure of an acute angle from the measure of an obtuse angle. Will the difference be less than 90°? Why? Draw angles to help you explain.

25. Add the measure of an acute angle to the measure of a right angle. Will the sum be less than 180°? Why? Draw angles to help you explain.

26. Subtract the measure of a right angle from the measure of an obtuse angle. What can you say about the difference? Why? Draw angles to help you explain.

GEOMETRY

12-5 Perpendicular and Parallel Lines

Objective
Name and draw perpendicular and parallel lines.

Look at the walls of your classroom.

Understand the Situation

- What is the measure of the angle where the wall and floor meet?
- Do the walls across from each other ever meet?

Perpendicular lines are two lines that meet, or **intersect**, to form right angles. **Parallel lines** are lines that never intersect in a flat space, or **plane. Skew lines** are lines in 3-dimensional space that never intersect and are not parallel.

Examples Name each pair of lines as perpendicular, parallel, or skew.

1. $\overleftrightarrow{FE}, \overleftrightarrow{DA}$
   parallel
   Write $\overleftrightarrow{FE} \parallel \overleftrightarrow{DA}$.

2. $\overleftrightarrow{AE}, \overleftrightarrow{AD}$
   perpendicular
   Write $\overleftrightarrow{AE} \perp \overleftrightarrow{AD}$.

3. $\overleftrightarrow{FE}, \overleftrightarrow{AB}$
   skew

Try This Name each pair of lines as parallel, perpendicular, or skew. Use the drawing above.

a. $\overleftrightarrow{FD}, \overleftrightarrow{AD}$
b. $\overleftrightarrow{EA}, \overleftrightarrow{FC}$
c. $\overleftrightarrow{BC}, \overleftrightarrow{AD}$

Examples Use a compass and a straightedge to draw perpendicular and parallel lines.

4. $\overleftrightarrow{WX} \perp \overleftrightarrow{YZ}$
   Draw $\overleftrightarrow{WX}$.

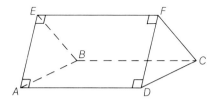

Open a compass the length of $\overleftrightarrow{WX}$. Put the compass on W. Make an arc. Put the compass on X. Make an arc. Draw a line through the points where the arcs intersect. Label the points Y and Z.

5. $\overleftrightarrow{JK} \parallel \overleftrightarrow{LM}$
   Draw $\overleftrightarrow{JK}$.

Draw $\overleftrightarrow{JL} \perp \overleftrightarrow{JK}$.
Draw $\overleftrightarrow{LM} \perp \overleftrightarrow{JL}$.
$\overleftrightarrow{JK} \parallel \overleftrightarrow{LM}$

340 CHAPTER 12

Try This Draw these lines.
  d. $\overleftrightarrow{BC} \parallel \overleftrightarrow{DE}$ e. $\overleftrightarrow{FG} \perp \overleftrightarrow{HI}$ f. $\overleftrightarrow{SR} \parallel \overleftrightarrow{TU}$ g. $\overleftrightarrow{OP} \perp \overleftrightarrow{UV}$

Exercises

Name each pair of lines as perpendicular, parallel, or skew.

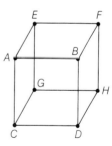

1. $\overleftrightarrow{BF}, \overleftrightarrow{EG}$
2. $\overleftrightarrow{EA}, \overleftrightarrow{EF}$
3. $\overleftrightarrow{GH}, \overleftrightarrow{CD}$
4. $\overleftrightarrow{AB}, \overleftrightarrow{BF}$
5. $\overleftrightarrow{GE}, \overleftrightarrow{AB}$
6. $\overleftrightarrow{HF}, \overleftrightarrow{EG}$
7. $\overleftrightarrow{GH}, \overleftrightarrow{EF}$
8. $\overleftrightarrow{EG}, \overleftrightarrow{BD}$
9. $\overleftrightarrow{EF}, \overleftrightarrow{CD}$

Draw these lines.

10. $\overleftrightarrow{AB} \parallel \overleftrightarrow{CD}$ 11. $\overleftrightarrow{RS} \perp \overleftrightarrow{UV}$ 12. $\overleftrightarrow{f} \parallel \overleftrightarrow{s}$ 13. $\overleftrightarrow{a} \perp \overleftrightarrow{t}$

Use this map to answer Exercises 14–17.

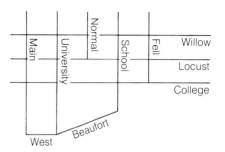

14. Is Normal Avenue perpendicular to School Street?
15. What streets are parallel to Willow Street?
16. Does Fell Avenue intersect College Avenue?
17. West Beaufort and School Street form what kind of angle?

Mixed Skills Review Choose the best estimate. Tell if the estimate is high or low.

18. 25% of 38   A. 5   B. 10   C. 20
19. 51% of 73   A. 35   B. 40   C. 30

Write in lowest terms.

20. $\frac{6}{8}$ 21. $\frac{12}{3}$ 22. $\frac{3}{9}$ 23. $\frac{8}{12}$ 24. $\frac{3}{15}$ 25. $\frac{7}{28}$

26. How many students voted for mysteries?

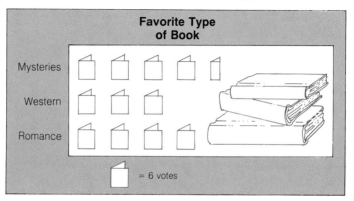

GEOMETRY

12-6 Polygons

Objective
Name polygons. State if a polygon is regular.

Chai cuts shapes out of colored glass to make window hangings.

Understand the Situation

- How many sides does the blue piece of glass have?
- How many sides does the yellow piece of glass have?
- Do any of the shapes have a curved side?

These shapes are polygons. A **polygon** is a plane, closed figure formed by segments. A polygon is named by the number of segments, or sides, it has.

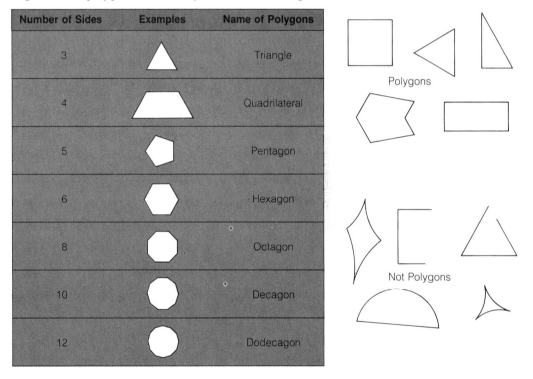

Number of Sides	Examples	Name of Polygons
3	△	Triangle
4	⬓	Quadrilateral
5	⬠	Pentagon
6	⬡	Hexagon
8	⬢	Octagon
10	◯	Decagon
12	◯	Dodecagon

A **regular polygon** is a polygon with all sides and all angles **congruent.** Congruence marks (≅) show that lines and angles are congruent.

These sides have the same length.
Write $\overline{AB} \cong \overline{AC} \cong \overline{CB}$.

These angles have the same measure.
Write $\angle ABC \cong \angle ACB \cong \angle BAC$.

Examples

Name each polygon. Tell if it is regular.

1.

   ABDCE is a regular pentagon. It has five congruent sides and five congruent angles.

2.

   LMN is a triangle. It is not regular. It has three sides of different length.

3.

   PQRS is a regular quadrilateral. It has four congruent sides and four congruent angles.

4.

   QRSTUV is a regular hexagon. It has six congruent sides and six congruent angles.

Try This

Name each polygon. Tell if it is regular.

a.   b.   c.   d.

Exercises

Name each polygon. Tell if it is regular.

1.   2.   3.   4.

5.   6.   7.   8.

9.   10. (pentagon) 11. (trapezoid) 12.

Group Activity

13. Draw a pentagon. Trade with other group members. Measure the angles. Add the total measure for each pentagon. Make a list of the totals. What do you find?

14. Draw a hexagon. Trade with other group members. Measure the angles. Add the total measure of all the angles for each hexagon. Make a list of the totals. What do you find?

Calculator

15. The sum of the angle measures of a regular 9-sided polygon is 1260°. What is the measure of one angle?

16. The sum of the angle measures of a regular dodecagon is 1800°. What is the measure of one angle?

GEOMETRY **343**

12-7 Triangles

Objective
Name special triangles.

Lonnie made gift packages out of cardboard triangles. For each package, he used four triangles, all with sides of the same length.

Understand the Situation
- What shape is each piece of cardboard?
- How many triangles does it take to make one package?
- Are all the sides of one triangle the same length?

Lonnie's triangles are **congruent** because they are the same size and shape. You can name different types of triangles.

Name of Triangle	Description	Example
Equilateral triangle	has three congruent sides	
Isosceles triangle	has only two congruent sides	
Scalene triangle	has no congruent sides	
Right triangle	has a right angle	

Examples
Name each type of triangle.

1.
   isosceles triangle

2.
   scalene triangle

3.
   right triangle

4.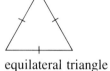
   equilateral triangle

Try This
Name each type of triangle.

a.

b.

c.

d.

Exercises

Name each type of triangle.

1.
2.
3.
4.
5.
6.
7.
8.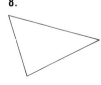

9. Use your protractor to measure the angles of an equilateral triangle. How do they compare?

10. Use your protractor to measure the angles of an isosceles triangle. How do they compare?

11. Use your protractor to measure the angles of a scalene triangle. How do they compare?

12. How do the sides and the angles of an equilateral, scalene, and isosceles triangle compare?

Group Activity

13. Draw two different triangles. Trade them with other group members. Use your protractor to measure the angles of the two triangles. Use a calculator to add the three angle measures for each triangle. Make a list of the totals. What do you find?

14. List examples of triangles in your school. Tell if each is equilateral, right, isosceles, or scalene.

Practice Through Problem Solving

15. Copy and cut out this figure. Make one cut and rearrange the pieces to form an equilateral triangle.

Problem Solving

16. How many triangles can you find in this figure?

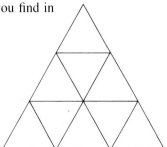

GEOMETRY **345**

12-8 Quadrilaterals

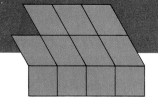

Objective
Name special quadrilaterals.

Armen is making a mosaic out of colored tiles. Each tile has four sides. Some of the tiles have four right angles.

Understand the Situation
- How many sides do Armen's tiles have?
- Are all the sides the same length?
- Are all the tiles the same shape?

These four-sided polygons are **quadrilaterals.** You can name special types of quadrilaterals.

Name of Quadrilateral	Description	Example
Parallelogram	has two pairs of parallel sides	
Rectangle	is a parallelogram with four right angles	
Rhombus	has four congruent sides	
Square	is a rhombus with four right angles	
Trapezoid	has only one pair of parallel sides	

Examples Name each type of quadrilateral.

1.
rhombus

has four congruent sides

2.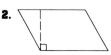
parallelogram

has two pairs of parallel sides

3.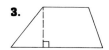
trapezoid

has only one pair of parallel sides

4.
rectangle

has two pairs of parallel sides and four right angles

Try This Name each type of quadrilateral.

a.    b.   c.   d.

346 CHAPTER 12

Exercises

Name each type of quadrilateral.

1.
2.
3.
4.
5.
6.
7.
8.

Draw and name each type of quadrilateral.

9. with only one pair of parallel sides
10. with four congruent sides and angles
11. with only two congruent sides and two congruent angles
12. with two pairs of parallel sides and four right angles
13. with two pairs of parallel sides and two pairs of congruent angles
14. with four congruent sides and two pairs of congruent angles
15. with no congruent sides and angles but with only one pair of parallel sides
16. with two pairs of parallel sides, with two pairs of congruent angles, and with four congruent sides

Group Activity

17. Draw three different quadrilaterals. Trade them with members of your group. Use your protractor to measure the angles of each. Use a calculator to add the angle measures. Make a list of the totals. What do you find?

Show Math

18. Draw and cut out of cardboard a quadrilateral that has no congruent sides and no congruent angles. Use it to trace and cut 12 copies out of paper. Can you arrange the quadrilaterals in a mosaic? Try again with a different quadrilateral. What do you find?

Problem Solving

19. How many squares can you find in this five-by-five square?

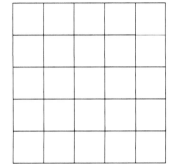

12-9 Perimeter

Objective
Find perimeters.

Laurie plans to build a fence around her garden. The hardware store sells fencing that costs $6.50 per foot. How many feet of fencing does Laurie need to buy?

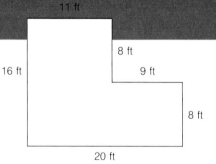

Laurie's Garden Plan

Understand the Situation

- How can Laurie find the amount of fencing she needs?
- How can she find the total cost?

The distance around a polygon is called the **perimeter**. You can find the perimeter of any polygon by adding the lengths of its sides.

Polygon	Formula	Symbols	Picture
Rectangle	$P = 2l + 2w$	l = length w = width P = perimeter	
Square	$P = 4s$	s = side P = perimeter	

Examples

1. Find the perimeter of Laurie's garden.

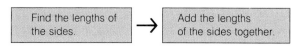

11, 8, 9, 8, 20, and 16 $11 + 8 + 9 + 8 + 20 + 16 = 72$

The perimeter of Laurie's garden is 72 ft. She would need 72 ft of fencing.

2.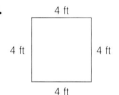

$P = 4s = 4 \times 4 = 16$ ft

3.

$P = 8 \times 2 = 16$ in.

4.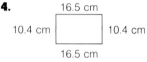

Use $P = 2l + 2w$.
$P = 2(16.5) + 2(10.4)$
$P = 53.8$ cm

348 CHAPTER 12

Try This Find the perimeter.

a.
b.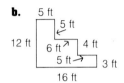
c.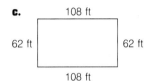
d.

Exercises

Find the perimeter.

1.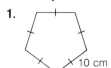
2.
3.
4.
5.
6.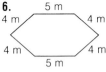
7.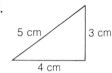
8.
9.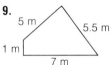
10.
11.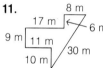
12.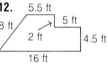

Mixed Applications

13. Kint Lok wants to trim the top of his dining room walls with a strip of wallpaper. His dining room measures 10 ft by 14 ft. How many feet of wallpaper should Kint Lok buy?

14. Albina wants to buy some fencing to put around her rectangular garden. Her garden measures 32 ft by 50 ft. How many 25-foot rolls of fencing will she need?

15. Alicia wants to frame the poster at the right. The frame she chose costs $4.75 per foot. What will the frame cost?

16. Jeffrey needs to buy chicken wire to fence in the area shown. Chicken wire costs 25¢ per foot. What will the wire cost?

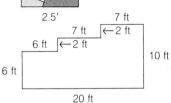

Mental Math

17. The perimeter of a large square is 400 ft. What is the length of one side?

18. The perimeter of a rectangle is 30 ft. Two of the sides are 10 ft long. How long are the other two sides?

GEOMETRY **349**

12-10 Circles and Circumferences

Objective
Find the circumference.

Thais estimates costs for road repair. She uses a trundle wheel to measure the length of sections of road. She finds the distance around the trundle wheel and multiplies by the number of complete turns.

A **radius** is a segment from the center to any point on the circle.

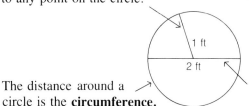

The distance around a circle is the **circumference.**

A **diameter** is a segment that has both endpoints on the circle and that passes through the center.

You cannot measure a circle's circumference with a straight measuring tool. But you can find its circumference if you measure the radius or diameter and use a number called pi.

Pi (π) is the ratio of the circumference of a circle to its diameter. π is about $\frac{22}{7}$, or about 3.14. The circumference of a circle is:

π times the diameter: $C = \pi d$,
or 2π times the radius: $C = 2\pi r$.

You may want to use a calculator for work with π formulas.

Examples

1. Find the circumference of the trundle wheel.

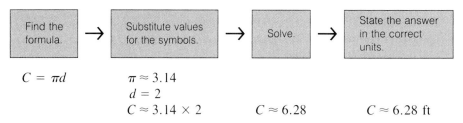

2. Find the circumference of this circle.

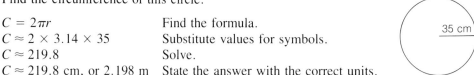

$C = 2\pi r$ — Find the formula.
$C \approx 2 \times 3.14 \times 35$ — Substitute values for symbols.
$C \approx 219.8$ — Solve.
$C \approx 219.8$ cm, or 2.198 m — State the answer with the correct units.

Try This Find the circumference of each circle.

a. b. c. d.

Exercises

Find the circumference of each circle.

1. 2. 3. 4.

5. 6. 7. 8.

9. 10. 11. 12.

Estimation Estimate the circumference of each circle.

13. 14. 15. 16.

17. 18. 19. 20.

Mixed Applications

21. Thais measured a section of road that was 100 turns of her trundle wheel. How long was this section?

22. The radius of the free-throw circle on a basketball court is 1.8 m. What is the circumference?

23. The radius of a spool is 2 cm. How long a piece of thread does it take to wrap once around the spool?

24. The diameter of a discus circle is 8 ft, 2.5 in. What is the circumference of the circle?

25. Harry used a trundle wheel that had a diameter of 1 m. How many turns of the wheel would it take to measure the length of a 100-m dash?

26. The circumference of a circle is 25π meters. What is the radius of the circle?

GEOMETRY **351**

12-11 Consumer Math: Budgets and Circle Graphs

Objective
Make a budget.
Make a circle graph.

A **budget** is a plan for using money. This circle graph shows Larry's budget. The sections of the graph show how he plans to use his money. The size of each section shows how much of the money he plans to use for each purpose.

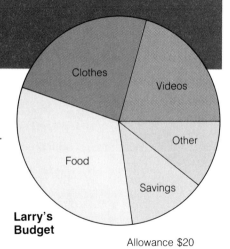

Larry's Budget
Allowance $20

Understand the Situation

- How much allowance does Larry get?
- For what item will he spend the most money?
- For what items will he spend the least money?

Example Draw a circle graph for Ali's weekly budget.
To draw a circle graph, follow each step:

savings $2
food $6
clothes $3
movies $6
other $1

(A) Find the total amount of money.

(B) Write each item's amount as a percent of the total. Round to the nearest percent.

(C) Multiply each percent by 360° to find the angle measures. Round to the nearest degree.

(D) Draw a circle.

(E) Draw each angle. Use the center of the circle as the vertex.

(F) Label each section.   **(G)** Title the circle graph.

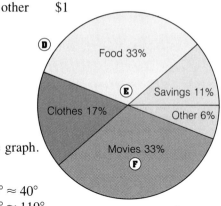

Ali's Budget **(G)**

(A)	**(B)**	**(C)**
Savings $2	2 ÷ 18 ≈ 11%	0.11 × 360° ≈ 40°
Food $6	6 ÷ 18 ≈ 33%	0.33 × 360° ≈ 119°
Clothes $3	3 ÷ 18 ≈ 17%	0.17 × 360° ≈ 61°
Movies $6	6 ÷ 18 ≈ 33%	0.33 × 360° ≈ 119°
Other $1	1 ÷ 18 ≈ 6%	0.06 × 360° ≈ 22°
Total $18		

Try This Draw a circle graph for each budget.

a. June's income = $12 a week
Food = $2, Movies = $4,
Clothes = $5, Savings = $1

b. Cheryl's income = $24 a week
Savings = $1, Food = $4, Car = $14,
Clothes = $2, Entertainment = $3

Exercises

Draw circle graphs for each student's budget.

Student	1. Luis	2. Fiona	3. Chris	4. Dee
Weekly Income	$18.00	$36.00	$12.00	$50.00
Savings	$ 2.00	$ 6.00	$ 1.00	$25.00
Food	$ 5.00	$ 9.00	$ 4.00	$ 5.00
Clothes	$10.00	$16.00	$ 1.00	$10.00
Videos	$ 0.00	$ 4.00	$ 5.00	$ 2.00
Other	$ 1.00	$ 1.00	$ 1.00	$ 8.00

Find each answer. Use the budgets listed in Exercises 1–4.

5. Which student spends the greatest percent on videos?

6. Which student spends the least percent on food?

7. Which student saves one half of his or her income?

8. Which student spends the greatest percent on clothes?

9. Which student spends the most money on clothes?

10. Which student spends the greatest percent on other things?

Use these circle graphs to answer Exercises 11–17.

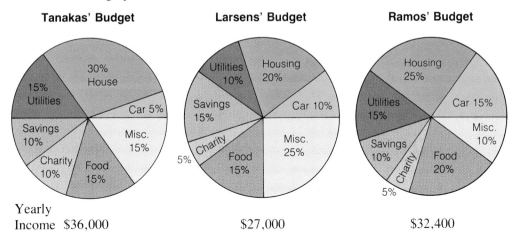

Yearly Income $36,000 $27,000 $32,400

11. Which family spends the least percent on their car?
12. Which family spends the greatest percent on housing and utility costs?
13. Which family saves the most money?

Estimation

14. How much would you estimate that the Tanaka family spends per year on food?

15. How much would you estimate that the Larsen family spends per year on housing?

16. Estimate how much money the Ramos family spends per year on food.

17. Estimate how much money the Ramos family spends on their car.

GEOMETRY **353**

12-12 Problem Solving: Determining Reasonable Answers

Objective
Decide if an answer is reasonable.

- SITUATION
- DATA
- PLAN
- ANSWER
- CHECK

Phar wants to build a pen for his rabbits. He plans to buy 35 ft of fencing. Is this amount of fencing reasonable?

(Figure: pen with sides 12 ft, 9 ft, 13 ft, 4 ft, 2 ft, 8 ft)

Example

Decide if the given answer is reasonable without solving the problem. If it is not reasonable, explain why not.

Is the amount of fencing that Phar plans to buy reasonable?

Phar should buy an amount of fencing equal to the perimeter of the pen.

8 ft + 2 ft + 4 ft + 9 ft + 12 ft + 13 ft ≈ 46 ft   Estimate.

No, 35 ft of fencing is not enough to make a pen as Phar plans.

Try This

Decide if the given answer is reasonable without solving the problem. If it is not reasonable, explain why not.

Problem: The diameter of a circular table is 5 ft. What is the circumference?
Answer: The circumference is about 31.4 ft.

Exercises

Decide if the given answer is reasonable without solving the problem. If it is not reasonable, explain why not.

1. *Problem:* The radius of a circular glass window is 8 in. What is the circumference of the window?
   Answer: The circumference is about 200.96 in.

2. *Problem:* A poster measures 2 ft by 3 ft. How many feet of frame would you need to frame the poster?
   Answer: You would need 10 ft of frame.

3. *Problem:* The perimeter of a square is 49.2 cm. What is the length of one side?
   Answer: The length of one side is 16.1 cm.

4. *Problem:* The circumference of a circle is 48π m. What is the radius of the circle?
   Answer: The radius is 12 m.

12-13 Problem Solving: Group Decisions

Objective
Make group decisions.

- SITUATION
- DATA
- PLAN
- ANSWER
- CHECK

The owner of a new restaurant opening near your school has asked your class to help her make some basic decisions. She is struggling with the problem of how to make the best use of the floor space. She is also trying to plan a menu and to calculate how many people to hire. She would like you to present your written recommendations. Include a scale drawing showing a possible floor plan for the restaurant.

Facts to Consider

- The restaurant floor space is 36 ft long and 30 ft wide.
- The kitchen and storage area should use about 25% of the floor space.
- One possibility for your scale drawing is a scale of $\frac{1}{4}$ in. to 1 ft. You are free to use any scale you wish.
- The restaurant owner would like to arrange the tables to seat the greatest number of customers possible.
- Tables should not be so close that customers complain about crowding or lack of privacy in conversations.
- The owner would like a complete explanation of the price of at least one meal on the menu.

Plan and Make a Decision

- Tell what you are being asked to do.
- How can you take best advantage of the floor space?
- What different seating arrangements can you think of? Would a drawing or model help you to visualize your ideas? What are the advantages and disadvantages of each plan?
- How will you settle differences among group members about the type of food to offer at the restaurant?
- On what will you base your menu prices?
- What factors will you consider when deciding how many people to hire?

Share Your Group's Decision

- Present your plan to the class.
- Explain the reasoning and calculations that lead to your final decision.
- Explain what problem-solving strategies you used to reach your decision.

Suppose
- Suppose your floor space was 70 ft by 60 ft. How would this affect your decisions?
- Suppose the kitchen and storage area should use about 40% of the floor space. How would this affect your decision?

12-14 Enrichment: Constructing Regular Polygons

Objective
Construct regular polygons.

Mrs. Lester is designing a quilt with regular hexagons. Carla's job is to cut the pieces. How can she draw the pattern?

Example Construct a regular hexagon using a straightedge and a compass.

Draw a circle with the desired radius.

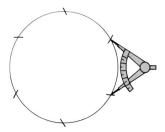

Keep the compass set for the radius. Mark off 6 arcs on the circumference of the circle.

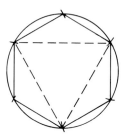

Connect these 6 points to form a regular hexagon.

Notice that you can connect every other point to form an equilateral triangle.

Try This Construct a regular hexagon in a circle with a radius of 3 cm.

Exercises

Construct a regular hexagon in a circle with the given radius.

1. $r = 1\frac{1}{4}$ in. **2.** $r = 2$ in. **3.** $r = 25$ mm

4. Construct two different equilateral triangles in a circle with a diameter of 3 in.

Construct the figure. Shade to make a quilt design.

5.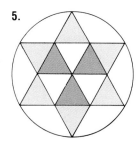

6.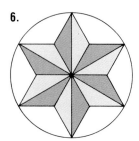

356 CHAPTER 12

12-15 Calculator: Finding Missing Measures

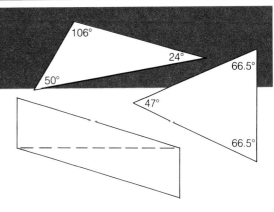

Objective
Use a calculator to find the missing measure of an angle in a polygon.

Find the sums of the measures of the angles in the two triangles. Draw another triangle any size you choose. Measure the angles. Find the sum of the measures.

The sum of the measures of the angles of any triangle is 180°.

Look carefully at the drawing of the quadrilateral. What is the sum of measures of its angles?

Exercises

Use a calculator to find the missing angle measures.

1. trapezoid

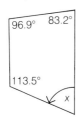

2. obtuse triangle

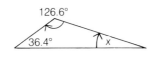

3. right triangle

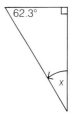

4. rhombus

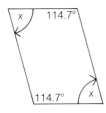

5. isosceles triangle

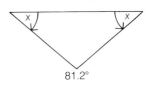

6. acute triangle

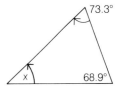

7. equilateral triangle

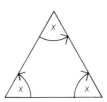

8. parallelogram

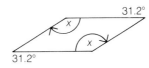

9. isosceles triangle

GEOMETRY

CHAPTER 12 REVIEW

Use the drawing to name the following.
1. three line segments
2. three rays
3. an acute angle
4. a right angle
5. an obtuse angle
6. parallel lines
7. perpendicular lines
8. three points

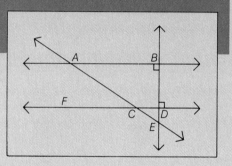

Match the name to the polygon. You may use more than one name.

9.   10.   11.

12.  13. 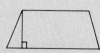 14.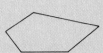

15. 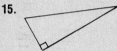 16.   17.

A. hexagon
B. pentagon
C. right triangle
D. square
E. isosceles triangle
F. trapezoid
G. parallelogram
H. equilateral triangle
I. octagon
J. rectangle
K. scalene triangle
L. regular polygon

Use the figures A–D to complete the sentence with <, >, or =.

A. 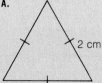   B. (circle, 2 cm)   C. 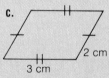   D.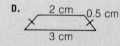

18. The perimeter of A ____ the perimeter of D.
19. The perimeter of C ____ the circumference of B.
20. The perimeter of C ____ the perimeter of D.
21. Give an example of skew lines in your classroom.

Draw and label the lines, line segments, and angles.
22. $\overline{AB} \perp \overline{XY}$
23. point T on $\overleftrightarrow{d}$
24. m∠FGH = 60°
25. $\overline{MN} \parallel \overline{OP}$
26. $\overleftrightarrow{g} \parallel \overleftrightarrow{h}$
27. $\overleftrightarrow{UV} \perp \overleftrightarrow{ST}$
28. point R on $\overline{CD}$
29. m∠RST = 140°

30. Draw a circle graph of the Rockets team budget. They will spend $50 for equipment, $100 for uniforms, $25 for insurance, and $75 for transportation.

CHAPTER 12 TEST

1. How many games are needed to finish a single-elimination softball tournament with 64 teams?

Use the drawing to answer items 2–11.

2. Name five points.
3. Name six line segments
4. Name two lines.
5. Name three rays.
6. Name four angles.
7. Name a right angle.

Name the pair of lines as perpendicular, parallel, or skew.

8. $\overleftrightarrow{AB}, \overleftrightarrow{DC}$
9. $\overleftrightarrow{DE}, \overleftrightarrow{AE}$
10. $\overleftrightarrow{BC}, \overleftrightarrow{ED}$
11. $\overleftrightarrow{AB}, \overleftrightarrow{BC}$

Draw and label the lines, line segments, and angles.

12. $m\angle ROY = 25°$
13. point R on $\overleftrightarrow{MN}$
14. $m\angle KLM = 110°$
15. point Q on $\overleftrightarrow{m}$
16. $\overline{MN} \perp \overline{PQ}$
17. $\overleftrightarrow{JK}$
18. $\overleftrightarrow{x} \parallel \overleftrightarrow{z}$
19. $\overleftrightarrow{BC} \perp \overleftrightarrow{DE}$

Use the drawing of triangles. Measure the angle. Name the angle as right, acute, or obtuse.

20. $\angle LMN$
21. $\angle R$
22. $\angle Y$
23. $\angle TUV$

Name the triangle as right, equilateral, isosceles, or scalene.

24. $\triangle QRS$
25. $\triangle LMN$
26. $\triangle TUV$
27. $\triangle XYZ$

Name the polygon. Tell if it is regular.

28.
29.

Name the type of quadrilateral. Find the perimeter.

30.
31.

Find the circumference.

32. $r = 2.5$ km
33. $d = 7$ yd
34. Draw a circle graph for Gina's party budget. She plans to spend $20 for decorations, $32 for refreshments, $18 for prizes, and $6 for invitations.

35. Is the given answer reasonable? If not, explain why not. *Problem:* The perimeter of a square is 36.4 cm. What is the length of one side? *Answer:* The length of one side is 4.1 cm.

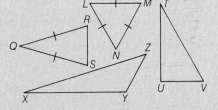

GEOMETRY 359

CHAPTER 12 CUMULATIVE REVIEW

1. Find the total interest.
   $p = \$750$
   $r = 1.9\%$ per month
   $t = 9$ months
   A. $1282.50
   B. $128.25
   C. $355.26
   D. $12,825

2. Find the amounts of broccoli, cheese, and walnuts you will use for 6 servings.
   A. $5\frac{1}{3}$ c broccoli, $1\frac{2}{3}$ c cheese, $\frac{1}{2}$ c walnuts
   B. $6\frac{1}{2}$ c broccoli, 2 c cheese, $\frac{3}{8}$ c walnuts
   C. 6 c broccoli, $1\frac{7}{8}$ c cheese, $\frac{9}{16}$ c walnuts
   D. 4 c broccoli, $1\frac{1}{4}$ c cheese, $\frac{3}{8}$ c walnuts

   Broccoli Supreme
   Serves 8 people.
   8 c chopped broccoli
   $2\frac{1}{2}$ c grated cheese
   $\frac{3}{4}$ c walnuts
   1 c onions, chopped
   1 egg, beaten
   1 tsp salt
   $\frac{1}{2}$ tsp pepper

3. Change the unit.
   24 fl oz = ___ c
   A. 3 c, 15 oz
   B. 3 c
   C. 3 c, 2 oz
   D. 2 c, 8 oz

4. Name the angle.
   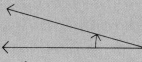
   A. obtuse   B. acute
   C. right   D. scalene

5. 84 is 120% of what number?
   A. 70
   B. 0.7
   C. 700
   D. 36

6. Divide.
   $4\frac{5}{6} \div 2$
   A. $2\frac{5}{12}$
   B. 3
   C. $2\frac{1}{2}$
   D. $9\frac{2}{3}$

7. Name the triangle.
   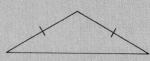
   A. equilateral   B. isosceles
   C. scalene   D. right

8. Change the unit.
   42.63 mm = ___ m
   A. 42,630
   B. 4.263
   C. 0.04263
   D. 4263

9. Subtract.
   $\frac{7}{9} - \frac{4}{9}$
   A. 3
   B. $\frac{1}{3}$
   C. $\frac{11}{9}$
   D. 0

10. Divide.
    $42\overline{)360}$
    A. 8 R24
    B. 9
    C. 8 R7
    D. 8 R6

11. Gil can ride a bike 6 mi in the same time Sam can ride 4 mi. They want to ride 9 mi and finish together. How much of a head start should Sam have?
    A. 3 mi
    B. 4 mi
    C. 5 mi
    D. 6 mi

12. Name the polygon.

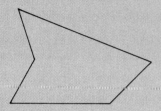

   A. triangle
   B. pentagon
   C. hexagon
   D. square

13. Tell if the situation calls for an exact answer or an estimate.
   A customer in a restaurant is figuring a 15% tip for the waiter.
   A. exact answer
   B. estimate

14. Find the circumference of a circle with these measurements.
   $r = 6$ cm
   A. 18.84 cm
   B. 3.768 cm
   C. 37.68 cm
   D. 21.84 cm

15. A long distance call on a pay phone costs 75¢ for the first 3 min and 21¢ for each additional minute. Jana has $2.50 in change. How long can she talk?
   A. 14 min
   B. 9 min
   C. 12 min
   D. 11 min

16. Use compatible numbers to write in lowest terms.
   $\frac{12}{42}$
   A. $\frac{3}{4}$
   B. $\frac{1}{2}$
   C. $\frac{2}{6}$
   D. $\frac{2}{7}$

17. A farmer had three kinds of animals, of which $\frac{1}{2}$ were cows, $\frac{1}{4}$ were horses, and the rest were pigs. He had 9 pigs. How many animals were on his farm?
   A. 26 animals
   B. 36 animals
   C. 28 animals
   D. 32 animals

18. Find the perimeter of a rectangle with these measurements.
   length = 11.5 in.
   width = 6.2 in.
   A. 35.4 in.
   B. 71.3 in.
   C. 58.4 in.
   D. 38.4 in.

19. Multiply.
   $\frac{4}{5} \times \frac{2}{3}$
   A. $1\frac{1}{5}$
   B. $\frac{2}{5}$
   C. $\frac{8}{15}$
   D. $\frac{3}{4}$

20. Name the quadrilateral.

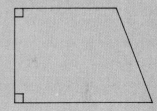

   A. trapezoid
   B. parallelogram
   C. rhombus
   D. rectangle

21. Compare. Use <, >, or =.
   5.0348 ☐ 50.348
   A. <
   B. >
   C. =

22. Write as a standard numeral.
   $(0.2)^4$
   A. 0.8
   B. 0.008
   C. 0.08
   D. 0.0016

23. Solve.
   $\frac{8}{c} = \frac{10}{42}$
   A. 40
   B. 33.6
   C. 32
   D. 33

Chapter 13 Overview

Key Ideas
- Use the problem-solving strategy *make a table*.
- Use tree diagrams.
- Find permutations and combinations.
- Find simple, experimental, and compound probability.
- Find odds.
- Use samples to make predictions.
- Find the mean, median, and mode.
- Collect and analyze data.

Key Terms
- counting principle
- frequency table
- histogram
- outcome
- simple probability
- independent
- random sample
- mode
- permutation
- compound probability
- odds
- mean
- range
- probability
- experimental probability
- expectation
- median

Key Skills

Add.

1. $85 + 34$
2. $67 + 34.35$
3. $56.2 + 19.5$
4. $40.8 + 9.6$
5. $26 + 6.87$
6. $42.05 + 4.5$
7. $45 + 278$
8. $23 + 503$
9. $307 + 185.3$
10. $125.4 + 37.6$
11. $324.07 + 546.3$
12. $78.0006 + 45.06$
13. $23.4 + 45.7 + 18.6$
14. $45.06 + 34.78 + 56.97$
15. $23.3 + 90.09 + 65.76$

Multiply.

16. $24 \times \frac{1}{4}$
17. $48 \times \frac{1}{6}$
18. $64 \times \frac{2}{3}$
19. $60 \times \frac{5}{12}$
20. $42 \times \frac{5}{6}$
21. $124 \times \frac{3}{4}$
22. $360 \times \frac{4}{5}$
23. $400 \times \frac{5}{8}$

Solve.

24. $\frac{3}{4} = \frac{x}{28}$
25. $\frac{2}{3} = \frac{x}{27}$
26. $\frac{5}{6} = \frac{25}{x}$
27. $\frac{5}{8} = \frac{45}{x}$
28. $\frac{x}{16} = \frac{5}{6}$
29. $\frac{x}{21} = \frac{20}{42}$
30. $\frac{3}{x} = \frac{18}{30}$
31. $\frac{4}{x} = \frac{24}{42}$

PROBABILITY AND STATISTICS 13

13-1 Problem Solving: Learning to Use Strategies

Objective
Use the strategy make a table.

- **SITUATION**
- **DATA**
- **PLAN**
- **ANSWER**
- **CHECK**

The zoo opened at 9 a.m. There were 5 people waiting to enter. At 9:10 a.m., 6 more people entered. At 9:20 a.m., 7 more people entered. At 9:30 a.m., another 8 people entered. If this rate continues, at what time will a total of 110 people have entered?

Some problems are easier to solve if you **make a table**.

Example Solve.

If this rate continues, at what time will a total of 110 people have entered the zoo?

Make a table. Use the data that is given in the problem.

Time	9:00	9:10	9:20	9:30
Number of People	5	6	7	8
Total Number of People	5	11	18	26

5 + 6   11 + 7   18 + 8

Fill in the table until you find the solution.

Time	9:00	9:10	9:20	9:30	9:40	9:50	10:00	10:10	10:20	10:30	10:40
Number of People	5	6	7	8	9	10	11	12	13	14	15
Total Number of People	5	11	18	26	35	45	56	68	81	95	**110**

If this rate continues, it will be 10:40 a.m. when 110 people have entered the zoo.

Try This Solve.

a. The zoo has 12 monkey areas in a row. There are 3 monkeys in the first area. There are 5 monkeys in the second area. There are 7 monkeys in the third area. If this rate continues, how many monkeys are there in the twelfth area?

b. Six men cleaned 6 cages in 6 h. Four women cleaned 4 cages in 4 h. How many cages can 12 men and 12 women clean in 12 h?

Exercises

Solve. Use one or more of the problem-solving strategies.

Problem-Solving Strategies
- Guess and check.
- Choose an operation.
- Make an organized list.
- Act it out.
- Find a pattern.
- Write an equation.
- Use logical reasoning.
- Draw a picture or diagram.
- **Make a table.**
- Use objects.
- Work backward.
- Solve a simpler problem.

1. Nola led a scavenger hunt. A team earned 3 points for finding 11 items, 6 for finding 12, 9 for finding 13, and so on. Qui's team earned 42 points. How many items did they find?

2. Morris bought 2 tickets to the amusement park for 3 times the cost of a T-shirt. He bought a T-shirt for twice the cost of lunch. He spent $6 on lunch. How much did he spend all together?

3. Littlefield has a city softball league. The league spent a total of $3240 on uniforms for all 12 teams. It cost $216 for every dozen uniforms. How many players are there on each team?

4. The Lincoln Park Zoo hosted 1,853,973 more people than the San Diego Zoo. The San Diego Zoo hosted 141,027 more people than the National Zoological Park. The Kings Island Zoo hosted 100,000 fewer people than the National Zoological Park. How many people visited each zoo?

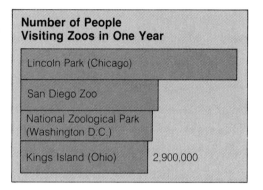

5. Carlita Romero runs a chicken ranch. On Tuesday, 1 egg hatched. On Wednesday, 4 eggs hatched. On Thursday, 9 eggs hatched. On Friday, 16 eggs hatched. If this rate continues, on what day of the week will 144 eggs hatch?

Suppose

6. Suppose in the example on p. 364 that 10 people were waiting to enter at 9 a.m. At 9:10 a.m., 15 people entered. At 9:20 a.m., 20 people entered. At 9:30 a.m., 25 people entered. If this rate continues, at what time will a total of 100 people have entered?

Evaluating Solutions

7. *Problem:* The Heart Association asked people to donate $5. The number of people who gave $5 each day is shown in the table. If this rate continues, how much money will be donated after 10 days?

   Darren filled in the table as shown. Was he correct? Why or why not?

Day	1	2	3	4	5	6	7	8	9	10
Number of People	1	3	6	10	15	20	25	30	35	40
Total Amount	$5	$20	$50	$100	$175	$195	$230	$260	$295	$335

13-2 Counting Problems

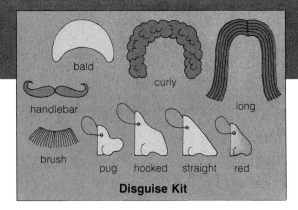

Disguise Kit

Objective
Make a tree diagram. Use the counting principle.

The Costume Shop is selling a disguise kit. How many different disguises can you make using a mustache and a wig?

Understand the Situation
- What are the choices for mustaches?
- How many choices are there for wigs?
- How many disguises can you make using a brush mustache and a wig?

A **tree diagram** shows all possible **outcomes,** or combinations.

Example 1. Make a tree diagram. List all possible outcomes of mustaches and wigs.

List each mustache.   *List each wig.*   *List possible outcomes.*

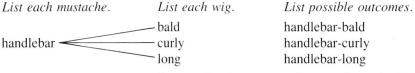

handlebar — bald, curly, long → handlebar-bald, handlebar-curly, handlebar-long

Draw a line between each mustache and wig.

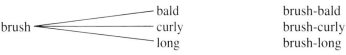

brush — bald, curly, long → brush-bald, brush-curly, brush-long

Try This Make a tree diagram. List all possible outcomes.
   a. of wigs and noses   **b.** of mustaches and noses

The **counting principle** is a way to find the number of possible outcomes. If you have m choices and n choices, then $m \times n$ gives the number of possible outcomes.

Example 2. Find the number of possible outcomes from the disguise kit.

mustaches × wigs × noses   Use the counting principle.
    2   ×   3   ×   4   = 24   Multiply the number of choices.

There are 24 possible disguises.

Try This Find the number of possible outcomes.
   c. The Costume Shop has added to the kit a Mohawk wig and a freckled nose. Find the number of disguises you can make now.

Exercises

Make a tree diagram. List all the possible outcomes.

1. List all possible outcomes for a restaurant and a movie theater. Choose one of each.

Dinner	Movie Theater
Alfredo's Pizza	Regency
Mel's Drive-in	Festival
Betty's Barbecue	Park

2. List all possible outcomes for a frozen yogurt. Choose one flavor, one topping, and one sauce.

Flavors	Toppings	Sauces
Vanilla	Walnuts	Carob
Strawberry	Fresh Berries	Peach
Lemon		

3. List all possible outcomes for a school election. Choose one president, one secretary, and one treasurer.

President	Secretary	Treasurer
Martinez	Wu	Lopez
Goldman	Henderson	Kline
Tran	O'Neil	

Find the number of possible outcomes.

4. Pizza-To-Go has these choices for pizza. Suppose you choose only one topping. Find the number of different kinds of pizza you can order.

Size	Crust	Topping
small	regular	cheese
medium	thick	peppers
large		mushrooms
jumbo		pepperoni
		onions
		olives

5. At Suza's Cafeteria, there are these choices on the menu. Find the number of different meals that include one soup, one main dish, and one vegetable.

Soup	Main Dish	Vegetables
Clam Chowder	Fish	Peas
Onion	Chicken	Corn
	Lamb Chops	Carrots
	Roast Beef	

6. At Sal's Salad Bar, you have these choices. How many different kinds of salads can you make with one choice of lettuce, one topping, and one dressing?

Lettuce	Toppings	Dressings
Spinach	Tomatoes	French
Iceberg	Olives	Italian
Romaine	Onions	Blue Cheese
	Beans	

Talk Math

7. Tell what information is given by a tree diagram that is not given by the counting principle. Why do you not always need to use a tree diagram?

13-3 Permutations

Objective
Find permutations using factorials.

Jerry, Rick, and Kirra are having their picture taken for the school paper. There are three places to stand. Rick is tallest. The photographer places him in the middle. She next puts Jerry on Rick's right.

Understand the Situation

- In how many different places can one student stand?
- After she has placed Rick, in how many places can the photographer put Jerry?
- In how many places can Kirra stand now?

A **permutation** is an ordered arrangement of the objects in a set.
You can find the number of permutations of n objects by multiplying.

Examples

1. Find the number of permutations of how Jerry, Rick, and Kirra can stand in three places.

There are three positions and three people.

Number of People for First Position	Number of People Left for Second Position	Number of People Left for Third Position	
↓	↓	↓	
3 ×	2 ×	1	= 6

There are 6 permutations of how Jerry, Rick, and Kirra can stand.

You can use the symbol $n!$ (n factorial) to represent a number of permutations. $n!$ represents $n \times (n-1) \times (n-2) \times \ldots \times 1$.

2. Find the number of permutations of four books on a shelf.

$n = 4 \quad n! = 4! = 4 \times 3 \times 2 \times 1 = 24$

There are 24 permutations of four books on a shelf.

Try This

a. Find the number of permutations of six friends standing in line for a movie.

b. Find the number of permutations of five bikes in a bike rack.

CHAPTER 13

Exercises

Solve.

1. A clothing shop has 3 mannequins in its display window. Find the number of ways the clothing shop can place them in a row.

2. The barber shop's hat rack has 4 places for hats. Find the number of permutations of 4 hats on the rack.

3. Sal, Kirby, Shari, Glenn, Paula, and Miko are taking batting practice in an empty lot. Find the number of batting orders possible.

4. Jeff's apartment building has 5 parking places. Find the number of permutations of 5 parked cars.

5. Brenda's band plans to play 8 songs in a set. In how many different orders could the band play the 8 songs?

6. Mikey only likes a few foods. He likes to eat them in different orders. In how many different orders can he eat a meal of soup, salad, potatoes, turkey, and fruit?

7. Alvin, Tracy, Kip, Kyoko, Jackie, Miguel, and Ellen ran a 100-m race. How many different ways could they finish the race?

8. Eight cars compete for the 5 parking places at Jeff's building. How many permutations are there of the 8 cars in the 5 places? [Hint: Multiply $n \times (n - 1) \times (n - 2)$. . . . Use only as many factors as there are places.]

Calculator

9. Find the greatest factorial that you can display on your calculator.

Talk Math

10. René says that 8! is the same as $8 \times 7!$. Is she correct? Tell why or why not.

Problem Solving

11. Rocky filed race entries for a 10K race. On the first day, he got 4 entries in the mail. On the second day, he got 8 entries. On the third day, he got 16 entries. If this rate continues, how many entries will he get on the seventh day?

Mixed Skills Review

Write each fraction as a decimal and as a percent.

12. $\frac{5}{8}$ 13. $\frac{3}{4}$ 14. $\frac{1}{5}$ 15. $\frac{5}{6}$ 16. $\frac{4}{9}$ 17. $\frac{2}{3}$

Estimate.

18. 24% of 12   19. $1\frac{1}{6} + 2\frac{9}{10} + 1\frac{3}{5}$ 20. $3589 \div 61$ 21. $622 + 589$

Find each answer.

22. $\frac{1}{8} \times \frac{1}{8}$ 23. $\frac{4}{9} \times \frac{4}{9}$ 24. $\frac{2}{7} \times \frac{2}{7}$ 25. $\frac{3}{5} \times \frac{3}{5}$ 26. $\frac{3}{4} \times \frac{3}{4}$

PROBABILITY AND STATISTICS

13-4 Simple Probability

Objective
Find simple probability.

Dee holds two red marbles and one blue marble behind her back. She puts one of the marbles in her right hand. What are the chances that the marble is blue?

Understand the Situation

- How many marbles did Dee have behind her back?
- How many of these marbles were blue?
- Is the marble in her right hand more likely to be red or blue?

You can find the **probability (P)** of Dee's choosing a blue marble.

$$\frac{\text{probability}}{\text{of an event}} = \frac{\text{number of ways the event can occur } (x)}{\text{number of possible outcomes } (y)}$$

You can write a probability as a fraction. The fraction cannot be less than 0 or more than 1. If the event cannot occur, the probability is 0. If the event is certain to occur, the probability is 1. You can think of a probability as x (events) out of y (outcomes). Each outcome must be **equally likely** to occur.

$$P(\text{event}) = \frac{x}{y} = x \text{ out of } y$$

Example Find the probability of Dee's picking a blue marble out of two red marbles and one blue marble.

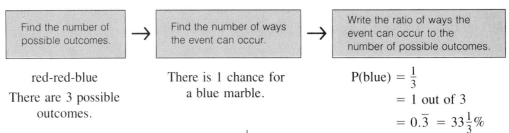

red-red-blue

There are 3 possible outcomes.

There is 1 chance for a blue marble.

$P(\text{blue}) = \frac{1}{3}$
$= 1 \text{ out of } 3$
$= 0.\overline{3} = 33\frac{1}{3}\%$

The probability of picking a blue marble is $\frac{1}{3}$, or 1 out of 3.

Try This

a. Find the probability of Dee's picking a red marble.
b. Find the probability of Dee's picking a yellow marble.
c. A jar contains 2 red marbles, 1 blue marble, and 3 yellow marbles. Find the probability of picking one red marble without looking.

Exercises

In his pocket, Donald has 6 quarters with the following mint dates: 1984, 1986, 1983, 1984, 1985, 1984. He has only these 6 coins in his pocket. He pulls out a quarter to buy a newspaper. Find the probability of Donald's picking each coin.

1. P(1986)
2. P(1984)
3. P(1985)
4. P(odd year)
5. P(even year)
6. P(divisible by 3)
7. P(<1990)
8. P(>2000)
9. P(1982)

10. In a grocery bag there are 7 cans of chicken-flavored cat food, 4 cans of fish-flavored cat food, and 6 cans of liver-flavored cat food. If you pick a can without looking, what is the probability that it will be liver-flavored?

11. A manufacturer tested 100 radios. Ninety-five radios passed the test. Five radios did not pass. Based on these results, what is the probability that the next radio manufactured will not pass the test?

Number Sense

12. Arrange the games in order from the greatest to the least chance of the underdog's winning.

Game	Favorite	Underdog	Probability of Underdog's Winning
1	Bulldogs	Cougars	$\frac{1}{60}$
2	Warriors	Lions	1 out of 75
3	Patriots	Bears	0.45
4	Spartans	Hawks	25%
5	Knights	Pirates	$\frac{2}{7}$

Sharon has 10 same-sized mugs in her cupboard. Two mugs are striped. Three mugs have penguins on them. Four mugs have city names on them. One mug has her name on it. Sharon takes one mug without looking. Use the number line to plot the probability of Sharon's picking each type of mug.

13. P(striped)
14. P(penguin)
15. P(city names)

Group Activity

16. Write the name of each student in your group on an index card. Fold the cards in half. Put all the cards in a shoe box.

    A. Have one person in your group draw a name without looking. What is the probability that the student will draw his or her own name?

    B. Put the name back. Have another student draw a name. What is the probability that it is a female's name? A male's name?

    C. Put the name back. Have another student draw a name. What is the probability that the student drawn has black hair? Blond hair? Red hair?

    D. Put the name back. Have another student draw a name. What is the probability that the student drawn has brown eyes? Blue eyes?

13-5 Experimental Probability

Objective
Find experimental probability.

Suppose you toss a toothpaste tube cap. What are the chances that the cap will land on its side? On its large end? On its small end? You can experiment to find data for probabilities.

Toss the toothpaste tube cap 100 times. Tally the results in a table like the one shown. Find the total of each possible outcome.

Possible Outcomes		
On its side	Large end down	Small end down
Total		

Write the ratio of the total of each possible outcome to the total number of trials. Each ratio gives an **experimental probability** of that outcome.

Suppose the cap landed on its side 75 times out of 100 trials. The experimental probability of the cap's landing on its side would be 75 to 100, or 75 out of 100. You could also write the probability as $\frac{3}{4}$, 3 out of 4, 0.75, or 75%.

Exercises

1. Would you say that each outcome is equally likely?

2. Are you certain that the cap will land at least once on the small end?

3. Is it highly likely that the cap will not land on the small end in 100 trials?

4. Is it possible that the cap will never land on the small end in 100 trials?

5. If you did this experiment again, do you think your results would be exactly the same? Repeat the experiment. How close are the results to the first experiment?

6. After doing this experiment, Jason said that the probability that the cap would land on its side is about 70%. Do you agree? Why or why not?

7. Suppose you tossed two toothpaste caps. Is it possible that both caps would land on the small end? On the large end? On the side?

8. Suppose you tossed two toothpaste caps. Is it highly likely that both caps would land on the small end? On the large end? On the side?

9. Toss a plastic bottle cap 100 times. Make a table like the one on p. 372. Keep a tally of the results. Find the experimental probability that the cap will land on its side, on its top, and on its bottom.

   A. Are these outcomes equally likely?
   B. Is it highly likely that the cap will never land on its side in 100 tosses?
   C. Is it possible that the cap will land on its side in 100 tosses?
   D. How could you change the probability by using a different type of bottle cap?

10. Place two red marbles and one blue marble of the same size in a bag. Without looking, pick two marbles from the bag. Replace the marbles. Repeat this process 100 times. Keep a record of how many times the colors of the marbles are different and how many times the colors are the same. Find the experimental probabilities of different and same.

   A. Are the outcomes *different* and *same* equally likely?
   B. Is it possible that you will never pick two marbles with the same color in 100 trials?
   C. Is it possible to have 10 straight outcomes when the marbles are the same color? Is it likely?
   D. Suppose you added one blue marble. Are the outcomes *different* and *same* equally likely now?

Group Activity

11. Design your own experiment in which you find the experimental probability of some outcomes. Do the experiment and decide if the outcomes are equally likely.

12. Think of an experiment in which the outcomes are equally likely. Do the experiment. Show why you believe the outcomes are equally likely.

Problem Solving

13. Horace Ginn owns a toy store. One day, he sold 28 marbles in the first hour. He sold 25 marbles in the second hour. He sold 22 marbles in the third hour. If this rate continues, what is the total number of marbles he will sell after eight hours?

14. Paige, Francisca, Len, and Kio traded marbles. Paige gave 7 marbles to Francisca. Paige received 13 marbles from Len and 12 marbles from Kio. She had 48 marbles after they finished trading. How many marbles did Paige have before they started trading?

15. Vera and Matt bought one bag of marbles. They took turns dividing the marbles. Vera chose 1 marble, then Matt chose 2 marbles. Vera chose 3 marbles, then Matt chose 4 marbles. When they were finished, Matt had 8 more marbles than Vera. How many marbles were in the bag before they started choosing?

13-6 Compound Probability

Objective
Find compound probability.

Jean is a star basketball player for Pikesville High. In a play-off game, she must make two free throws to win the game. What is the probability that Jean will make the two shots?

Jean Bowman's Scoring Probability
Field Goals—13 out of 25
Free Throws—7 out of 10

Understand the Situation

- About how many free throws does Jean make out of every 10 attempts?
- About how many free throws does she make out of every 100 attempts?
- If Jean makes the first free throw, does that affect her chance of making the second free throw?

Compound probability gives the probability for more than one event. The events are **independent** if one event does not affect the outcome of the other.

If **A** and **B** are **independent events**, then **P(A and B) = P(A) × P(B)**.

Examples

1. Find the probability that Jean will make both free throws.

   The two free throws are independent events. Jean makes 7 out of every 10 free throws. The probability of her making any given free throw is $\frac{7}{10}$.

   P(two free throws) = P(one free throw) × P(one free throw)
   $= \frac{7}{10} \times \frac{7}{10} = \frac{49}{100}$

   The probability that Jean will make the two free throws is $\frac{49}{100}$, or 49 out of 100. The probability is about $\frac{1}{2}$, or 1 out of 2.

2. Michael Jordan makes about 17 out of 20 free throws. What is the probability that he will sink two free throws in a row?

   P(two free throws) = P(one free throw) × P(one free throw)
   $= \frac{17}{20} \times \frac{17}{20} = \frac{289}{400}$

   The probability that Jordan will make two free throws in a row is $\frac{289}{400}$. The probability is about $\frac{3}{4}$, or 3 out of 4.

Try This

a. What is the probability that Jean will make two field goals in a row?

b. Larry Bird makes about 9 out of 10 free throws. What is the probability that he will make two free throws in a row?

Exercises

There are 6 blue marbles, 3 red marbles, 5 green marbles, and 1 white marble all the same size in a box. In another box, there are 2 red blocks, 5 green blocks, 4 white blocks, and 6 blue blocks all the same size. Suppose you take a marble and a block from the boxes without looking. Find each probability.

1. P(blue marble and blue block)
2. P(green marble and white block)
3. P(white marble and red block)
4. P(red marble and red block)
5. P(red marble and green block)
6. P(blue marble and white block)
7. Ralph Sampson makes about 34 out of 50 free throws. In a key game, Ralph must make two free throws for his team to win. What is the probability that his team will win the game?
8. Brad Davis makes 2 out of every 5 three-point field goals. Find the probability that Brad will make four straight three-point field goals. Give your answer as a percent.
9. Alvin Robertson makes 15 out of every 29 field goals he shoots. Find the probability that Alvin will make three field goals in a row. Give your answer as a percent.

Data Hunt

10. Find the probability of a professional or college basketball player of your choice making a free throw. Then find the probability of the player making three free throws in a row. Give your answer as a percent.

Group Activity

11. Toss three coins together 20 times. Keep a record of each result. What is the experimental probability of tossing three heads at once? Add a fourth coin. What is the experimental probability of tossing four heads at once?

Mixed Skills Review

Solve.

12. $\frac{3}{8} = \frac{n}{16}$
13. $\frac{18}{y} = \frac{9}{2}$
14. $\frac{1}{5} = \frac{8}{w}$
15. $\frac{f}{6} = \frac{10}{12}$
16. $\frac{7}{14} = \frac{s}{4}$

Find each answer.

17. $35 + 24 + 36 + 28 + 26$
18. $42 + 40 + 38 + 41 + 43$
19. $210 \div 6$
20. $252 \div 7$
21. $260 \div 8$
22. $336 \div 10$
23. $\frac{1}{600} \times 800$
24. $\frac{1}{500} \times 650$
25. $300 \times \frac{1}{250}$
26. $300 \times \frac{1}{180}$

Round to the given place.

27. $5.09 dime
28. 0.318 tenths
29. 452 thousands
30. $28.45 dollar
31. 7 tens
32. 0.546 ones

Write the ratio.

33. 21 km in 7 h
34. 6 wins to 4 losses
35. 3 oranges for 39¢

13-7 Odds

Objective
Find odds.
Find probabilities given odds.

Odds are ratios that compare the ways an event can and cannot occur. These odds gave the ratio of the chances of winning the NCAA basketball tournament to the chances of not winning it in 1987.

Odds of Winning the 1987 NCAA Men's Tournament

North Carolina	1:2	Syracuse	1:12
Indiana	2:5	Oklahoma	1:15
UNLV	1:3	Kansas	1:20
Alabama	1:5	Notre Dame	1:30
Georgetown	1:5	Providence	1:30
Iowa	1:8	Louisiana State	1:40
DePaul	1:8	Duke	1:50
Florida	1:12	Wyoming	1:1000

Understand the Situation

- What were the odds of Alabama's winning?
- Based on these odds, which team had the worst chance of winning?

You can find the odds in favor of and against an outcome.

$$\text{odds in favor of an outcome} = \frac{\text{number of ways it can occur}}{\text{number of ways it cannot occur}}$$

$$\text{odds against an outcome} = \frac{\text{number of ways it cannot occur}}{\text{number of ways it can occur}}$$

Example

1. Butch has 6 golf tees in his pocket. Two are red. One is yellow. Three are white. He picks a tee without looking. What are the odds in favor of Butch's picking a red tee?

$$\frac{\text{number of ways he can pick red}}{\text{number of ways he cannot pick red}} = \frac{2 \text{ red tees}}{1 \text{ yellow} + 3 \text{ white tees}} = \frac{2}{4}$$

The odds of Butch's picking a red tee are $\frac{2}{4}$, or $\frac{1}{2}$, 1 to 2, or 1:2.

Try This

a. Punky has 5 pairs of running socks. Two pairs are yellow. Three pairs are blue. She picks a pair without looking. What are the odds against Punky's picking a yellow pair?

You can state probabilities if you know the odds for an event.

Example

2. What was the probability that North Carolina would win the NCAA tournament? Use the odds listed in the table.

The table shows that the odds were 1:2 that North Carolina would win.

$$\frac{\text{number of ways it could occur}}{\text{number of possible outcomes}} = \frac{1 \text{ way they would win}}{2 \text{ ways they would not win} + 1 \text{ way they would win}} = \frac{1}{3}$$

Try This

b. What was the probability that Georgetown would win the NCAA tournament? Use the odds listed in the table.

Exercises

A bag contains the blocks listed in this table. Without looking, a student picks a block. Find the odds.

Blocks, equal size
5 red
3 green
2 blue
1 white

1. odds against drawing a red block
2. odds in favor of drawing a blue block
3. odds in favor of drawing a green block
4. odds against drawing a blue block
5. odds in favor of drawing a white block
6. odds against drawing a green block

Without looking, a student draws one block from the same bag of blocks. Find the probability.

7. P(red block)
8. P(green block)
9. P(blue block)
10. P(white block)
11. P(white block or red block)
12. P(green block or blue block)
13. P(not a blue block)
14. P(not a red block)

Use the odds listed in the NCAA table on p. 376.

15. What was the probability that UNLV would win?
16. What was the probability that Providence would win?

Mixed Applications

17. The table on p. 376 is based on one writer's point of view. A different writer gave Iowa a 60% chance of winning the NCAA tournament. Write this writer's prediction as odds for Iowa's winning. (Hint: First write the probability as a fraction.)

Number Sense

18. A sportscaster gave the probability of Oklahoma's winning as 15%. Was her point of view different from that of the writer who made the table on p. 376? Write the 15% probability as odds in favor of Oklahoma's winning.

Group Activity

19. Use the list of Favorite Sports to Watch. Write your favorite sports on an index card. Make a tally of the group's favorite sports. Fold each card in half. Put the cards in a box.

    A. Have one student in your group draw a card without looking. What are the odds against picking football?
    B. Put the card back. Have another student draw a card. What are the odds in favor of picking tennis?
    C. Make up your own questions about odds in favor of or against picking a sport.

Favorite Sports to Watch
Football
Baseball
Gymnastics
Tennis
Basketball
Swimming

PROBABILITY AND STATISTICS

13-8 Expectations

Objective
Evaluate expectations.

To raise money and gain members for the club, the Bike Club held a raffle for a bike. They sold 500 tickets. In other years they found that about 1 out of 20 ticket buyers joined the club. How many new members could the club expect from the raffle?

Understand the Situation
- How many tickets did the club sell?
- What is the probability that a ticket buyer will join the club?

An **expectation** is a number or quantity that can be expected as an outcome. To find the expectation, multiply the number of possible events by the probability of an event.

Examples

1. How many new members can the Bike Club expect to gain from the raffle?

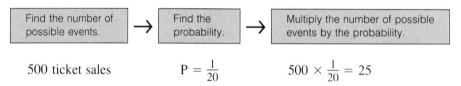

500 ticket sales $P = \frac{1}{20}$ $500 \times \frac{1}{20} = 25$

The Bike Club can expect to gain 25 new members.

2. A farmer raising tomatoes planted 180 seeds. The seed company gave a 95% probability that each seed would sprout. How many plants can the farmer expect from this planting?

180 seeds $\times \frac{95}{100} = 171$ The farmer can expect 171 plants from 180 seeds.

Try This

a. Mr. Lomax planted 700 seeds of a new kind of bean. In earlier trials he found that only about 1 out of 35 of the seeds actually sprouted. How many plants can Mr. Lomax expect from this planting?

b. The Bike Club now has 40 members. Usually 65% of their members join the 50-km ride each year. How many members can the club expect at this year's ride?

c. Mr. Paoli knows that $\frac{2}{3}$ of the people who buy fruit at his roadside fruit stand buy a jar of honey as well. He has about 150 customers each weekend. How many jars of honey can Mr. Paoli expect to sell on a weekend?

Exercises

Find the expectation.

1. Sales records show that 25% of the customers at Pizza-To-Go buy a pepperoni pizza. The shop sells 240 pizzas on an average day. Find the number of pepperoni pizzas the shop expects to sell today.

2. The New Sound shop held a drawing for a CD player. Two-thirds of the entries were from old customers. The rest of the entries were from new customers. How many new customers can the shop expect from 2550 entries?

3. Gail Veseley estimates that about 1 out of every 6 customers in her shop buys a poster. She plans a special sale on posters to draw about 1800 customers to her store. How many posters can Gail expect to sell during the sale?

4. K&S Advertising places radio and TV ads for many shops at Bay City Mall. The mall serves an area of about 240,000 people. Find the number of customers K&S expects to reach by radio, by TV, and by both radio and TV.

Probability of Reaching Customers	
By radio	1 out of 3
By TV	1 out of 4
By both radio and TV	1 out of 20

Number Sense

5. The Bike Club must sell at least 150 tickets to break even on their raffle. The members find that they sell 1 ticket for each 10 people they ask. In order to break even, how many people must the members ask to buy a ticket?

Estimation

Estimate the expectation.

6. A store owner finds that 35% of the browsers in her store actually buy something. If she can get 1300 people to walk through her store in a week, about how many sales can the store owner expect?

7. A large community fund drive gets a contribution from 1 out of 11 people they call. This year the volunteers will telephone 12,500 people. How many contributions can they expect?

Mixed Application

8. The 54 members of the Range Riders Bike Club can bring guests to their annual 50-km ride and barbeque. Use the graph to find how many people the cooks can expect to feed at the barbeque.

Bring a Friend!

PROBABILITY AND STATISTICS **379**

13-9 Consumer Math: Using Samples

Objective
Use samples to make predictions.

Highland Drinks, Inc. is testing a new drink called *Coconut Breeze*. Cynthia Lora is in charge of testing in Hillsboro, a town of 25,000 people. She asked a **sample** of the people in Hillsboro to taste three unlabeled drinks. Cynthia will compare the number of tasters who say they like *Coconut Breeze* to the total number of tasters. This way she can predict how many people in the town are likely to buy the new drink. This is a chart of people's choices in the taste test.

Cynthia wanted a sample of many different buyers in Hillsboro. She held the taste test in five different parts of the town at several times of day. Cynthia expects that testing a variety of people will give her a better prediction.

Which drink do you like best?

Drink	Number of People
A	80
B	25
C	20

Many companies use sampling to predict how popular their products will be. Television ratings are also based on samples.

Understand the Situation
- How many people live in Hillsboro?
- How many people tested the drinks?
- How many people liked drink A best?
- Why was it important for Cynthia to test many different people?

You can use proportions to make predictions from samples.

Example Predict how many people in Hillsboro would like drink A.

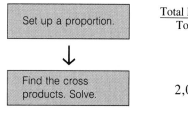

$$\text{Total liked drink A} \rightarrow \frac{80}{125} = \frac{n}{25{,}000} \leftarrow \text{Prediction for Hillsboro}$$
$$\text{Total polled} \rightarrow \qquad\qquad \leftarrow \text{Total people}$$

$$80 \times 25{,}000 = n \times 125$$
$$2{,}000{,}000 = n \times 125$$
$$2{,}000{,}000 \div 125 = n \times (125 \div 125)$$
$$16{,}000 = n$$

Cynthia predicted that 16,000 people out of 25,000 would like drink A.

Try This Solve.

a. Predict how many people in Hillsboro would like drink B.

b. Predict how many people in Hillsboro would like drink C.

Exercises

1. There are about 18,000 possible car buyers in Haskell. The table shows a random sample from each of three regions. Predict how many people in Haskell would buy each model.

2. Baxter, Inc. is testing a new shampoo product. They took a random sample of consumers in Solano County. There are about 2,516,000 consumers in the county. Predict how many people in the county would like each product.

Which car would you buy?

Model	Region I	Region II	Region III
A	31	22	12
B	19	28	38

Which shampoo do you like better?

Region I	Region II
Product A: 75	Product A: 39
Product B: 25	Product B: 61

Region III	Region IV
Product A: 46	Product A: 56
Product B: 54	Product B: 44

In Jackson County, a polling group expects 20,000 people to vote for supervisor. The group questioned a random sample of voters. They put the results in a table.

3. Predict who will win the election.
4. Predict how many votes Evans will receive.
5. Predict how many votes García will receive.
6. Predict how many votes Jarona will receive.
7. Predict how many votes Tanaka will receive.
8. Do you think Evans is highly likely to receive more votes in the election than García? Explain.

Poll—Supervisor Election

If the election were held today, for whom would you vote?

Evans	82
García	80
Jarona	40
Tanaka	110
Other, no view	38
Total	350

Group Activity

9. Think of a topic of current interest to students in your school. Design a survey to predict the views of the whole school. Report to the class your survey results, your predictions, the size of your sample, and how you made sure the sample was random.

Mixed Skills Review

Find the answer. Use mental math.

10. $\frac{1}{2} + \frac{3}{4}$
11. 25% of 28
12. $59 + $2
13. $1\frac{5}{6} - 1\frac{1}{6}$

Evaluate.

14. $(y + 10) \div 2$ for $y = 12$
15. $(23 + w) \div 2$ for $w = 15$

13-10 Problem Solving: Working with Data

Objective
Make a frequency table and histogram.

- SITUATION
- **DATA**
- PLAN
- ANSWER
- CHECK

A sit-up contest for high school girls was held at the fitness center. This table shows the number of sit-ups for each person. To help organize the data, Andy made a **frequency table**.

Sit-ups

55	61	62	61	60	54	57
58	62	63	60	62	62	62
57	58	54	55	56	57	58
59	61	62	64	61	63	65
62	60	65	61	64	60	61
62	63	61	58	59	54	55
60	62	64	65	66	64	66
56	65	63	55	60	57	59
60	64	66	67	66	64	65
57	61	62	57	61	60	54

Understand the Situation
- What was the least number of sit-ups?
- What was the greatest number of sit-ups?
- How many girls did exactly 60 sit-ups?

Example

1. Make a frequency table for the first two rows of data.

   (A.) List each number that appears from least to greatest.

   (B.) Mark a tally next to a number each time it occurs.

   (C.) Count the tallies. List the frequency.

Sit-ups	Tally	Frequency
(A) 54	I (B)	1 (C)
55	I	1
57	I	1
58	I	1
60	II	2
61	II	2
62	IIII	5
63	I	1

Try This   **a.** Complete the frequency table for all of the sit-up data.

Andy used a histogram to represent the set of data. A **histogram** is a bar graph of frequencies.

Example

2. How many girls did exactly 55 sit-ups?

   There were 4 girls who did 55 sit-ups.

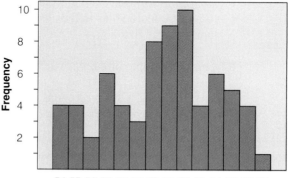

Sit-up Contest

Try This   **b.** How many girls did exactly 65 sit-ups?

Exercises

Make a frequency table and a histogram. Then answer the question.

1. the pulse rates of several members before a workout at the fitness center:

   76, 78, 68, 74, 80, 79, 68, 75, 74, 79, 73, 74, 81, 75, 72, 80, 69, 68, 72, 70, 76, 80, 68, 69, 70, 71, 75

   How many more members had a pulse rate of 68 than of 73?

2. the systolic blood pressure of several members before a workout:

   121, 130, 126, 131, 136, 127, 124, 133, 130, 126, 127, 125, 130, 125, 130, 129, 130, 126, 124, 131, 129, 131

   How many members had a systolic blood pressure between 123 and 128?

3. the number of push-ups done by a group of high school boys:

   75, 62, 70, 68, 68, 70, 69, 65, 72, 74, 60, 65, 69, 68, 71, 72, 69, 65, 68, 63, 68, 74, 61, 73, 68, 66, 68

   How many boys did more than 67 push-ups?

4. the weights in pounds of several men after a diet/workout program:

   167, 175, 180, 178, 172, 169, 178, 181, 173, 175, 181, 176, 169, 173, 178, 181, 174, 176, 182, 179

   Suppose more men entered the diet/workout program. The new men's weights doubled the number who weighed 178 lb and tripled the number who weighed 176 lb. How many new men entered the program?

This histogram shows the heights of basketball players on the fitness center team.

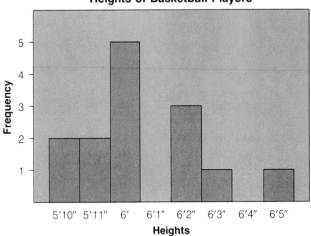

Heights of Basketball Players

5. How many players are 5 ft 11 in.?

6. How many players are 6 ft 2 in.?

7. How tall are the greatest number of players?

8. The center on the team is the tallest player. How tall is he?

9. The point guard on the team is one of the shortest players. How tall is he?

Suppose 10. Suppose the team lost a player whose height is 6 ft 2 in. They gained two players, one whose height is 6 ft 3 in. and one whose height is 6 ft. Make a histogram that shows the heights of the players on the new team.

Data Hunt 11. Find the pulse rates, for one minute, of each student in the class. Make a frequency table and a histogram of the data.

13-11 Mean and Mode

Objective
Find mean and mode.

An article stated that the average teenager watched 24 h of television per week. The students in Mr. Tan's class wanted to know if the class's average was that high. What was the class's average number of TV viewing hours per week?

Hours of TV Viewing by Students in Mr. Tan's Class Week of May 1–7

14, 10, 28, 12, 21, 3, 18, 9.5, 21, 8, 12.5, 9, 15.5, 4.5, 10, 13, 6.5

Understand the Situation

- How many students in Mr. Tan's class kept track of their TV viewing?
- What could be meant by the term *average teenager*?

The **mean** is the average of a set of numbers. You can use the mean as a single number to represent a set of data.

Example 1. Find the mean of the data on TV viewing.

| Find the sum of the numbers. | → | Find the number of addends. | → | Divide the sum by the number of addends. |

14 + 10 + 28 + 12 + 21 + 3 + 18 + 9.5 + 21 + 8 + 12.5 + 9 + 15.5 + 4.5 + 10 + 13 + 6.5 = 215.5

There are 17 addends.

215.5 ÷ 17 = 12.67647 ≈ 12.7

Round to the nearest tenth.

The mean number of TV viewing hours for the class was about 12.7 h per week.

Try This

a. Find the mean TV viewing hours of the students' younger brothers and sisters. Hours in a week: 14, 12.5, 17, 21, 8, 20, 9, 24, 19, 6, 15.5, 13, 7.5, 12, 19, 10, 12

The **mode** of a set of data is the value that occurs most often. There is no mode if no number occurs more than any other number in the set of data.

Example 2. Find the mode of this data.

TV Sets per Household

2, 1, 1, 3, 2, 2, 3, 1, 2, 2, 4, 0, 1, 2, 2, 3, 2, 1

Count the number of times each value occurs.
The 1 occurs 5 times. The 2 occurs 8 times. The 3 occurs 3 times. The 4 occurs 1 time. The value 2 occurs most often. The mode is 2 TV sets.

384 CHAPTER 13

Try This

b. Students in another school surveyed their class on the number of TV sets per household. Find the mode.

TV Sets per Household

2, 3, 1, 3, 2, 2, 3, 3, 4, 3, 2, 1, 3, 2, 4, 2, 3, 2, 3, 1, 1

Exercises

Find the mean.

1. **Hours Spent Watching Movies on VCR for One Week**

   4, 3.5, 0, 1.5, 4, 1.5, 0, 0, 6, 3.5, 8, 2, 3, 4.5, 0, 1.5, 2, 0

2. **Hours Spent Watching Sports on TV for One Week**

   0, 5.5, 6, 1, 2, 3.5, 0, 1.5, 0, 1.5, 3, 6.5, 8, 1, 0, 3, 4, 7, 0, 4, 2.5

3. **Hours Spent Watching News Programs for One Week**

   6, 1, 5, 4.5, 1.5, 8, 1.5, 7, 6.5, 3.5, 2.5, 6, 0, 4, 8, 1, 2.5

4. Estimate the mean number of hours students spent watching Saturday morning TV. Check your estimate.

5. Find the mode of the number of hours students spent watching Saturday morning TV.

Survey of Elementary School Students

Number of hours spent watching Saturday morning TV

1, 2, 2.5, 2, 0, 5, 3.5, 3, 5, 3, 4, 2, 5, 0, 3, 2, 6, 1.5, 5, 0, 3, 5, 2.5, 3, 5, 1

6. Six-year-old Teresa watches two half-hour shows each Saturday morning. Compare her TV viewing to the students in the survey.

7. Find the mode and the mean of these sizes of TV sets in the students' homes. Can you buy a set in the mean size?
   19 in., 24 in., 24 in., 13 in., 19 in., 24 in., 5 in., 13 in., 19 in., 24 in., 19 in., 13 in., 19 in.

8. Find the mode and the mean number of households watching a mini-series on four nights. 32,570,000; 33,490,000; 32,240,000; 32,490,000

9. Use the data in Exercise 2. Find the mode. Compare it to the mean. Use the data for only those people who did watch sports on TV. Find the mean. How does the mean change?

10. Find the mode and the mean of the number of people watching TV each night.

11. A company plans to advertise a new product on TV for 4 nights in a row. When should the company advertise? Use the data in the chart.

Mean Number of People in U.S. Watching TV

Night	Number of People
Sunday	108,100,000
Monday	93,800,000
Tuesday	93,800,000
Wednesday	92,400,000
Thursday	92,900,000
Friday	92,400,000
Saturday	90,000,000

13-12 Range and Median

Objective
Find range and median.

Several students in Ms. Martinez' class work after school.

Work After School
Hours per Week
8, 10, 12, 6, 8, 9, 14, 18, 12, 17, 19, 23, 19, 13, 12, 9, 12, 14, 15, 13, 22

Understand the Situation

- How many students work after school?
- What is the least number of hours worked?
- What is the greatest number of hours worked?

The **range** is the difference between the greatest and least numbers in a set of data. The **median** is the middle value when you arrange the numbers in order.

Examples

1. Find the range and median of the number of hours students work after school.

   To find the range, find the least and greatest numbers. Subtract the least number from the greatest number. The greatest number is 23. The least number is 6.

   $23 - 6 = 17$ The range is 17 h.

   To find the median, put the list in order from least to greatest.

   6, 8, 8, 9, 9, 10, 12, 12, 12, 12, **13**, 13, 14, 14, 15, 17, 18, 19, 19, 22, **23**
   least value   middle value   greatest value

   Find the middle value. There are 21 values. The eleventh value is the middle value. The median is 13.

2. Find the range and median of these values.
   14, 15, 16, 17, 17, 20

   $20 - 14 = 6$ The range is 6.

   If there is an even number of values, the median is the mean of the two middle values. 14, 15, **16, 17**, 17, 20 $\frac{16 + 17}{2} = 16.5$

   The median is 16.5 h.

Try This

a. Some students worked after school at fast-food restaurants. Find the range and median of the hours they worked.
   12, 9, 10, 8, 15, 20, 15, 16, 10, 6, 17, 22, 18, 16

b. Some students worked for retail businesses. Find the range and the median of their hourly pay rates. $3.85, $4.25, $3.75, $5.00, $4.10, $5.50, $3.75, $4.00, $4.25, $3.85, $4.50, $4.15, $4.00

Exercises

Find the range and the median for each set of data.

1. Number of hours per week worked by students at gas stations:
   14, 21, 15, 17, 14, 23, 18, 13.5, 19, 18, 22, 27, 30, 21, 16, 19.5

2. Number of hours per week worked by students in retail businesses:
   19, 18, 22, 27, 30, 21, 16, 19.5, 23, 12, 18, 20, 22, 29, 28, 15

3. Weekly income of students working after school:
   $35, $76, $76, $24, $45, $65, $80, $67, $65, $28, $100, $98, $45, $87, $90, $65, $125, $65, $76, $86, $90, $95, $45, $85, $65

4. Number of employers in local malls who hire students:
   12, 18, 23, 28, 28, 17, 19, 21, 10, 19, 10, 15, 12, 25, 30, 18, 23, 17

5. Use the data in Exercise 1. What could happen to make the range greater?

6. Use the data in Exercise 2. The students who were working 18 h per week now work 24 h per week. How does this affect the range?

7. At the Ranch House Restaurant, people who bus tables earn from $3.75 to $5 per hour. Waiters earn from $4.15 to $5.40 per hour. Which job has the greater salary range?

8. Use the data in Exercise 1. Suppose the students working 14 h or less quit their jobs. How would the median change?

9. Use the data in Exercise 3. Give two ways that the median income for after-school jobs might change to a lower value.

Talk Math

10. Suppose you learn that student jobs pay from $3.75 to $8 per hour. The median pay is $4.10. What does this tell you?

11. The pay ranges at job B and job Q are both $4. The median pay at job Q is $5.75. The median pay at job B is $4.50. How could you explain this?

12. Which do you think better represents the weekly salaries of 10 students, the mean or the median? Why?

Weekly Salaries of 10 Students Working Summer Jobs
$150, $125, $135, $170, $100, $125, $425, $145, $115, $125

Mental Math

13. Find the range of the data in Exercise 12.

14. Find the range of tips made by Ruth for 10 days of work busing tables.
    $20, $18, $18.50, $28, $21, $26, $15, $23.50, $19, $21

Data Bank

15. Find the range and median of the data on Minimum Hourly Wage Rates listed in the Data Bank on p. 493.

Show Math

16. Make a bulletin board display of newspaper articles that use *range* and *median*. Highlight the words wherever they appear. Report on one article to the class.

13-13 Problem Solving: Data Collection and Analysis

Objective
Collect and analyze data.

- SITUATION
- DATA
- PLAN
- ANSWER
- CHECK

In the late 1930s, television was first introduced in the U.S. Today most homes have at least one TV set. Many groups have gathered statistics on TV viewing habits in the U.S. These statistics show the average number of hours watched by TV viewers who are from 12 to 17 years old.

Hours of TV Viewing							
per person per day Viewers 12–17 years old							
	1980	1981	1982	1983	1984	1985	1986
Feb.	3:33	3:40	3:15	3:36	3:47	3:46	3:42
Mar.	3:03	3:07	3:03	3:35	3:31	3:16	3:13
Apr.	3:17	2:56	3:08	3:40	3:25	3:21	3:04
May	2:45	2:43	2:35	3:05	3:01	2:47	2:45
Jul.	3:43	3:28	3:39	4:05	3:53	3:54	4:08
Oct.	3:01	2:47	3:13	3:21	2:48	3:06	2:50
Nov.	3:17	2:55	3:13	3:33	3:13	3:22	3:00
Dec.	3:23	3:06	3:31	3:36	3:24	3:24	3:06

Work in groups to complete the following.

Situation
Analyze the data on viewing habits in the chart. Use the data to create a graph that will help people spot trends in viewing. Predict the average number of hours of TV viewing per day by 12–17 year olds in 1987 and 1988.

Data Collection
Survey your classmates about their TV viewing habits. Find how many hours per month each student in the class watches TV. Compare the data you collect to the data in the national survey. Include your findings in your report.

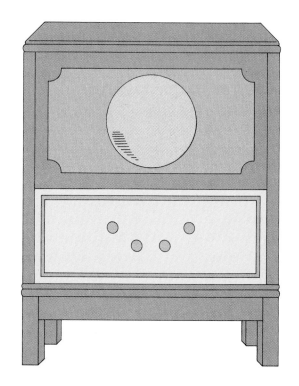

Facts to Consider

- The statistics in the table were gathered in a national survey.
- The data represents an average number of TV viewing hours per person per day over a one-month period.
- The data is written as hours and minutes.
  4:21 means 4 h and 21 min.
- The data covers over-the-air cable television viewing.
- When this survey was completed, there were only two statistics available for 1987:
  February   3:08
  March   2:52

Share Your Group's Decision

- Present your graph and your report to the class.
- Compare your findings with other groups.

Suppose Suppose you survey people outside of your class. Will this data change your report?

Mixed Skills Review Find the answer.

1. $8000 - 4336$
2. 584×398
3. $\frac{1}{3} + \frac{5}{6}$
4. $\$4650 \div 100$
5. $\frac{6}{7} \div \frac{3}{14}$

Decide if you need an exact answer or an estimate.

6. You are a reporter for your school newspaper. You are writing an article about the basketball game last night. You report how many points each player scored in the game.

7. You are a reporter for your school newspaper. You are writing an article about the basketball game last night. You report how many fans attended the game.

Compare. Use $<$, $>$, or $=$.

8. $15\% \square 0.2$
9. $\frac{2}{3} \square 0.66$
10. $\frac{5}{6} \square \frac{8}{9}$
11. $\frac{1}{5} \square 20\%$

PROBABILITY AND STATISTICS

13-14 Enrichment: Stem-and-Leaf Diagrams

Objective
Make stem-and-leaf diagrams.

Fanny's Bait Shop held a contest for the heaviest bluefish caught off the Hatteras Inlet.

Bluefish Contest
weights caught to nearest pound
12, 15, 21, 18, 19, 13, 12, 15, 13, 16, 21, 25, 22, 30, 12, 25, 9, 10, 8, 10, 13, 16, 27, 7, 26, 17, 8, 16, 19, 20, 13

Understand the Situation
- What was the weight of the heaviest fish caught?
- Were there many fish over 10 lb caught this year?

Fanny uses a **stem-and-leaf diagram** to organize the data.

Example Make a stem-and-leaf diagram of the weights. How many bluefish over 20 lb were caught?

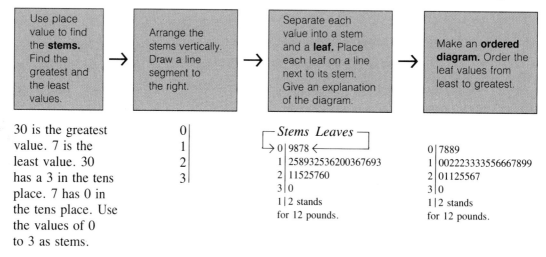

30 is the greatest value. 7 is the least value. 30 has a 3 in the tens place. 7 has 0 in the tens place. Use the values of 0 to 3 as stems.

Nine bluefish over 20 lb were caught.

Exercises

Make an ordered stem-and-leaf diagram of the set of data. Then answer the question.

1. the lengths in inches of brown trout caught by the Tri-City fishing club

   18, 24, 22, 19, 17, 25, 18, 20, 11, 9, 11, 13, 21, 19, 30, 23, 18, 17, 15, 25, 28, 21, 22, 19, 21, 19, 24, 20, 19, 14, 19, 26, 23

   What length was caught most often?

2. the weights in pounds of Atlantic cod caught from a commercial fishing boat

   10, 24, 22, 19, 18, 25, 18, 21, 20, 15, 13, 21, 19, 22, 18, 17, 20, 11, 13, 22, 24, 19, 18, 21, 22, 13, 19, 15, 20, 14, 12, 19, 20, 22

   What were the lightest and heaviest weights?

13-15 Computer: Arrays

Objective
Identify elements in arrays.

An **array** is a set of variables. You can identify an element of an array by its subscript, or the element in parentheses.

A(0), A(1), A(2), . . . , A(8) is an array of 9 elements.

There are two kinds of arrays. A **one-dimensional** array needs only one number to identify an element in that array. A **two-dimensional** array needs two numbers: a row number and a column number. To write the location of an element in a two-dimensional array, write the name of the array and then write the row number and the column number in parentheses.

Example Identify the 2 in this array.

$$A \begin{bmatrix} 3 & 2 \\ 4 & 1 \end{bmatrix} \quad \underset{\substack{\text{name of} \\ \text{array}}}{\nearrow} \underset{\substack{\text{row} \\ \text{number}}}{A(1,2)} \underset{\substack{\text{column} \\ \text{number}}}{\nwarrow}$$

Try This Identify the element in array Q.

   a. 1   **b.** 2   **c.** 3   **d.** 5 $Q \begin{bmatrix} 1 & 3 & 5 \\ 6 & 2 & 4 \end{bmatrix}$

Exercises

Identify the element in array R.

1. 5	2. 13	3. 11	4. 45
5. 20	6. 10	7. 64	8. 29

$$R \begin{bmatrix} 22 & 46 & 33 & 45 \\ 64 & 13 & 28 & 31 \\ 11 & 23 & 29 & 10 \\ 27 & 20 & 9 & 5 \end{bmatrix}$$

Identify the element in array T.

9. red	10. pink	11. green
12. silver	13. purple	14. black

$$T \begin{bmatrix} \text{red} & \text{orange} & \text{blue} \\ \text{black} & \text{purple} & \text{green} \\ \text{silver} & \text{pink} & \text{yellow} \end{bmatrix}$$

Find the element in array S.

15. S(2,2)	16. S(1,3)	17. S(2,3)
18. S(1,1)	19. S(2,1)	20. S(1,2)

$$S \begin{bmatrix} 1 & 4 & 2 \\ 3 & 5 & 0 \end{bmatrix}$$

Write Your Own Problem

21. Write a 3-by-5 array with friends' names. Have a classmate identify elements in your array.

PROBABILITY AND STATISTICS

CHAPTER 13 REVIEW

Choose the one that does not belong in each group. Explain why not.

1. A. probability
   B. permutation
   C. odds in favor
   D. odds against

2. A. compound
   B. experimental
   C. simple
   D. expectation

3. A. sample
   B. counting principle
   C. outcomes
   D. tree diagram

4. A. mean
   B. probability
   C. median
   D. mode

Match each question with the method that gives the answer.

5. How many ways can I choose a soup and drink from a choice of 5 soups and 3 drinks?
6. What is the probability of the Cowboys winning three games in a row?
7. How many people in the county will vote for Wong?
8. Are the Houston Astros likely to win the World Series?
9. How many ways can I line up seven books?

A. permutations
B. experimental probability
C. counting principle
D. odds in favor
E. median
F. compound probability
G. random sample

Complete each sentence.

10. If A and B are independent events, then P(A and B) = _____.
11. If there are m choices and n choices, then _____ gives the number of possible outcomes.
12. The probability of an event is the number of ways the event can occur divided by _____.
13. The odds in favor of an outcome is the number of ways it can occur divided by _____.

Match.

14. 2, 5, 5, 6
15. 3, 5, 7, 9
16. 1, 3, 4, 4
17. 2, 2, 6, 8
18. 1, 3, 6, 6
19. 2, 2, 3, 9

A. mean = 4
C. median = 5
E. median = 2
G. median = 4

B. range = 3
D. mode = 6
F. mean = 6
H. mean = 5

CHAPTER 13 TEST

1. Aretha is selling raffle tickets. One ticket costs $1. Three tickets cost $2. Seven tickets cost $3. At this rate, how many tickets cost $7?
2. Make a tree diagram. Find all the possible outcomes for lunch and a museum.
   Lunch: Sally's Salad Bar, Sam's Sandwich Shop, Pete's Pizza Parlor
   Museum: Modern Art, Contemporary Design, American Indian
3. Find the number of possible outcomes for dinner and a movie theater.
   Dinner: José's Barbecue, The Burger Pitt, Japanese Gardens
   Movie Theater: The Grand, The Park, The Guild, The Festival
4. There are 6 hats on a rack. Find the number of permutations of the hats.
5. Out of the first 200 ballots, 140 people voted for Appel. 60 people voted for Mitkin. What is the probability that the next vote will be for Appel?
6. Suppose you toss a coin 500 times. About how many times should it land *heads*?
7. Kathy bowls a 200 game 1 in every 8 games. What is the probability that she will bowl two 200 games in a row?

A bag contains the following marbles: 3 red, 2 blue, 5 green, and 6 black. Without looking, Mabel picks a marble. Find the odds.

8. in favor of picking a green marble
9. against picking a black marble

10. Lee's Kennels have about 1 champion dog for every 24 puppies they raise. Suppose they raise 96 pups. How many champions can they expect?
11. A marketing assistant ran a taste test with a random sample of consumers in Adrian, a town of about 20,000 people. Predict about how many people would like Product B.

Product	Number of People Who Liked Product
A	50
B	82
C	68

12. Make a frequency table and a histogram for the set of data. Then answer the question.
    test scores for Mr. Romero's history class:
    83, 90, 82, 88, 92, 96, 75, 88, 79, 90, 88, 86, 82, 79, 75, 96, 95, 84, 82, 85, 70, 98, 88, 80, 82, 65, 100, 85, 88
    How many more students scored an 88 than an 82?
13. Find the mean and the mode of the data. Round to the nearest hundredth.
    amount of rain in inches each month for Houston, Texas, in 1985:
    3.57, 3.54, 2.68, 3.54, 5.10, 4.52, 4.12, 4.35, 4.65, 4.05, 4.03, 4.04
14. Find the range and median of the data.
    the number of tornadoes each month:
    0, 2, 11, 17, 34, 22, 10, 5, 124, 2, 0, 5

CHAPTER 13 CUMULATIVE REVIEW

1. Robert arrived at the museum after Victoria. Aiko did not arrive last. Ines arrived before Aiko. Charlie and Ines were not the first people to arrive. Who was the third person to arrive?
   A. Ines
   B. Charlie
   C. Robert
   D. Aiko

2. Add. Use mental math.
   $4\frac{1}{2} + 6 + 3\frac{1}{2}$
   A. 15
   B. 14
   C. 13
   D. $14\frac{1}{2}$

3. Betsy prints her own photos. Three out of every 4 photos come out right. She printed 2 photos today. What is the probability that both of the photos came out right?
   A. $\frac{3}{4}$
   B. $\frac{27}{64}$
   C. $\frac{9}{16}$
   D. $\frac{4}{7}$

4. 156 is 60% of what number?
   A. 93.6
   B. 2.6
   C. 260
   D. 190

5. How many toothpicks would you need to make a pattern that is 10 squares long?
   A. 30 toothpicks
   B. 41 toothpicks
   C. 40 toothpicks
   D. 31 toothpicks

6. Suds, Inc. is testing new soaps. They took a random sample of consumers in Lifka, a town of 16,000 people. Predict how many people would like Product A.
   A. 6000 people
   B. 5200 people
   C. 5500 people
   D. 5000 people

Product	Number of People Who Liked Product
A	65
B	90
C	45

7. Write $\frac{22}{6}$ as a mixed number or a whole number.
   A. 4
   B. $3\frac{2}{3}$
   C. $3\frac{3}{6}$
   D. $4\frac{1}{3}$

8. Write as a decimal. 115%
   A. 11.5
   B. 115
   C. 0.115
   D. 1.15

9. Estimate. Round to the nearest hundred.
   539 + 649
   A. 1200
   B. 1190
   C. 1100
   D. 1170

10. Multiply.
    357 × 244
    A. 87,108
    B. 3570
    C. 87,008
    D. 86,108

11. Divide.
    72.8 ÷ 100
    A. 728
    B. 0.728
    C. 7.28
    D. 7280

12. In a shopping bag there are 5 packets of dill seeds, 2 packets of oregano seeds, and 8 packets of basil seeds. Each packet is the same size. Suppose you pick a packet without looking. What is the probability that it will be a packet of basil seeds?
    A. $\frac{1}{4}$
    B. $\frac{7}{15}$
    C. $\frac{8}{15}$
    D. $\frac{5}{8}$

13. Compare. Use <, >, or =.
    0.16 □ $\frac{1}{5}$
    A. <
    B. >
    C. =

14. Solve.
    $\frac{n}{9} = \frac{4}{12}$
    A. 5
    B. 4
    C. 2
    D. 3

15. Add.
    $1\frac{1}{5} + 6\frac{2}{3}$
    A. $7\frac{3}{8}$
    B. $7\frac{13}{15}$
    C. $7\frac{1}{3}$
    D. $7\frac{11}{15}$

16. Decide which operation you would use to solve this problem.
    Problem: Fusako earned △ for □ days of work. About how much money is this per day?
    A. addition
    B. subtraction
    C. multiplication
    D. division

17. Name two lines in this drawing.
    A. $\overleftrightarrow{AB}, \overleftrightarrow{x}$
    B. $\overrightarrow{CD}, \overleftrightarrow{y}$
    C. $\overleftrightarrow{EF}, \overleftrightarrow{z}$
    D. $\overrightarrow{BC}, \overleftrightarrow{x}$

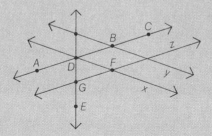

18. Change the unit.
    24 ft = _____ yd
    A. 288
    B. 72
    C. 2
    D. 8

19. Subtract.
    9.003 − 0.678
    A. 8.435
    B. 8.325
    C. 9.435
    D. 9.325

20. Divide.
    $\frac{5}{6} \div \frac{2}{10}$
    A. $\frac{1}{6}$
    B. $\frac{7}{16}$
    C. $4\frac{1}{6}$
    D. $1\frac{7}{8}$

21. Find the perimeter of a rectangle with these measurements.
    $l = 20$ cm, $w = 12$ cm
    A. 32 cm
    B. 240 cm
    C. 52 cm
    D. 64 cm

Chapter 14 Overview

Key Ideas

- Use the problem-solving strategy *use logical reasoning*.
- Find the area of squares, rectangles, parallelograms, triangles, and circles.
- Find the area of irregular-shaped figures.
- Find the cost of carpeting.
- Identify space figures: prisms, pyramids, cones, and cylinders.
- Find surface area.
- Find the volume of prisms, cylinders, pyramids, and cones.
- Understand logical meanings of the words *and*, *or*, and *not*.

Key Terms

- logical reasoning
- square
- square units
- height
- square yard
- pyramid
- sphere
- edge
- rectangle
- parallelogram
- triangle
- cube
- cylinder
- surface area
- space figure
- area
- base
- circle
- prism
- cone
- volume

Key Skills

Multiply.

1. 3.14×7
2. 15.2×7.2
3. 5.8×3.02
4. 0.6×0.8
5. 6×0.07
6. 23.5×5.8
7. 3.4×1.8
8. 9.08×0.33
9. $\frac{1}{2} \times 34 \times 8$
10. $\frac{1}{2} \times 4.5 \times 3$
11. $\frac{1}{2} \times \frac{3}{4} \times 6$
12. $\frac{1}{2} \times 5.5 \times 6.5$

Evaluate.

13. $l \times w$ for $l = 5$ and $w = 9$
14. s^2 for $s = 7$
15. $b \times h$ for $b = 4$ and $h = 5.5$
16. $\frac{1}{2}bh$ for $b = 8$ and $h = 5$
17. $3.14 \times r^2$ for $r = 6$
18. $3.14 \times r^2$ for $r = 2.5$
19. s^3 for $s = 2.2$
20. lwh for $l = 3$, $w = 4.6$, $h = 5.5$
21. $3.14r^2h$ for $r = 3$ and $h = 8$
22. $\frac{4}{3} \times 3.14 \times r^3$ for $r = 1.2$

AREA AND VOLUME

14

14-1 Problem Solving: Learning to Use Strategies

Objective
Use the strategy logical reasoning.

- SITUATION
- DATA
- **PLAN**
- ANSWER
- CHECK

Look at the pile of cubes. This is called a **space figure**. The pile is filled in on the back. How many cubes are there?

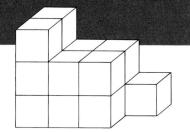

You can answer this question using **logical reasoning**.

Example

1. How many cubes are in the stack?

   Think about pulling the stack apart. What would each part look like? Count the cubes in each part. Add the total number of cubes. There are 15 cubes.

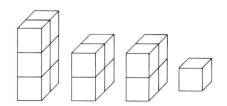

Try This How many cubes are in each stack?

a.   b.   c.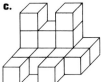

Examples What shapes and how many of each shape make up the faces of the solid figure?

2. 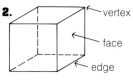   Dotted lines show hidden **edges**.

6 squares

3.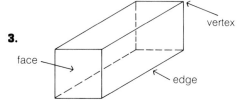

2 squares, 4 rectangles

Try This What shapes and how many of each shape make up the faces of the solid figure?

d.    e.

398 CHAPTER 14

Exercises

Solve. Use one or more of the problem-solving strategies.

Problem-Solving Strategies
- Guess and check.
- Choose an operation.
- Make an organized list.
- Act it out.
- Find a pattern.
- Write an equation.
- **Use logical reasoning.**
- Draw a picture or diagram.
- Make a table.
- Use objects.
- Work backward.
- Solve a simpler problem.

1. How many cubes are in this stack?

2. How many cubes are in this stack?

3. Oliver had two part-time jobs. He made a total of $90 every two weeks. One job paid twice as much as the other. How much money did Oliver earn for each job every two weeks?

4. Vanita has two pairs of pants, three shirts, and two sweaters. How many different outfits can she make?

5. What shapes and how many of each shape make up the faces of the solid figure?

6. How many cubes are in this stack?

7. Find the number of games that you would need to finish a single elimination soccer tournament with 18 teams.

8. What shapes and how many of each shape make up the faces of the solid figure?

   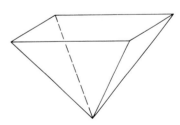

Show Math   9. Draw a picture to show how this solid figure looks when viewed from the top, the back, and the right side.

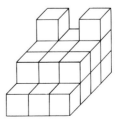

AREA AND VOLUME **399**

14-2 Area of Squares and Rectangles

Objective
Find the area of squares and rectangles.

Jan plans to buy tile to cover her porch floor. Each tile measures one foot on a side. How many tiles does she need to cover the floor?

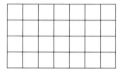

The number of **square units** it takes to cover a region is the **area** of the region.

In Jan's case, the square unit is one square foot. When you find the number of tiles, you will find the area in square feet.

Example

1. Find the number of tiles Jan will need for her porch.

   Count the squares in the picture. Jan will need 28 tiles to cover the porch. The area of the porch is 28 square feet, or 28 sq ft, or 28 ft².

Try This Find the area.

a.    b. 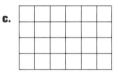   c.

You can use a formula to find the area of a rectangle and a square.

Polygon	Formula	Symbols	Picture
Rectangle	$A = l \times w$, or lw	A = area l = length w = width	w, l
Square	$A = s \times s$, or s^2	A = area s = side	s

Examples Find the area of the square or the rectangle.

2. $l = 12$ ft, $w = 8$ ft
   $A = l \times w$
   $A = 12 \times 8$
   $A = 96$
   $A = 96$ sq ft, or 96 ft²

3. $s = 10$ cm
   $A = s^2$
   $A = 10^2$
   $A = 100$
   $A = 100$ sq cm, or 100 cm²

Try This Find the area of the square or the rectangle.

d. $s = 9$ cm   e. $l = 9.4$ in.   f. $l = 18$ m   g. $s = 6$ ft
              $w = 8.5$ in. $w = 10$ m

Exercises

Find the area.

1.   2.  3.

Find the area of the square or the rectangle.

4. $l = 21$ ft
   $w = 5$ ft

5. $s = 7.5$ ft

6. $l = 5$ ft
   $w = 2\frac{1}{2}$ ft

7. $l = 0.7$ m
   $w = 0.2$ m

8.
9.
10.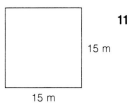
11.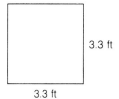

Show Math Draw rectangles that have the following areas.

12. 32 cm^2 13. 48 in.^{2} 14. 36 in.^{2} 15. 100 cm^2

Suppose 16. Suppose the length and width of this rectangle were twice as long. How much larger would the area be?

Talk Math 17. Tell two different ways you could find the area of this room.

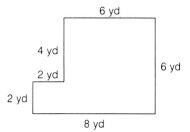

Problem Solving 18. The worker who measured the room for carpet left out some of the dimensions. Find the area.

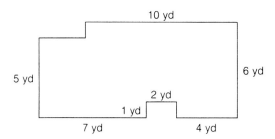

14-3 Area of Parallelograms

Objective
Find the area of parallelograms.

Les is making a metal sculpture using pieces shaped like parallelograms. He buys metal by the square foot. What is the area of the piece of metal in his plan?

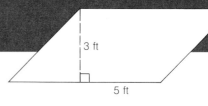

Base = 5 ft   Height = 3 ft

Understand the Situation

- What is the length of the base of the parallelogram?
- What is the height of the parallelogram?

You can take a parallelogram and make it into a rectangle. Cut off the end. Fit it onto the other end. Both figures have the same area. You can find the area of a parallelogram using a formula like the one for the area of a rectangle.

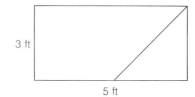

Polygon	Formula	Symbols	Picture
Parallelogram	$A = b \times h$, or bh	A = area b = base h = height	

Examples

1. Find the area of the parallelogram in Les' plan.
   $A = b \times h$
   $A = 5 \times 3$
   $A = 15$
   The area of the parallelogram is 15 sq ft, or 15 ft^2.

2. Find the area of a parallelogram with $b = 40$ ft and $h = 12$ ft.
   $A = b \times h$
   $A = 40 \times 12$
   $A = 480$
   $A = 480$ sq ft, or 480 ft^2

Try This   Find the area of the parallelogram.

a.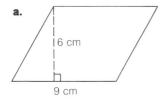

b.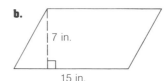

c. $b = 16.3$ cm
   $h = 2.7$ cm

d. $b = 2\frac{1}{2}$ ft
   $h = 1\frac{1}{4}$ ft

Exercises

Find the area of the parallelogram.

1.

2.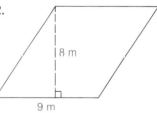

3.

4.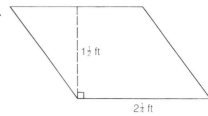

5.

6.

7. $b = 61$ ft
   $h = 7$ ft

8. $b = 15.2$ cm
   $h = 7.2$ cm

9. $b = 1.2$ in.
   $h = 0.7$ in.

10. $b = 3.4$ ft
    $h = 1.8$ ft

11. $b = 6.2$ m
    $h = 3.3$

12. $b = 2\frac{1}{2}$ ft
    $h = 3\frac{1}{2}$ ft

13. $b = 1\frac{1}{5}$ in.
    $h = 2\frac{1}{3}$ in.

14. $b = 3.5$ m
    $h = 4.2$ m

Calculator Use your calculator to find the area of each parallelogram. Round your answer to the nearest thousandth.

15. $b = 4.013$; $h = 5.796$
16. $b = 6.051$; $h = 4.927$
17. $b = 3.021$; $h = 5.821$

Show Math 18. Draw and label two different parallelograms with an area of 24 square units.

Estimation Estimate the area of each figure in terms of square units. 1 square unit: ☐

19. 20. 21.

AREA AND VOLUME **403**

14-4 Area of Triangles

Objective
Find the area of triangles.

Tomas needs to order paint for the gable end of his house. He needs to find the area. What is the area?

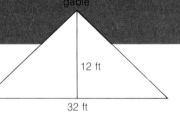

Understand the Situation
- What shape is the gable end?
- How high is the gable end?
- How wide is the base of the gable end?

The formula for the area of a triangle is related to the formula for the area of a parallelogram. Each triangle has the same base and height as the parallelogram. Each triangle has one half its area.

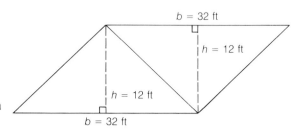

Polygon	Formula	Symbols	Picture
Triangle	$A = \frac{1}{2} \times b \times h,$ or $\frac{1}{2}bh$	A = area b = base h = height	

Examples

1. What is the area of the gable end of Tomas' house?

   $A = \frac{1}{2} \times b \times h$
   $A = \frac{1}{2} \times 32 \times 12$
   $A = 192$

   The area of the gable end is 192 sq ft, or 192 ft².

2. Find the area of a triangle with $b = 3$ ft and $h = 4$ ft.

   $A = \frac{1}{2} \times b \times h$
   $A = \frac{1}{2} \times 3 \times 4$
   $A = 6$
   $A = 6$ sq ft, or 6 ft²

Try This Find the area of the triangle.

a. 6 ft, 3 ft

b. 8 in., 6 in.

c. $b = 12$ cm
$h = 6$ cm

d. $b = 6.2$ m
$h = 5.2$ m

404 CHAPTER 14

Exercises

Find the area of the triangle.

1.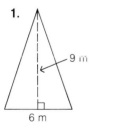
2.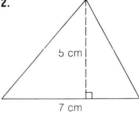
3.
4.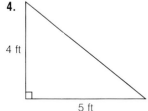
5.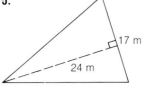
6.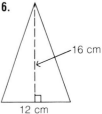

7. $b = 4$ ft
   $h = 2$ ft
8. $b = 6$ cm
   $h = 3$ cm
9. $b = 2\frac{1}{2}$ in.
   $h = 4\frac{1}{4}$ in.
10. $b = 15$ ft
    $h = 6$ ft
11. $b = 7.8$ m
    $h = 1.1$ m
12. $b = 2.4$ m
    $h = 6.82$ m
13. $b = 1\frac{1}{4}$ in.
    $h = 2\frac{1}{3}$ in.
14. $b = 6.3$ cm
    $h = 2.5$ cm

Mixed Applications

15. Rhonda has to order bricks for a triangular gable end with a height of 16 ft and a base of 44 ft. Six bricks cover a square foot. About how many bricks does she need?

16. Susanna is buying paint to cover a triangular-shaped area with a height of 12 ft and a base of 44 ft. One gallon of the paint covers 200 sq ft. About how many gallons does she need to buy to cover the area?

Mixed Skills Review

Find the balance. Elliot's starting checking account balance is $464.27.

17. On May 2, Elliot wrote a check for $14.97. On May 4, he deposited $78.01.

Find the greatest common factor for each pair of numbers.

18. 4, 10
19. 21, 45
20. 52, 78
21. 18, 38
22. 60, 15

Find the perimeter.

23. rectangle: $l = 14$ ft
    $w = 12$ ft
24. square: $s = 9$ ft
25. rectangle: $l = 4$ cm
    $w = 1.5$ cm

14-5 Area of Circles

Objective
Find the area of circles.

Sarah has to mix fertilizer for a flower bed. The box gives amounts for areas in square feet. What is the area of Sarah's flower bed?

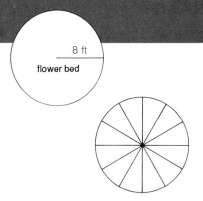

Understand the Situation
- What does Sarah need to find?
- What shape is the flower bed?
- What measure is shown?

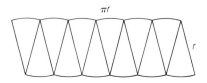

You can see how the area of a circle is found. Cut the circle into wedges. Arrange the wedges to form a shape like a parallelogram. The height is r and the base is $\frac{1}{2}$ the circumference, $\frac{1}{2}(2\pi r)$, or πr. The area, $\pi \times r \times r$, is approximately the same as the circle.

Polygon	Formula	Symbols	Picture
Circle	$A = \pi \times r \times r$, or πr^2	A = area r = radius	

Examples

1. Find the area of Sarah's flower bed.
   Use 3.14 for π.
   $A = \pi r^2$
   $A \approx 3.14 \times 8^2$
   $A \approx 200.96$
   The area is about 200.96 sq ft, or 200.96 ft².

2. Find the area of a circle with $d = 12$ in. Use 3.14 for π.
   $A = \pi r^2$
   $A \approx 3.14 \times 6^2$ ($r = d \div 2$)
   $A \approx 3.14 \times 36$
   $A \approx 113.04$
   $A \approx 113.04$ sq in., or 113.04 in.²

Try This Find the area. Use 3.14 for π. Round to the nearest hundredth.

a.   b.   c. $r = 7$ ft   d. $d = 3$ ft

406 CHAPTER 14

Exercises

Find the area. Use 3.14 for π. Round to the nearest hundredth.

1.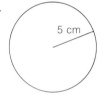
2.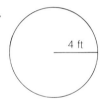
3.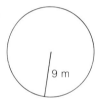

4.
5.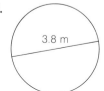
6.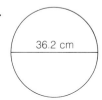

7. $r = 1.7$ cm
8. $r = 0.5$ in.
9. $d = 2.2$ m
10. $r = 5\frac{1}{4}$ ft
11. $d = 14$ ft
12. $d = 26$ in.
13. $r = 2\frac{1}{2}$ ft
14. $d = 9$ m

Calculator Use your calculator to check Belinda's work. If you find a mistake, give the correct answer.

> Name **Belinda**
>
> Find the area of each circle. Use 3.14 for π.
> Round your answer to the nearest hundredth.
>
> 15. $r = 4.12$ cm   16. $r = 2.4$ m   17. $r = 5.6$ ft   18. $r = 8.07$ in.
>
> $A \approx 12.94$ cm² $A \approx 18.09$ m² $A \approx 98.47$ ft² $A \approx 204.5$ in.²

Mixed Skills Review

Estimate.

19. $\frac{14}{34} \times 22$
20. $\$49.03 + \$3.85 + \$26.57$
21. $324 \div 8$

Name each polygon.

22.
23.
24.
25.

Write using exponents. Then write as a standard numeral.

26. $14 \times 14 \times 14 \times 14$
27. $(3)(3)(3)(3)(3)$
28. 7.6 cubed

14-6 Consumer Math: Buying Carpet

Objective
Change square feet to square yards.

- SITUATION
- DATA
- PLAN
- ANSWER
- CHECK

Joan's living room measures 15 ft by 12 ft. Carpet is sold in square yards. How many square yards of carpet would it take to cover her floor?

Understand the Situation

- What are the length and width of Joan's living room?
- In what unit of measure is her living room given?
- In what unit of measure is carpet sold?

You need to change square feet to square yards to find the amount of carpet. The picture shows why there are 9 square feet in 1 square yard.

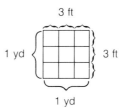

You can count 9 square feet in the square yard.
3 ft × 3 ft = 9 ft^2

Example Find the number of square yards of carpet that Joan needs.

You can work this problem two ways.

A. Find the number of square feet. Change the square feet to square yards.

$A = l \times w$
$A = 15 \times 12$
$A = 180$ ft^2
$180 \div 9 = 20$ yd^2

B. Find the room's area in square yards. Change feet to yards.

$l = 15$ ft $= 5$ yd
$w = 12$ ft $= 4$ yd
$A = l \times w$
$A = 5 \times 4$
$A = 20$ yd^2

Joan's living room will take 20 yd^2 of carpet. Joan may need more carpet if she has to match a pattern.

Try This Solve.

a. Find the area in square yards of a room that measures 12 ft by 8 ft.

b. Find the area in square yards of a room that measures 9 ft by 12 ft.

Exercises

Find the area of each room in square yards.
1. 9 ft by 9 ft
2. 15 ft by 18 ft
3. 10 ft by 14 ft
4. 9 ft by 15 ft
5. 12 ft by 16 ft
6. 20 ft by 18 ft
7. 12 ft by 12 ft
8. 15 ft by 15 ft

Mixed Applications

Solve.

9. How many square yards of carpet would it take to cover this room's floor? The carpet does not have a pattern.

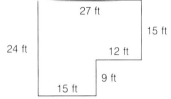

10. How many square yards of carpet would it take to cover this room's floor? The carpet does not have a pattern.

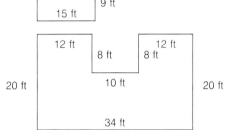

Problem Solving

11. How many cubes are in this stack?

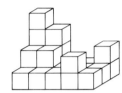

Group Activity

12. Work with your group to design a garden for a local senior citizens center. The center has given you an area that measures 6 yd by 8 yd. They want $\frac{1}{3}$ of the area planted with flowers, $\frac{1}{4}$ with grass, $\frac{1}{6}$ with small shrubs, $\frac{1}{8}$ with fruit trees, and $\frac{1}{8}$ with a path of flat stones. Draw your plan. Label each area in square feet.

Mixed Skills Review

Write each in words.
13. 7432
14. $659.83
15. $8.09
16. 20,405

Change these units.
17. 15 yd = _____ ft
18. 18 fl oz = _____ c
19. 3.2 T = _____ lb
20. 4000 m = _____ km

Find each answer.
21. $63.4 \div 6$
22. $\frac{1}{9} \times 23$
23. $0.35 + 6.4$
24. $\frac{3}{4} \div 5\frac{1}{6}$

14-7 Space Figures

Objective
Count faces, edges, and vertices. Name space figures.

Look at the space figures shown. Which has five **faces,** eight **edges,** and five **vertices**?

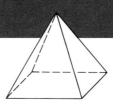

Example

1. Tell how many faces, edges, and vertices this prism has.

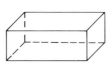

Faces = 6
Edges = 12
Vertices = 8

Try This   Tell how many faces, edges, and vertices each prism has.

a.   b.    c.

These drawings show some common space figures. Name prisms and pyramids for the shapes of their bases. A prism has two congruent parallel polygons as bases.

Prisms

  Triangular   Cube   Rectangular

Pyramids

                    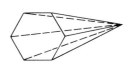

  Triangular   Square   Hexagonal

Cylinders,
Cones,
and Spheres

  Cylinder   Cone   Sphere

Example

2. Name the space figure.

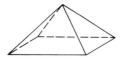

square pyramid

410   CHAPTER 14

Try This Name each space figure.

d.  e. 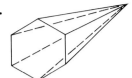 f.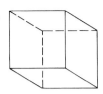

Exercises

Tell how many faces, edges, and vertices each figure has.

1. 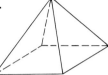 2.   3.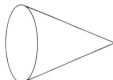

4.   5.   6.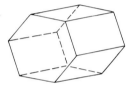

Name each space figure.

7.   8.   9.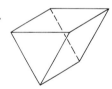

10. 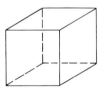 11.  12.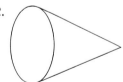

13. What space figure has exactly four triangular faces?
14. What space figure has three rectangular faces?
15. What space figure has one curved rectangular face and two circular faces?
16. What space figure has two hexagonal and six rectangular faces?

Talk Math 17. Describe a space figure to other members of the class. Let the class members draw that figure. Compare their drawings with the shape you described.

14-8 Surface Area of Rectangular Prisms

Objective
Find the surface area of rectangular prisms.

Esther needs to rustcoat this sculpture. What is the surface area of the sculpture?

Understand the Situation

- What is the shape of the sculpture?
- What is the length of each edge?
- How many faces does the sculpture have?

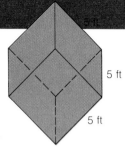

The **surface area** of a space figure is the sum of the areas of each of its faces.

Examples

1. Find the surface area of Esther's sculpture.

   To see all the surfaces, draw a pattern of the cube.

   Each face of the cube is a square. Find the area of each face.

   $A = s^2 = 5^2 = 25$
   $A = 25 \text{ ft}^2$

   There are 6 faces on the cube.

   Multiply to find the surface area of the cube.

   Surface area = $25 \text{ ft}^2 \times 6 = 150 \text{ ft}^2$

2. Find the surface area of the rectangular prism.

   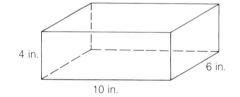

   Find the area of each face.

   Top: $10 \times 6 = 60 \text{ in.}^2$ Front: $10 \times 4 = 40 \text{ in.}^2$ Left side: $6 \times 4 = 24 \text{ in.}^2$
   Bottom: $10 \times 6 = 60 \text{ in.}^2$ Back: $10 \times 4 = 40 \text{ in.}^2$ Right side: $6 \times 4 = 24 \text{ in.}^2$

   Add the areas. $60 + 60 + 40 + 40 + 24 + 24 = 248$ The surface area is 248 in.^2

Try This   Find the surface area.

a.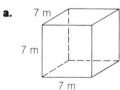

b.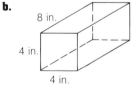

c.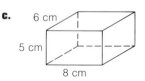

Exercises

Find the surface area.

1.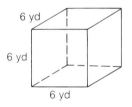
2.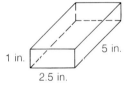
3.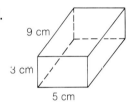
4.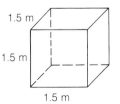
5.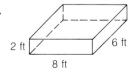
6.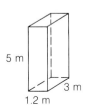
7.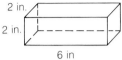
8.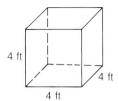
9.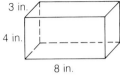

Mixed Applications

10. Edna wants to paint the walls and ceiling of her room. One quart of paint covers 50 sq ft. How many quarts of paint will she need?

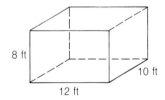

11. A contractor is building forms for pouring cement. How many square feet of plywood will it take to line the inside walls but not the top?

12. The surface area of the cube shown is 24 in.² How long is an edge of the cube?

Problem Solving

Decide if you would need to find the perimeter or the area for Exercises 13–16.

13. How many bricks will I need to make my patio?
14. How many boards will I need to put a fence around my garden?
15. How many bricks will I need to make an edge around my flower bed?
16. How much cardboard will I need to build a box?

AREA AND VOLUME   **413**

14-9 Volume of Rectangular Prisms

Objective
Find the volume of rectangular prisms.

Kenji is packing puzzle blocks into boxes. He can line the bottom of a box with 4 blocks on one edge and 5 blocks on another. He can pack the blocks 3 layers high. In all, how many blocks can he fit into a box?

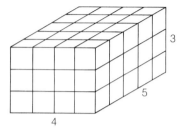

Understand the Situation

- What shape are the puzzle blocks?
- What shape is the box?
- How many blocks will fit on the bottom layer?
- How many layers will there be?

The volume of a space figure is the number of **cubic units** it takes to fill the figure. There is a formula to find the volume of rectangular prisms.

Space Figure	Formula	Symbols	Picture
Rectangular prism	$V = l \times w \times h$, or lwh	V = volume l = length w = width h = height	
Cube	$V = s \times s \times s$, or s^3	V = volume s = side	

Examples

1. Find the volume of Kenji's box.

   $V = l \times w \times h$
   $V = 5 \times 4 \times 3$
   $V = 60$
   $V = 60$ cubic units

2. Find the volume of a cube with an edge of 5 in.

   $V = s^3$
   $V = 5^3$
   $V = 125$
   $V = 125$ in.^{3}

Try This Find the volume of the rectangular prism.

a.
7 in.
9 in.
4 in.

b.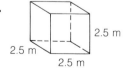
2.5 m
2.5 m
2.5 m

c. $l = 17$ ft
$w = 2$ ft
$h = 8$ ft

Exercises

Find the volume of the rectangular prism.

1.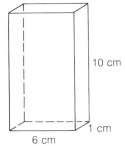

2.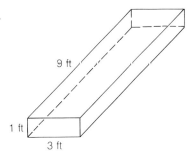

3.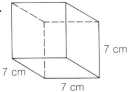

4.

5.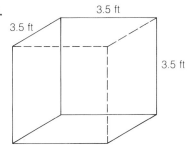

6.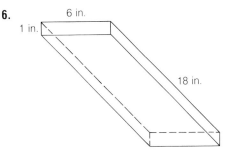

7. $l = 9$ cm
   $w = 3$ cm
   $h = 3$ cm

8. $l = 4$ m
   $w = 2.2$ m
   $h = 1.5$ m

9. $s = 6$ yd

10. $s = 4$ in.

11. $l = 10$ ft
    $w = 10$ ft
    $h = 5$ ft

12. $l = 2.5$ m
    $w = 1.8$ m
    $h = 1$ m

13. $s = 10$ in.

14. $s = 2.5$ cm

Data Hunt 15. Find the volume of your classroom.

Mental Math 16. The volume of a rectangular box is 24 ft³. Divide the box in half. How many cubic feet does each of the half boxes contain?

17. Ms. Yee has a cube that has a volume of 27 cubic units. What is the length of one edge of the cube?

Suppose 18. A cube has an edge of 1 unit. Suppose you doubled the length of each edge. What would happen to the volume of the cube?

AREA AND VOLUME

14-10 Volume of Cylinders

Objective
Find the volume of cylinders.

Take two pieces of $8\frac{1}{2}$ in. by 11 in. paper and some tape. Make two cylinders with a different height and radius. Do the cylinders have the same volume?

Understand the Situation

- Are the heights of the two cylinders the same?
- Are the lengths of the radius the same?
- How can you check your answer about the volumes?

You can check the volumes of the cylinders in more than one way. Fill one of the cylinders with dry rice or beans. Put the same amount of rice or beans into the other cylinder. Which has the greater volume?

You can use a formula to find the volume of a cylinder.

Space Figure	Formula	Symbols	Picture
Cylinder	$V = \pi r^2 \times h$, or $\pi r^2 h$	V = volume r = radius h = height	

Examples Find the volume of the cylinder. Use 3.14 for π.

1. $1\frac{5}{8}$ in., $8\frac{1}{2}$ in.
$V = \pi r^2 h$
$V \approx 3.14 \times 1.625 \times 8.5$
$V \approx 3.14 \times 2.64 \times 8.5$
$V \approx 70.64$
$V \approx 70.64$ in.^{3}

2. $r = 1\frac{1}{4}$ in.
$h = 11$ in.
$V = \pi r^2 h$
$V \approx 3.14 \times (1.25)^2 \times 11$
$V \approx 3.14 \times 6.25 \times 11$
$V \approx 53.97$
$V \approx 53.97$ in.^{3}

Try This Find the volume of the cylinder. Use 3.14 for π.

a. 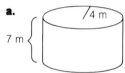 4 m, 7 m

b. 2 ft, 8 ft

c. $r = 2$ in.
$h = 1.5$ in.

d. $r = 3$ m
$h = 5$ m

CHAPTER 14

Exercises

Find the volume of the cylinder. Use 3.14 for π.

1.
2.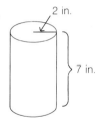
3.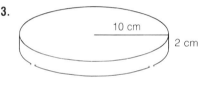

4.
5.
6.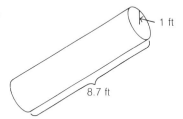

7. $r = 2$ in.
   $h = 4$ in.
8. $r = 4.5$ m
   $h = 10$ m
9. $r = 10$ m
   $h = 4.5$ m
10. $r = 3.2$ cm
    $h = 4$ cm
11. $r = 8$ ft
    $h = 12$ ft
12. $r = 1.2$ cm
    $h = 5$ cm
13. $r = 3$ in.
    $h = 2.5$ in.
14. $r = 5$ m
    $h = 3.2$ m

Mixed Applications

15. Wilma found an old pipe in a vacant lot. What was its volume?

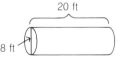

16. The Castillas have a small pool in their backyard. What is the volume of the pool?

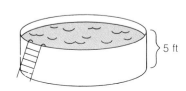

Data Hunt

17. If the Castillas fill this pool 4 ft deep, what would be the weight of the water in the pool? (Hint: Find how much one cubic foot of water weighs.)

Number Sense

18. The Wagger family wants to build a new storage tank that will hold twice as much as the one shown. Dan says the new tank should have twice the old radius and twice the old height. Lenny says that it should just have twice the old radius. Ken says that it should just have twice the old height. Who is right? Show why.

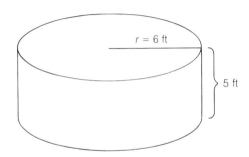

14-11 Volume of Pyramids and Cones

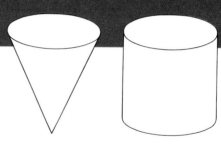

Objective
Find the volume of pyramids and cones.

Take a cone-shaped water cup. Use paper to make a cylinder that has the same base and the same height as the cone. How much greater is the volume of the cylinder?

Understand the Situation

- How do the heights of the cylinder and the cone compare?
- How do the areas of the bases of the cylinder and the cone compare?
- How can you compare the volumes of the cylinder and the cone?

You can compare the volumes using dry rice or beans. Fill the cone and use it to fill the cylinder. Do this until the cylinder is full. The volume of a cone is $\frac{1}{3}$ the volume of a cylinder with the same height and base area. The volume of a pyramid is equal to $\frac{1}{3}$ the volume of a prism with the same base area and height. You can use a formula to find the volume of a cone and a pyramid.

Space Figure	Formula	Symbols	Picture
Cone and Pyramid	$V = \frac{1}{3} \times B \times h$, or $\frac{1}{3}Bh$	V = volume B = area of the base h = height	(cone and pyramid with height h)

Examples Find the volume of the cone or pyramid. Use 3.14 for π.

1. $V = \frac{1}{3}Bh$

$V \approx \frac{1}{3} \times (3.14 \times 3^2) \times 5$

$V \approx 47.1$ cubic units

2. $r = 4$ cm; $h = 6$ cm

$V = \frac{1}{3}Bh$

$V \approx \frac{1}{3} \times (3.14 \times 4 \times 4) \times 6$

$V \approx 100.48$ cm^3

3. $V = \frac{1}{3}Bh$

$V = \frac{1}{3} \times (3 \times 3) \times 7$

$V = 21$ cubic units

4. $l = 4$ cm; $w = 5$ cm; $h = 6$ cm

$V = \frac{1}{3}Bh$

$V = \frac{1}{3} \times (4 \times 5) \times 6$

$V = 40$ cm^3

Try This Find the volume of the cone or pyramid. Use 3.14 for π.

a.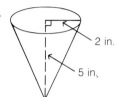

b. $r = 3$ in.
$h = 2$ in.

c.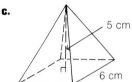

d. $l = 6$ ft
$w = 2$ ft
$h = 2$ ft

Exercises

Find the volume of the cone or pyramid. Use 3.14 for π.

1.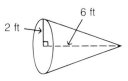
2.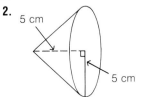
3.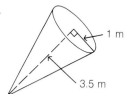

4. $r = 2$ cm
$h = 7$ cm

5. $r = 1.5$ ft
$h = 3$ ft

6. $r = 3$ m
$h = 4$ m

7. $r = 2$ in.
$h = 1.5$ in.

8.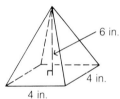
9.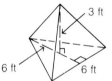
10.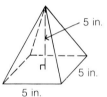

11. $l = 2$ in.
$w = 2$ in.
$h = 3$ in.

12. $l = 10$ ft
$w = 8$ ft
$h = 9$ ft

13. $l = 4.2$ cm
$w = 3$ cm
$h = 5$ cm

14. $l = 5$ m
$w = 5$ m
$h = 3$ m

Estimation 15. Leon had a paper cone with a radius of 1 in. and a height of 4 in. He guessed that it would hold 8 in.^{3} of water. How would you estimate this volume? Estimate your own answer.

Data Bank 16. The Great Pyramid in Egypt is a square pyramid. What is its volume in cubic feet? See the Data Bank on p. 493.

Suppose 17. Suppose that you doubled the height of a cone. What would that do to the volume of the cone? Suppose that you doubled the radius of a cone. What would that do to the volume? Suppose that you doubled both the height and the radius. What would that do to the volume?

18. Suppose that you doubled the length, width, and height of a pyramid. What would that do to the volume of the pyramid?

AREA AND VOLUME

14-12 Problem Solving: Using Logic

Objective
Use the logic of *and*, *or*, and *not*.

- **SITUATION**
- **DATA**
- **PLAN**
- **ANSWER**
- **CHECK**

The words *and*, *or*, and *not* play an important role in problem solving. Look at the meanings in the table. Try the following problems.

Word	Logical Meaning
and	Each part must be true.
or	At least one part must be true.
not	No part can be true.

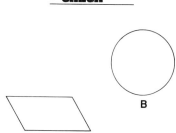

Examples Name the shapes.

1. All daps have four vertices or are circles. What shapes are daps?

   Shapes A, B, and D are daps. A and D are quadrilaterals. B is a circle.

2. All foms are triangles and have two sides that are congruent. What shapes are foms?

   Shape C is the only fom. Shape E is a triangle, but it does not have two congruent sides. Shape F has two congruent sides, but it is not a triangle.

3. Zads do not have four vertices. What shapes are zads?

   Shapes B, C, E, and F are zads. They do not have four corner points.

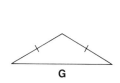

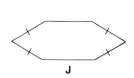

Try This Name the shapes.

a. All wips have segment sides and enclose a space. What shapes are wips?

b. All pils have a right angle or two congruent sides. What shapes are pils?

c. All cixes have an obtuse angle and are not triangles. What shapes are cixes?

Exercises

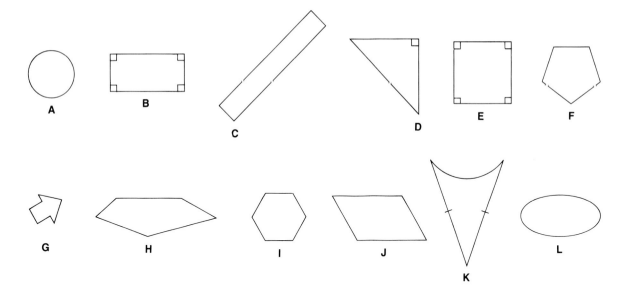

Name the shapes.

1. All baps have three vertices or no segment sides. What shapes are baps?
2. All guls have four vertices and at least one obtuse angle. What shapes are guls?
3. All dors have vertices and at least three segment sides. What shapes are dors?
4. All fuzes have a right angle or an obtuse angle. What shapes are fuzes?
5. All caxes have no vertices and no segment sides. What shapes are caxes?
6. All mifs have no segment sides and no right angles. What shapes are mifs?

Look at the following table. Use the information given to answer Exercises 7–12.

	Column 1	Column 2	Column 3	Column 4	Column 5
Row 1	Fred	Yoshi	June	Alan	Jos
Row 2	Shana	Julian	Lei	Rudra	Lei
Row 3	Alan	Lei	Shana	Sam	Anne
Row 4	Jos	Denny	Jos	Anne	Frank

7. Who shares a row and also shares a column with Julian?
8. Who shares a row and also shares a column with Fred?
9. Who does not share a row and does not share a column with Shana?
10. Who shares a row or shares a column with Alan?
11. Who shares a row and also shares a column with Alan?
12. Who shares a row or does not share a column with Lei?

AREA AND VOLUME **421**

14-13 Problem Solving: Developing Thinking Skills

Objective
Use problem-solving strategies.

- SITUATION
- DATA
- PLAN
- ANSWER
- CHECK

The volume of a crate is 140 ft³. The length of the crate is 3 ft longer than the width. The width is 1 ft shorter than the height. What are the length, width, and height of the crate?

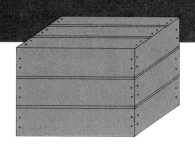

Exercises

Follow the *Thinking Actions*. Answer each question.

Try these **before** starting to write.

Thinking Actions Before
- Read the problem.
- Ask yourself questions to understand it.
- Think of possible strategies.

Try these **during** your work.

Thinking Actions During
- Try your strategies.
- Stumped? Try answering these questions.
- Check your work.

1. What is the volume of the crate?
2. What is the formula for finding the volume of the crate?
3. How much longer is the length than the width?
4. What strategy, or strategies, might help you to solve this problem?
5. Will making a table help you to solve this problem?
6. Can you guess the measurements and then check your guess?
7. If the width is 2 ft, what is the length?

8. Below are two correct solutions. Write the answer to the problem in a complete sentence. Name the strategies shown.

Solution 1

Width	Length	Height	Volume
2 ft ×	5 ft ×	3 ft =	30 ft³
3 ft ×	6 ft ×	4 ft =	72 ft³
4 ft ×	7 ft ×	5 ft =	140 ft³

Solution 2

W = 5 ft
L = 5 ft + 3 ft = 8 ft
H = 5 ft + 1 ft = 6 ft

5 ft × 8 ft × 6 ft = 240 ft³

too high

W = 4 ft
L = 4 ft + 3 ft = 7 ft
H = 4 ft + 1 ft = 5 ft

4 ft × 7 ft × 5 ft = 140 ft³

It checks.

Solve. Use one or more of the problem-solving strategies.

9. Mr. Liedholm saves coupons. Every day he cuts out 10 coupons. Every third day, he uses 3 coupons at the grocery store. Suppose he has 9 coupons on the first day. On what day will Mr. Liedholm have exactly 100 coupons?

Problem-Solving Strategies
- Guess and check.
- Choose an operation.
- Make an organized list.
- Act it out.
- Find a pattern.
- Write an equation.
- Use logical reasoning.
- Draw a picture or diagram.
- Make a table.
- Use objects.
- Work backward.
- Solve a simpler problem.

10. Toby bought 3 items at a department store. She paid with a $20 bill. The clerk gave her $0.80 change. What items did Toby buy?

Slippers	$7.75
Socks	$6.50
Handkerchief	$4.50
Sunglasses	$18.05
Belt	$6.95

11. All of the students in Ms. Holzer's class have brown, blond, red, or black hair. There are twice as many students with brown hair as with black hair. There are 4 more students with black hair than with blond hair. There are 4 more students with blond hair than with red hair. There are 2 students with red hair. How many students are in Ms. Holzer's class?

12. Jinn-Hwa, Leda, Van, and Dean play different instruments: an electric guitar, the drums, a piano, and a saxophone. Van does not play the saxophone. Jinn-Hwa cannot carry her instrument. Leda needs sticks to play her instrument. Who plays each instrument?

13. Nalini put olives around the top edges of a rectangular casserole. The casserole was 30 cm long and 20 cm wide. She put an olive every 5 cm. How many olives did she use?

14. Rory bought pencils at the stationery store. The pencils were on sale for 25¢ each. For every 3 pencils a customer bought, the store was giving away 1 pencil free. Rory spent $1.75. How many pencils did Rory get?

15. The area of a rectangular patio is 54 ft^2. The length is 3 ft longer than the width. What are the length and the width of the patio?

Solve It Another Way

16. Collin solved this problem using the strategy *make a table*. Solve this problem using a different strategy. *Problem:* Mrs. Simmons bought a piece of ribbon 12 ft long to make bows. Each bow needs a piece of ribbon 1 ft long. How many cuts will she have to make?

Feet	1	2	3	4	5	6	7	8	9	10	11	12
Cuts	0	1	2	3	4	5	6	7	8	9	10	11

AREA AND VOLUME

14-14 Enrichment: Spatial Visualization

Objective
See spatial objects as two-dimensional figures.

Mario looked at a box and wondered what a flattened model would look like.

A solid figure like the box is called a **polyhedron.** Each of the polygons that form the solid is called a **face.** The sides of the polygons are called **edges.**

Understand the Situation
- What kind of polygon are the top and bottom of the box?
- How many edges do you need to cut to make the box flat?
- How many rectangles does it take to make the box?

Examples

1. Look at the polyhedron. Draw a picture of a flattened model.

Count the number of faces. Imagine cutting along enough edges to flatten the figure.

2. Draw a picture of the polyhedron.

Exercises

Look at the polyhedron. Draw a picture of a flattened model.

1. 　　2. 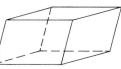　　3.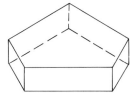

Draw a picture of the polyhedron.

4. 　　5.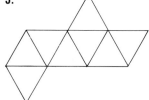

14-15 Calculator: Irregular Areas

Objective
Use a calculator to find irregular areas.

Frank Toledo owns a ranch. He grows crops to provide feed for livestock. What is the area of his wheat fields?

You can use a calculator to help you find the answer.

Example Find the area of Frank's wheat fields.
Divide the fields into sections.
Find the area of each section.
Area A: 3.2 km × 5.8 km = 18.56 km²
Area B: 1.4 km × 4.7 km = 6.58 km²
Add the areas.
The area of the wheat fields is 25.14 km².
18.56 km² + 6.58 km² = 25.14 km²

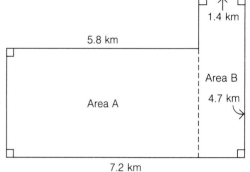

Exercises

Find the area of the farm. Use 3.14 for π.

1.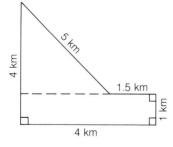

2.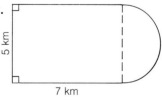

3.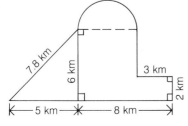

4.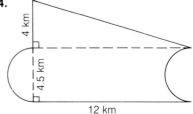

AREA AND VOLUME **425**

CHAPTER 14 REVIEW

Match.
1. volume of a cone
2. area of a parallelogram
3. area of a triangle
4. volume of a cylinder
5. area of a circle
6. volume of a rectangular prism

A. lwh
B. πr^2
C. s^3
D. $\pi r^2 h$
E. $\frac{1}{2}bh$
F. $\frac{1}{3}Bh$
G. $2\pi r$
H. bh

Choose the one that does not belong in each group. Explain why not.

7. A. B. 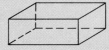 C. D.

8. A. B. C. 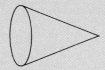 D.

9. A. B. 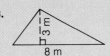 C. D.

10. A. B. C. D.

Write the missing word.
11. Area is to square units as volume is to _____ units.
12. Circle is to cone as polygon is to _____.
13. Square foot is to square yard as 1 is to _____.
14. Points are to vertices as segments are to _____.
15. The three statements about a cube are scrambled. Unscramble each to state a true relationship.
    A. length of a side = 24 in.^{2}
    B. surface area = 8 in.^{3}
    C. volume = 2 in.

CHAPTER 14 TEST

Solve.

1. How many blocks are in this stack?

Find the area. Use $\pi = 3.14$. Round to the nearest tenth.

2.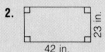

3.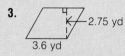

4.

5.

6.

7.

Find the area of each room in square yards.

8. living room: 24 ft by 36 ft
9. bedroom: 16 ft by 9 ft
10. den: 12 ft by 10 ft

Tell how many faces, edges, and vertices each figure has.

11.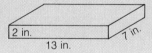

12.

13.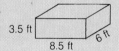

Find the surface area.

14.

15.

16.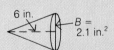

Find the volume. Round to the nearest whole number.

17.

18. (cylinder, 13 ft, 2.5 ft)

19. (cone, 6 in., $B = 2.1$ in.^{2})

20. (7.5 m, 2 m, 2 m)

Name the shapes.

21. All rets have at least 4 vertices and congruent sides. What shapes are rets?
22. All gams have no vertices or less than 4 sides. What shapes are gams?

A   B   C   D   E   F

AREA AND VOLUME

CHAPTER 14 CUMULATIVE REVIEW

1. Find the median for this data.
   number of hours per week worked by students at the library: 22, 15, 16.5, 12, 17, 14.5, 21, 18, 14, 16, 26, 24, 19, 23, 27, 11
   A. 22 h
   B. 18 h
   C. 16 h
   D. 17.5 h

2. Find the area of this parallelogram.
   $b = 3\frac{1}{4}$ ft
   $h = 2\frac{1}{2}$ ft
   A. $8\frac{1}{8}$ ft^2
   B. $6\frac{3}{4}$ ft^2
   C. $7\frac{1}{2}$ ft^2
   D. $8\frac{3}{4}$ ft^2

3. Add.
   45.09 + 39.89
   A. 84.88
   B. 74.88
   C. 84.98
   D. 74.98

4. Name the polygon. Tell if it is regular.
   A. dodecagon, not regular
   B. regular pentagon
   C. octagon, not regular
   D. rectangle, not regular

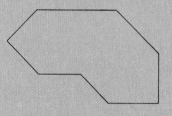

5. Find the area of a circle with $r = 6.6$ cm. Use 3.14 for π.
   A. ≈41.448 cm^2
   B. ≈20.724 cm^2
   C. ≈136.778 cm^2
   D. ≈120.70 cm^2

6. Find the mean of this data. hours spent listening to music in a week: 10, 16, 24, 18, 6, 9, 12, 22, 7, 13, 19, 14, 5, 7
   A. 12 h
   B. 16 h
   C. 14 h
   D. 13 h

7. Find the volume. Use 3.14 for π.
   A. ≈266.9 ft^3
   B. ≈907.46 ft^3
   C. ≈181.492 ft^3
   D. ≈264.8 ft^3

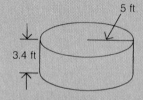

8. An art gallery has 7 photographs hung in a row on one wall. Find the number of permutations of 7 photographs.
   A. 5040
   B. 14
   C. 64
   D. 49

9. Name the space figure.
   A. cylinder
   B. rectangular prism
   C. triangular pyramid
   D. hexagonal pyramid

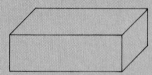

10. Zvi has 8 different hats. Two are blue. Five are yellow. One is red. He picks a hat without looking. What are the odds against Zvi picking a yellow hat?
    A. 1:5
    B. 2:5
    C. 3:5
    D. 5:3

11. All pils have a right angle or two congruent sides. What shapes are pils?
    A. A, B, C
    B. D
    C. A, B, C, D
    D. B, C, D

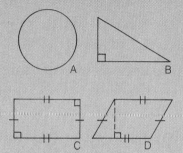

12. At Pablo's Potato Filler, you have these choices. How many different kinds of potatoes can you make with one choice of size, one topping, and one sauce?
    A. 28 potatoes
    B. 24 potatoes
    C. 33 potatoes
    D. 27 potatoes

Size	Topping	Sauce
Medium	Tomatoes	Cheese
Jumbo	Mushrooms	Tomato
	Olives	Chili
	Bacon bits	

13. Write the decimal as a fraction in lowest terms.
    0.72
    A. $\frac{72}{100}$
    B. 7.5
    C. $\frac{18}{25}$
    D. $\frac{9}{125}$

14. What shapes and how many of each shape do you need to build this space figure?
    A. 5 squares, 4 triangles
    B. 2 squares, 4 rectangles
    C. 4 triangles, 1 square
    D. 5 rectangles, 4 triangles

15. Subtract.
    $122\frac{3}{8} - 71\frac{5}{9}$
    A. $50\frac{59}{72}$
    B. $41\frac{3}{8}$
    C. $51\frac{13}{72}$
    D. 42

16. Change the unit.
    563 cm = _____ m
    A. 56.3
    B. 5.63
    C. 5630
    D. 0.563

17. Divide.
    $8\frac{1}{6} \div 2\frac{2}{3}$
    A. $4\frac{1}{9}$
    B. $4\frac{1}{4}$
    C. $2\frac{10}{93}$
    D. $3\frac{1}{16}$

18. Multiply.
    9.16 × 4.2
    A. 37.32
    B. 36.32
    C. 38.472
    D. 38.468

Chapter 15 Overview

Key Ideas
- Use the problem-solving strategy *work backward*.
- Name and order integers.
- Add, subtract, multiply, and divide integers.
- Check budgets.
- Estimate with integers.
- State what data is missing.
- Make group decisions.

Key Terms
- positive
- negative
- integer
- budget

Key Skills

Add or subtract.

1. $85 - 34$
2. $67 + 34$
3. $56 - 19$
4. $40 - 14$
5. $126 + 56$
6. $37 - 18$
7. $45 + 278$
8. $23 + 503$
9. $307 - 234$
10. $809 - 506$
11. $1548 - 372$
12. $3006 - 451$
13. $(34 + 28) - 16$
14. $(56 - 28) + 16$
15. $(52 - 18) - 15$
16. $(8 + 79) + 65$

Multiply or divide.

17. 16×4
18. 21×8
19. $21 \div 7$
20. 8×43
21. $24 \div 3$
22. $45 \div 5$
23. 105×7
24. 86×123
25. $64 \div 8$
26. 45×23
27. $60 \div 12$
28. $48 \div 4$
29. $23 \times 5 \times 12$
30. $125 \div 5$
31. $626 \div 4$
32. $12 \times 12 \times 34$

INTEGERS 15

15-1 Problem Solving: Learning to Use Strategies

Objective
Use the strategy work backward.

- SITUATION
- DATA
- **PLAN**
- ANSWER
- CHECK

You can solve many problems **working backward** from the answer to the question. A detective must often work backward to solve a crime.

Example Solve.

Owen collects stamps. He has 60 fewer Spanish stamps than French stamps. He has 70 stamps that are either Spanish or Irish. He has 30 Irish stamps. How many stamps does Owen have from each country? How many stamps does Owen have all together?

Solve by working backward.

Data in the Problem	*Work Backward*
There are 60 fewer Spanish stamps than French stamps.	$40 + 60 = 100$ French stamps ↑
↓	
70 stamps are either Spanish or Irish.	$70 - 30 = 40$ Spanish stamps ↑
↓	
There are 30 Irish stamps.	30 Irish stamps

Owen has 30 Irish stamps, 40 Spanish stamps, and 100 French stamps. He has 170 stamps all together.

Try This Solve.

Pam Curry owns a pizza parlor. One Monday night, she found that there were 3 times as many orders for sausage pizza as there were for ham pizza. The number of orders for ham pizza was twice the number for pepperoni pizza. There were 20 orders for pepperoni pizza. How many orders did she have for each type of pizza? How many orders did she have all together?

Exercises

Solve. Use one or more of the problem-solving strategies.

Problem-Solving Strategies
- Guess and check.
- Choose an operation.
- Make an organized list.
- Act it out.
- Find a pattern.
- Write an equation.
- Use logical reasoning.
- Draw a picture or diagram.
- Make a table.
- Use objects.
- **Work backward.**
- Solve a simpler problem.

1. Ana Cortez manages the Red River Diner. One morning, she found that 4 times as many people had whole milk as had nonfat milk on their cereal. Only half as many people had 2% milk as had whole milk. Eight people had 2% milk. How many people had each type of milk on their cereal? How many people had milk with cereal?

2. Su-Lin had $2 to spend on school supplies. What are the different ways she could spend her money? Use the price list shown.

   School Supplies
   Erasers 50¢ each
   Pencils 25¢ each
   Pens $1 each

3. In Bay City High School's last basketball game, Swifty scored 10 more points than Brad. Brad scored 6 points less than Manuel. Manuel scored 10 points less than Butch. Butch scored 11 times as many points as Palmer. Palmer scored only 3 points. How many points did each player score?

4. How many games do you need to finish a single-elimination softball tournament with 20 teams?

5. How many different kinds of pizza with 3 toppings each can you order?

   Pizza Palace Special
   Your Choice of Any Three Toppings
   - olives
   - ham
   - sausage
   - mushroom

6. The freshman officers at Sonora High School asked students what color class T-shirt they liked best. Twice as many students liked purple as liked pink. One fourth as many students liked grey as liked purple. Six more students liked black than liked grey. Fifteen students liked grey. How many students liked pink?

Write Your Own Problem

7. Write a problem that you could solve working backward. Use data from this chart.

   Ardena's Stamps
   15 German
   10 Japanese
   5 Australian

8. Write a story problem that this picture would help you solve. Then solve the problem.

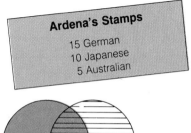

15-2 Positive and Negative Numbers

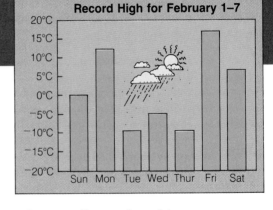

Objective
Name and order integers.

Positive numbers are numbers above 0 on this scale. **Negative numbers** are numbers below 0 on this scale. Read $^-7$ as *negative seven*.

Understand the Situation

- What is the lowest recorded temperature?
- What is the highest recorded temperature?

Opposites are numbers like 3 and $^-3$. They are the same distance from 0 but on opposite sides of it. All whole numbers and their opposites make up the set of **integers.**

Examples Write an integer to represent the statement.

1. Death Valley is 282 ft below sea level. $^-282$
2. The Texas Commerce Tower is 1002 ft high.   1002
3. Randy overdrew her account by $60. $^-60$

Try This Write an integer to represent the statement.

a. The stunt man fell 416 ft. **$^-416$**   b. The temperature was 15° below zero.

One integer is greater than another integer if it is farther to the right, or above, on the number line.

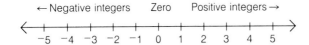

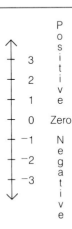

Examples Compare. Use < or >.

4. $^-7 \square ^-3$ $^-7 < ^-3$ $^-3$ is to the right of $^-7$ on the number line.

5. $2 \square ^-4$ $2 > ^-4$ $^-4$ is to the left of 2 on the number line.

Try This Compare. Use < or >.
c. ⁻5 ☐ ⁻13   d. 2 ☐ ⁻7   e. 0 ☐ 9   f. ⁻4 ☐ ⁻1

Exercises

Write an integer to represent each.
1. a bank deposit of $425
2. 456 ft below sea level
3. 5 min before lift-off
4. 15 points ahead in a basketball game
5. 6 h ago
6. a loss of 10 yd
7. 40 min after leaving
8. a gain of 16 lb
9. 40° below zero
10. a stock decline of 5 points

Compare. Use < or >.
11. ⁻6 ☐ ⁻9
12. 0 ☐ ⁻8
13. 9 ☐ ⁻2
14. ⁻34 ☐ ⁻12
15. ⁻198 ☐ ⁻200
16. 7 ☐ ⁻12
17. ⁻3201 ☐ ⁻34
18. ⁻900 ☐ ⁻899
19. 5678 ☐ 5099
20. 59 ☐ 0
21. ⁻99 ☐ 99
22. ⁻63 ☐ ⁻75
23. ⁻89 ☐ ⁻54
24. ⁻6 ☐ ⁻1
25. 0 ☐ ⁻90
26. ⁻18 ☐ ⁻81
27. 12 ☐ ⁻43
28. 2 ☐ ⁻100

Number Sense

Arrange these integers in order from least to greatest.
29. 9870, ⁻198, 78, ⁻3201, 0, ⁻87
30. 12, ⁻90, 200, 89, ⁻199, 32
31. 67, 42, ⁻981, ⁻41, 950, 39
32. ⁻79, 54, ⁻2, ⁻97, 78, 6
33. ⁻103, 80, 914, ⁻72, ⁻85, 611
34. ⁻1, ⁻26, 38, ⁻47, 5, 12
35. 20, ⁻31, 57, ⁻62, ⁻1, 2
36. 0, 10, ⁻64, ⁻3, ⁻10, 3

Group Activity

37. List the ways in which people use negative integers in your school. Check such areas as bookkeeping, science, and sports.

Calculator

38. Calculators represent negative numbers in different ways. Enter each of the following keystrokes on your calculator. How does your calculator represent negative numbers?
    A. [−] [3]   Does the display read ⁻3?
    B. [3] [−]   Does the display read ⁻3?
    C. [±] [3]   Does the display read ⁻3?
    D. [3] [±]   Does the display read ⁻3?

INTEGERS

15-3 Adding Integers

Objective
Add integers.

In the third quarter of a football game, Ted gained 3 yd on one play and lost 6 yd on the next play. What number represents the total number of yards Ted gained or lost on the two plays?

Understand the Situation
- Does the word *gained* suggest a positive or negative direction?
- Does the word *lost* suggest a positive or negative direction?
- What two integers do you want to add?

You can use a number line to add integers.

Example
1. Add. Use a number line. $3 + (^-6)$

Find the first addend on the number line. → If the second addend is positive, move that many units to the right. If the second addend is negative, move that many units to the left. → The number at which you stop is the sum.

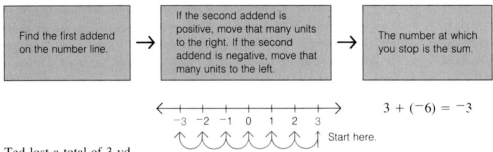

$3 + (^-6) = ^-3$

Ted lost a total of 3 yd.

Try This
Add. Use a number line.
a. $^-5 + 7$ b. $3 + (^-3)$ c. $^-4 + (^-2)$ d. $3 + 5$

The sum of two opposite numbers is always zero. $4 + (^-4) = 0$

Here are the rules for adding integers.

Rule		Examples
Numbers with the same sign	Add the number of units. Use the same sign.	$3 + 3 = 6$ $^-3 + ^-3 = ^-6$
Numbers with different signs	Subtract the number of units. Use the sign of the number farther from zero.	$6 + ^-3 = 3$ $^-6 + 3 = ^-3$

CHAPTER 15

Examples Add.

2. $^-6 + (^-8)$
$^-6 + (^-8) = ^-14$ $^-6$ and $^-8$ have the same sign. Add. $6 + 8 = 14$
Use the same sign. The sum is negative.

3. $3 + (^-5)$
$3 + (^-5) = ^-2$ 3 and $^-5$ have different signs. Subtract. $5 - 3 = 2$
$^-5$ is farther from 0 than 3. Use the negative sign. The sum is negative.

Try This Add.

e. $6 + 13$ **f.** $^-9 + 5$ **g.** $7 + (^-11)$ **h.** $^-12 + (^-6)$

Exercises

Add. Use a number line.

1. $6 + (^-4)$ **2.** $7 + (^-3)$ **3.** $^-5 + 8$ **4.** $^-8 + (^-3)$
5. $^-8 + (^-4)$ **6.** $2 + (^-2)$ **7.** $^-4 + 5$ **8.** $^-9 + 6$
9. $8 + 3$ **10.** $10 + (^-8)$ **11.** $^-5 + 5$ **12.** $3 + (^-2)$

Add.

13. $^-17 + 9$ **14.** $^-19 + 5$ **15.** $^-20 + 20$ **16.** $^-17 + 8$
17. $^-15 + (^-12)$ **18.** $^-21 + (^-5)$ **19.** $^-31 + (^-12)$ **20.** $^-19 + (^-11)$
21. $45 + (^-34)$ **22.** $36 + (^-21)$ **23.** $21 + (^-67)$ **24.** $14 + (^-54)$
25. $t = ^-37 + (^-16)$ **26.** $k = ^-62 + (^-45)$ **27.** $23 + (^-23) = f$
28. $^-43 + 280 = g$ **29.** $^-67 + 189 = b$ **30.** $a = ^-54 + 320$
31. $^-102 + (^-45) = d$ **32.** $^-225 + 59 = k$ **33.** $(^-67) + 56 + (^-11)$
34. $8 + (^-97) + (^-45)$ **35.** $(^-89) + 46 + 89$

Mixed Applications

36. Jackie owed Lynn $45. On Friday she borrowed $15 more from her. In the next two weeks, she paid Lynn $12 on each of three different days. How much does Jackie owe Lynn now?

37. In a football game, Jiro gained 54 yd in the first quarter, lost 16 yd in the second quarter, lost 12 yd in the third quarter, and gained 41 yd in the fourth quarter. What was his total yardage for the game?

Mixed Skills Review Find each answer.

38. $56 + 57$ **39.** $108 - 91$ **40.** 54×5 **41.** $83 + 299$

42. $120 \div 15$ **43.** 17×11 **44.** $52 \div 13$ **45.** $231 - 177$

INTEGERS

15-4 Subtracting Integers

Objective
Subtract integers.

Han is building a fence around his yard. He wants each fence post to be 6 ft above ground and 2 ft below ground. How long should each post be?

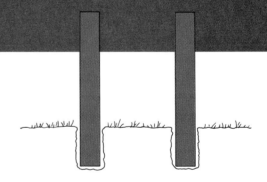

Understand the Situation

- What integer can you use to show the length of the post above the ground?
- What integer can you use to show the length of the post below the ground?

You can use a number line to find the difference.

Example Subtract. Use a number line.

1. $6 - (^-2)$

| Find the **minuend** on the number line. | → | Add the opposite of the **subtrahend**. | → | The number at which you stop is the difference. |

$6 - (^-2)$ $\quad\quad 6 - (^-2)\;\; 6 + 2 \quad\quad 6 - (^-2) = 8$

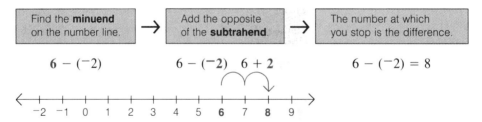

Each post should be 8 ft long.   ↑ Start here.

Try This Subtract. Use a number line.

a. $5 - 3$ **b.** $7 - (^-2)$ **c.** $^-3 - 2$ **d.** $^-8 - (^-6)$

Here is the rule for subtracting integers.

Rule	Examples
Find the opposite of the subtrahend. Add the opposite to find the difference.	$3 - (^-5) = 3 + 5 = 8$ $^-4 - 6 = ^-4 + {^-6} = {^-10}$

Example Subtract.

2. $^-9 - (^-7)$
   $^-9 + 7 = {^-2}$ The opposite of $^-7$ is 7.

438   CHAPTER 15

Try This Subtract.

e. $8 - (^-6)$ f. $^-7 - 2$ g. $^-12 - (^-3)$ h. $17 - 32$

Exercises

Subtract. Use a number line.

1. $5 - 7$
2. $6 - 10$
3. $3 - 12$
4. $7 - 13$
5. $^-8 - 6$
6. $^-9 - 4$
7. $^-1 - 9$
8. $^-4 - 12$
9. $^-4 - (^-5)$
10. $^-7 - (^-6)$
11. $^-8 - (^-10)$
12. $^-12 - (^-15)$

Subtract.

13. $^-10 - 10$
14. $^-9 - 9$
15. $^-23 - 23$
16. $^-87 - 87$
17. $17 - (^-17)$
18. $32 - (^-32)$
19. $43 - (^-43)$
20. $18 - (^-18)$
21. $12 - 19 = n$
22. $z = 23 - 34$
23. $a = 89 - 109$
24. $76 - 89 = p$
25. $0 - 12 = b$
26. $t = 0 - 41$
27. $76 - (^-98) = v$
28. $c = ^-23 - (^-23)$

29. Find the difference between $^-46$ and $^-46$.

30. Take away 208 from 98.

Evaluate.

31. $^-34 - y$ for $y = 0$
32. $z - (^-123)$ for $z = 122$
33. $^-69 - u$ for $u = ^-31$
34. $^-84 - v$ for $v = ^-58$

Mixed Applications

35. On a winter day, Thelma saw on TV that the temperature in the suburbs was $^-11°C$. Downtown the temperature was $^-6°C$. What was the temperature difference?

36. The deepest point in the Atlantic Ocean is the Puerto Rico Trench. It is 8605 m below sea level. The deepest point in the Pacific Ocean is the Mariana Trench. It is 10,924 m below sea level. How much higher is the bottom of the Puerto Rico Trench?

Calculator Use your calculator to find each answer. Use the ± key.

37. $3 + (^-15) + 67 - 18$
38. $^-34 + (^-18) - 16 + 123$
39. $45 + (^-17) - (^-32) + 81$
40. $^-56 + 23 + (^-18) - 7$
41. $^-67 + 82 - 23 - 45$
42. $^-12 + (^-12) - (^-12) + 12$
43. $^-12 + 33 + (^-49) - 6 + 82$
44. $1 + (^-2) + (^-4) + 8 - (^-12)$
45. $^-94 - 16 + (^-5) - 10 - (^-57)$
46. $28 - 43 - (^-17) + 8 - 31$

Data Hunt

47. Find the difference between the highest and lowest record temperatures in your state.

15-5 Consumer Math: Checking a Budget

Objective
Check a budget.

- SITUATION
- DATA
- PLAN
- ANSWER
- CHECK

Nick set a budget for his week's income. He babysits for a neighbor. He planned to make $10 a day for five days. Was Nick under or over his budget? How much?

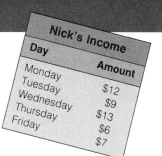

Understand the Situation
- How much did Nick plan to make each day?
- Did he make more or less than he planned on Tuesday?

You can use integers to show how much Nick was under or over his budget.

Example Find how much Nick was under or over his budget.

Make a table to show how much Nick was under or over each day's goal.

Day	Monday	Tuesday	Wednesday	Thursday	Friday
Difference	12 − 10 = 2	9 − 10 = ⁻1	13 − 10 = 3	6 − 10 = ⁻4	7 − 10 = ⁻3

Add the integers. $2 + (^-1) + 3 + (^-4) + (^-3) = ^-3$

Nick was $3 under his weekly budget.

Try This Carla made a weekly budget. She wanted to spend only $5 a day. She spent $2, $6, $1, $5, $7, $6, and $3 for the week. Find how much Carla was under or over her budget.

Exercises

1. This table shows Michelle's income for 5 days of yard work. She planned to make $24 a day. Was she under or over her budget? How much?

Day	1	2	3	4	5
Income	$19	$27	$26	$28	$22

2. This table shows Tyson's expenses for one week. He planned to spend only $6 a day. Was he under or over his budget? How much?

Days	1	2	3	4	5	6	7
Expenses	$4	$7	$5	$2	$6	$8	$9

15-6 Estimating Sums

Objective
Estimate sums.

Chandra keeps track of records bought and sold for a used-record store. She tries to keep a mental total of the increase or decrease of the store's cash. How much did the store's cash amount increase or decrease for these entries?

Records	Bought	Sold
3 Beatles	$6	$12
1 Miles Davis	$2	—
1 U2	—	$5
2 Bach	—	$9
4 Rolling Stones	$11	—

Understand the Situation
- What integer can you use to represent the entry for the Miles Davis album?
- What operation can you use to find the amount of increase or decrease in cash?

Example Estimate the amount of increase or decrease in cash for the record store.

| Find numbers that are almost opposites. | → | The sum of opposite numbers is zero. Cross out the pairs of numbers whose sum is about zero. | → | Add the remaining integers. |

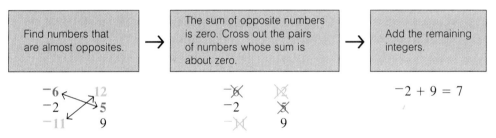

$$-2 + 9 = 7$$

The amount of increase in cash was about $7.

Try This In one hour, the record store bought records for $4, $7, $9, and $13. They sold records for $12, $11, and $5. Estimate the amount of increase or decrease in cash.

Exercises

Estimate.
1. $^-14 + 22 + 25 + 15 + (^-21) + 19$
2. $33 + (^-42) + 36 + 32 + (^-34) + (^-35)$
3. $^-109 + 59 + 82 + (^-60) + 110$
4. $212 + (^-200) + (^-222) + 201 + 228$
5. $500 + (^-498) + (^-501) + 489 + 508$
6. $851 + (^-849) + (^-839) + 840 + (^-822)$

INTEGERS

15-7 Multiplying Integers

Objective
Multiply integers.

The elevator in Lorene's office building rises and falls at the rate of 10 ft per second. How many feet does the elevator fall in 3 s?

10 ft, 1 s
10 ft, 1 s
10 ft, 1 s

Understand the Situation

- What integer can you use to represent the distance the elevator falls in one second?
- For how many seconds has the elevator traveled?
- What operation do you need to find the answer?

You can multiply to find how far the elevator has traveled.

Examples

1. Find how far the elevator has traveled.

   Look at the above diagram. The elevator has traveled down 3 groups of 10 ft, or 30 ft.

   $3 \times (^-10) = ^-30$

2. Find how far the elevator will fall in 0 s. The elevator travels down 0 ft in 0 s.

   $0 \times (^-10) = 0$

3. Look at this pattern. What is $^-2 \times ^-1$?

   As each factor decreases by 1, each product increases by 2.

   $^-2 \times ^-1 = 2$

   $^-2 \times 3 = ^-6$
   $^-2 \times 2 = ^-4$
   $^-2 \times 1 = ^-2$
   $^-2 \times 0 = 0$
   $^-2 \times ^-1 = \square$

Here are the rules for multiplying with integers.

Rule	Examples
The product of two integers with different signs is negative.	$3 \times (^-5) = ^-15$ $^-2 \times 4 = ^-8$
The product of two integers with the same sign is positive.	$7 \times 2 = 14$ $^-5 \times (^-2) = 10$
The product of any integer and zero is zero.	$0 \times 6 = 0$ $^-9 \times 0 = 0$

Try This Multiply.

a. $^-3 \times (^-8)$
b. $16 \times (^-4)$
c. $^-12 \times 6$
d. $^-21 \times (^-8)$

Exercises

Multiply.

1. $9 \times (^-8)$
2. $7 \times (^-7)$
3. $8 \times (^-3)$
4. $6 \times (^-9)$
5. $^-10 \times 8$
6. $^-9 \times 8$
7. $^-11 \times 7$
8. $^-3 \times 12$
9. $^-120 \times 5$
10. $^-23 \times 8$
11. $^-32 \times 6$
12. $^-16 \times 9$
13. 10×34
14. $0 \times {}^-56$
15. $27 \times (^-17)$
16. $42 \times (^-21)$
17. $9 \times (^-7) = t$
18. $n = {}^-8 \times 6$
19. $^-7 \times (^-4) = r$
20. $s = {}^-8 \times (^-4)$
21. $m = {}^-6 \times (^-12)$
22. $^-7 \times (^-20) = b$
23. $^-47 \times 34 = q$
24. $107 \times (^-5) = j$
25. Find the product of $^-87$ and $^-33$.
26. Find what $^-508$ times 0 equals.

Evaluate.

27. ^-63r for $r = 112$
28. $s(^-98)$ for $s = 45$
29. $78m$ for $m = {}^-26$
30. ^-102n for $n = {}^-307$

Mixed Applications

31. An elevator was traveling down at the rate of 20 ft per second. How far would the elevator travel in 6 s?
32. An elevator was traveling down at the rate of 20 ft per second. How far would the elevator travel in 12 s?
33. A scuba diver went to a depth of 220 ft below sea level. A submarine went 8 times as deep. How deep did the submarine go?
34. Lynette dug a hole that was 15 in. deep. A tractor dug a hole that was 6 times as deep. How deep was the hole that the tractor dug?

Number Sense

Tell what the sign of the product would be if these were the signs of the factors.

35. $+ + {}^- + {}^-$
36. $+ {}^- {}^- {}^-$
37. $+ {}^- + {}^- +$
38. $^- {}^- {}^- + {}^-$
39. $^- + + {}^- + {}^-$
40. $^- {}^- + {}^- + {}^- + +$
41. $+ + {}^- {}^- + + {}^- +$

Write Math

42. Write a rule for finding the sign of a product without multiplying first.

Mental Math

Find each answer. Use mental math.

43. $^-40 + 120 + 60 + (^-30) + (^-20)$
44. $^-20 + 50 + (^-100) + 60 + 20$
45. $300 + (^-120) + (^-80) + 250 + (^-50)$
46. $^-25 + (^-175) + 130 + 50 + (^-180)$
47. $50 + 100 + (^-20) - 30 + 200$
48. $5 \times (^-2) + (10 - 5) - (^-20) + 10$

INTEGERS

15-8 Dividing Integers

Objective
Divide integers.

Cass, Susan, and Fred lost a total of $75 when the Star Band concert tour went bankrupt. Each lost the same amount of money. How much money did each lose?

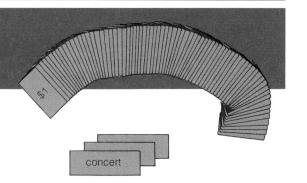

Understand the Situation
- How many people lost money?
- What number can you use to represent the total amount of money lost?
- What operation do you need to find the answer?

You can divide to find the answer.

Examples

1. How much did Cass, Susan, and Fred each lose?

   Divide the money into an equal number of groups.

   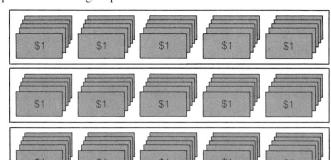

   $^-75 \div 3 = {}^-25$

   Each person lost $25.

2. Divide. $^-25 \div ({}^-5)$

   If $^-25 \div ({}^-5) = x$,
   then $\quad {}^-5 \times x = {}^-25$.
   $\qquad\qquad x = 5$
   So, $^-25 \div ({}^-5) = 5$.

Here are the rules for dividing integers.

Rule	Examples
The quotient of two integers with the same sign is positive.	$15 \div 5 = 3$ $^-10 \div ({}^-5) = 2$
The quotient of two integers with different signs is negative.	$20 \div ({}^-5) = {}^-4$ $^-12 \div 4 = {}^-3$

Try This

Divide.

a. $42 \div ({}^-7)$ b. $^-42 \div ({}^-7)$ c. $^-72 \div ({}^-9)$ d. $55 \div ({}^-5)$

Exercises

Divide.

1. $36 \div (^-6)$
2. $28 \div (^-7)$
3. $26 \div (^-2)$
4. $48 \div (^-8)$
5. $72 \div (^-12)$
6. $45 \div (^-5)$
7. $27 \div (^-9)$
8. $63 \div (^-7)$
9. $^-56 \div (^-8)$
10. $^-32 \div 16$
11. $^-64 \div 8$
12. $^-49 \div 7$
13. $^-88 \div 11$
14. $^-84 \div 7$
15. $^-45 \div 3$
16. $^-40 \div 5$
17. $80 \div (^-20) = r$
18. $e = ^-34 \div (^-2)$
19. $^-39 \div (^-3) = w$
20. $l = ^-56 \div (^-7)$
21. $f = ^-121 \div 11$
22. $0 \div (^-7) = h$
23. $v = 124 \div (^-4)$
24. $^-153 \div (^-9) = i$

25. What is $^-189$ divided by 21?
26. Divide $^-104$ by $^-13$.

Evaluate.

27. $^-126 \div b$ for $b = ^-18$
28. $c \div ^-42$ for $c = 0$
29. $\frac{q}{-5}$ for $q = 65$
30. $\frac{-256}{k}$ for $k = ^-16$

Mental Math

Solve. Use mental math.

31. The 4 members of the Zoom-Boom Band spent $160 on outfits. Each member paid the same amount. How much money did each member pay?

32. Charmaine, Jack, Val, Fran, and Teri spent $150 on tickets for a concert. Each ticket was the same price. How much money did each person spend?

Problem Solving

33. The temperature on Wednesday was 5° higher than the temperature on Tuesday. Thursday's temperature was 10° lower than Wednesday's temperature. The temperature on Friday was 2° higher than the temperature on Thursday. Friday's temperature was 62°F. What was the temperature each day?

Mixed Skills Review

Find each answer.

34. $678 + 490$
35. $^-12 + 24$
36. $509 - 388$
37. $4 \times (^-11)$
38. $2400 \div 60$
39. $^-38 - 24$
40. $37 + (^-13)$
41. $^-36 \div (^-9)$

Choose a calculation method that you would use to solve each problem. Tell why.

42. The attendance for three concerts was 23,405, 22,672, and 25,993. What was the total attendance for the three concerts?

43. Gil said he could eat 5 pieces of pizza. Molly said she could eat 3 pieces. Bea said she could eat 4 pieces. How many pizzas should they buy if each pizza is cut into 6 pieces?

15-9 Exploring Operations with Integers

Objective
Explore operations with integers.

Meki and Maria worked on the statistics for the football team. They found the average total yards gained by the quarterbacks per game. There were 10 games in the season. Here is how they found the average total yards:

Season Totals		
Yards	Joe	Mike
Rushing	565	−256
Passing	1762	210

Joe
```
   565
 +1762
 ─────
  2327
```

```
        232.7
    ┌────────
 10 │ 2327.0
      20
      ──
       32
       30
       ──
        27
        20
        ──
         70
         70
         ──
          0
```

Mike
```
  −256
 + 210
 ─────
   −46
```

```
       −4.6
    ┌──────
 10 │ −46.0
      40
      ──
       60
       60
       ──
        0
```

Exercises

1. Why did Meki and Maria first add the number of yards rushing and passing?
2. What would these sums represent?
3. Why is Mike's sum a negative number?
4. Why did they divide the sum for each player by 10?
5. What does the quotient in each problem represent?
6. Find the average: 17-yd loss, 9-yd gain, 4-yd gain, 8-yd loss, 7-yd gain.
7. Find each player's average rushing gain per game and average passing gain per game.

Estimation Estimate. Tell if each answer seems reasonable or not reasonable.

8. $-89.2 + (-23.4) = -11.26$
9. $56\frac{3}{4} \times 2 = -113.5$
10. $40.48 \div (-2.3) = -17.6$
11. $-3\frac{3}{4} + 2\frac{1}{2} = -6\frac{1}{4}$
12. $-1500 \times 200 = -300{,}000$
13. $3.128 \times (-10) = -31.28$

Calculator Find each answer. Round each product to the nearest tenth.

14. $-23.56 + 37.87$
15. -54.29×2.2
16. $37.45 - (-19.71)$
17. $-55.02 \div 13.1$
18. $-43.51 - (-13.78)$
19. $36.01 \times (-4.4)$
20. $226.32 \div (-18.4)$
21. $-48.39 + 37.81$
22. $-23.5 \times (-6.3)$

Hanna worked in a quality control position in a factory. Her job was to measure the length of each pipe to find if it was over or under the correct measurement. She found the following measurements for 10 pipes.

-0.01 cm $+0.20$ cm -0.17 cm $+0.14$ cm -0.09 cm
-0.12 cm -0.23 cm $+0.47$ cm -0.22 cm -0.17 cm

23. What was the average error for the ten objects?
24. What was the greatest error for pipes over the correct measurement?
25. What was the greatest error for pipes under the correct measurement?

Number Sense The sum and product of two integers is shown. Find the two integers.

26. ☐ ☐
    sum: 1
    product: -12
27. ☐ ☐
    sum: -7
    product: 10
28. ☐ ☐
    sum: -5
    product: -24

Mixed Skills Review Estimate.

29. $598 + 322$
30. 54×8
31. $2413 \div 39$
32. $723 - 519$
33. $34 + 33 + 31 + 29 + 52 + 49$
34. $98 + 149 + 101 + 151 + 99 + 102$

Find each answer. Use mental math.

35. $49 + 2$
36. 50% of 84
37. $12\frac{3}{4} + 8\frac{1}{4}$
38. 560×5
39. 25% of 100
40. $10\frac{2}{3} - 6\frac{1}{3}$
41. $\$15.25 - \$4.50$
42. $1800 \div 60$

Compare. Use $<$, $>$, or $=$.

43. $3.002 \;\square\; 30.01$
44. $\frac{3}{4} \;\square\; 0.75$
45. $66\% \;\square\; \frac{2}{3}$
46. $0.22 \;\square\; \frac{2}{9}$
47. $\frac{4}{5} \;\square\; \frac{12}{15}$
48. $\frac{1}{6} \;\square\; 17\%$
49. $7.2\% \;\square\; 0.72$
50. $0.07 \;\square\; 0.11$

15-10 Problem Solving: Missing Data

Objective
Decide what data is missing in a problem.

- SITUATION
- **DATA**
- PLAN
- ANSWER
- CHECK

Petroil Oil Company drilled an oil well 1475 ft deep. They estimated that the oil deposit was $^-1148$ ft. Did Petroil drill the well the correct depth?

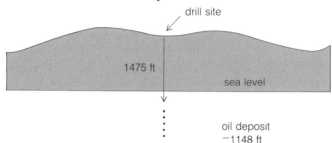

Some problems may have missing data so that you are unable to solve them.

Example Solve. If the problem has missing data, state what data is missing.

Did Petroil drill the well the correct depth?

Decide what data you need to solve the problem.

elevation of drill site − depth of oil deposit = well depth

Substitute the known data. Is any data missing?

elevation of drill site − ($^-1148$ ft) = 1475 ft

The elevation of the drill site is missing.

Try This Solve. If the problem has missing data, state what data is missing.

Another drill site was 179 ft above sea level. Petroil drilled 1850 ft. Did they reach an oil deposit?

Exercises

Solve. If the problem has missing data, state what data is missing.

1. Petroil drilled a well 1400 ft deep to an oil deposit. The drill site was located 30 mi outside Stanton. What was the depth of the oil deposit below sea level?

2. A drill site was 243 ft above sea level. An oil deposit was $^-1000$ ft. How far did Petroil have to drill to reach oil?

3. The longest well that Petroil had to drill was 3000 ft. How much greater is this than the shortest well that Petroil had to drill?

4. Petroil estimated that an oil deposit at one drill site was between $^-613$ ft and $^-748$ ft. How far would Petroil have to drill to reach oil?

448 CHAPTER 15

15-11 Problem Solving: Group Decisions

Objective
Make group decisions.

- SITUATION
- DATA
- PLAN
- ANSWER
- CHECK

Ned has trouble managing his budget. He goes over it each week. Here are his expenses and budget for the last four weeks. See what you can do to help Ned manage his budget.

Item	Weekly Budget	Week 1	Week 2	Week 3	Week 4
Clothes	$ 5.00	$ 7.23	$ 4.16	$18.25	$14.60
Food	4.00	2.23	6.18	3.76	4.15
Entertainment	11.00	18.25	10.00	14.75	16.24
Transportation	5.00	2.50	4.75	12.75	3.45
Savings	5.00	0.00	0.00	0.00	0.00
Tapes	9.00	8.00	3.00	9.00	9.00
Other	6.00	15.79	18.22	10.15	2.44

Facts to Consider
- You only have Ned's records to study.
- You should not have Ned try to earn more money.
- Ned's parents will not lend him any money.

Plan and Make a Decision
- Describe what your group must do.
- How would you find if Ned is within his budget or not?
- How could a calculator help?
- How could graph paper help?
- Help Ned manage his budget. What changes would you suggest?

Share Your Group's Decision
- Present your plan to the class.
- Did other groups find a similar solution? How did the solutions differ?

Suppose
- Ned found a part-time job. He made an extra $10 per week. What changes would you make in his budget?

Mixed Skills Review

Find the answer.

1. $36 \div (^-9)$
2. $499 + 625$
3. $3000 - 232$
4. 36×39
5. 2.3×100
6. $\$208 + \$110$
7. $72.38 \div 100$
8. $15 \times (^-3)$
9. $4.002 - 0.9$
10. 5.6×2.05

15-12 Enrichment: Wind Chill Factor

Objective
Use a wind chill table.

A combination of cold and wind makes you feel colder than the temperature alone. The effect of wind on the temperature is called the **wind chill factor**. The Wind Chill table shows what the temperature feels like for a given temperature and wind speed.

Wind Speed in mph	Wind Chill — Temperature in Degrees Fahrenheit									
	35	30	25	20	15	10	5	0	−5	−10
0	35	30	25	20	15	10	5	0	−5	−10
5	33	27	21	16	12	7	0	−5	−10	−15
10	22	16	10	3	−3	−9	−15	−22	−27	−34
15	16	9	2	−5	−11	−18	−25	−31	−38	−45
20	12	4	−3	−10	−17	−24	−31	−39	−46	−53
25	8	1	−7	−15	−22	−29	−36	−44	−51	−59
30	6	−2	−10	−18	−25	−33	−41	−49	−56	−64
35	4	−4	−12	−20	−27	−35	−43	−52	−58	−67

Examples

1. Find the wind chill temperature. The actual temperature is 20°F. The wind is 5 mph.

   Find 20°F in the top row of the table. Read down the column to the row beginning with 5 mph. The wind chill temperature is 16°F.

2. Find the actual temperature. The wind chill temperature is −15°F. The wind is 10 mph.

   Find the row beginning with 10 mph in the table. Read across until you find −15°F. Look at the top of the column. The actual temperature is 5°F.

3. Find the wind speed. The wind chill temperature is 8°F. The actual temperature is 35°F.

   Find 35°F. Read down the column until you find 8°F. Read across to the wind speed column. The wind speed is 25 mph.

Try This

a. Find the wind chill temperature. The actual temperature is 10°F. The wind speed is 5 mph.

b. Find the actual temperature. The wind chill temperature is −22°F. The wind speed is 25 mph.

c. Find the wind speed. The actual temperature is 25°F. The wind chill temperature is 2°F.

Exercises

1. Find the wind chill temperature. The actual temperature is 35°F. The wind speed is 35 mph.

2. Find the wind speed. The actual temperature is 25°F. The wind chill temperature is −10°F.

3. Find the wind chill temperature. The actual temperature is 30°F. The wind speed is 10 mph.

4. Find the actual temperature. The wind chill temperature is −49°F. The wind speed is 30 mph.

5. Find the actual temperature. The wind chill temperature is −31°F. The wind speed is 20 mph.

15-13 Calculator: Integers

Objective
Use a calculator to solve problems with integers.

This graph shows the temperature ranges for five European cities. What is the range in temperature for Copenhagen, Denmark?

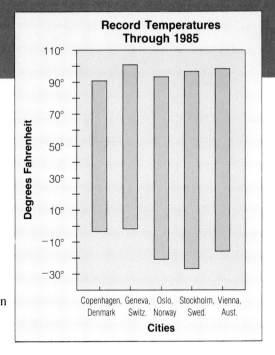

Example Use a calculator to find the range in temperature for Copenhagen, Denmark.

91°F − (−3°F) =   Find the difference
highest   lowest   between the highest and the lowest temperatures.

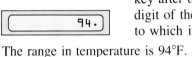

Press the change sign key after the last digit of the number to which it applies.

The range in temperature is 94°F.

Try This Use a calculator to find the range in temperature.

a. Geneva, Switz.
b. Oslo, Norway
c. Stockholm, Sweden
d. Vienna, Austria

Exercises

Use a calculator to find the range in temperature.

1. Texas
2. Mississippi
3. Oregon
4. Idaho
5. Alaska
6. Hawaii
7. Florida
8. Alabama
9. Kentucky
10. Colorado
11. California
12. Illinois

Data Bank

13. Use a calculator to find the range in temperature for Boise, Idaho. Use the Data Bank on p. 488. **113°**

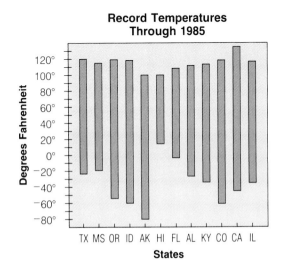

INTEGERS **451**

CHAPTER 15 REVIEW

Answer true or false.
1. $^-5 > {^-3}$ _____
2. $50 > {^-10}$ _____
3. $^-17 < {^-21}$ _____
4. $3 + {^-2} = 1$ _____
5. $10 + (^-10) = {^-20}$ _____
6. $(^-32) + (^-17) = 15$ _____

Write an integer to represent each statement.
7. 5 hours ago _____
8. 125° above zero _____
9. 23 points ahead _____
10. a loss of 16 yards _____

Arrange the integers in order from least to greatest.
11. $^-29, 57, {^-3}, {^-20}, 18$
12. $23, {^-23}, {^-52}, 0, 17$

Match.
13. $^-8 - 7$
14. $^-2 - (^-6)$
A. $^-30$
B. 33
15. $^-25 - 5$
16. $32 - 17$
C. 15
D. $^-15$
17. $0 - (^-30)$
18. $18 - (^-15)$
E. $^-63$
F. $^-72$
19. $^-21 \times 3$
20. $12 \times (^-6)$
G. 30
H. 4

Fill in the blank.
21. $25 \div (^-5) =$ _____
22. $^-15 \div 3 =$ _____
23. $^-51 \div (^-3) =$ _____

Choose the true statement.
24. Javier made a weekly budget. He wanted to spend only $6 a day. He spent $3, $2, $8, $1, $10, $2, and $5 for a week.
   A. He is over budget by $11.
   B. He is over budget by $17.
   C. He is under budget and can buy an $11 tape.
   D. He is under budget but has just enough left for $5 movie ticket.

25. The record store is doing an accounting report covering the months of June, July, August, and September.

	June	July	August	September
Purchases	$1000	$800	$1200	$600
Sales	$700	$1200	$1100	$900

   A. There was an overall increase of $500 from June to September.
   B. There was an overall decrease of $300 from June to September.
   C. There was an overall increase of $1000 from June to September.
   D. There was an overall increase of $300 from June to September.

CHAPTER 15 TEST

1. Fred has 9 coins. He has twice as many pennies as nickels. He has half as many quarters as dimes. He has one more quarter than nickels. How many of each type of coin does he have?

Write an integer to represent each statement.

2. a $37 loss
3. 15 bonus points on a quiz

Compare. Use < or >.
4. 104 □ 1004
5. ⁻89 □ ⁻91
6. 3 □ ⁻200

Add.
7. 11 + (⁻7)
8. ⁻5 + (⁻16)
9. 42 + (⁻93)
10. ⁻101 + 47

Subtract.
11. 8 − 17
12. ⁻13 − 53
13. ⁻27 − (⁻12)
14. ⁻45 − (⁻45)

15. This table shows Nadine's expenses for one week. She planned to spend only $4 a day. Was she under or over her weekly budget? How much?

Day	1	2	3	4	5	6	7
Expenses	$2	$5	$7	0	$1	$9	$5

Estimate.
16. 18 + (⁻38) + 55 + 37 + (⁻17)
17. 654 + (⁻655) + (⁻621) + 621 + 673

Multiply.
18. 4 × (⁻7)
19. (⁻9) × 15
20. (⁻21) × (⁻11)
21. (⁻34) × 34

Divide.
22. 36 ÷ (⁻4)
23. 96 ÷ (⁻16)
24. ⁻121 ÷ (⁻11)
25. ⁻322 ÷ 23

26. Jake and Sam are running backs for rival football teams. The table shows each player's rushing yardage for each quarter of the championship game. Find each player's total yardage gain for the game.

Quarter	1	2	3	4
Jake's Yardage	16	31	⁻7	23
Sam's Yardage	⁻4	35	⁻2	59

27. Solve. If the problem has missing data, state what data is missing.

The temperature at 12:00 p.m. on Tuesday was ⁻4°C. The temperature at 12:00 p.m. on Wednesday was 4° warmer than the temperature at 2:00 p.m. on Tuesday. What was the temperature at 12:00 p.m. on Wednesday?

CHAPTER 15 CUMULATIVE REVIEW

1. What is the difference between the day's high and low for Fairfield?
   A. 23°
   B. 29°
   C. 25°
   D. 35°

   Record Low and High Temperature for April 30

City	Low	High
Dumas	59°F	82°F
Fairfield	54°F	79°F
Pearsall	48°F	83°F
Littlefield	55°F	84°F

2. Write an integer to represent this statement. The temperature dropped 45 degrees.
   A. 45°
   B. −45°
   C. 0.45°
   D. −0.45°

3. Find the area of a triangle with these measures.
   $b = 18$ cm
   $h = 12$ cm
   A. 21.6 cm²
   B. 108 cm
   C. 21.6 cm
   D. 108 cm²

4. Divide.
   $^-63 \div (^-9)$
   A. 7
   B. 567
   C. $^-7$
   D. 54

5. For the past three weeks, Pablo has biked a total of 170 mi. He biked 30 mi more the second week than the first week. He biked 38 mi more the third week than the second week. How many miles did he bike the second week?
   A. 24 mi
   B. 54 mi
   C. 68 mi
   D. 44 mi

6. Estimate.
   $^-12 + 36 + 25 + (^-37) + 10 + 22$
   A. 150
   B. 40
   C. 130
   D. 50

7. Measure this angle.
   A. 45°
   B. 135°
   C. 60°
   D. 90°

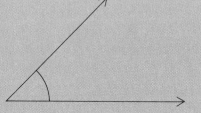

8. Find the scale measurement for 22 ft. Use a scale of $\frac{1}{2}$ in. = 10 ft.
   A. $1\frac{1}{4}$ in.
   B. $2\frac{1}{2}$ in.
   C. $1\frac{1}{10}$ in.
   D. $\frac{3}{4}$ in.

9. Find the volume of a pyramid with these measures.
   $B = 6$ cm²
   $h = 10$ cm
   A. 20 cm³
   B. 60 cm³
   C. 30 cm³
   D. 80 cm³

10. Loren made a weekly budget. He wanted to spend only $8 a day. He spent $10, $2, $4, $9, $8, $5, and $12 for the week. How much was Loren under or over his budget?
    A. over $56
    B. over $50
    C. under $6
    D. over $8

11. Write the fraction as a decimal.
    $2\frac{3}{5}$
    A. 2.6
    B. 2.4
    C. 0.26
    D. 2.35

12. Divide.
    623 ÷ 14
    A. 43 R7
    B. 43 R12
    C. 45
    D. 44 R7

13. Add.
    ⁻8 + 12
    A. ⁻4
    B. 16
    C. 4
    D. ⁻20

14. Read the scale.
    A. 384.5 g
    B. 340 g
    C. 385 g
    D. 380.5 g

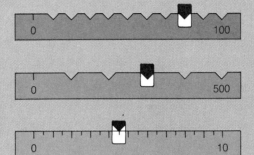

15. The highway patrol checked how fast cars were traveling. Out of 168 cars, 5 times as many cars were going the speed limit as were speeding. How many cars were speeding?
    A. 30 cars   C. 33.6 cars
    B. 32 cars   D. 28 cars

16. Find the area of a square with this measure.
    s = 9 ft
    A. 18 ft²
    B. 81 ft²
    C. 36 ft²
    D. 6561 ft²

17. Find the volume of a cube with this measure.
    s = 6.2 m
    A. 238.328 m³
    B. 38.44 m³
    C. 18.6 m³
    D. 38.44 m

18. Add.
    $3\frac{14}{15} + 6\frac{7}{15}$
    A. $9\frac{21}{30}$ B. $9\frac{2}{3}$
    C. $9\frac{7}{15}$ D. $10\frac{2}{5}$

19. Find the greatest common factor.
    12, 47
    A. 2
    B. 3
    C. 6
    D. 1

20. Subtract.
    12 − (⁻3)
    A. 9
    B. 15
    C. ⁻15
    D. ⁻9

21. Multiply.
    ⁻42 × (⁻8)
    A. 326
    B. 336
    C. ⁻336
    D. ⁻456

22. Find the surface area of a cube with this measure.
    s = 6 m
    A. 216 m³
    B. 216 m²
    C. 36 m²
    D. 36 m³

23. Find the sales tax.
    $8.49 + $0.98
    A. $0.51
    B. $0.06
    C. $0.56
    D. $0.57

6% SALES TAX		
Transaction	Tax	Transaction
.01– .10	.00	8.42– 8.58
.11– .22	.01	8.59– 8.74
.23– .39	.02	8.75– 8.91
.40– .56	.03	8.92– 9.08
.57– .73	.04	9.09– 9.24
.74– .90	.05	9.25– 9.41
.91–1.08	.06	9.42– 9.58
1.09–1.24	.07	9.59– 9.74
1.25–1.41	.08	9.75– 9.91
1.42–1.58	.09	9.92–10.08

Chapter 16 Overview

Key Ideas
- Use the problem-solving strategy *guess and check*.
- Write algebraic expressions using variables.
- Solve equations using addition, subtraction, multiplication, and division.
- Solve equations with multiple steps.
- Solve problems using formulas.
- Use the Pythagorean Theorem.
- Find squares and square roots.
- Read and graph ordered pairs.
- Collect and analyze data.

Key Terms
- algebraic
- check
- ordered pair
- expression
- formula
- coordinate grid
- equation
- square
- component
- solution
- square root
- hypotenuse
- origin

Key Skills

Add or subtract.

1. $23 + 8$
2. $74 - 8$
3. $82 + 25$
4. $28 - 16$
5. $36 + 35$
6. $31 + 85$
7. $146 - 56$
8. $147 - 66$
9. $543 + 248$
10. $3508 - 776$

Multiply or divide.

11. 35×3
12. $32 \div 8$
13. 76×4
14. 23×8
15. $63 \div 7$
16. 125×6
17. $165 \div 5$
18. $260 \div 4$
19. 75×20
20. 66×34

ALGEBRA 16

16-1 Problem Solving: Learning to Use Strategies

Objective
Use the strategy guess and check.

- SITUATION
- DATA
- **PLAN**
- ANSWER
- CHECK

Mario just received 5 new videos. He now has 17 videos. How many videos did he have before he received the new videos?

You can use the **guess and check** strategy to find the number of videos.

Example Solve by guessing and checking.

1. How many tapes did Mario have before he received the new videos?

v = the number of Mario's videos	Use a variable to represent the unknown quantity.
$v + 5 = 17$	Write an equation to represent the problem.
$10 + 5 = 15$ $15 < 17$	Guess 10. Check. 10 is too small.
$12 + 5 = 17$ $17 = 17$ ✔	Guess 12. Check. It is correct.

Mario had 12 tapes before he received the new videos.

Try This Solve by guessing and checking.

a. $x + 3 = 24$ **b.** $y + 12 = 47$ **c.** $w + 5 = 18$ **d.** $t + 13 = 25$

Example Solve by guessing and checking.

2. Diego sold 16 tickets to the school dance. He needed to sell 37 more tickets to sell all of his tickets. How many tickets did he have to start?

t = the total number of tickets	Use a variable to represent the unknown quantity.
$t - 16 = 37$	Write an equation.
$55 - 16 = 39$ $39 > 37$	Guess 55. Check. 55 is too big.
$53 - 16 = 37$ $37 = 37$ ✔	Guess 53. Check. It is correct.

Diego had 53 tickets to start.

Try This Solve by guessing and checking.

e. $c - 14 = 66$ **f.** $z - 21 = 34$ **g.** $t - 15 = 34$ **h.** $s - 25 = 105$

Exercises

Solve. Use one or more of the problem-solving strategies.

Problem-Solving Strategies
- **Guess and check.**
- Choose an operation.
- Make an organized list.
- Act it out.
- Find a pattern.
- Write an equation.
- Use logical reasoning.
- Draw a picture or diagram.
- Make a table.
- Use objects.
- Work backward.
- Solve a simpler problem.

1. Bess got 5 pairs of earrings for her birthday. She now has 13 pairs. How many pairs of earrings did she have before her birthday?

2. In how many different orders can you play these notes on a piano?

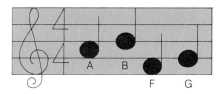

3. Tobias spent 15¢ in change for a pencil. He now has 37¢ in change. How much change did he have before he bought the pencil?

4. Replace the letters with the digits 0–9 to write a correct multiplication problem.

$$\begin{array}{r} AB \\ \times AB \\ \hline CAB \\ BD \\ \hline EAB \end{array}$$

5. Maude numbered each page in her journal. She wrote 137 digits. How many pages long was her journal?

6. Lyle, Valerie, Sean, and Serena sold pom-poms at the track meet. Lyle sold twice as many pom-poms as Valerie. Valerie sold 2 less than Sean. Sean sold 5 more than Serena. Serena sold 9 pom-poms. How many pom-poms did Lyle sell?

Write Your Own Problem

Write a word problem that you could solve using the equation.

7. $w + 15 = 27$ 8. $p - 9 = 16$ 9. $b + 10 = 13$ 10. $s - 11 = 15$

11. Write a problem that you could solve using the strategy *guess and check*. Use the data in the graph.

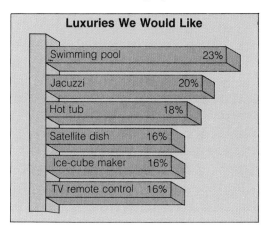

ALGEBRA **459**

16-2 Writing Algebraic Expressions Using Variables

Objective
Write algebraic expressions using variables.

You can use addition, subtraction, multiplication, and division to write word expressions as algebraic expressions.

What other words can you write in this chart?

+	−	×	÷
sum increase	difference take away	times product	divide ?

Example Write four word expressions for each algebraic expression.

1. $m + 7$

   Some possible answers:
   the sum of m and 7 m increased by 7
   7 added to a number m 7 more than m

Try This Write four word expressions for each algebraic expression.

a. $r - 3$ b. $6y$ c. $w \div 9$ d. $t + 5$ e. $\frac{q}{4}$

Examples Write an algebraic expression for each sentence.

2. Vera scored 17 points more than Bryce.

   b = Bryce's points   Use a variable to represent the number of points Bryce has.
   $b + 17$ Write an expression to represent the number of points Vera has.

3. The tree was twice as high as the house.

   h = house   Use a variable to represent the height of the house.
   $2h$ Write an expression to represent the height of the tree.

Try This Write an algebraic expression for each sentence.

f. Lionel was 5 years younger than Dan.
g. Sara had 4 times as much money as Jenny.
h. Fred weighed $\frac{1}{3}$ as much as Douglas.
i. Naomi had 2 more magazines than Mei.

Exercises

Write four word expressions for each algebraic expression.

1. $f + 4$
2. $t - 5$
3. $8w$
4. $w \div 3$
5. $\frac{c}{3}$
6. $r - 1$
7. $7z$
8. $h + 2$
9. $p \div 4$
10. $\frac{d}{5}$

Write an algebraic expression for each sentence.

11. Harry was twice as fast as Jarrod.
12. Archie was 2 years older than Tai.
13. Belinda was only half as tall as Janice.
14. Cindi had 4 more problems correct than Freddie.
15. Jackson had 3 times as many hits as Al.
16. Don's time for the run was 10 seconds less than Jo's.
17. The length of the board was $\frac{1}{9}$ as long as the pipe.
18. Clara sold 5 times as many programs as Brett.

Write Your Own Problem

Write a sentence like the ones in Exercises 11–18 for each algebraic expression.

19. $t - 4$
20. $y + 8$
21. $j \div 3$
22. $5b$
23. $\frac{w}{2}$

Number Sense

Evaluate each expression. Arrange in order from least to greatest.

24. $m + 5$, $2m \div 3$, $4m - 2$, m, $m + 7$ for $m = 6$
25. $j - 2$, $3j$, $2j + 10$, $j + 7$, j for $j = 5$
26. $k \div 2$, $6k + 2$, $k - 16$, k, $k \times k$ for $k = 4$
27. $m + 4$, $9 - m$, $2m$, $4m + 2$, $5m \div 25$ for $m = 5$

Problem Solving

28. Match each picture with the correct algebraic expression.

    b
    $b - 1$
    $b + 3$
    $3b$
    $b \div 2$

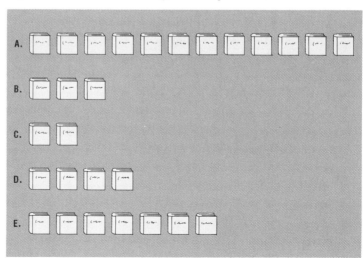

ALGEBRA

16-3 Solving Equations with Addition

Objective
Solve one-step equations with addition.

Adrian received a check for $37. After he deposited it, his account had a balance of $617. How much was the balance before the deposit?

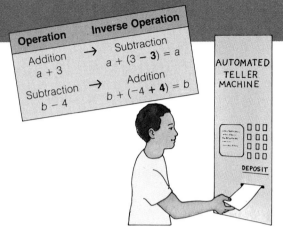

Understand the Situation

- How much money did Adrian receive?
- What does it mean *to deposit a check*?
- What was the new balance?

You can solve this problem using an equation. A value for the variable that makes an equation true is a **solution.**

Use **inverse operations** to find solutions. Addition and subtraction are inverse operations. Each operation will undo the action of the other.

Read the addition example in the table. Note that subtracting 3 will undo the action of adding 3. Always use the same operation on both sides of an equation to keep the sides equal.

Example Solve.

How much was Adrian's balance before he deposited the check?

b = the old balance	Use a variable to represent the unknown quantity.
$b + 37 = 617$	Write an equation to represent the problem.
$(b + 37) - 37 = 617 - 37$	Find the value of b. Use the inverse
$b + (37 - 37) = 580$	operation to get b alone on one side of the
$b + 0 = 580$	equation. Use the same operation on both
$b = 580$	sides of the equation.
$580 + 37 = 617$	Check. Replace the variable in the equation
$617 = 617$ ✔	with your answer.

Adrian's balance was $580 before he deposited the check.

Try This Solve.

a. $c + 14 = 38$
b. $w + 289 = 315$
c. $t + 16 = {}^-91$
d. $p + 15 = {}^-7$

Exercises

Solve.

1. $x + 3 = 12$
2. $p + 9 = 16$
3. $m + 13 = 38$
4. $s + 19 = 43$
5. $p + 23 = 54$
6. $y + 33 = 89$
7. $h + 12 = 8$
8. $z + 8 = 3$
9. $k + 19 = 11$
10. $l + 7 = {}^-2$
11. $n + 8 = {}^-6$
12. $a + 10 = {}^-3$
13. $c + 13 = {}^-22$
14. $w + 17 = {}^-34$
15. $p + 21 = {}^-45$
16. $21 = y + 4$
17. $19 = t + 7$
18. $32 = x + 9$
19. ${}^-7 = 4 + v$
20. $57 = 72 + f$

Mixed Applications

Solve.

21. Freda added 8 oz of oil to some gasoline for her lawnmower. The scale on the container showed that the container now held 128 oz. How many ounces of gasoline were in the container before Freda added oil?

22. Jian spent a total of $124 shopping for school clothes. One shirt he bought cost $15. What was the total cost of the other clothes he bought?

23. Denise uses the equation $14{,}000 = L + 7950$ to find the approximate length of a par-70 golf course. L stands for the length of the golf course in yards. Find the approximate length of the par-70 course.

24. The record temperature shift in one day was in Browning, Montana, in 1916. The temperature rose 100°F from morning to afternoon. The afternoon high was 44°F. What was the morning low?

Write Math

25. Write an equation that has a solution of 7.
26. Write an equation that has a solution of 0.
27. Write an equation that has a solution of ${}^-3$.
28. Write an equation that has a solution of ${}^-5$.

Mixed Skills Review

Find the greatest common factor for each pair of numbers.

29. 86, 14
30. 17, 41
31. 27, 68
32. 35, 90
33. 24, 36

Find the answer.

34. $\$34.96 - \$5.48$
35. 30% of 65
36. $4\frac{2}{3} \times 3\frac{1}{2}$
37. $16\overline{)342}$

Find the area.

38. square
    $s = 12.4$ m

39. 16.3 in. / 9.5 in.

40. 4 cm / 4 cm

41. rectangle
    $l = 3\frac{1}{2}$ ft
    $w = 7$ ft

16-4 Solving Equations with Subtraction

Objective
Solve one-step equations with subtraction.

Tara spent $14 at the sporting goods sale. She had $3 left in her wallet. How much money did she have to start?

Understand the Situation
- How much money does Tara have now?
- How much money did she spend?
- Does she have more or less money now than at the start?

You can solve Tara's problem using an equation with subtraction.

Example Solve.

How much money did Tara have to start?

d = amount of money Tara had to start	Use a variable to represent the unknown quantity.
$d - 14 = 3$	Write an equation to represent the problem.
$(d - 14) + 14 = 3 + 14$ $d + (^-14 + 14) = 17$ $d + 0 = 17$ $d = 17$	Find the value of d. Use the inverse operation to get d alone on one side of the equation. Use the same operation on both sides of the equation.
$17 - 14 = 3$ $3 = 3$ ✓	Check. Replace the variable in the equation with your answer.

Tara had $17 to start.

Try This Solve.

a. $c - 14 = 38$ **b.** $w - 28 = 31$ **c.** $r - 16 = 91$ **d.** $s - 23 = 44$

Exercises

Solve.

1. $t - 6 = 16$
2. $x - 9 = 17$
3. $y - 17 = 31$
4. $x - 42 = 28$
5. $h - 12 = 51$
6. $y - 56 = 76$
7. $m - 23 = {}^-12$
8. $k - 34 = {}^-19$
9. $x - 65 = {}^-48$
10. $y - 33 = {}^-48$
11. $m - 29 = {}^-77$
12. $y - 42 = {}^-54$
13. $15 = y - 8$
14. ${}^-5 = y - 2$
15. $11 = t - 9$
16. $17 = x - 13$
17. ${}^-15 = y - 25$
18. $x - ({}^-21) = 7$
19. $n - ({}^-8) = 22$
20. ${}^-18 = v - 5$

Mixed Applications

Solve.

21. Kenji bought a weight bench for $99 off the regular price. He paid $375. What was the regular price?
22. Mitzi skied a downhill course in 56.2 s. This was 1.7 s faster than her old record. What was the time of her old record?
23. Bernardo played 67 tennis matches. He won 36 of them. How many matches did he lose?
24. Regina made three equal payments of $36 on her bike. She still owed $246. How much was the total cost of the bike?

Mental Math

Solve. Use mental math.

25. $w - 5 = 24$
26. $t - 10 = 49$
27. $b - 99 = 200$
28. $f - 19 = 60$

Problem Solving

29. The difference in age between Felix and Felicia is 5 years. The sum of their ages is 23 years. How old are Felix and Felicia?

Mixed Skills Review

Write each improper fraction as a mixed number or a whole number.

30. $\frac{24}{16}$
31. $\frac{32}{3}$
32. $\frac{86}{12}$
33. $\frac{67}{67}$
34. $\frac{20}{5}$
35. $\frac{0}{8}$

Find the answer.

36. $\frac{3}{6} = \frac{18}{n}$
37. $\$17 - \$3.65$
38. ${}^-8 \times ({}^-16)$
39. $9.36 \div 10$

Arrange in order from least to greatest.

40. $0.4, \frac{4}{8}, 1.3, \frac{7}{3}, 2$
41. $\frac{29}{4}, \frac{74}{10}, 7.2, \frac{45}{6}, 7$
42. $5, \frac{27}{5}, 4\frac{10}{5}, 4.9$
43. $9.5, \frac{100}{12}, 8\frac{1}{2}, \frac{85}{9}, 7.79$
44. $\frac{32}{100}, 0.267, \frac{3}{8}, 0.312, \frac{13}{32}$
45. $3, \frac{50}{16}, 2.4, \frac{29}{14}, 2.9$
46. ${}^-4, 0, {}^-5, 2.5, {}^-25$
47. ${}^-12, 20, {}^-49, {}^-62, 4$
48. $97, {}^-98, 94, {}^-46, 89$

49. 9.7 cm, 90.7 mm, 970 mm, 9.7 m, 0.907 cm

ALGEBRA

16-5 Solving Equations with Multiplication or Division

Objective
Solve one-step equations with multiplication or division.

You have already learned that addition and subtraction are inverse operations. In order to get the variable alone on one side of an equation, you must use the correct inverse operation.

Operation		Inverse Operation
Addition $a + 3$	→	Subtraction $a + (3 - 3) = a$
Subtraction $b - 4$	→	Addition $b + (-4 + 4) = b$
Multiplication $c \times 5$, or $5c$	→	Division $c \times (5 \div 5) = c$, or $\frac{5c}{5} = c$
Division $d \div 6$, or $\frac{d}{6}$	→	Multiplication $d \times (\frac{1}{6} \times 6) = d$, or $\frac{d}{6} \times 6 = d$

You can solve equations with multiplication or division using inverse operations.

Read the multiplication example in the table. Note that dividing by 5 will undo the action of multiplying by 5. In the division example, see that multiplying by 6 will undo the action of dividing by 6.

Examples Solve.

1. $5x = 30$

   $5x = 30$
   $\frac{5x}{5} = \frac{30}{5}$
   $x = 6$

   Use the inverse operation to get x alone on one side of the equation. Use the same operation on both sides of the equation.

   $5(6) = 30$
   $30 = 30$ ✓

   Check. Replace the variable in the equation with your answer.

2. $\frac{m}{10} = 20$

   $\frac{m}{10} = 20$
   $\frac{m}{10} \times 10 = 20 \times 10$
   $m = 200$

   Use the inverse operation to get m alone on one side of the equation. Use the same operation on both sides of the equation.

   $\frac{200}{10} = 20$
   $20 = 20$ ✓

   Check. Replace the variable in the equation with your answer.

Try This Solve.

a. $6v = 48$ b. $9k = 342$ c. $\frac{n}{7} = 63$ d. $\frac{1}{14}w = 42$

CHAPTER 16

Exercises

Solve.

1. $3x = 12$
2. $5y = 30$
3. $8t = 72$
4. $-5y = 40$
5. $-6x = 72$
6. $-8x = 64$
7. $12m = -48$
8. $9p = 36$
9. $7t = -56$
10. $-12x = -84$
11. $-15y = -30$
12. $-14y = -42$
13. $23y = -69$
14. $18x = -54$
15. $16x = -48$
16. $9.1k = 546$
17. $\left(\frac{1}{2}\right)h = 49$
18. $\frac{x}{3} = 7$
19. $\frac{x}{4} = 3$
20. $\frac{m}{2.3} = 115$

21. Find the number r that when divided by 7 gives a quotient of 17.
22. Find the number s that when multiplied by 6 gives a product of 252.
23. Find the number t that when divided by 13 gives the quotient of 12.
24. Find the number n that when multiplied by $\frac{1}{3}$ gives a product of 12.

Calculator Solve and check. Round to the nearest hundredth.

25. $567y = 697.41$
26. $\frac{x}{10.8528} = 0.238$
27. $159x = 37{,}524$
28. $\frac{y}{11.7} = 0.9$

Number Sense

29. Write an equation that has a solution between 4 and 5.
30. Write an equation that has a solution between 0 and -1.

Group Activity

31. Use a tape measure to measure in centimeters the height of each group member. Next measure in centimeters the distance from each student's hip bone to the floor. Divide the measure of a student's height by the measure of the student's hip bone height. Is each answer close to 1.618? This ratio is called the "golden ratio." Find other examples of pairs of measurements with this ratio.

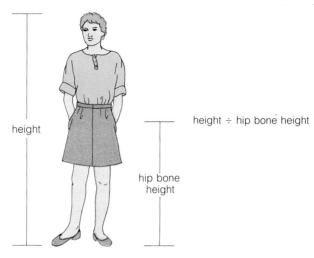

ALGEBRA

16-6 Solving Multiple-Step Equations

Objective
Solve multiple-step equations.

Caspar had a coupon for 25¢ off on one can of chili. He bought 7 cans for $3.11. How much did each can cost originally?

Understand the Situation
- How many cans did Caspar buy?
- How much did he spend?
- How much did he save by using the coupon?

You can write an equation with more than one operation for the situation.

Examples Solve.

1. How much did each can of chili cost originally?

   c = cost of one can of chili   Use a variable to represent the unknown quantity.

   $7c - 0.25 = 3.11$ Write an equation for the problem.

   $7c + (^-0.25 + 0.25) = 3.11 + 0.25$ Use inverse operations to get the variable alone on one side of the equation.
   $$7c = 3.36$$
   $$\frac{7c}{7} = \frac{3.36}{7}$$
   $$c = 0.48$$

   $7(0.48) - 0.25 = 3.11$ Check. Replace the variable in the equation with your answer.
   $3.36 - 0.25 = 3.11$
   $3.11 = 3.11$ ✔

   Each can of chili originally cost 48¢.

2. Mrs. Hamada bought a used refrigerator for $225. She paid $45 down and the rest in 9 equal monthly payments. How much was each monthly payment without interest?

   m = amount of each payment   Use a variable.
   $9m + 45 = 225$ Write an equation.
   $9m + (45 - 45) = 225 - 45$ Subtract 45 from both sides.
   $$9m = 180$$
   $$\frac{9m}{9} = \frac{180}{9}$$ Divide both sides by 9.
   $$m = 20$$

   $9(20) + 45 = 225$ Check.
   $180 + 45 = 225$
   $225 = 225$ ✔

   Each monthly payment was $20.

Try This Solve.

a. $4x + 7 = 35$ b. $5f - 8 = 47$ c. $17 = 2v - 9$ d. $225 = 4z + 65$

Exercises

Solve.

1. $8y + 4 = 68$
2. $5p + 6 = 31$
3. $30 = 3t + 6$
4. $72 = 7x + 9$
5. $13 = 3m + 4$
6. $9y + 6 = 51$
7. $^-5y + 6 = 16$
8. $^-8x + 4 = 28$
9. $35 = ^-4y + 7$
10. $86 = ^-9x + 5$
11. $55 = ^-7y + 6$
12. $^-4t + 11 = 43$
13. $4x - 6 = 34$
14. $33 = 3t - 9$
15. $15 = 6y - 3$
16. $0.5y - 0.7 = 4.8$
17. $\left(\frac{1}{3}\right)m + 5 = 41$
18. $87 = ^-8y - 9$
19. $\left(\frac{1}{2}\right)l - 6 = 42$
20. $^-6.8 = 1.1y + 0.9$

Mixed Applications

Solve.

21. Sue bought 10 cans of juice concentrate for $4.65. She gave the clerk a discount coupon for 25¢ off the total price. What did each can of concentrate originally cost?

22. Zack bought 14 cans of soup and one gallon of milk. The milk cost $1.95. He paid a total of $9.65. How much did each can of soup cost?

23. Jerry bought a $164 radio with a $20 down payment. He agreed to make 6 equal monthly payments. What is the amount of each monthly payment without interest?

24. Jackie's parents bought a $1200 video camera with a $120 down payment. They agreed to make 30 equal monthly payments. How much was each monthly payment without interest?

Mental Math

Which of the following equations would it be easy for you to solve using mental math? Solve those equations.

25. $x + 2 = 7$
26. $5x = 325$
27. $x - 5 = 12$
28. $3x - 7 = 353$
29. $4y - 10 = 484$
30. $x + 9 = 14$
31. $\left(\frac{1}{2}\right)m + 25 = 49$
32. $\left(\frac{1}{2}\right)n = 4$

Suppose

Suppose in the second example on p. 468 that Mrs. Hamada's installment plan was as shown. Find the amount of each monthly payment.

33. Down payment: $10
    Equal monthly payments: 15

34. Down payment: $57
    Equal monthly payments: 6

Problem Solving

35. The school built a new gym and locker rooms. Lydia's job was to put numbers on all the girls' lockers. The lockers were to be numbered consecutively beginning with 1. Lydia had to put up one digit at a time. She used 342 digits in all. How many lockers were built in the new girls' locker room?

16-7 Consumer Math: Solving Problems Using Formulas

Objective
Solve consumer problems using formulas.

- SITUATION
- DATA
- PLAN
- ANSWER
- CHECK

Alonzo paid $3 for parking. How long did he park in the lot?

Parking Rates
- $1 for first hour
- $0.50 for each additional 20 min

Understand the Situation
- How much does it cost to park for the first hour?
- How much does each additional 20 min of parking cost?

Examples

1. How long did Alonzo park?

   c = cost of parking
   g = number of groups of 20 min after the first hour

   Use variables to represent the amounts that can change.

   $c = 1 + 0.50g$ Write the fomula.
   $3 = 1 + 0.50g$ Substitute the known variables.
   $3 - 1 = (1 - 1) + 0.50g$ Solve.
   $2 = 0.50g$
   $\frac{2}{0.50} = \frac{0.50g}{0.50}$
   $4 = g$ Alonzo parked 4 groups of 20 min after the first hour.
   He parked for 2 h 20 min.

2. Mel's Bowling Alley charges $0.75 for renting shoes and $0.50 for each game. How many games could Tanya bowl for $6.25?

   c = total cost of bowling
   g = number of games bowled

   Use variables for the amounts that can change.

   $c = 0.75 + 0.50g$ Write the formula.
   $6.25 = 0.75 + 0.50g$ Substitute the known variables.
   $6.25 - 0.75 = (0.75 - 0.75) + 0.50g$ Solve.
   $5.50 = 0.50g$
   $\frac{5.50}{0.50} = \frac{0.50g}{0.50}$
   $11 = g$ Tanya could bowl 11 games for $6.25.

Try This

a. Ina paid $5 for parking. How long did she park?

b. How many games can Neal bowl for $4.75? He rented shoes.

Exercises

Solve. Use the formulas in the examples on p. 470.

1. Georgia paid $8.50 for parking. How long did she park?
2. How much would it cost to rent shoes and to bowl 10 games?
3. Eduardo parked his van for 24 h. How much money did he pay for parking?
4. How many games can Mimi bowl for $2.25? She rented shoes.
5. How much would it cost to park a car for 10 h?
6. Pat paid $3.75 to rent shoes and to bowl. How many games did Pat bowl?

Solve.

7. $6l + 0.25 = 3.25$
8. $4x + 0.17 = 0.97$
9. $-2n + 0.10 = 0$
10. $5b - 0.35 = 3.15$
11. $7z - 0.13 = 0.78$
12. $9p - 0.07 = 2$
13. $\frac{v}{8} = 2.56$
14. $\frac{y}{9} = 0.07$
15. $\frac{5t}{8} = 0.4$
16. $\frac{n}{-3} = 4.8$
17. $\frac{f}{-3} = 1.08$
18. $\frac{u}{5.5} = 2$
19. $\frac{2m}{3} = 7.2$
20. $\frac{3b}{5} = 0.9$

The formula for the stopping distance of a car is $D = \frac{S^2}{30f}$. D is distance in feet. S is speed in miles per hour. f is the drag factor of the road. Round to the nearest foot.

21. Find the stopping distance for a car going 55 miles per hour on each of the road surfaces listed in the table.

Road Surface	Drag Factor
Dry asphalt	0.75
Wet asphalt	0.6
Snow	0.3
Ice	0.1

Data Hunt Find the formula to change degrees Celsius to degrees Fahrenheit. Give the temperature in Fahrenheit. Round to the nearest degree.

22. 20°C
23. 12°C
24. 5°C
25. 0°C
26. ⁻4°C
27. ⁻10°C

Mixed Skills Review

Write using exponents. Then write as a standard numeral.

28. $2 \times 2 \times 2 \times 2 \times 2$
29. $5 \times 5 \times 5 \times 5$
30. $(0.3) \times (0.3) \times (0.3)$

Write each fraction or mixed numeral as a decimal.

31. $3\frac{4}{5}$
32. $\frac{5}{2}$
33. $\frac{1}{3}$
34. $\frac{24}{33}$
35. $9\frac{8}{11}$
36. $1\frac{5}{8}$

Find the answer.

37. 54 qt = ____ gal
38. 8.4 m = ____ cm
39. $64.26 \div 12.6$
40. 42×47

ALGEBRA

16-8 Squares and Square Roots

Objective
Find squares and square roots.

Gilda is building a square pen for her calf. The area of the pen should be 100 sq ft. What should be the length of each side?

Understand the Situation
- What shape will the pen be?
- What should the area of the pen be?
- Will the four sides of the pen be the same length?

To find the length of each side, find the square root. The **square root** of a number is a number that is multiplied by itself to get that number.
Write $\sqrt{a}$ to represent the positive square root of a.

Example Find the square root.

1. Find the length of each side of the pen. Find $\sqrt{100}$.

s = one side of the pen
$s^2 = 100$ *Think: What number times itself equals 100?*
$10^2 = 100$, so $\sqrt{100} = 10$ The length of each side of the pen is 10 ft.

Try This Find the square root.

a. $\sqrt{25}$ b. $\sqrt{49}$ c. $\sqrt{64}$ d. $\sqrt{81}$ e. $\sqrt{144}$ f. $\sqrt{400}$

You can look up squares and square roots of larger numbers in the table on p. 495.

Examples Find the square or square root. Use the table shown.

2. 13^2

n	n^2	$\sqrt{n}$
12	144	3.464
13	169	3.606

Find 13 under n in the table.
Look across under n^2 in the table.
This number is the square.
$13^2 = 169$

3. $\sqrt{48}$

n	n^2	$\sqrt{n}$
47	2209	6.856
48	2304	**6.928**

Find 48 under n in the table.
Look across under $\sqrt{n}$ in the table.
This number is the square root.
$\sqrt{48} = 6.928$

Try This Find the square or square root. Use the table on p. 495.

g. 23^2 h. $\sqrt{50}$ i. 34^2 j. $\sqrt{24}$ k. 19^2 l. $\sqrt{37}$

Exercises

Find the square root.

1. $\sqrt{16}$
2. $\sqrt{9}$
3. $\sqrt{36}$
4. $\sqrt{121}$
5. $\sqrt{4}$
6. $\sqrt{900}$
7. $\sqrt{2500}$
8. $\sqrt{10,000}$
9. $\sqrt{1600}$
10. $\sqrt{3600}$
11. $\sqrt{8100}$
12. $\sqrt{4900}$

Find the square or square root. Use the table on p. 495.

13. 17^2
14. 28^2
15. 16^2
16. 15^2
17. 42^2
18. 61^2
19. 87^2
20. 35^2
21. 57^2
22. 63^2
23. 75^2
24. 90^2
25. $\sqrt{256}$
26. $\sqrt{441}$
27. $\sqrt{841}$
28. $\sqrt{2704}$
29. $\sqrt{784}$
30. $\sqrt{2025}$
31. $\sqrt{6889}$
32. $\sqrt{1024}$
33. $\sqrt{1764}$
34. $\sqrt{2601}$
35. $\sqrt{3249}$
36. $\sqrt{7225}$

Mixed Applications

37. The area of a square is 2116 square units. What is the length of one side of the square?

38. The area of a square is 6400 square units. What is the length of one side of the square?

39. The area of a circle is 28.26 square units. What is the radius of the circle?

40. A given number t is squared. Five is added to the square to get a sum of 54. What is the value of t?

Talk Math

41. Look at this list of square numbers.

    1, 4, 9, 16, 25, 36, 49, 64, 81, 100, ...

    Describe some of the patterns you see.

Calculator

42. Find the square root of each number.

    $\sqrt{0.01} = $ _____ $\sqrt{0.1} = $ _____ $\sqrt{1} = $ _____
    $\sqrt{10} = $ _____ $\sqrt{100} = $ _____ $\sqrt{1000} = $ _____

    Can you find a pattern? Guess the positive square root of 0.001 and 100,000. Check your guess with your calculator.

Estimation Estimate the square root.

43. $\sqrt{10}$
44. $\sqrt{24}$
45. $\sqrt{98}$
46. $\sqrt{50}$
47. $\sqrt{37}$
48. $\sqrt{120}$
49. $\sqrt{142}$
50. $\sqrt{15}$
51. $\sqrt{80}$
52. $\sqrt{63}$

Problem Solving

53. The Texas Star Bank received 11 new computers. The bank must divide the computers among three branches. Each branch is to receive at least one computer. In how many ways can the 11 computers be divided among the three branches?

16-9 Pythagorean Theorem

Objective
Use the Pythagorean Theorem.

Fumiko wanted to know how far it was to sail from the dock to her house across the lake.

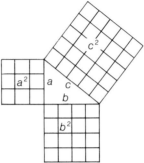

If you know the lengths of two sides of a right triangle, you can find the length of the third side. Use the Pythagorean Theorem to find the missing length.

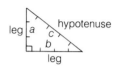

The **Pythagorean Theorem** states that in any right triangle, the sum of the areas of the squares on the legs is equal to the area of the square on the **hypotenuse**. You can write this rule as a formula: $a^2 + b^2 = c^2$.

Examples Find the missing length.

1. Find the distance from the dock to Fumiko's house.

   $a^2 + b^2 = c^2$ Use the Pythagorean Theorem.
   $3^2 + 4^2 = c^2$ Substitute variables with known quantities.
   $9 + 16 = c^2$ Solve.
   $25 = c^2$
   $\sqrt{25} = \sqrt{c^2}$ Find the square root of both sides.
   $5 = c$

   The distance from the dock to Fumiko's house is 5 mi.

2.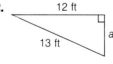

   $a^2 + b^2 = c^2$
   $a^2 + 12^2 = 13^2$
   $a^2 + 144 = 169$
   $a^2 + (144 - 144) = 169 - 144$
   $a^2 = 25$
   $\sqrt{a^2} = \sqrt{25}$
   $a = 5$ ft

3.

   $a^2 + b^2 = c^2$
   $9^2 + b^2 = 15^2$
   $81 + b^2 = 225$
   $(81 - 81) + b^2 = 225 - 81$
   $b^2 = 144$
   $\sqrt{b^2} = \sqrt{144}$
   $b = 12$ m

Try This   Find the missing length.

a. 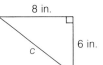   b. 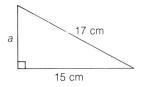   c.

Exercises

Find the missing length.

1. 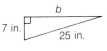   2. 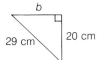   3.

4. 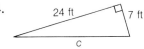   5.   6.

Find the missing length. Use a calculator or the table on p. 495. Round to the nearest thousandth.

7. $a = 10$ cm
$b = \square$
$c = 26$ cm

8. $a = 2$ in.
$b = 7$ in.
$c = \square$

9. $a = \square$
$b = 2$ ft
$c = 3$ ft

10. $a = 6$ m
$b = 8$ m
$c = \square$

11. $a = \square$
$b = 6$ cm
$c = 12$ cm

12. $a = 16$ mi
$b = \square$
$c = 18$ mi

13. $a = 10$ yd
$b = 10$ yd
$c = \square$

14. $a = \square$
$b = 20$ in.
$c = 25$ in.

Mixed Applications

Solve. Round to the nearest thousandth.

15. Antonia sailed 5 mi east. Then she sailed 5 mi north. How far, along a straight line, was she from her starting point?

16. A ramp to a truck is 10 ft long. The distance, along the ground, from one end of the ramp to the truck is 8 ft. How far above ground is the ramp at the truck?

Practice Through Problem Solving

17. Draw a square by repeating the shape of the triangle shown.

ALGEBRA **475**

16-10 Graphing Ordered Pairs

Objective
Graph ordered pairs on a coordinate grid.

The key on the map says that Austin is at (7, D). To find Austin, find 7 on the horizontal scale and D on the vertical scale. Find the square where the sections intersect.

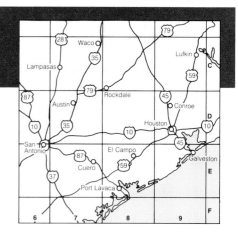

Understand the Situation

- What number would you use to find Rockdale?
- What letter would you use to find Cuero?

An **ordered pair** is a pair of letters or numbers used on a **coordinate grid**. The **first component** (x) tells you how far to move left or right from the **origin**, or (0, 0). The **second component** (y) tells you how far to move up or down from the origin.

Example

1. Name the ordered pair for points A, B, and C.

   The ordered pairs are: $A(1, 3)$, $B(2, 4)$, and $C(5, 2)$.

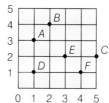

Try This Name the ordered pair for each point. Use the grid in Example 1.
   a. D **b.** E **c.** F

You can graph ordered pairs (x, y) on coordinate grids.

Examples Graph and label these points on a coordinate grid.

2. $A(4, 5)$ 3. $B(^-1, 3)$
4. $C(^-4, ^-2)$ 5. $D(4, ^-2)$

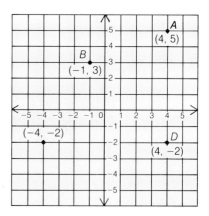

Try This Graph and label these points on a coordinate grid.
   d. $A(1, 5)$ **e.** $B(^-2, 4)$ **f.** $C(^-3, ^-5)$ **g.** $D(3, 5)$

CHAPTER 16

Exercises

Name the ordered pair for each point.

1. A
2. B
3. C
4. D
5. E
6. F
7. G
8. H
9. I
10. J
11. K
12. L

Graph and label these points on a coordinate grid.

13. $A(6, 3)$
14. $B(^-2, 5)$
15. $C(3, ^-5)$
16. $D(8, ^-2)$
17. $E(5, 8)$
18. $F(^-6, ^-7)$

Name the letter for each ordered pair.

19. $(1, 5)$
20. $(^-2, 4)$
21. $(6, ^-3)$
22. $(2, ^-4)$
23. $(5, 5)$
24. $(^-7, 4)$

Show Math Copy and connect the following ordered pairs in order on a coordinate grid. What does the figure look like?

25. $(0, 0), (7, 0), (9, 2), (7, 2), (7, 3),$
    $(6, 3), (5, 2), (0, 1), (0, 0)$

26. $(0, 0), (9, 0), (7, 2), (6, 2), (6, 6),$
    $(4.5, 8), (3, 6), (3, 2), (2, 2), (0, 0)$

Data Bank 27. Use the map on p. 493 to plan a trip. Visit at least five cities. Describe your trip using ordered pairs. Trade your plan with a classmate. Find what cities your classmate planned to visit.

Problem Solving 28. You are at $(1, 5)$ on the coordinate grid shown. Go straight to $(5, 3)$. How far is the distance if each grid is one square mile?

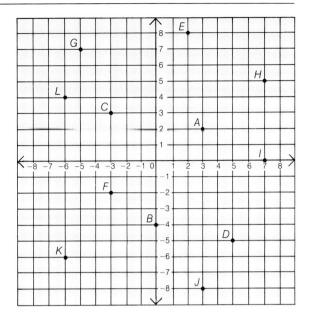
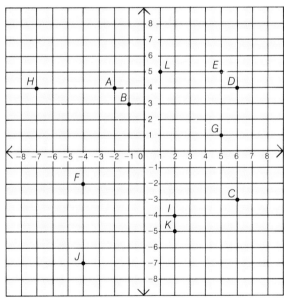
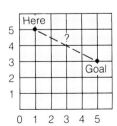

ALGEBRA **477**

16-11 Problem Solving: Determining Reasonable Answers

Objective
Decide if an answer is reasonable.

- SITUATION
- DATA
- PLAN
- ANSWER
- CHECK

Jonelle wants to save $350 to begin dental assistant training. She plans to save $27 each month. She uses the equation $27 \times m = \$350$ to find how many months it will take to reach her goal.

Jonelle used a calculator to solve the equation.

The calculator display showed | 9450. |

Is this a reasonable answer to the problem?

Understand the Situation
- What does the variable m represent?
- What operation should Jonelle use to solve the equation?

An estimate can help you decide if the answer to a problem is reasonable.

Examples Decide if the answer is reasonable without solving the problem. If it is not reasonable, tell why not.

1. Jonelle wants to find *for how many months* she needs to save. She guesses $m = 12$ months. 27×12 is about 30×10, or $300. $300 is close to $350 so the answer should be close to 12. 9450 months is not reasonable.

   To solve $27 \times m = \$350$ for m, *divide* both sides of the equation by 27.

2. *Problem:* Jonelle's other costs for training will be $49 for books, $52.50 for uniforms, $45 for laboratory supplies, and $10 for a locker. What additional amount should she save?

   Answer: | 156.60 | $49 + \$52.50 + \$45 + \$10 \approx (\$50 \times 3) + \$10$
   Jonelle should save about $160. The answer is reasonable.

Try This Decide if the answer is reasonable without solving the problem. If it is not reasonable, tell why not.

a. *Problem:* Lin had a balance of $126.04. He wrote two checks for $61 and $28.39. He deposited $43.57.
   What is Lin's new balance? Answer: | 80.22 |

b. *Problem:* Find the interest if a savings account of $500 earns 5% for 2 years. Answer: | 500. |

478 CHAPTER 16

Exercises

Decide if the answer is reasonable without solving the problem. If it is not reasonable, tell why not.

1. *Problem:* Sal is saving to buy a set of car mechanic tools. He works 16 hours each week at a garage. His pay is $4.50 per hour. If he saves half of his pay, how much will Sal have saved in 10 weeks?

   Answer: `360.`

2. *Problem:* Eric is saving $21 each month to buy a new bike. How long will it take to save $137?

   Answer: 10 months

3. *Problem:* Sophie is making a banner that will go completely around a circular bandstand. The diameter of the bandstand is 30 ft. What length of fabric must she use to go around the circumference of the bandstand?

   Answer: Sophie must use 75 ft of fabric.

4. *Problem:* You are the treasurer for the Soccer Club this year. Draw a circle graph to show the sources of income for the club.

   Answer:

   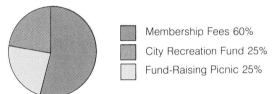

   - Membership Fees 60%
   - City Recreation Fund 25%
   - Fund-Raising Picnic 25%

5. *Problem:* In April the balance in Chieko's checking account was $47.73. She had a service charge of $4. She deposited $32.90. She wrote checks for $9.63, $22, and $19.75. What is Chieko's new balance?

   Answer: `25.25`

Suppose

6. *Problem:* Suppose Jonelle needed to save $338. She planned to save $13 each week. How long would it take her to save $338?

   Answer: It would take Jonelle about half a year.

Write Your Own Problem

7. Write a problem. Use the data in the table. Write an answer that is reasonable or that is not reasonable. Trade problems with a classmate.

   Training Courses Offered at
   Rio Verde Junior College
   Fee: $65 per semester
   Word Processing—2 semesters
   Technical Drawing—2 semesters
     (plus $5 storage drawer key)
   Travel Agent—3 semesters
   Commercial Art—4 semesters
     (plus $5 storage drawer key)

16-12 Problem Solving: Data Collection and Analysis

Objective
Collect and analyze data.

- SITUATION
- DATA
- PLAN
- ANSWER
- CHECK

The Speedy Subscription Service is planning a sales drive for Midland. The Sales Department has assembled data describing the ages of subscribers to four of the most popular magazines from the last sales drive in Midland.

SUBSCRIPTION FORM
Please enter a subscription to _____ magazine for _____ issues.
Subscription Price _____
Name _____
Address _____
City _____
State/Zip _____
☐ Payment enclosed
Charge to _____ Signature _____
☐ MC ☐ Visa Exp. Date _____
Account No. _____

Data Summary Sheet

The following data gives for each magazine the percent of subscribers that fall within each age group. In addition, the table gives the annual subscription rate, the number of issues, the percent of the subscription price that Speedy gets as a commission for taking the order, and the present number of subscriptions in Midland.

Sports Pix

Ages	21–30	31–40	41–50	51–60	61–up
% of Subscribers	54	24	12	6	4

$52/26 issues—40% commission
1614 subscriptions

News Alive

Ages	21–30	31–40	41–50	51–60	61–up
% of Subscribers	18	23	34	13	12

$78/52 issues—30% commission
2314 subscriptions

Flowers and Stuff

Ages	21–30	31–40	41–50	51–60	61–up
% of Subscribers	12	23	32	19	14

$36/12 issues—10% commission
812 subscriptions

Craft World

Ages	21–30	31–40	41–50	51–60	61–up
% of Subscribers	12	37	34	13	4

$18/12 issues—12% commission
746 subscriptions

Work in groups to complete the following.

Situation

Your group is to choose two more magazines to add to the list for the sales drive and choose one from the present list to drop. The new magazines must be like one of the four already on the list: a sports magazine, a news magazine, a garden magazine, or a crafts magazine.

Data Collection

Collect data from your group and two other groups on the ages of their parents and what magazines they currently get. Get a listing of parents' hobbies and interests.

Facts to Consider

This is the breakdown of Midland's population by age at present.

Under 5 years	7.8%
5 to 17 years	19.8%
18 to 24 years	15.1%
25 to 31 years	11.2%
32 to 48 years	11.3%
49 to 55 years	11.4%
56 to 64 years	11.1%
65 and older	12.3%

A recent survey has shown that, on the average, the money that an adult in Midland might spend on a new magazine subscription is $30 per year.

Share Your Group's Plan

- Explain your group's plan to the class. Use a table or a chart as a visual aid. Tell how you made decisions and why you think that your plan is a good one.
- How can you compare two groups' plans? How would you tell which was better?

Suppose

- Suppose that the number of young people in Midland in the 21–40 age range doubled. What effect would that have on your decision?
- Suppose that the number of people in Midland in the 51–up age range doubled. What effect would that have on your decision?
- Suppose that this had been the age breakdown for people in Midland.

Under 5 years	6.2%
5 to 17 years	13.1%
18 to 24 years	14.1%
25 to 31 years	11.4%
32 to 48 years	12.7%
49 to 55 years	13.8%
56 to 64 years	11.3%
65 and older	17.4%

Would this have made a difference in your decision? If so, how?

16-13 Enrichment: Reading a Grid Map

Objective
Read a grid map.

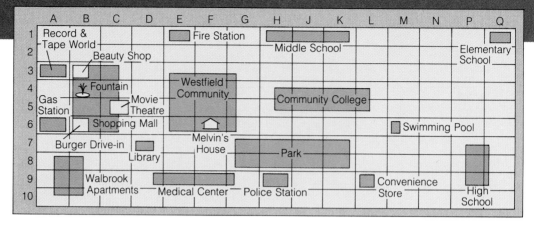

Melvin made a grid map of his town to help his new neighbor Donna find her way around. What are the coordinates for the swimming pool?

Examples

1. Locate the swimming pool on the map. Look for the horizontal and vertical coordinates of the grid. The pool is in the square with coordinates M6.

2. Melvin said that he knew what place was at any pair of coordinates Donna could name. Donna tested him with the coordinates A3. Melvin said the Record & Tape World store was located there. Was he correct?

   Find A on the horizontal axis and move down to 3 on the vertical axis. The Record & Tape World store is in square A3. Melvin was right.

Try This

a. Give the coordinates for the library.　　b. What is found in square E1?

Exercises

Give the coordinates.

1. fire station　　2. police station　　3. convenience store　　4. Melvin's house

Give all the coordinates for all the squares for the place shown on the map.

5. high school　　6. community college　　7. middle school　　8. Walbrook Apartments

What is located in the square?

9. B4　　10. A6　　11. B6　　12. B3　　13. C5

14. G7 to K7 and G8 to K8　　15. B3 to B6 and C3 to C6

16-14 Computer: Squares and Square Roots

Objective
Use a computer program to find squares and square roots.

This program will find squares and square roots of positive numbers on your computer. The symbol $^\wedge$ means *is raised to a power of.* $^\wedge 2$ squares the value of the number or the value of the variable written before it. You can also write the square of a number, n, as $n * n$. To find the square root of a number, write SQR followed by the number in parentheses.

```
10 INPUT ``Number'';N
20 PRINT ``Square is'';N^2;`` , Square root is'';SQR(N)
30 GOTO 10
```

Example

Write the number as a square using $^\wedge$. Find the value of the square.

1. 5

$5^\wedge 2 = 25$ Write the symbol $^\wedge 2$ after the number to be squared.

Try This

Write the number as a square using $^\wedge$. Find the value of the square.

a. 3   **b.** 2   **c.** 0   **d.** 4   **e.** 6   **f.** 8

Example

Write the square root of the number using SQR(n). Find the square root.

2. 9

SQR(9) = 3   Write the number in parentheses after SQR.

Try This

Write the square root of the number using SQR(n). Find the square root.

g. 4   **h.** 25   **i.** 16   **j.** 100   **k.** 144   **l.** 36

Exercises

Write the number as a square using $^\wedge$. Find the value of the square.

1. 1   **2.** 7   **3.** 9   **4.** 10   **5.** 12   **6.** 13

Write the square root of the number using SQR(n). Find the square root.

7. 1   **8.** 64   **9.** 49   **10.** 121   **11.** 81   **12.** 169

A **Pythagorean triple** is a set of three numbers that satisfies the Pythagorean Theorem. For example, 3, 4, and 5 form a Pythagorean triple: $3^2 + 4^2 = 5^2$. Decide if the set of numbers forms a Pythagorean triple.

13. 6, 8, 10   **14.** 5, 12, 13   **15.** 4, 4, 16   **16.** 12, 35, 37

CHAPTER 16 REVIEW

Match.
1. $c + 5 = {}^-12$ A. 28
2. $z - 25 = 80$ B. $^-4$
3. $w - 8 = 15$ C. 8
4. $a - 2 = {}^-6$ D. $^-25$
5. $p + 12 = 23$ E. $^-17$
6. $^-9 = b + 16$ F. 119
7. $25 = 17 + b$ G. 105
8. $y - ({}^-6) = 34$ H. 23
9. $^-13 = k - 7$ I. 11
10. $m - 29 = 90$ J. $^-6$

Write the missing number.
11. If $2x = 10$, then $x = $ _____ .
12. If $8z = {}^-72$, then $z = $ _____ .
13. If $^-3y = {}^-12$, then $y = $ _____ .
14. If $x \div {}^-2 = 6$, then $x = $ _____ .

Fill in the box with the correct number or letter.

15. $15 = 6t - 3$
    $\square = 6t$
    $3 = t$

16. $4x + 8 = {}^-24$
    $\square x = {}^-32$
    $x = \square$

17. $^-5k - 10 = {}^-45$
    $^-5k = \square$
    $\square = 7$

18. The difference between 50 and n is written:
    A. $50n$
    B. $50 - n$
    C. $n = 50$
    D. $50 \div n$

19. Seventy-two increased by 2 is written:
    A. $72 \div 2$
    B. 72×2
    C. $72 - 2$
    D. $72 + 2$

20. The sum of the square root of 64 and the square root of 121 is:
    A. 8
    B. 11
    C. 19
    D. 3

21. The difference between the square root of 81 and the square of 3 is:
    A. 78
    B. 6
    C. 0
    D. 3

22. Find the missing length in this right triangle.
    $a = 24$ $b = 32$ $c = $ _____

CHAPTER 16 TEST

1. Clem got 8 cassette tapes for his birthday. Now he has 42 tapes. How many tapes did he have before his birthday?

Write three word expressions for each algebraic expression.

2. $q + 9$
3. $v - 4$
4. $12c$
5. $\frac{h}{7}$

Solve.

6. $a + 12 = {}^-9$
7. ${}^-14 = b - 6$
8. $\frac{w}{12} = {}^-5$
9. ${}^-4u = {}^-64$
10. $9z - 7 = {}^-97$
11. $0.6 - 0.3c = 0.12$

12. Parking costs $1 for the first hour and $0.75 for each additional half hour. Amy parked her car for $5\frac{1}{2}$ hours. How much did she pay for parking?

Solve.

13. $7y - 0.45 = 0.25$
14. $\frac{p}{-6} = {}^-3.7$
15. $3k + 0.33 = {}^-6$

Find the answer. Use the table on p. 495.

16. $\sqrt{169}$
17. $\sqrt{225}$
18. $\sqrt{2601}$
19. $\sqrt{5776}$
20. 33^2
21. 57^2
22. 86^2
23. 94^2

Find the missing length in the right triangle. Round to the nearest thousandth.

24.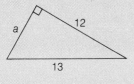
25.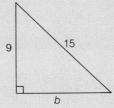
26.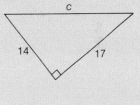

Graph and label these points on a coordinate grid.

27. $(3, 0)$
28. $({}^-4, 1)$
29. $({}^-2, {}^-3)$

Decide if the answer is reasonable without solving the problem. If it is not reasonable, tell why not.

30. *Problem:* Rona planned to fence an odd-shaped garden plot. The sides measured 12 ft, 5 ft, 9 ft, 11 ft, and 15 ft. How much fencing should she buy?

    Answer: Rona should buy 41 ft of fencing.

CHAPTER 16 CUMULATIVE REVIEW

1. A fruit stand sold 5 times as many apples as peaches. It sold half as many oranges as apples. The stand sold 50 oranges. How many apples and peaches did it sell?
   A. 100 apples, 20 peaches
   B. 100 apples, 80 peaches
   C. 500 apples, 50 peaches
   D. 100 apples, 250 peaches

2. Find the square root. $\sqrt{121}$
   A. 12
   B. 14
   C. 11
   D. 61

3. The hat store has 15 hat displays in a row. There are 12 hats in the first display. There are 16 hats in the second display. There are 20 hats in the third display. If this rate continues, how many hats are there in the fifteenth display?
   A. 60 hats
   B. 72 hats
   C. 56 hats
   D. 68 hats

4. Write an algebraic expression for this sentence. Enrique is 6 years younger than Camila.
   A. $6 + c$
   B. $6 - c$
   C. $c - 6$
   D. $\frac{6}{c}$

5. Find the area in square yards of a room that measures 21 ft by 18 ft.
   A. 13 yd^2
   B. 378 ft^2
   C. 42 yd^2
   D. 13 ft^2

6. Solve.
   $\frac{x}{6} = 582$
   A. 97
   B. 3492
   C. 3482
   D. 3060

7. Name the pair of lines as parallel, skew, or perpendicular.
   A. parallel
   B. skew
   C. perpendicular

8. Divide.
   $107 \overline{)535}$
   A. 5 R7
   B. 5 R5
   C. 4 R25
   D. 5

9. Multiply.
   $^{-}19 \times (^{-}19)$
   A. $^{-}371$
   B. 361
   C. 38
   D. $^{-}38$

10. Find the percent of increase or decrease. 52 to 234
    A. 305% increase
    B. 182% increase
    C. 3.5% decrease
    D. 350% increase

11. Divide.
    $2\frac{3}{4} \div \frac{5}{8}$
    A. $4\frac{4}{5}$
    B. $2\frac{1}{8}$
    C. $\frac{2}{5}$
    D. $4\frac{2}{5}$

12. Solve.
    $3x - 11 = 22$
    A. 11
    B. $3\frac{2}{3}$
    C. 33
    D. $4\frac{1}{3}$

13. Name the ordered pair for point A.
    A. $(5, {}^-2)$
    B. $(2, 5)$
    C. $(4, {}^-2)$
    D. $(5, {}^-1)$

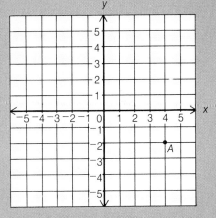

14. Estimate. 99 is about what percent of 392?
    A. 250%
    B. 25%
    C. 400%
    D. 99%

15. On one floor of an apartment building, 58% of the space is for apartments. Sixteen percent is for storage. What percent of the space is left for hallways?
    A. 74%
    B. 26%
    C. 7.4%
    D. 4.2%

16. Solve.
    $37 = h - ({}^-8)$
    A. 29
    B. 18
    C. 45
    D. ${}^-29$

17. Mel had to paint a triangular deck. The base of the triangle was 15 ft. The height of the triangle was 8 ft. What area would Mel paint?
    A. 15 ft^2
    B. 60 ft^2
    C. 23 ft^2
    D. 120 ft^2

18. Solve.
    $p + 26 = {}^-41$
    A. ${}^-67$
    B. 15
    C. 67
    D. 25

19. Find the missing length of the right triangle.
    $a = \square$ cm
    $b = 24$ cm
    $c = 25$ cm
    A. 49 cm
    B. ${}^-1$ cm
    C. 7 cm
    D. 1 cm

20. Add. Use mental math.
    $299 + 3$
    A. 296
    B. 2993
    C. 292
    D. 302

21. Estimate.
    $33 + 31 + 28 + 29$
    A. 120
    B. 90
    C. 150
    D. 100

22. Multiply.
    3.05×1000
    A. 0.305
    B. 0.0305
    C. 30.5
    D. 3050

DATA BANK

Record Temperature and Snowfall for 1985

T = trace — no data

Station	Temperature (°F)				Sleet or Snow		
	Highest	Date	Lowest	Date	Total (in.)	Greatest in 24 h	Date
Albany, NY	92	8/15	−7	12/26	50.7	6.5	3/4
Albuquerque, NM	100	7/6	5	2/1	7.1	2.0	2/3
Anchorage, AK	76	6/30	−10	2/22	—	5.6	4/8
Asheville, NC	93	6/5	−16	1/21	8.5	2.4	2/12
Atlanta, GA	99	6/6	−8	1/21	1.9	1.5	2/11
Baltimore, MD	99	8/14	−6	1/21	10.9	3.7	1/31
Barrow, AK	68	8/2	−42	2/25	22.5	1.6	9/21
Birmingham, AL	100	6/5	−6	1/21	0.3	0.3	2/12
Bismarck, ND	100	7/7	−34	2/4	51.5	9.4	11/25
Boise, ID	101	7/9	−12	2/4	41.6	6.0	11/28
Boston, MA	93	8/15	6	1/21	27.2	5.9	2/5
Buffalo, NY	91	7/25	−10	1/21	168.2	16.9	1/19
Burlington, VT	89	8/14	−15	12/19	82.0	9.2	3/4
Charleston, SC	100	6/2	6	1/21	T	T	4/28
Charleston, WV	96	8/13	−15	1/21	49.2	7.8	2/12
Chicago, IL	99	9/7	−27	1/20	38.8	7.5	1/1
Cincinnati, OH	92	9/9	−21	1/20	28.8	6.2	2/12
Cleveland, OH	92	7/8	−18	1/20	74.2	8.1	2/13
Columbus, OH	92	9/8	−19	1/20	43.8	6.3	2/12
Concord, NH	93	8/15	−19	2/4	55.5	10.4	3/4
Dallas, TX	106	9/1	7	2/2	5.1	2.4	2/1
Denver, CO	98	7/7	−15	1/31	67.5	8.7	9/28
Des Moines, IA	101	1/8	−20	1/20	45.3	10.9	12/1
Detroit, MI	91	9/7	−15	2/3	60.9	5.5	3/3
Dodge City, KS	18	6/8	−7	1/12	26.5	6.4	1/9
Duluth, MN	85	7/13	−13	1/19	110.0	15.0	9/2
Fairbanks, AK	83	6/3	−48	2/21	54.4	5.4	11/16
Fresno, CA	107	7/3	29	12/14	0.0	T	4/11
Galveston, TX	95	9/1	20	1/21	T	T	1/12
Grand Rapids, MI	92	7/13	−12	2/9	88.1	9.1	2/11
Hartford, CT	97	8/14	−3	2/4	27.2	4.4	2/5
Helena, MT	101	7/9	−27	11/23	45.5	7.2	12/11
Honolulu, HI	93	9/16	54	2/1	0.0	0.0	—
Houston, TX	102	9/1	16	1/21	1.7	1.0	1.12
Huron, SD	105	7/7	−30	12/18	80.5	18.3	3/3
Indianapolis, IN	91	7/14	−22	1/20	29.8	4.9	2/10
Jackson, MS	99	6/4	2	1/21	1.7	1.4	2/1
Jacksonville, FL	100	6/4	7	1/21	T	T	1/20
Kansas City, MO	98	7/13	−14	1/20	27.9	7.6	1/9
Lander, WY	96	7/10	−22	1/31	100.9	22.7	11/12
Little Rock, AR	102	8/1	−2	1/20	11.3	6.3	2/1
Los Angeles, CA	97	7/1	39	12/12	0.0	0.0	—
Louisville, KY	96	7/14	−16	1/20	13.7	5.5	2/1
Marquette, MI	86	7/13	−27	2/1	269.3	25.8	12/1
Memphis, TN	97	6/4	−4	1/20	20.7	8.1	1/3

DATA BANK

Record Temperature and Snowfall for 1985

T = trace   — no data

Station	Temperature (°F)				Sleet or Snow		
	Highest	Date	Lowest	Date	Total (in.)	Greatest in 24 h	Date
Miami, FL	98	6/4	30	1/22	0.0	0.0	—
Milford, UT	104	6/6	−11	2/5	63.6	20.4	3/27
Milwaukee, WI	98	7/8	−25	1/20	64.6	10.9	1/1
Minneapolis, MN	102	6/8	−25	1/19	91.9	14.7	3/31
Mobile, AL	99	6/6	3	1/21	T	T	2/11
Moline, IL	97	9/7	−21	1/20	35.7	8.8	1/1
Nashville, TN	98	7/14	−17	1/21	18.3	6.7	1/1
Newark, NJ	97	8/15	−8	1/21	21.6	3.9	2/5
New Orleans, LA	94	6/5	14	1/21	0.4	0.4	1/20
New York, NY	95	8/15	−2	1/21	19.5	5.7	2/5
Nome, AK	83	7/19	−28	2/20	52.0	3.6	5/6
Norfolk, VA	97	9/3	−3	1/21	4.3	2.3	1/20
Oklahoma City, OK	104	5/30	1	1/20	6.7	2.8	12/12
Omaha, NE	99	6/8	−18	1/20	23.5	6.0	3/30
Philadelphia, PA	93	8/14	−6	1/21	17.8	3.5	1/17
Phoenix, AZ	115	8/24	28	2/1	0.1	0.1	12/11
Pittsburgh, PA	92	8/13	−18	1/21	45.4	4.8	4/8
Portland, ME	91	8/15	−9	2/4	49.1	9.9	3/4
Portland, OR	101	7/19	13	11/24	9.8	2.2	2/4
Providence, RI	93	8/15	2	1/21	24.8	7.6	2/5
Raleigh, NC	95	7/10	−9	1/21	4.1	2.4	1/28
Rapid City, SD	104	7/7	−19	12/2	66.9	10.4	3/2
Reno, NV	101	6/18	4	11/14	31.2	15.4	11/9
Richmond, VA	100	8/14	−6	1/21	9.6	4.2	1/20
Rochester, NY	91	8/13	−19	1/21	99.8	6.3	3/4
St. Louis, MO	97	7/14	−18	1/20	12.1	2.9	12/19
Salt Lake City, UT	102	7/5	−9	2/1	74.7	10.7	12/7
San Antonio, TX	103	9/2	14	2/2	15.9	13.2	1/12
San Diego, CA	98	10/3	38	12/13	T	T	11/12
San Francisco, CA	99	7/1	33	12/12	0.0	0.0	—
Sault Ste. Marie, MI	85	8/9	−26	2/2	174.3	12.4	12/1
Savannah, GA	104	6/2	3	1/21	0.0	0.0	—
Seattle, WA	93	7/19	10	11/23	24.9	7.8	11/21
Shreveport, LA	103	8/31	10	1/21	4.8	4.4	2/1
Sioux City, IA	103	6/8	−18	12/18	33.6	10.3	3/30
Spokane, WA	100	7/9	−21	11/23	61.4	8.1	11/8
Springfield, MO	101	7/12	−13	1/20	25.9	4.6	1/19
Syracuse, NY	91	8/15	−7	2/4	124.2	10.8	1/22
Tampa, FL	99	6/5	21	1/21	0.0	0.0	—
Washington, DC	98	8/15	−4	1/21	10.7	3.9	1/17
Wilmington, DE	94	8/14	−14	1/21	16.7	3.9	1/25

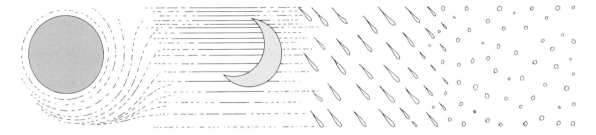

DATA BANK

Mileage Chart

Atlanta						
1115	Boston					
722	1009	Chicago				
2244	3082	2113	Los Angeles			
896	219	831	2849	New York		
2515	3174	2205	425	2996	San Francisco	

Gravitational Pull of Planets and the Moon

Multiply Earth weight by the given factor to find weight on another planet or the moon.

Planet/Moon	Factor
Mercury	0.37
Venus	0.88
Mars	0.38
Jupiter	2.64
Saturn	1.15
Uranus	1.15
Neptune	1.12
Pluto	1.04
Moon	0.17

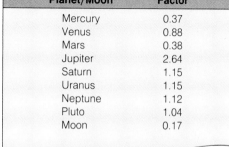

Sports Equipment

Item	Diameter	Weight
soccer ball	$8\frac{3}{5}$ in.	14–16 oz
softball	$3\frac{39}{50}$ in.	$6\frac{1}{4}$–7 oz
tennis ball	$2\frac{9}{16}$ in.	$2\frac{1}{32}$ oz
volleyball	$8\frac{59}{100}$ in.	$9\frac{1}{2}$ oz
handball	$1\frac{7}{8}$ in.	$2\frac{3}{10}$ oz
golf ball	$1\frac{17}{25}$ in.	$1\frac{31}{50}$ oz
bowling ball	$8\frac{3}{5}$ in.	$15\frac{5}{19}$ lb
baseball	$2\frac{7}{23}$ in.	$5\frac{3}{16}$ oz
basketball	$9\frac{4}{17}$ in.	20–22 oz
hockey puck	3 in.	$5\frac{1}{2}$ oz

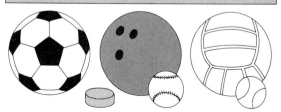

Nails

Penny size	Length (in.)	Common nails per pound	Casing nails per pound	Finishing nails per pound
2d	1	875	1000	1350
3d	$1\frac{1}{4}$	550	625	850
4d	$1\frac{1}{2}$	300	450	600
6d	2	175	225	300
8d	$2\frac{1}{2}$	100	150	200
10d	3	65	95	125
16d	$3\frac{1}{2}$	45	70	—
20d	4	30	—	—

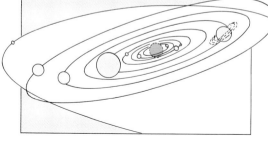

DATA BANK

Gold Medalists in Men's High Jump Olympic Games 1932–1984

Year	Medalist	Height
1932	Duncan McNaughton, Canada	6 ft $5\frac{5}{8}$ in.
1936	Cornelius Johnson, United States	6 ft $7\frac{13}{16}$ in.
1948	John L. Winter, Australia	6 ft 6 in.
1952	Walter Davis, United States	6 ft 8.32 in.
1956	Charles Dumas, United States	6 ft $11\frac{1}{4}$ in.
1960	Robert Shavlakadze, USSR	7 ft 1 in.
1964	Valery Brumel, USSR	7 ft $1\frac{3}{4}$ in.
1968	Dick Fosbury, United States	7 ft $4\frac{1}{4}$ in.
1972	Yuri Tarmak, USSR	7 ft $3\frac{3}{4}$ in.
1976	Jacek Wszola, Poland	7 ft $4\frac{1}{2}$ in.
1980	Gerd Wessig, E. Germany	7 ft $8\frac{3}{4}$ in.
1984	Dietmar Mogenburg, W. Germany	7 ft $8\frac{1}{2}$ in.

How an Average Adult Spends a Day

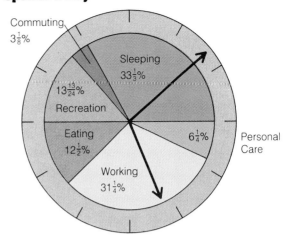

Commuting $3\frac{1}{8}$%
Sleeping $33\frac{1}{3}$%
Recreation $13\frac{13}{24}$%
Eating $12\frac{1}{2}$%
Personal Care $6\frac{1}{4}$%
Working $31\frac{1}{4}$%

Blueprint

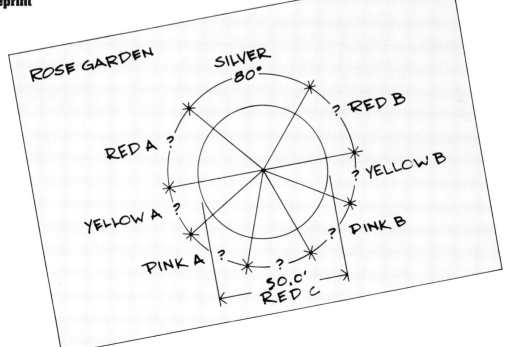

DATA BANK

6% Sales Tax Schedule

Transaction	Tax	Transaction	Tax	Transaction	Tax	Transaction	Tax
.01– .10	.00	6.42– 6.58	.39	12.92–13.08	.78	19.42–19.58	1.17
.11– .22	.01	6.59– 6.74	.40	13.09–13.24	.79	19.59–19.74	1.18
.23– .39	.02	6.75– 6.91	.41	13.25–13.41	.80	19.75–19.91	1.19
.40– .56	.03	6.92– 7.08	.42	13.42–13.58	.81	19.92–20.08	1.20
.57– .73	.04	7.09– 7.24	.43	13.59–13.74	.82	20.09–20.24	1.21
.74– .90	.05	7.25– 7.41	.44	13.75–13.91	.83	20.25–20.41	1.22
.91–1.08	.06	7.42– 7.58	.45	13.92–14.08	.84	20.42–20.58	1.23
1.09–1.24	.07	7.59– 7.74	.46	14.09–14.24	.85	20.59–20.74	1.24
1.25–1.41	.08	7.75– 7.91	.47	14.25–14.41	.86	20.75–20.91	1.25
1.42–1.58	.09	7.92– 8.08	.48	14.42–14.58	.87	20.92–21.08	1.26
1.59–1.74	.10	8.09– 8.24	.49	14.59–14.74	.88	21.09–21.24	1.27
1.75–1.91	.11	8.25– 8.41	.50	14.75–14.91	.89	21.25–21.41	1.28
1.92–2.08	.12	8.42– 8.58	.51	14.92–15.08	.89	21.42–21.58	1.29
2.09–2.24	.13	8.59– 8.74	.52	15.09–15.24	.91	21.59–21.74	1.30
2.25–2.41	.14	8.75– 8.91	.53	15.25–15.41	.92	21.75–21.91	1.31
2.42–2.58	.15	8.92– 9.08	.54	15.42–15.58	.93	21.92–22.08	1.32
2.59–2.74	.16	9.09– 9.24	.55	15.59–15.74	.94	22.09–22.24	1.33
2.75–2.91	.17	9.25– 9.41	.56	15.75–15.91	.95	22.25–22.41	1.34
2.92–3.08	.18	9.42– 9.58	.57	15.92–16.08	.96	22.42–22.58	1.35
3.09–3.24	.19	9.59– 9.74	.58	16.09–16.24	.97	22.59–22.74	1.36
3.25–3.41	.20	9.75– 9.91	.59	16.25–16.41	.98	22.75–22.91	1.37
3.42–3.58	.21	9.92–10.08	.60	16.42–16.58	.99	22.92–23.08	1.38
3.59–3.74	.22	10.09–10.24	.61	16.59–16.74	1.00	23.09–23.24	1.39
3.75–3.91	.23	10.25–10.41	.62	16.75–16.91	1.01	23.25–23.41	1.40
3.92–4.08	.24	10.42–10.58	.63	16.92–17.08	1.02	23.42–23.58	1.41
4.09–4.24	.25	10.59–10.74	.64	17.09–17.24	1.03	23.59–23.74	1.42
4.25–4.41	.26	10.75–10.91	.65	17.25–17.41	1.04	23.75–23.91	1.43
4.42–4.58	.27	10.92–11.08	.66	17.42–17.58	1.05	23.92–24.08	1.44
4.59–4.74	.28	11.09–11.24	.67	17.59–17.74	1.06	24.09–24.24	1.45
4.75–4.91	.29	11.25–11.41	.68	17.75–17.91	1.07	24.25–24.41	1.46
4.92–5.08	.30	11.42–11.58	.69	17.92–18.08	1.08	24.42–24.58	1.47
5.09–5.24	.31	11.59–11.74	.70	18.09–18.24	1.09	24.59–24.74	1.48
5.25–5.41	.32	11.75–11.91	.71	18.25–18.41	1.10	24.75–24.91	1.49
5.42–5.58	.33	11.92–12.08	.72	18.42–18.58	1.11	24.92–25.08	1.50
5.59–5.74	.34	12.09–12.24	.73	18.59–18.74	1.12	25.09–25.24	1.51
5.75–5.91	.35	12.25–12.41	.74	18.75–18.91	1.13	25.25–25.41	1.52
5.92–6.08	.36	12.42–12.58	.75	18.92–19.08	1.14		
6.09–6.24	.37	12.59–12.74	.76	19.09–19.24	1.15		
6.25–6.41	.38	12.75–12.91	.77	19.25–19.41	1.16		

DATA BANK

Federal Minimum Hourly Wage Rate, 1950–1988

Year	Rate
1950–1955	$0.75
1956–1960	1.00
1961–1962	1.15
1963–1966	1.25
1967	1.40
1968–1973	1.60
1974	2.00
1975	2.10
1976–1977	2.30
1978	2.65
1979	2.90
1980	3.10
1981–1988	3.35

The Great Pyramid

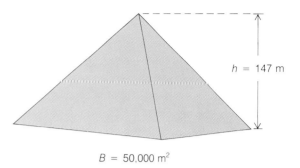

$h = 147$ m

$B = 50{,}000$ m^2

Map of Texas

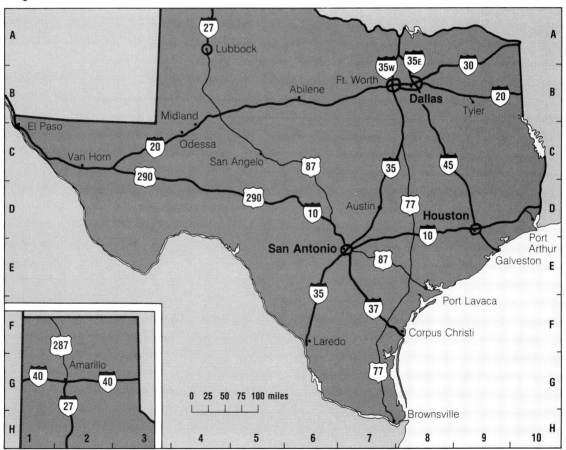

DATA BANK

Symbols

<	is less then	π	pi ($\approx$ 3.14)
>	is greater than	P(x)	probability of the outcome x
=	is equal to	-5	negative 5
$\neq$	is not equal to	$0.\overline{6}$	repeating decimal
$\leq$	is less than or equal to	x^2	x squared
$\geq$	is greater than or equal to	x^3	x cubed
$\approx$	is approximately equal to	$\sqrt{}$	square root
$\sim$	is similar to	$\overleftrightarrow{AB}$	line AB
$\cong$	is congruent to	$\overrightarrow{AB}$	ray AB
'	foot (feet)	$\overline{AB}$	segment AB
"	inch (inches)	$\angle ABC$	angle ABC
$a{:}b, \frac{a}{b}$, a to b	ratio of a to b	$m\angle ABC$	measure of angle ABC
%	percent	$\overleftrightarrow{AB} \perp \overleftrightarrow{CD}$	line AB perpendicular to line CD
°F	degrees Fahrenheit	$\overleftrightarrow{AB} \parallel \overleftrightarrow{CD}$	line AB parallel to line CD
°C	degrees Celsius		

Measurements

Time

24 hours (h) = 1 day (d)   60 minutes (min) = 1 hour   60 seconds (s) = 1 minute

Customary Units of Measure

Length

1 foot (ft) = 12 inches (in.)
1 yard (yd) = 3 feet = 36 inches
1 mile (mi) = 1760 yards = 5280 feet

Capacity

1 cup (c) = 8 fluid ounces (fl oz)
1 pint (pt) = 2 cups = 16 fluid ounces
1 quart (qt) = 2 pints = 4 cups
1 gallon (gal) = 4 quarts = 8 pints

Weight

1 pound (lb) = 16 ounces (oz)
1 ton (T) = 2000 pounds

Temperature

water boils: 212 degrees Fahrenheit (°F)
normal body temperature: 98.6°F
room temperature: 68°F
water freezes: 32°F

Metric Units of Measure

Length

1 kilometer (km) = 1000 meters (m)
1 hectometer (hm) = 100 meters
1 dekameter (dam) = 10 meters
1 decimeter (dm) = 0.1 meter
1 centimeter (cm) = 0.01 meter
1 millimeter (mm) = 0.001 meter

Capacity

1 kiloliter (kL) = 1000 liters (L)
1 milliliter (mL) = 0.001 liter

Mass

1000 kilogram (kg) = 1 ton (t)
1 kilogram (kg) = 1000 grams (g)
1 milligram (mg) = 0.001 gram

Temperature

water boils: 100 degrees Celsius (°C)
normal body temperature: 37°C
room temperature: 20°C
water freezes: 0°C

DATA BANK

Formulas

Perimeter
Rectangle $P = 2l + 2w$, or $2(l + w)$
Square $P = 4s$

Circumference
Circle $C = \pi d$, or $2\pi r$

Area
Rectangle $A = l \times w$, or lw
Square $A = s \times s$, or s^2
Parallelogram $A = b \times h$, or bh
Triangle $A = \frac{1}{2} \times b \times h$, or $\frac{1}{2}bh$
Circle $A = \pi \times r \times r$, or πr^2

Volume
Rectangular prism $V = l \times w \times h$, or lwh
Cube $V = s \times s \times s$, or s^3
Cylinder $V = \pi r^2 \times h$, or $\pi r^2 h$
Cone and pyramid $V = \frac{1}{3} \times B \times h$, or $\frac{1}{3}Bh$

Other
Simple interest $i = prt$
Distance $d = rt$
Pythagorean Theorem $a^2 + b^2 = c^2$

Squares and Square Roots

N	N²	√N	N	N²	√N	N	N²	√N
1	1	1	35	1225	5.916	69	4761	8.307
2	4	1.414	36	1296	6	70	4900	8.367
3	9	1.732	37	1369	6.083	71	5041	8.426
4	16	2	38	1444	6.164	72	5184	8.485
5	25	2.236	39	1521	6.245	73	5329	8.544
6	36	2.449	40	1600	6.325	74	5476	8.602
7	49	2.646	41	1681	6.403	75	5625	8.660
8	64	2.828	42	1764	6.481	76	5776	8.718
9	81	3	43	1849	6.557	77	5929	8.775
10	100	3.162	44	1936	6.633	78	6084	8.832
11	121	3.317	45	2025	6.708	79	6241	8.888
12	144	3.464	46	2116	6.782	80	6400	8.944
13	169	3.606	47	2209	6.856	81	6561	9
14	196	3.742	48	2304	6.928	82	6724	9.055
15	225	3.873	49	2401	7	83	6889	9.110
16	256	4	50	2500	7.071	84	7056	9.165
17	289	4.123	51	2601	7.141	85	7225	9.220
18	324	4.243	52	2704	7.211	86	7396	9.274
19	361	4.359	53	2809	7.280	87	7569	9.327
20	400	4.472	54	2916	7.348	88	7744	9.381
21	441	4.583	55	3025	7.416	89	7921	9.434
22	484	4.690	56	3136	7.483	90	8100	9.487
23	529	4.796	57	3249	7.550	91	8281	9.539
24	576	4.899	58	3364	7.616	92	8464	9.592
25	625	5	59	3481	7.681	93	8649	9.644
26	676	5.099	60	3600	7.746	94	8836	9.695
27	729	5.196	61	3721	7.810	95	9025	9.747
28	784	5.292	62	3844	7.874	96	9216	9.798
29	841	5.385	63	3969	7.937	97	9409	9.849
30	900	5.477	64	4096	8	98	9604	9.899
31	961	5.568	65	4225	8.062	99	9801	9.950
32	1024	5.657	66	4356	8.124	100	10,000	10
33	1089	5.745	67	4489	8.185			
34	1156	5.831	68	4624	8.246			

DATA BANK

Deposit Slip

DEPOSIT		Dollars	Cents
Cash	Currency		
	Coin		
Date _____	Checks		
_____			
Sign here for cash received			
Arrow Bank of Texas			
392 Main Street	Total		
Lufkin, TX 75901	Less Cash Received		
⑈095101340⑈ ⋯1649⋮61403⑈	Net Deposit		

Check Register

NUMBER	DATE	DESCRIPTION OF TRANSACTION	(−) PAYMENT	✓ T	(−) FEE (IF ANY)	(+) DEPOSIT	BALANCE

Check

```
                                                              174
                              _____ 19 ____      75-134/951
 Pay to the
 Order of  _____ | $ [_____]
                                                   Dollars
    TO   Arrow Bank
         of Texas
         392 Main Street
         Lufkin, TX 75901

 For _____

 ⑈095101940⑈ 499000067118011⬛  0001
 INTERCHECKS INC.—SPECIAL DUPLICATE
```

EXTRA MIXED SKILLS REVIEW

Chapter 1

A. Find the answer.
1. $8 - 3$
2. 4×7
3. $5 + 9$
4. $18 \div 6$
5. $12 + 8$
6. $10 \div 2$
7. 8×3
8. $11 - 6$
9. $7 \div 7$
10. $7 + 7$
11. $7 - 7$
12. 7×7
13. Sara spent $11 at the grocery store, $3 at the drugstore, and $8 at the record shop. How much money did she spend in all?
14. Tod started the day with $10. He spent $5 for some socks and $2 for a magazine. How much money did he have left?
15. Find the total of 7 and 4 and 3.
16. What is 5 subtracted from 12?
17. What is 9 multiplied by 6?
18. How much is 15 divided by 5?
19. What is 27 divided by 3?
20. What is 4 times as much as 8?

B. Find the answer.
1. $2 + 7$
2. 3×6
3. $12 \div 4$
4. $11 - 0$
5. $10 - 2$
6. $11 + 4$
7. 8×6
8. $10 \div 5$
9. 9×9
10. $9 \div 9$
11. $9 + 9$
12. $9 - 9$
13. How much is 35 divided by 7?
14. Find the difference between 7 and 4.
15. What is 14 divided into 2 equal parts?
16. Find the total of 2 and 6 and 4.
17. What is 5 times as much as 4?
18. Find the product of 2 and 10.

Solve. Use this list of numbers: 20, 16, 4, 8, 24, 12.

19. What is the least number?
20. What is the greatest number?
21. Find a number that is 5 times as much as one of the other numbers.
22. What number divides evenly into all the other numbers?

C. Find the answer.
1. $4 \div 4$
2. 5×5
3. $10 + 10$
4. $11 - 3$
5. 6×6
6. $6 - 6$
7. $6 \div 6$
8. $6 + 6$
9. $4 + 5$
10. $18 \div 9$
11. 5×3
12. $10 - 1$
13. Find the product of 7 and 4.
14. Find the difference between 8 and 5.
15. Twelve people joined a club. They each paid $2 dues. How much money did the new members pay all together?
16. Four baseball fans shared 20 baseball cards equally. How many cards did each fan get?

EXTRA MIXED SKILLS REVIEW

Chapter 2

A. Arrange in order from least to greatest.
1. 1, 0.9, 1.9, 0.19, 1.09
2. 0.054, 0.45, 0.045, 0.0544, 0.4
3. 1234, 1324, 1342, 1243, 1224
4. 79.1, 78.9, 70.91, 7.091, 9.701

Round to the nearest hundred.
5. 507
6. 663
7. 98
8. 918
9. 16,792

Round to the nearest tenth.
10. 42.01
11. 7.924
12. 16.078
13. 292.96
14. 3.45

Estimate.
15. 210 × 12
16. 475 + 308
17. 1261 − 729
18. 194 × 58

B. Write as a standard numeral or in money notation.
1. forty-six hundredths
2. twelve thousand, seven hundred two
3. eighty-nine dollars and thirty-four cents
4. nine million, fourteen thousand, six

Round to the given place.
5. $0.70 dollar
6. 23.4324 thousandth
7. 1863.009 tenth
8. 0.997 hundredth

Compare. Use <, >, or =.
9. 5678 ☐ 5879
10. 4.602 ☐ 40.62
11. 93.71 ☐ 93.710
12. 0.008 ☐ 0.08

Describe the action. Tell which operation you would use to solve the problem.
13. Mr. Li sold ◯ tools from a rack of △ tools. How many tools were left?
14. Jack is ☐ inches tall and Jerry is ◯ inches tall. How much taller is Jack?

C. Estimate.
1. 218 ÷ 29
2. 642 ÷ 12
3. 171 ÷ 8
4. 413 ÷ 50

Write the word name.
5. 147,900
6. 3512
7. $82.15
8. $632.41

Does the situation call for an exact answer or an estimate? Tell why.
9. A newspaper editor is finding the population of a city to write an article.
10. The tax director is finding the population of the city to write a budget for the next year.
11. You are checking your bank balance after writing a check.
12. You are finding the cost of a coat on sale at $\frac{1}{3}$ off.

Chapter 3

A. Find the answer.
1. 173 − 110
2. 0.84 + 3.57
3. 491 + 68
4. 62.8 − 29.4
5. 84.9 + 27
6. 407 − 352
7. 529 + 108
8. 480 − 237

Describe the action. Tell which operation you would use to solve the problem.
9. How many tables would you need to seat △ students at an awards dinner? Each table seats ◯.
10. ☐ students from each of △ classes came to the dinner. How many students attended?

Choose the calculation method you would use to find the answer. Tell why.
11. 100 + 642 + 3
12. 38.4 + 9.007 + 25
13. 204 − 92

B. Find the answer.
1. 5,149 + 27,462
2. 471.6 − 0.0385
3. $10.47 − 6.82
4. 15,608 − 9,939
5. 21.09 − 14.347
6. 91 + 742 + 609
7. 12 − 7.4896
8. 22.745 + 303

Find the answer. Use mental math.
9. $36.50 + $4.12
10. 96 − 4
11. 18¢ + 3¢
12. $17.43 − $6
13. 404 + 72
14. 5 + 28
15. $5 + $2.25
16. 119 − 2

Decide if the answer is reasonable without solving the problem. If it is not reasonable, tell why not.
17. *Problem:* Add $12.15, $37.42, and $19.95. What is the total?
    Answer: $69.52
18. *Problem:* How much fencing will enclose a square pen 4 yards on a side?
    Answer: 14 yards

C. Estimate.
1. 672 − 203
2. 987 + 120
3. 555 + 282
4. 3245 − 1996
5. 42 + 819 + 287
6. 567 + 29 + 541
7. 871 − 236
8. 38 + 61 + 75

Describe the action. Tell which operation you would use to solve the problem.
9. Bike tires cost $△. That is $☐ more than they cost last year. What was last year's price?
10. J.J. has saved $☐ for a motorbike. The bike costs $△. How much more money does J.J. need to save?

Find the answer.
11. $100 − $45.95
12. 7907 − 519
13. 30,182 + 7,929
14. 8088 − 679
15. 74 + 951 + 0.4
16. 336 − 12.08
17. 63.041 − 0.296
18. 25.046 − 16.7

EXTRA MIXED SKILLS REVIEW

Chapter 4

A. Multiply.
1. 0.0721×10
2. 690×1000
3. 8.33×100
4. 1000×94
5. 10×452.7
6. 201.3×100
7. 1000×0.4721
8. 5.5×100

Find the area.
9. $l = 42$ inches, $w = 10$ inches
10. $l = 64$ feet, $w = 32$ feet

Write using exponents and as a standard numeral.
11. $4 \times 4 \times 4 \times 4$
12. $(0.7)(0.7)(0.7)$
13. 3.1 cubed
14. 12 squared

Find the answer.
15. 550×2
16. 164×96
17. $\$910 - \$73$
18. $5.24 - 0.78$

B. Estimate.
1. $3200 + 2952 + 792 + 3031$
2. $104 + 85 + 98 + 105$
3. $31 + 78 + 33 + 80$
4. $17 + 21 + 29 + 19 + 32$

Write as a standard numeral.
5. 10^4
6. 8^2
7. $(0.3)^3$
8. $(2.5)^2$
9. 7^4
10. 5^5

Find the product or sum. Use mental math.
11. 609×6
12. $467 + 22$
13. $31 + 24 + 40$
14. 81×3

Find the answer.
15. $493 + 831 + 64$
16. 47×247
17. 46×8.01
18. 83×97

C. Estimate. Is the exact answer greater or less than the given number? Use $<$ or $>$.
1. $421 \times 704 \; \square \; 30{,}000$
2. $296 \times 141 \; \square \; 30{,}000$
3. $807 \times 539 \; \square \; 440{,}000$

Describe the action. Tell which operation you would use to solve the problem.
4. The Junior Class wants to raise $\$\triangle$ by selling dance tickets at $\$\bigcirc$ each. How many tickets must they sell?
5. $\triangle$ crafts students will share $\square$ pounds of clay equally. How much clay will each student have?

Find the answer.
6. 44×6
7. 78×7
8. 309×62
9. 89×14
10. $\$14.94 + \$6.02$
11. $71 - 0.69$
12. $\$39 \times 1.5$
13. 132×3.2

EXTRA MIXED SKILLS REVIEW

Chapter 5

A. Estimate.
1. 39 + 41 + 16 + 38
2. 596 + 82 + 611
3. 2941 + 618
4. 374 + 22 + 61

Choose the calculation method you would use to find the answer. Tell why.
5. 213 × 40
6. 972.345 ÷ 10
7. 16 × 50.78
8. 24 + 13.6

Solve.
9. How many $10 bills could you get for $126.35? How much money is left?
10. How many $10 bills could you get for $4290? How much money is left?

Find the answer.
11. 2719 ÷ 8
12. 29,785 ÷ 35
13. 2646 ÷ 42
14. 812 ÷ 14
15. 79)5056
16. 18)234
17. 12)3171.6
18. 5)1046.5

B. Find the answer.
1. 12,942 + 6,798
2. 34.02 − 9.74
3. 10.213 − 0.789
4. 4578 + 9046
5. 12.5 × 3.6
6. 17,334 ÷ 54
7. 3620.7 ÷ 9
8. 201.7 × 43
9. You have a 3467-pound truckload of sand. How many 100-pound sacks can you fill with this sand?
10. A wagon train of pioneers traveled about 17 miles per day for 100 days. How far did they travel?

Find the answer. Use mental math.
11. 0.431 ÷ 10
12. 708.942 ÷ 10
13. 678 ÷ 100
14. 12,586 ÷ 10
15. 240 + 15
16. 23 − 4
17. 560 + 120
18. 175 − 25

C. Find the answer.
1. $34.95 × 5
2. $16.07 × 12
3. $58.43 × 16
4. $904.60 × 100
5. 7749 ÷ 63
6. 21 × 7.79
7. 94 × 0.213
8. 1785 ÷ 21
9. Suppose 12 people share $36.72 equally. How much money will each receive?
10. Suppose Gary saves $6.50 per week. How long will it take him to save $234?

Find the answer. Use mental math.
11. 72 × 3
12. 41 × 5
13. 6 × 31
14. 10 × 24

Choose the calculation method you would use to find the answer. Tell why.
15. 469 × 243
16. $946.12 ÷ 10
17. 209 × 72
18. 839 ÷ 37.9

EXTRA MIXED SKILLS REVIEW

Chapter 6

A. Solve. Use the graph for Exercises 1–4.
1. Did enrollment increase or decrease between 1986 and 1987?
2. When was the greatest increase in enrollment?
3. How much did enrollment change between 1985 and 1987?
4. Use the data in the line graph to make a pictograph. Let 🕴 = 1000 students.

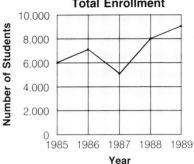

Big Pine School District Total Enrollment

Find the answer.
5. 0.007×6
6. $0.089 + 4.2$
7. $1.62 - 0.972$
8. $25.992 \div 7.6$

B. Find the answer.
1. $12{,}642 \div 1000$
2. $29{,}304 \div 100$
3. 7821×1000
4. 26×100

Find the answer. Use mental math.
5. 31×4
6. $150 + 275$
7. $\$17.92 - \$0.41$
8. 6×54

9. Gym students were shooting baskets to try to make 5 baskets in a row. The table shows how many trials it took the students to reach their goal. Make a vertical bar graph of the data.

Number of Students	Number of Trials
5	1
19	2
40	3
26	4
15	5 or more

C. Find the answer.
1. $0.103 - 0.08$
2. $2.8 + 309 + 68$
3. $\$97.45 \times 78$
4. $84{,}409 + 632$

Solve. Use the bar graph for Exercises 5–8.
5. Estimate the traffic from 7 to 8 p.m.
6. When is the heaviest traffic?
7. Estimate the traffic from 4 to 6 p.m.
8. Which time period has more traffic: 4 to 5 p.m. or 7 to 8 p.m.?

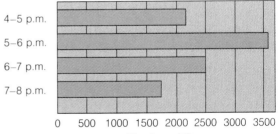

Rush Hour Traffic on the B Street Bridge

Chapter 7

A. Estimate.
1. 9.7 × 806
2. 17)3600
3. 291 × 39
4. 4759 ÷ 122

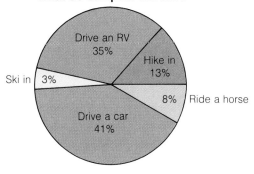

Travel to the Campsite
What do campers choose?

Solve. Use the circle graph for Exercises 5–8.
5. What is the most popular way for campers to travel?
6. What percent more campers hike than ride horses?
7. What percent of campers hike, ski, or ride a horse?
8. Suppose the skiers drove cars. Then what percent of the campers would travel in cars?

B.
1. Draw a horizontal bar graph for this data from the Running Club.
   5 members run 0–12 mi/wk,
   17 members run 12–20 mi/wk,
   8 members run more than 20 mi/wk
2. Draw a line graph for this data.
   Average number of Texas tornadoes for the first six months of the year.
   January, 1; February, 3; March, 8; April, 18; May, 36; June, 18

Write using exponents and as a standard numeral.
3. 3 × 3 × 3 × 3
4. 100 cubed
5. 9 × 9 × 9
6. 29 squared
7. 5 to the third power
8. 7 × 7 × 7 × 7 × 7 7^5,

Compare. Use <, >, or =.
9. 2 ft □ 21 in.
10. 81.6 km □ 816 m
11. 1 qt □ 32 fl oz
12. 91 mL □ 0.91 L
13. 18 fl oz □ 2 c
14. 1 mi □ 5200 ft

C. Change the unit.
1. 13 ft = _____ in.
2. 4 qt = _____ pt
3. 4 lb = _____ oz
4. 4 qt = _____ c
5. 2 gal = _____ pt
6. 2.5 T = _____ lb

Write as a standard numeral.
7. 14^2
8. 3^3
9. 9^2
10. 7 squared
11. 10^5
12. $(1.3)^2$
13. $(2.05)^2$
14. 100^4
15. $(1.2)^4$
16. 10^3
17. 4 cubed
18. 6^5

Arrange in order from least to greatest.
19. 176, 16.7, 1.76, 167, 17.6
20. 4 cm, 41 mm, 4 km, 40 cm, 4.0 m

EXTRA MIXED SKILLS REVIEW

Chapter 8

A. Solve. Use the pictograph for Exercises 1–4.
1. Which class read the most books?
2. Which class read about 5 books each?
3. Estimate the total number of books a class of 30 ninth graders read.
4. Estimate the total number of books a 9th grader and a 12th grader read.

Average Number of Books Read During the Summer by Students

Write in lowest terms.

5. $\frac{10}{5}$ 6. $\frac{40}{62}$ 7. $\frac{100}{60}$

8. $\frac{4}{32}$ 9. $\frac{12}{8}$ 10. $\frac{17}{34}$

B. Estimate the likely temperature.
1. weather for water skiing   **A.** 30°C   **B.** 10°C   **C.** 25°F   **D.** 52°F
2. weather for snow skiing   **A.** 20°C   **B.** 56°F   **C.** 45°C   **D.** 15°F
3. oven for baking bread   **A.** 100°C   **B.** 400°F   **C.** 212°F   **D.** 60°C

Arrange in order from least to greatest.

4. $1\frac{1}{4}, \frac{7}{8}, \frac{4}{9}, \frac{9}{4}, 2\frac{4}{9}$

5. 3 mL, 33 L, 0.303 kL, 30 mL, 3.3 L

Measure to the nearest $\frac{1}{4}$ in.

6. 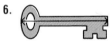 7.  8.

C. Find the answer.
1. 20% + 41% 2. 94% − 62% 3. 27 × 12 4. 78.2 ÷ 1.7

Compare. Use <, >, or =.
5. 4 yd ☐ 10 ft 6. 4.02 mm ☐ 402 cm 7. 0.5 t ☐ 500 kg

Estimate.
8. Which fraction is about $\frac{1}{2}$?   **A.** $\frac{8}{4}$ **B.** $\frac{14}{27}$ **C.** $\frac{15}{20}$
9. Which fraction is about $\frac{2}{3}$?   **A.** $\frac{67}{100}$ **B.** $\frac{24}{48}$ **C.** $\frac{15}{10}$

EXTRA MIXED SKILLS REVIEW

Chapter 9

A. Give the likely measure: inches, feet, yards, or miles.
1. the length of a school hallway
2. the length of the Ohio River
3. the distance around the moon
4. the depth of a cup
5. Write the fraction of figures that are shaded.
6. Write the fraction of triangles that are shaded.
7. Write the fraction of squares that are shaded.
8. Write the fraction of circles that are shaded.
9. Write the fraction of figures that are arrows.

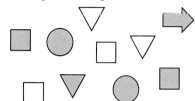

B. Find the reciprocal.
1. $\frac{1}{12}$
2. 13
3. $\frac{5}{24}$
4. $\frac{84}{101}$
5. $\frac{22}{1}$
6. $\frac{5}{8}$

Write as a decimal.
7. $\frac{4}{11}$
8. $\frac{1}{5}$
9. $\frac{91}{10}$
10. $\frac{2}{3}$
11. $12\frac{1}{3}$
12. $\frac{437}{1000}$

Read the scale or gauge.
13.
14.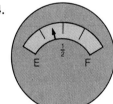
15.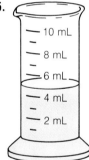

C. Find the answer.
1. $\frac{2}{3} \times \frac{4}{9}$
2. $\frac{5}{6} \div \frac{1}{2}$
3. $\frac{3}{4} \times 5\frac{1}{12}$
4. $\frac{11}{12} \div 6$

Write as a fraction or a mixed number.
5. 0.75
6. 2.4
7. 9.69
8. 0.534
9. 0.109
10. 1.5

Write using exponents.
11. (0.4)(0.4)(0.4)
12. 71 × 71
13. 3.6 to the fifth power

Chapter 10

A. Write as a fraction or a mixed number.
1. 0.25
2. 2.6
3. 8.35
4. 0.725
5. 0.303
6. $2.\overline{6}$

Find the answer.
7. $8 \div \frac{2}{9}$
8. $11 \div \frac{1}{4}$
9. $\frac{4}{5} \times 3\frac{1}{2}$
10. $\frac{7}{8} + \frac{3}{4}$
11. $\frac{1}{2} - \frac{2}{5}$
12. $\frac{11}{15} - \frac{3}{5}$
13. $\frac{7}{10} \div 3$
14. $\frac{9}{4} + \frac{9}{10}$

Write equivalent fractions.
15. $\frac{3}{4} = \frac{\square}{16} = \frac{\square}{20}$
16. $\frac{11}{15} = \frac{\square}{30} = \frac{\square}{45}$
17. $\frac{21}{14} = \frac{42}{\square} = \frac{63}{\square}$

B. Find the answer.
1. $\frac{12}{13} + \frac{4}{39}$
2. $\frac{8}{9} \div 4$
3. $\frac{3}{7} \times 3\frac{1}{2}$
4. $\frac{5}{9} - \frac{2}{8}$
5. $8105 + 810.5$
6. $38{,}271 \times 6$
7. 9340×8
8. 0.0721×6

Write as a decimal.
9. $\frac{29}{100}$
10. $\frac{15}{20}$
11. $\frac{7}{12}$
12. $\frac{16}{1000}$
13. $\frac{7}{8}$
14. $\frac{13}{26}$

Write using exponents.
15. $(0.6)(0.6)(0.6)$
16. 4.3 to the fourth power
17. 13×13

Find the answer. Use mental math.
18. $4\frac{2}{3} + 1\frac{1}{3} + 6$
19. $19\frac{2}{5} - 7\frac{1}{5}$
20. $\frac{4}{5} \times \frac{1}{10}$
21. $\frac{8}{15} \div 2$

C. Change the unit.
1. 2.3 km = ____ m
2. 3 gal = ____ qt
3. 6.5 lb = ____ oz
4. 5 pt = ____ c
5. 43.2 cm = ____ mm
6. 143 g = ____ kg

Estimate.
7. $4026 \div 4.8$
8. $12 \div 3.1$
9. $55 \div 8.9$
10. $333 \div 4.3$
11. $2943 - 671$
12. $809 - 128$
13. $\$10.45 - \$6.95$
14. $\$42 - \$18.12$

Find the answer.
15. $\frac{1}{4} \times \frac{3}{4}$
16. $\frac{1}{4} \div \frac{3}{4}$
17. $\frac{3}{4} - \frac{1}{8}$
18. $\frac{3}{4} + 4\frac{9}{16}$
19. 23×56.4
20. 2.29×1000
21. $7 \times 98{,}423$
22. $\$124.05 - \$9$

EXTRA MIXED SKILLS REVIEW

Chapter 11

A. Write as a fraction.
1. 0.625
2. 90%
3. $12\frac{1}{2}\%$
4. 125%
5. $0.\overline{3}$

Compare. Use <, >, or =.
6. $\frac{1}{16}$ ☐ 0.16
7. $\frac{1}{3}$ ☐ 0.34
8. 9.5 ☐ 95%
9. 0.6 ☐ 60%
10. 12% ☐ 0.12
11. 0.100 ☐ 1%
12. 0.147 ☐ 1.47%
13. $\frac{4}{11}$ ☐ $0.\overline{36}$

Find the answer.
14. 0.25×132
15. $\frac{5}{28} \div \frac{1}{7}$
16. $\frac{23}{24} \div \frac{2}{3}$
17. $\frac{3}{5}$ of 12

B. Write as a percent.
1. 0.3
2. $\frac{1}{4}$
3. 0.429
4. 6
5. $\frac{3}{4}$

Find the answer.
6. $29.2\% - 14\%$
7. $75\% + 16\%$
8. $11.2\% - 5.3\%$
9. $12\frac{1}{2}\% + 9\%$

10. Estimate what percent of the number line each letter represents.

11. Estimate what fraction of the number line each letter represents.

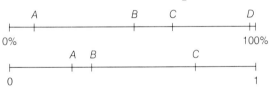

C. Write as a decimal.
1. 89%
2. 4.2%
3. 139%
4. $37\frac{1}{2}\%$
5. 400%

Find the missing length. The figures are similar.

6.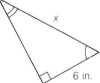
7.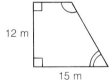
8.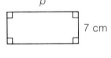
9.

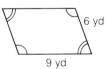

Chapter 12

A. This 10 by 10 grid shows the layout of the theater club storeroom. Give the percent of the space for each.

1. costume closet
2. props
3. makeup chest
4. hats and wigs
5. open space
6. shoes
7. both props and makeup
8. What is the difference between the percent of space for props and for costumes?

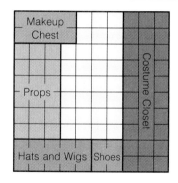

Find the answer. Use mental math.
9. 75% of 24
10. 50% of 218
11. 25% of 84
12. 60% of 25

B. Solve.
1. 23.4 is □% of 26.
2. 75% of □ is 6.
3. 33.8 is 52% of □.
4. 20% of □ is 25.
5. 90 is □% of 150.
6. 1% of □ is 0.57.
7. 27 is $33\frac{1}{3}$% of □.
8. 100% of □ is 318.
9. 50 is □% of 200.

Write the ratio.
10. 75 right to 32 left
11. 7 wins to 4 losses
12. 57 in to 57 out
13. 15 red to 12 blue
14. 21 open to 11 shut
15. 1 up to 145 down

Estimate.
16. 15% of $21.95
17. 20% of 178
18. 48% of 32
19. 73% of $39.50

C. Use a ruler and the map. Estimate the distances.
1. Rock Springs to Laramie
2. Casper to Cheyenne
3. Jackson to Casper
4. Sheridan to Rock Springs
5. Cheyenne to Jackson
6. Cody to Sheridan

State if the ratios are equal. Use = or ≠.
7. $\frac{11}{12}$ □ $\frac{35}{36}$
8. $\frac{10}{13}$ □ $\frac{30}{39}$
9. $\frac{23}{48}$ □ $\frac{5}{10}$
10. $\frac{2}{3}$ □ $\frac{5}{12}$
11. $\frac{17}{4}$ □ $\frac{18}{5}$

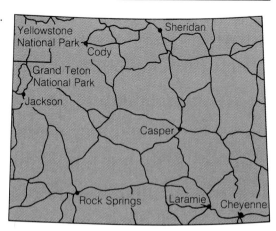

Wyoming Scale: $\frac{3}{8}$ in. = 60 mi

Chapter 13

A. Solve.

1. A long distance call costs $1.50 for 4 minutes. At this rate, how long can you talk for $4.50?

2. Four pounds of grapes cost $2.60. At this rate, what weight of grapes can you buy for $3.90?

3. A word processor earns $62 in 8 h. What is the hourly wage?

4. The sales tax on a dress that costs $96 is $5.28. At this rate, what will be the tax on a $74 dress?

Name the polygon. Tell if it is regular.

5.   6.  7. 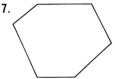 8.

B. Name the type of quadrilateral.

1.   2.  3. 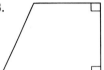 4.

5. Estimate what fraction of the number line each letter represents.

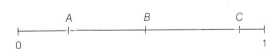

C. Find the answer.

1. $\frac{12}{7}$ of 35
2. $\frac{3}{16}$ of 29
3. 12% of 96
4. $\frac{5}{9}$ of 63
5. $\frac{2}{5}$ of 7
6. 9% of 1246
7. 28% of 60
8. $\frac{1}{3}$ of 51

Find the circumference or the perimeter. Use 3.14 for π.

9. 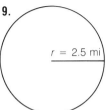 10. 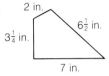 11. 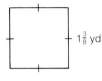 12.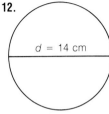

EXTRA MIXED SKILLS REVIEW

Chapter 14

A. A bag holds 7 blue blocks, 4 green blocks, and 2 red blocks. You draw blocks from the bag without looking. You replace the blocks after each draw.

1. Find the probability of picking a blue block two times in a row.
2. Find the odds against picking a green block.
3. Find the odds for picking a red block.
4. Find the probability of picking a green block.

Draw and label each of the following.

5. point X on $\overleftrightarrow{KJ}$
6. m∠QRS = 80
7. $\overline{BC} \perp \overline{OP}$
8. $\overleftrightarrow{XY} \parallel \overleftrightarrow{FG}$

Find the answer.

9. $\frac{7}{8} \div \frac{1}{2}$
10. $\frac{14}{15} \div \frac{2}{3}$
11. 9% of 2
12. 21% of 16

B.

1. Find the number of possible meals. You choose one each from 2 soups, 4 salads, and 5 sandwiches.
2. Find the number of permutations of the orders in which 5 runners could finish a race.
3. Find the range and median for this set of data. *Career batting averages:* .217, .259, .302, .291, .264, .289, .255, .261, .259
4. Find the mean and the mode for this set of data. *Miles per gallon of gas:* 21, 17, 26, 16, 23, 22, 23, 25, 19, 20, 21

Compare. Use <, >, or =.

5. 63% □ $\frac{2}{3}$
6. 1.2 g □ 120 mg
7. 12.5% □ $\frac{1}{8}$
8. 5274 ft □ 1 mi

Draw the following angles. Name each angle as right, acute, or obtuse.

9. m∠ABC = 80
10. m∠TUV = 10
11. m∠LMN = 130
12. m∠DEF = 30

C. Measure the angle.

1.
2.
3.

4. Estimate what percent of the number line each letter represents.

Give the fraction for the shaded area. Then write the fraction as a percent.

5.
6.
7.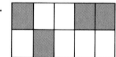
8.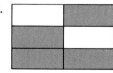

EXTRA MIXED SKILLS REVIEW

Chapter 15

A. Name the figure. Give the number of faces.

1.
2.
3.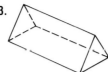
4.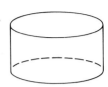

Find the volume. Use 3.14 for π.

5. cone:
   $B = 8.4$ ft^2
   $h = 4.1$ ft

6. rectangular prism:
   $l = 5$ m, $w = 6$ m
   $h = 7$ m

7. pyramid:
   $B = 5$ yd^2
   $h = 9$ yd

8. cylinder:
   $r = 7$ mm
   $h = 8$ mm

Find the area. Use 3.14 for π.

9. square:
   $s = 4$ yd

10. triangle:
    $b = 2$ in., $h = 12$ in.

11. parallelogram:
    $b = 5$ m, $h = 4.2$ m

12. circle:
    $d = 29$ mi

B. Name the figure. Give the number of vertices.

1.
2.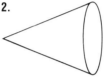
3.
4.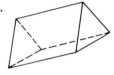

5. Six out of 10 cyclists wear helmets. There were 45 trailbikers riding in Big Trails Park last weekend. How many of these cyclists would you expect to see wearing helmets?

6. A ranger noticed that 1 out of 7 of the cars visiting the park were from out of state. On Sunday there were 343 cars at the park. What is the expectation for the number of out-of-state cars?

C. Find the answer.

1. $17 \times \frac{2}{3}$
2. 2% of 39
3. $2\frac{1}{3} \times 54$
4. $\frac{3}{4} \div 16$

5. Find the perimeter of the rectangle.
   $l = 29$ ft 6 in. and $w = 12$ yd

6. Find the circumference of the circle. Use 3.14 for π. $r = 12.7$ km

7. Find the volume of the cone. Use 3.14 for π. $r = 4$ in. and $h = 9$ in.

8. Find the surface area of the cube.
   $s = 6.2$ cm

Choose the correct formula.

9. simple interest
10. volume of a cone
11. area of a circle
12. volume of a cube
13. area of a triangle
14. area of a parallelogram

Formulas	
$A = \frac{1}{2}bh$	$V = \frac{1}{3}Bh$
$i = prt$	$V = s^3$
$V = lwh$	$A = \pi r^2$
$A = bh$	$A = s^2$

Chapter 16

A. Find the answer. Use mental math.

1. $2\frac{2}{3} + 7\frac{1}{3} + 3\frac{2}{5}$
2. $10\frac{8}{9} + \frac{3}{4} + 7\frac{1}{9}$
3. 25% of 84
4. 40% of 200

Find the area. Use 3.14 for π.

5.
72 mm
109 mm

6.
$r = 5$ mi

7.
23 cm
42.6 cm

8.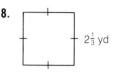
$2\frac{1}{3}$ yd

Estimate.

9. $39 \div 4$
10. $743 \div 81$
11. $2864 \div 72$
12. $1693 \div 604$

B. Compare. Use <, >, or =.

1. 15 ☐ ⁻15
2. 250 ☐ 2.5%
3. 45% ☐ $\frac{4}{5}$
4. ⁻12 ☐ 0
5. 19% ☐ 1.9
6. ⁻84 ☐ ⁻21
7. 3 ☐ 300%
8. 66% ☐ 0.6

Find the answer.

9. $^-21 \div 7$
10. $416 - (^-7)$
11. $^-69 + (^-72)$
12. $13 \times (^-2)$

Find the circumference or perimeter. Use 3.14 for π.

13.
$d = 12$ mi

14.
21.7 cm
75.2 cm

15.
$s = 4\frac{1}{2}$ yd

16.
$r = 27.5$ mm

C. Find the answer.

1. $63 \times (^-5)$
2. $32 - (^-47)$
3. $^-84 \div 14$
4. $^-29 \times (^-7)$

Find the missing length. The figures are similar.

5.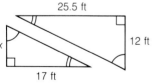
25.5 ft
12 ft
17 ft
x

6.
12 cm
16.2 cm
20.25 cm
y

7.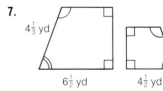
$4\frac{1}{3}$ yd
$6\frac{1}{2}$ yd
$4\frac{1}{2}$ yd
s

Pick one block without looking from: 2 red, 3 blue, 1 green, and 6 yellow blocks. Replace the block after it is drawn. Find the probability of picking:

8. green
9. red, then green
10. yellow or blue
11. blue

GLOSSARY

accuracy The quality of being without errors.

acute angle An angle that has a measure less than 90°.

addend A number added to other numbers to find a sum.

algebraic expression An expression that contains at least one variable, for example, b or $n + 7$.

angle ($\angle$) Two rays with a common endpoint called the vertex.

area The total number of square units it takes to cover a region.

average The sum of the values in a set divided by the number of values in that set.

axis (plural: axes) A reference line on a graph showing the scale.

balance The amount of money in a bank account.

bar graph A graph that uses vertical or horizontal bars to compare quantities.

base In exponential notation a^n, a is the base; the side of a geometric figure that is perpendicular to its height.

batting average In baseball the number of hits divided by the number of times at bat (written without a zero in the ones place, for example, .271).

baud rate The rate at which a modem transfers data.

bits per second The baud rate is measured in bits of data per second.

blueprint A white-on-blue photographic copy of an architectural or engineering design.

break apart To separate the digits of a number into place values of ones, tens, hundreds, etc., in order to do mental calculations.

budget A plan for using money.

calculation method The procedure used to carry out a mathematical process.

capacity The measure of the amount that a container will hold.

Celsius temperature scale (°C) The metric temperature scale, in which 0° is the freezing point of water and 100°C is the boiling point of water.

chart A group of related facts shown in the form of a table, diagram, or graph.

circle All the points in a plane that are a fixed distance from a point called the center.

circle graph A graph that uses parts of a circle to compare parts of a whole.

circumference The distance around a circle.

clustering When given addends are close in value to the same number, using that number for each addend in order to estimate the sum.

commission A percent of the money taken in on sales, often paid to a sales clerk or agent in addition to salary or wages.

common Shared by all members of a group.

compatible numbers Numbers close in value to given numbers that can be used instead of the given numbers in order to make a problem easier to estimate.

compensation Adding to or subtracting from a given number to make mental calculations easier, then adjusting the answer by the same amount.

component One of the letters or numbers in an ordered pair.

composite number Any whole number greater than 1 that has more than two factors.

compound probability The probability for more than one event.

cone A space figure with exactly one circular base and exactly one vertex, not in the same plane as the base.

congruent Having the same size and shape.

congruent angles Angles that have the same measure.

congruent segments Segments that have the same length.

GLOSSARY

coordinate grid A rectangular system of parallel and perpendicular lines used for graphing.

coordinates See *ordered pair*.

corresponding sides The matching sides in a pair of congruent or similar figures.

counting principle A way to find all possible outcomes for a set of related events. With m choices for one event and n choices for another event, $m \times n$ gives the number of possible outcomes.

cross products For fractions $\frac{a}{b}$ and $\frac{c}{d}$, the *cross products* are ad and bc.

cube A rectangular prism with six square faces.

cubic units Any units with six square faces, used to measure volume.

customary system The system of measurement that uses the inch, foot, yard, and mile as the basic units of length; the ounce, pound, and ton as units of weight; the fluid ounce, pint, quart, and gallon as units of capacity. Sometimes called the English system.

cylinder A space figure with two congruent, parallel, circular bases.

data Information from which conclusions can be inferred.

decagon A polygon with ten sides.

decimal A numeral that uses place value and a decimal point (.) to name a number, for example, 0.64 and 12.507.

deduction Any amount subtracted from gross earnings.

degree (°) The unit used to measure angles and temperature.

denominator The number of equal parts into which a whole is divided; in the fraction $\frac{a}{b}$, the denominator is b.

deposit An amount of money put into a bank account or put down as a pledge or partial payment.

deposit slip A form used to list the amount of money added to a bank account.

diagram A drawing that explains an object or idea by outlining its parts and their relationships.

diameter A segment that passes through the center of a circle and has endpoints on the circle.

difference The answer to a subtraction problem.

digits The basic symbols used to write numerals. In our system, the digits are 0, 1, 2, 3, 4, 5, 6, 7, 8, and 9.

discount A reduction from the usual price of an item.

dividend In a division problem, the number you divide.

divisor The number by which you divide another number.

dodecagon A polygon with twelve sides.

double bar graph A graph with two sets of data, each set represented by a set of bars.

double line graph A graph with two sets of data, each set represented by a line.

down payment The initial amount paid when buying an item on an installment plan.

edge The line at which two faces of a space figure meet.

elapsed time The amount of time that has passed from a starting time to an ending time.

equation A number sentence that is a statement of equality between two expressions.

equilateral triangle A triangle with all three sides the same length.

equivalent fractions Fractions that name the same number, for example, $\frac{1}{2}, \frac{2}{4}, \frac{5}{10}$.

estimate An approximation of the exact answer.

evaluate an algebraic expression To find the value of an expression by substituting values for the variables in an expression.

exact A precise answer or measurement.

expectation The expected outcome based upon the probability and the number of events.

experimental probability Probability based upon the results of an experiment.

exponent A raised number that tells how many times a value is multiplied by itself. In a^n, n is the exponent.

GLOSSARY

faces Any one of the surfaces of a geometric figure.

factor A number that is multiplied by another number to yield a product.

factorial The product of a certain series of consecutive whole numbers that begins with a certain number and decreases to 1.

Fahrenheit temperature scale (°F) The customary temperature scale, in which 32°F is the freezing point of water and 212°F is the boiling point of water.

figure See *geometric figure*.

flowchart A diagram showing the process of work through a set of operations.

formula A mathematical rule expressed in symbols.

fraction A numeral used to name the number of objects in a set or parts in a whole; any number expressed in the form $\frac{a}{b}$ where a is the numerator, b is the denominator, and $b \neq 0$.

frequency table A table that shows how many times an event occurs.

front-end estimation An estimate made by adding the leading digits of a number.

geometric figure A closed shape made up of line segments or planes.

gram (g) The basic unit of mass in the metric system.

graph A diagram that compares numerical data. Includes bar, line, circle, coordinate graphs, and pictographs.

greatest common factor (GCF) The greatest whole number that is a factor of two or more numbers.

grid See *coordinate grid*.

gross pay The total amount of earnings before any deductions.

height In any geometric figure, the perpendicular distance from a vertex to a base.

hexagon A polygon with six sides.

histogram A bar graph showing frequencies.

horizontal scale The scale used on the horizontal (side-to-side) axis of a graph.

hypotenuse The side opposite the right angle in a right triangle.

improper fraction A fraction whose numerator is greater than or equal to its denominator.

independent event In compound probability, an event that does not affect the outcome of other events.

installment plan A way of buying an item by paying part of the cost at first, and then paying part of the remaining cost each month until the full amount plus interest is paid.

integers The whole numbers and their opposites.

interest Money paid for the use of money, usually a percent, of the amount invested, loaned, or borrowed.

intersecting lines Lines that meet at only one point.

inverse operations Operations that reverse the action of each other. Addition will undo subtraction; multiplication will undo division.

invert To reverse an object or action; when dividing fractions, to reverse the position of the numerator and the denominator. See *reciprocal*.

isosceles triangle A triangle with at least two congruent sides.

least common denominator (LCD) The least common multiple of the denominators of two or more fractions. See *least common multiple*.

least common multiple (LCM) The least non-zero number that is a multiple of two or more given numbers.

line (↔) A set of points determined by two points, and extending endlessly in both directions.

line graph A graph using line segments to show the direction of change in data.

line segments Two endpoints and all the points between them.

liter (L) The basic unit of capacity in the metric system.

logic The science of correct reasoning.

logical reasoning The act of reasoning from a known principle to an unknown.

GLOSSARY

loop A step in a computer program that repeats an action.

lowest terms A fraction is in *lowest terms* when the only common factor of the numerator and the denominator is 1.

mass The measure of the quantity of matter in an object.

mean The average; the sum of a set of numbers divided by the number of members in the set.

measurement Finding length, mass, or capacity by comparison with a standard scale.

median The middle number in a set of numbers given in consecutive order.

mental math Working problems without using pencil and paper or a calculator; solving a problem "in your head."

meter (m) The basic unit of length in the metric system.

metric system The system of measurement that uses the meter as the basic unit of length, the liter as the basic unit of capacity, and the gram as the basic unit of mass.

minuend In a subtraction problem, the number from which you subtract.

mixed number A numeral that uses a whole number and a fraction to name a number, for example, $2\frac{3}{4}$.

mode The value that occurs most often in a set of data.

modem A device that changes computer data to electrical pulses.

money notation Writing a number using the $ or ¢ symbols; used to show an amount of money.

multiple The product of a number and another whole number.

multiple-step problem A problem that can be solved using two or more operations.

negative number A number that is less than zero. A raised minus sign is used to show the negative value, for example, $^-3$.

net pay The amount of money earned after deductions.

numerator The number of equal parts of a whole that are being taken; in the fraction $\frac{a}{b}$, the numerator is a.

numerical expression A single numeral, or two or more numerals used with operation symbols, for example, 8, 5 − 3, 4 × 6.

objective A goal or purpose.

obtuse angle An angle with measure greater than 90° and less than 180°.

octagon A polygon with eight sides.

odds In probability, the ratio comparing the ways an event can and cannot occur.

operation A mathematical process, for example, addition or division, that makes a change in a quantity.

opposites Numbers that are the same distance from 0, but on opposite sides of 0; and whose sum is 0, for example, 3 and $^-3$.

ordered pair A pair of numbers in a particular order that are the coordinates of a point in a grid.

origin The point where the horizontal and vertical axes of a grid intersect.

outcome One possible result of an event.

overtime Paid work done beyond the set working hours, usually at a greater rate of pay.

parallel lines (∥) Lines in the same flat space, or plane, that never intersect.

parallelogram A quadrilateral with two pairs of parallel sides.

pentagon A polygon with five sides.

percent (%) The ratio of a number to 100.

perimeter The sum of the lengths of the sides of a polygon.

permutation An arrangement of the objects in a set into a certain order.

perpendicular lines (⊥) Two lines that meet or intersect to form right angles.

pi (π) The ratio of the circumference of a circle to its diameter, equal to about 3.14.

pictograph A diagram or graph using picture-like symbols to represent data.

GLOSSARY

place value The value given to the position a digit occupies in a numeral. In the decimal system, each place of a numeral is ten times the value of the place to its right.

plane A surface of a space figure.

point An exact location in space.

polygon A closed figure formed by three or more segments in the same plane.

polyhedron A closed three-dimensional figure where each face is a polygon.

positive number A number with a value greater than zero.

power The number of times a base number is multiplied by itself as shown by an exponent. Read 10^3 as *ten to the third power*.

precision Being exact in measurement.

prime number A whole number greater than 1 that has exactly two factors, itself and 1.

principal An amount of money borrowed or loaned; amount upon which interest is computed.

prism A space figure that has rectangular sides and a pair of parallel congruent bases.

probability The ratio of the number of times a certain outcome can occur to the number of total possible outcomes.

problem-solving strategy A technique or approach that might be used to help solve a problem; for example, making a diagram, writing an equation, working backward.

product The answer to a multiplication problem.

proper fraction A fraction whose numerator is less than its denominator, for example $\frac{7}{8}$.

proportion A numerical sentence stating that two ratios are equal.

protractor An instrument that is used to measure angles.

pyramid A space figure that has three or more triangular sides and a polygon as a base.

Pythagorean Theorem A rule that states that the square of the hypotenuse of a right triangle is equal to the sum of the squares of the other two sides.

quadrilateral A polygon with four sides.

quotient The answer to a division problem.

radius Any segment that joins the center to a point on a circle.

random sample A group in which every member of the population has an equal chance of being included.

range The difference between the greatest and least numbers in a set of data.

rate A ratio used to compare two measures.

rate of interest The percent of principal charged as interest on a loan.

ratio A comparison of one number to another.

ray ($\rightarrow$) A part of a line that has one end point and continues endlessly in one direction.

reciprocal Two numbers are *reciprocals* if their product is 1; inverse.

rectangle A parallelogram with four right angles.

rectangular prism A prism with rectangular bases.

regroup To make a group of 10 from 1 of the next highest place value, or 1 from 10 of the next lowest place value. Example: 1 hundred can be regrouped into 10 tens; 10 ones can be regrouped into 1 ten.

regular polygon A polygon with all sides the same length and all angles the same measure.

remainder The number remaining when one number is divided by another; the answer to a subtraction problem.

repeating decimal A decimal with a group of digits that repeats endlessly. A bar is placed over the digits that repeat, for example, $52.\overline{6}$.

rhombus A parallelogram with four congruent sides.

right angle An angle having a measure of 90°.

right triangle A triangle that has a right angle.

round To express a number in even units by using the nearest multiple of a chosen place value. To round 47 to the nearest ten, use 50 as the nearest multiple of ten.

running balance The balance after adding or subtracting each entry in a checkbook.

scale A series of marks along a line at regular intervals used in measurement or graphs; the ratio between the dimensions used in a scale drawing or map and the actual dimensions.

GLOSSARY

scale drawing A drawing of an object with a fixed ratio of the dimensions in the drawing to the dimensions of the actual object.

scalene triangle A triangle with no sides the same length.

segment (—) Part of a line; two points and all the points between them.

similar figures Figures with the same shape but not necessarily the same size.

simple interest A percent of the principal only, earned or paid over a given period of time.

simple probability The probability for one event.

skew lines Lines in a three-dimensional space that never intersect and are not parallel.

solution The value for the variable that makes an equation true.

space figure A closed, three-dimensional figure.

sphere The set of all points in space at a fixed distance from a point called the center.

spreadsheet program A computer program that calculates numbers arranged in rows and columns as in a financial report.

square A quadrilateral with sides of equal length and four right angles; the product of a number multiplied by itself.

square root ($\sqrt{}$) If $a^2 = b$, then a is the square root of b.

square units Any units with sides of equal length and four right angles used to measure area.

standard numerals A number expressed in digits.

stem-and-leaf diagram A diagram that uses place value to arrange statistical values.

subtrahend In a subtraction problem, the number that you subtract.

sum The answer to an addition problem.

surface area The sum of the areas of the faces of a figure.

table An arrangement of related facts, usually rows and columns.

telecommunications The process of sending information.

terminating decimal A decimal with a fixed number of digits.

trapezoid A quadrilateral with at least one pair of parallel sides.

tree diagram A diagram that shows all possible outcomes of an event or a set of permutations.

triangle A polygon with three sides.

tuition The charge for instruction at a college or private school.

unit price The cost of one unit of an item.

unit rate The rate for a single quantity of a measure.

variable A letter or other symbol used to stand for a number in an expression or an equation.

vertex (plural: vertices) The point that the two rays of an angle have in common; the intersection of two sides of a polygon.

vertical scale The scale used on the vertical (top-to-bottom) axis of a graph.

volume The measure of a solid region in terms of cubic units.

whole number A number, 1, 2, 3, etc., that has no fractional or decimal parts.

withdrawal An amount of money taken out of a bank account.

TRY THIS ANSWERS

1-1
a. eighty-seven thousand, seventy-six
b. twenty-four and thirty-four hundredths
c. fifty-four hundredths
d. five hundred sixty-five dollars
e. eight dollars and sixty-eight cents
f. 470,243
g. 5.208
h. $1.49

1-2
a. <   b. <   c. <   d. >
e. 12, 12.03, 12.045, 12.3, 12.42
f. 0.003, 0.02, 0.03, 0.032, 0.23
g. 0.8, 0.9, 0.99, 1, 1.02, 9.0
h. 1834, 1839, 1842, 1883, 1983

1-3
a. 50   b. 1.5   c. 3400
d. 3.55   e. $7.00   f. $1.90

1-4
a. compare, subtraction
b. find the missing part, subtraction

1-5
a. 100   b. 30   c. 70
d. 600   e. 300   f. 500
g. 3000   h. 12,000   i. $2
j. $12   k. $7   l. $17

1-6
a. separate into a given number of groups, division
b. put together, multiplication

1-7
a. 2000   b. 2100   c. 5000
d. 180,000   e. 180,000   f. 1,200,000
g. 15,000   h. 42,000

1-8
a. A   b. B
Answers may vary.
(Exact answer given.)
c. 70 (68)   d. 5 (5.3)
e. 80 (≈81.44)   f. 60 (≈58.05)

1-9 a. and b. Answers may vary.

1-11 a.

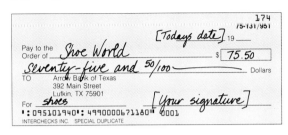

b.

2-1 a. 32 mi   b. 3 mi

2-2 a. 64   b. 1086   c. 2531
d. 63.09   e. 0.897   f. 3.852

2-3 a. 317   b. 374   c. 2847
d. 22.22   e. 173.3   f. 3.228

2-5 Answers may vary.
(Exact answer given.)
a. 700 (707)   b. 1500 (1502)
c. 1580 (1581)   d. 1140 (1135)
e. 550 (556)   f. 1100 (1103)
g. 10,900 (10,900)

2-6

DEPOSIT		Dollars	Cents
Date [Today's date]	Currency	8	00
	Coin	10	00
	24-953	25	00
Sign here for cash received	24-871	11	75
Arrow Bank of Texas 392 Main Street Lufkin, TX 75901	Total	54	75
	Less Cash Received	0	00
075101340:..1649:52149:	Net Deposit	54	75

TRY THIS ANSWERS

2-7

NUMBER	DATE	DESCRIPTION OF TRANSACTION	PAYMENT	✓	FEE (IF ANY)	DEPOSIT	BALANCE
	9/24	paycheck				26 80	205 67
							26 80
							232 47
349	9/27	Stuffed Shirts	21 24				21 24
							211 23

2-8 a. $3.50 b. $4.25 c. $65 d. 109 e. 51 f. 49 g. $5.50 h. $14.75

2-9 Answers may vary.
  a. mental math   b. calculator
  c. paper and pencil   d. mental math

2-10 a. reasonable
  b. not reasonable; if Fred bought his TV in 1960, the year would be 1978 not 1989

3-1 a. 6 choices b. 24 arrangements

3-2 a. 144 b. 10,818 c. 620 d. 19,432 e. 255 f. 2135

3-3 a. 9683 b. 12,065 c. 24,640 d. 115,828 e. 364,524

3-5 a. about 160 b. about 586

3-6 $165

3-7 a. 11.522 b. 10.8 c. 0.0156 d. 31.05 e. $72.75

3-8 a. 93.4 b. 4.56 c. 1758 d. 1400 e. 50.2 f. 78,940 g. 21,600 h. 9012

3-9 a. $298.23 b. yes, $1083.39

3-10 a. 5^2, 25 b. 4^3, 64 c. $(2.1)^3$, 9.261 d. 7^5; 16,807

3-11 a. 128 b. 63 c. 492 d. 45 e. 160

4-1 a. bike $80, helmet $20
  b. 1 quarter, 1 dime, 3 nickels

4-2 a. 42 b. 48 c. 116 d. 86 R4 e. 489 R2

4-3 a. 6 R5 b. 7 R26 c. 7 R12 d. 69 R10 e. 214 R8

4-4 a. 50 R3 b. 504 R1 c. 600 R3 d. 303 R14 e. 800 R12

4-6 2 min

4-8 a. 3.12 b. 2.53 c. 0.195 d. 1.047 e. 1.68

4-9 a. 4.56 b. 0.784 c. 3 d. 0.045

4-10 a. 3.2 b. 0.26 c. 13 d. 0.065

4-13 a. 21 b. 26 c. 3 d. $0.\overline{6}$

4-14 a. correct b. incorrect; 27,885 c. incorrect; 20,856

5-1 a. 15th day b. 25 days, $9

5-2 a. about 2000 b. about 3000 c. about 1500
  d.
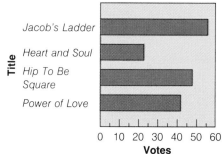
Favorite Huey Lewis Hits

5-3 a. 190 b. 150 c. 140 d. 80 e. 50
  f.
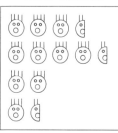
Sci-fi Classics

TRY THIS ANSWERS

5-4 a. about $22,000 b. about $35,000
c.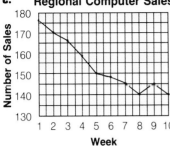

5-5 a. 17% b. 20% c. 63%
d. 8%

5-6 Answers may vary.
(Exact answer given)
a. 180 (180) b. 100 (101)
c. $44 ($44.27)

5-7 a. 207 b. 594 c. 181
d. 196 e. $129

5-8 a. The data is the same. In Graph II the space between amounts is greater, so the line is steeper.
b. It would go down in weeks 2 to 3.

5-9 B

6-1 $16

6-2 a. inches b. feet c. yards
d. miles e. 36 ft
f. 10 ft 10 in. g. 5 mi

6-3 a. a botanist measuring plant growth
b. Janet cutting ribbon for a hatband
c. $2\frac{1}{4}$ in. d. 1 in.

6-4 a. quarts b. fluid ounces
c. quarts d. gallons
e. 11 qt 1 pt f. 2 c 4 fl oz
g. 17 gal

6-5 a. ounces b. tons c. pounds
d. pounds e. 2T 1000 lb
f. 104 oz g. 1 lb 7 oz
h. 0.25 lb i. 5000 lb
j. 36 oz.

6-6 a. 2 lb 4 oz b. 3 lb 2 oz
c. 5 lb 5 oz d. 4 fl oz
e. 11 fl oz f. 16 fl oz

6-7 a. millimeters b. centimeters
c. kilometers d. meters

6-8 a. 960 mm b. 0.712 km
c. 43.8 cm d. 1.9 cm
e. 2,389,000 m f. 0.176 m

6-9 a. 0.0547 L b. 1900 mL
c. 0.0462 kL d. 2.384 kL
e. 2000 L f. 90 mL

6-10 a. 8230 g b. 5000 kg
c. 0.65 kg d. 6400 mg
e. 0.986 g f. 4 t

6-11 a. 2 cm b. 472.5 g c. 7 mL

6-12 a. B b. Fahrenheit c. Celsius

6-13 a. $V = lwh$ b. $A = \pi r^2$
c. $i = prt$ d. $67

7-1 a. about 49 min b. about 30 min

7-2 a. $\frac{3}{19}$ b. $\frac{16}{19}$ c. $\frac{2}{3}$

7-3 a. $\frac{6}{22}$ b. $\frac{4}{12}$ c. $\frac{35}{42}$
d. 21 e. 20 f. 80
g. 42 h. 64

7-4 a. 7 b. 4 c. 12 d. 15
e. 1 f. $\frac{2}{3}$ g. $\frac{1}{4}$ h. $\frac{7}{4}$
i. $\frac{16}{45}$ j. $\frac{3}{7}$ k. $\frac{3}{4}$

7-5 a. $\frac{21}{4}$ b. $\frac{17}{5}$ c. $\frac{19}{12}$ d. $\frac{17}{6}$
e. $\frac{35}{3}$ f. $\frac{35}{8}$ g. $1\frac{6}{7}$ h. $1\frac{1}{4}$
i. 5 j. $2\frac{2}{3}$ k. $2\frac{1}{3}$ l. 1

7-6 a. 16 b. 30 c. 27
d. 42 e. 36 f. 40

7-7 a. $\frac{3}{6}, \frac{2}{6}$ b. $\frac{3}{10}, \frac{4}{10}$ c. $\frac{7}{15}, \frac{20}{15}$
d. $\frac{15}{20}, \frac{4}{20}$ e. $\frac{4}{14}, \frac{7}{14}, \frac{5}{14}$

7-8 a. > b. < c. < d. = e. >
f. < g. = h. > i. >

TRY THIS ANSWERS

7-9   a. about $\frac{5}{8}$ filled   b. about $\frac{1}{8}$ filled   c. about $\frac{1}{2}$ filled

7-10   a. $\frac{1}{2}$ b. $\frac{2}{5}$ c. $\frac{3}{4}$ d. 1   e. $\frac{5}{6}$

7-11   a. 0.8   b. $0.\overline{3}$ c. 1.25   d. $2.\overline{4}$ e. 1.5   f. 0.125

7-12   a. $\frac{21}{25}$ b. $\frac{3}{1000}$ c. $\frac{1}{6}$ d. $10\frac{1}{3}$ e. $5\frac{3}{10}$

7-13   a. $35.63   b. 0   c. $265

7-15   a. 14 h 20 min   b. 1 h 15 min

8-1   Torr horse, Budd dog, Constance rabbit, Dee cat

8-2   a. $\frac{1}{6}$ b. $\frac{1}{2}$ c. 6   d. 1   e. $\frac{1}{16}$

8-3   a. 1   b. $2\frac{4}{9}$ c. 21   d. $5\frac{1}{7}$ e. $\frac{3}{8}$

8-4   a. $3096.80   b. $1697.62

8-5   a. $\frac{3}{2}$ b. $\frac{1}{21}$ c. $\frac{8}{5}$ d. $\frac{2}{7}$ e. 1   f. $\frac{4}{23}$

8-6   a. $1\frac{3}{32}$ b. $1\frac{1}{2}$ c. $\frac{1}{24}$ d. 12   e. $2\frac{1}{4}$ f. $\frac{1}{72}$ g. 1   h. 4   i. 24   j. 5000

8-8   a. about 16   b. about 36   c. about 54   d. about 18   e. about 12

8-11   a. $15   b. $12.60   c. $14.25   d. $10.05   e. $5.10   f. $9.90

9-1   15 students

9-2   a. $\frac{7}{9}$ b. $\frac{1}{3}$ c. $\frac{1}{2}$ d. $\frac{3}{5}$ e. $\frac{2}{3}$

9-3   a. $1\frac{1}{4}$ b. $6\frac{1}{2}$ c. 1   d. $5\frac{1}{5}$

9-4   a. $\frac{13}{15}$ b. $\frac{17}{24}$ c. $\frac{1}{2}$ d. $\frac{1}{20}$

9-5   a. $1\frac{5}{24}$ b. $6\frac{25}{28}$ c. $202\frac{2}{9}$ d. $5\frac{17}{30}$

9-6   a. 7   b. $8\frac{1}{2}$ c. $9\frac{1}{4}$ d. $3\frac{3}{8}$

9-7   a. $\frac{3}{2}$ b. $\frac{5}{3}$ c. $\frac{13}{9}$ d. $\frac{9}{5}$ e. $\frac{12}{7}$

9-8   a. $\frac{1}{3}$ b. $3\frac{13}{18}$ c. $1\frac{1}{2}$ d. $1\frac{5}{8}$

9-9   a. $9\left(8\frac{37}{40}\right)$ b. $2\left(2\frac{1}{6}\right)$ c. $4\frac{1}{2}\left(4\frac{107}{144}\right)$ d. $3\left(3\frac{1}{30}\right)$ e. $12\left(11\frac{20}{21}\right)$

9-10 $1\frac{5}{8}$ in.

9-13   a. 5 p.m. CST   b. 11 a.m. MST

9-14   Loop occurs from lines 10 to 30. The program will print
THE QUICK BROWN FOX JUMPED OVER THE LAZY DOG
repeatedly (until user presses CTRL C).

10-1   a. 20 h   b. 44 albums

10-2   a. $\frac{2}{1}$ b. $\frac{5}{3}$ c. $\frac{6}{5}$ d. $\frac{1}{7}$ e. $\frac{1}{7}$ f. $\frac{36}{1}$

10-3   a. $=$ b. $\neq$ c. $=$ d. $\neq$

10-4   a. $19.50   b. $10\frac{1}{2}$ h

10-5   a. 38.5 mi/gal   b. 0.4 kg/person

10-6   a. 40 oz for $1.79   b. 125 sq ft for 79¢   c. 5 cans for $1.55

10-7   a. 15 in.   b. 8.75 m

10-8   a. 18 ft   b. $\frac{3}{5}$ in.   c. $\frac{7}{10}$ in.   d. $2\frac{1}{2}$ in.   e. 1 in.   f. $\frac{21}{40}$ in.   g. $\frac{17}{40}$ in.

10-9   a. 9%   b. 20%   c. 13%

10-10   a. 0.63   b. 0.019   c. 0.08   d. 0.25   e. 1.468   f. 96%   g. 43.6%   h. 246%   i. 100%   j. $33\frac{1}{3}$%

10-11   a. 38%   b. 62.5%   c. $66\frac{2}{3}$%   d. 20%   e. 50%   f. $\frac{29}{100}$ g. $\frac{8}{25}$ h. $\frac{1}{40}$ i. $\frac{1}{80}$ j. $\frac{3}{50}$ k. $\frac{881}{1000}$

10-12   a. $>$ b. $<$ c. $>$ d. $=$

TRY THIS ANSWERS

10-13 **a.** not reasonable; both percents about 50%
    b. reasonable

11-1 **a.** 3 eggs   **b.** about 2 calendars
   c. $4284.75   **d.** about 2 buttons

11-2 **a.** 8 students   **b.** 93.8 lb
   c. 7 members

11-3 **a.** 40%   **b.** 80%

11-4 **a.** 50 seals   **b.** 50 seals

11-5 **a.** 12   **b.** 30   **c.** 40   **d.** 28

11-6 Answers may vary. (Exact answer given.)
   a. 30% (≈32.4%)   **b.** 90 (106.08)

11-7 **a.** $93.60   **b.** $206.25
   c. $61,875

11-8 **a.** 40% increase
   b. 12.5% decrease

11-9 **a.** $0.04   **b.** $1.05   **c.** $1.44

11-11 **a.** $189   **b.** $175.50

12-1 **a.** 51 matches   **b.** 20 oranges

12-2 Answers may vary.   **a.** A, C
   b. $\overleftrightarrow{AB}, \overleftrightarrow{DF}$ **c.** $\overline{CD}, \overline{CH}$
   d.–**g.**

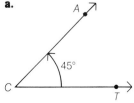

12-3 **a.** $\overrightarrow{ZA}, \overrightarrow{ZB}, \overrightarrow{ZC}, \overrightarrow{ZD}, \overrightarrow{ZE}, \overrightarrow{ZF}$
   b. Answers may vary. ∠EZB, ∠BZD, ∠DZF, ∠FZC, ∠CZA, ∠AZE
   c. 40°   **d.** 90°   **e.** 130°

12-4 **a.**

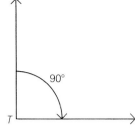

b.

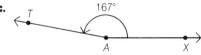

c.

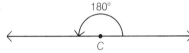

d.

e. acute   **f.** right   **g.** obtuse
h. acute

12-5 **a.** perpendicular   **b.** skew   **c.** parallel

d.

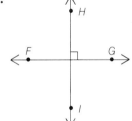

e.

f.

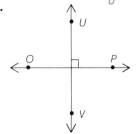

g.

TRY THIS ANSWERS

12-6 **a.** hexagon, not regular
b. regular triangle
c. regular octagon
d. dodecagon, not regular

12-7 **a.** equilateral **b.** right
c. isosceles **d.** scalene

12-8 **a.** trapezoid **b.** parallelogram
c. square **d.** rhombus

12-9 **a.** 15 cm **b.** 56 ft
c. 340 ft **d.** 20 in.

12-10 **a.** 18.84 cm **b.** 12.56 in.
c. 25.748 mi **d.** 219.8 cm

12-11 **a.**

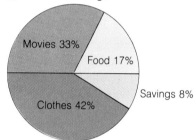

June's Budget

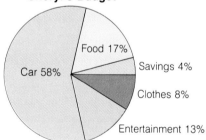

Cheryl's Budget

12-12 not reasonable; 5 ft × 3.14 ≈ 15 ft

12-14

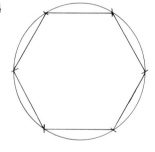

13-1 **a.** 25 monkeys **b.** 60 cages

13-2 **a.**

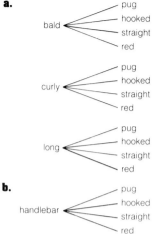

b. (handlebar, brush with pug/hooked/straight/red)

c. 40 disguises

13-3 **a.** 720 permutations **b.** 120 permutations

13-4 **a.** $\frac{2}{3}$ **b.** 0 **c.** $\frac{1}{3}$

13-6 **a.** $\frac{169}{625}$ **b.** $\frac{81}{100}$

13-7 $\frac{3}{2}$

13-8 **a.** 20 plants **b.** 26 members **c.** 100 jars

13-9 **a.** 5000 people **b.** 4000 people

13-10 **a.**

Sit-Ups

Number	Tally	Frequency
54	IIII	4
55	IIII	4
56	II	2
57	HHI I	6
58	IIII	4
59	III	3
60	HHI III	8
61	HHI IIII	9
62	HHI HHI	10
63	IIII	4
64	HHI I	6
65	HHI	5
66	IIII	4
67	I	1

b. 5 girls

TRY THIS ANSWERS

13-11 a. ≈14.1 h   b. 3

13-12 a. 16 h, 15 h   b. $1.75, $4.10

13-15 a. Q (1,1)   b. Q (2,2)
c. Q (1,2)   d. Q (1,3)

14-1 a. 21 cubes   b. 16 cubes   c. 15 cubes
d. 4 triangles, 1 rectangle
e. 5 squares, 4 triangles

14-2 a. 15 sq units   b. 9 sq units
c. 24 sq units   d. 81 cm²
e. 79.9 in.²   f. 180 m²
g. 36 ft²

14-3 a. 54 cm²   b. 105 in.²
c. 44.01 cm²   d. $3\frac{1}{8}$ ft²

14-4 a. 9 ft²   b. 24 in.²
c. 36 cm²   d. 16.12 m²

14-5 a. ≈28.26 m²   b. ≈12.56 in.²
c. ≈153.86 ft²   d. ≈7.07 ft²

14-6 a. $10\frac{2}{3}$ yd²   b. 12 yd²

14-7 a. Faces: 5   b. Faces: 8
    Edges: 9   Edges: 18
    Vertices: 6   Vertices: 12
c. Faces: 4   d. cylinder
    Edges: 6
    Vertices: 4
e. hexagonal pyramid   f. cube

14-8 a. 294 m²   b. 160 in.²   c. 236 cm²

14-9 a. 252 in.³   b. 15.625 m³   c. 272 ft³

14-10 a. ≈351.68 m³   b. ≈100.48 ft³
c. ≈18.84 in.³   d. ≈141.3 m³

14-11 a. ≈20.93 in.³   b. ≈18.84 in.³
c. 40 cm³   d. 8 ft³

14-12 a. G, I, J, K   b. G, I, J, K, L   c. J

14-15 11.75 km²

15-1 20 pepperoni, 40 ham, 120 sausage; 180 orders

15-2 a. ⁻416   b. ⁻15   c. >
d. >   e. <   f. <

15-3 a.
$-5 + 7 = 2$

b.
$3 + (-3) = 0$

c.
$-4 + (-2) = -6$

d.
$3 + 5 = 8$

e. 19   f. ⁻4   g. ⁻4   h. ⁻18

15-4 a.
$5 - 3 = 2$

b.
$7 - (-2) = 9$

c.
$-3 - 2 = -5$

d.
$-8 - (-6) = -2$

e. 14   f. ⁻9   g. ⁻9   h. ⁻15

15-5 $5 under

15-6 about $5 less

15-7 a. 24   b. ⁻64   c. ⁻72   d. 168

15-8 a. ⁻6   b. 6   c. 8   d. ⁻11

15-10 missing data: depth of oil deposit

15-12 a. 7°F   b. 15°F   c. 15 mph

15-13 a. 102°   b. 114°   c. 123°   d. 112°

16-1 a. 21   b. 35   c. 13   d. 12
e. 80   f. 55   g. 49   h. 130

TRY THIS ANSWERS

16-2 **a.** *r* decreased by 3, *r* less 3, 3 less than *r*, the difference when 3 is taken from *r*
b. 6 times *y*, the product of 6 and *y*, *y* multiplied by 6, the product of *y* and 6
c. *w* divided by 9, *w* separated into 9 equal groups, the quotient of *w* and 9, one ninth of *w*
d. the sum of *t* and 5, *t* increased by 5, 5 more than *t*, 5 added to *t*
e. *q* divided by 4, *q* separated into 4 equal groups, the quotient of *q* and 4, one fourth of *q*
Answers may vary. **f.** $d - 5$
g. $4j$ **h.** $\frac{1}{3}d$ **i.** $m + 2$

16-3 **a.** 24 **b.** 26 **c.** -107 **d.** -22

16-4 **a.** 52 **b.** 59 **c.** 107 **d.** 67

16-5 **a.** 8 **b.** 38 **c.** 441 **d.** 588

16-6 **a.** 7 **b.** 11 **c.** 13 **d.** 40

16-7 **a.** 3 h 40 min **b.** 8 games

16-8 **a.** 5 **b.** 7 **c.** 8 **d.** 9
e. 12 **f.** 20 **g.** 529
h. 7.071 **i.** 1156 **j.** 4.899
k. 361 **l.** 6.083

16-9 **a.** 10 in. **b.** 8 cm **c.** 40 yd

16-10 **a.** (1, 1) **b.** (3, 2) **c.** (4, 1)
d.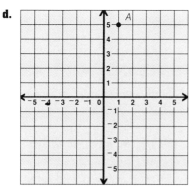
e.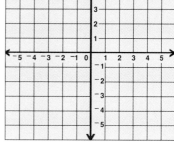
f.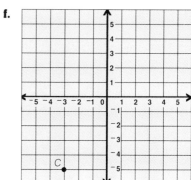
g.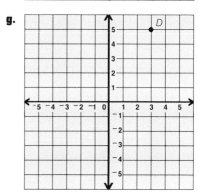

16-11 **a.** reasonable
b. not reasonable, $500 would be 50% per year

16-13 **a.** D7 **b.** fire station

16-14 **a.** 3^2, 9 **b.** 2^2, 4 **c.** 0^0, 0
d. 4^2, 16 **e.** 6^2, 36 **f.** 8^2, 64
g. SQR(4), 2 **h.** SQR(25), 5
i. SQR(16), 4 **j.** SQR(100), 10
k. SQR(144), 12 **l.** SQR(36), 6

INDEX

Act it out, 22, 182
Acute angles, 338
Addend, 68
Addition. See also *Sums*
    associative property of, 79
    calculator and, 131, 439
    estimation and, 12, 21, 40, 68, 127, 130, 254, 257
    inverse operations and, 240
    mental math and, 11, 44, 79, 131, 243, 250
    of customary units, 174
    of decimals, 34
    of fractions, 242, 246
    of integers, 436
    of mixed numbers, 244, 248
    of percents, 128, 284
    of time, 206
    of whole numbers, 34
    regrouping and, 251
    solving equations with, 462
    understanding, 10
    using number lines, 436
    writing zeros with, 34
Algebra
    evaluating expressions, 37, 63
    graphing on a coordinate grid, 476
    solving equations, 129, 225, 249, 462, 464, 466, 468
    solving multiple-step equations, 468
    variables, 7, 460
    Working with Variables, 7, 37, 63, 129, 225, 249
Algebraic expressions, 37, 63, 129, 225, 249, 460
Angles
    acute, 338
    calculator and, 343, 345, 357
    congruent, 280, 342, 344
    drawing, 338
    estimation and, 337
    measuring, 336

    mental math and, 339
    naming, 338
    obtuse, 338
    of polygons, 342, 343, 345, 347
    right, 338
Approximately equal to ($\approx$), 111
Area
    calculator and, 403, 407
    estimation and, 62, 403
    irregular, 425
    of a circle, 170, 406
    of a parallelogram, 402
    of a rectangle, 62, 170, 400
    of a square, 400
    of a triangle, 170, 404
    surface, 412
Average, 384. See also *Mean*

Balance, 52
Bank statements, reconciling, 52
Bar graphs, 122
    computer generated, 139
    double, 138
    histograms, 382
Base
    of a geometric figure, 402, 404, 410
    of an exponent, 78
Batting average, 200
Blueprints, 334
Break apart numbers, 79, 229
Budgets, 139, 352, 440

Calculation methods, choosing, 46, 227
Calculator
    addition and, 131, 439
    angle measures and, 343, 345, 357
    area of a circle and, 407
    area of a parallelogram and, 403
    as a calculation method, 46
    checking answers and, 63, 65, 73, 155, 199
    division and, 67, 98, 447, 467

    factorials and, 369
    fractions and, 198, 233
    integers and, 135, 139, 447, 451
    inverse operations and, 113
    irregular area and, 425
    maintaining a checking account and, 53
    measurement and, 153, 161, 175
    multiplication and, 65, 67, 75, 223, 447, 467
    order of operations and, 112
    patterns and, 75, 99, 473
    percent and, 287, 289, 307, 321
    place value and, 5, 7
    sales tax and, 321
    scale drawings and, 297
    solving equations and, 37, 65, 467
    square roots and, 473
    subtraction and, 37
    unit prices and, 279
Capacity
    customary units of, 152
    metric units of, 162
Careers, 14, 15, 20, 66, 72, 73, 150, 151, 163, 214, 224, 229, 232, 242, 243, 250, 324, 425, 478
Celsius, 168, 471
Centimeter, 158
Checking accounts, 24, 42, 43, 52, 53
Circle graphs, 128, 139, 352
    computer generated, 139
    mental math and, 129
Circles
    area of, 170, 406
    calculators and, 407
    circumference of, 350
    diameter of, 350
    estimation and, 351
    pi, 350
    radius of, 350

INDEX

Circumference, 350
  estimation and, 351
Clustering, 68
Commissions, 324
Common denominator, 193, 242, 244, 246, 252
Comparison
  of decimals, 6, 290–291
  of fractions, 194, 290–291
  of integers, 441
  of unit prices, 278, 279
  of percents, 290–291
  of whole numbers, 6
  on a number line, 434
Compatible numbers, 18, 130, 197, 226, 250, 314
Compensation, 131
Components, 476
Composite numbers, 82
Compound probability, 374
Computers, 25, 83, 139, 207, 261, 325, 391, 483
Cones, 410
  volume of, 418
Congruence, 280, 342
Congruent angles, 280, 342, 344
Congruent segments, 342
Consumer Math, 14, 15, 24, 42, 43, 52, 53, 66, 76, 77, 80, 98, 100, 102, 108, 110, 132, 156, 170, 196, 204, 220, 228, 229, 232, 256, 259, 278, 316, 320, 324, 352, 380, 388, 408, 440, 449, 462, 468, 470
Coordinate grids, 476
Coordinates, 476
Corresponding sides, 280
Counting, 44
Counting principle, 366
Cross products, 272
Cubes, 410
  of numbers, 78
  surface area of, 412
  volume of, 170, 414, 415
Cubic units, 414
Cubit, 149
Cup, 152
Current deductions, 76
Customary units
  addition of, 174
  comparing, 148

estimation and, 151, 153, 155
  of capacity, 152
  of length, 148
  of weight, 154
  measuring to the nearest quarter inch, 150
  reading scales, 156
  temperature, 168, 471
Cylinders, 410
  volume of, 416

Data Bank, 13, 78, 95, 155, 187, 226, 253, 287, 337, 387, 419, 451, 477, 488–496
Data Collection and Analysis, 80, 136, 294, 295, 388, 480
Data Hunt, 21, 25, 133, 149, 201, 257, 279, 289, 317, 375, 383, 415, 417, 439, 471
Decagon, 342
Decimals
  addition of, 34
  as fractions, 200
  as mixed numbers, 200
  as percents, 286–287, 290
  comparing and ordering, 6
  comparing with percents and fractions, 290
  division by 10, 100, or 1000, 106
  division by decimals, 108
  division by whole numbers, 104
  fractions as, 198
  mental math and, 201, 203
  mixed numbers as, 198
  multiplication of, 72, 74, 83
  multiplication by 10, 100, or 1000, 74
  multiplication using computers, 83
  on a number line, 287
  percents as, 286
  place value, 4
  reading, 4
  repeating, 198
  rounding, 8, 202
  subtraction of, 36
  terminating, 198
  writing, 4, 286

  writing zeros in, 7, 286
Decimeter, 158
Deductions, 76
Degrees
  of angles, 336, 338
  of temperature, 168
Dekameter, 106, 158
Denominator, 184
  common, 193, 242, 244, 246, 252
  least common, 193, 246, 248, 252
Deposit slips, filling out, 42
Deposits, 42
  outstanding, 52
Determining Reasonable Answers, 48, 67, 91, 95, 109, 130, 292, 354, 478
Diagram, 32
Diameter, 350
Differences. See *Subtraction*
Discount, 290, 291, 306, 307, 312
Dividend, 94
Division. See also *Quotients*
  by 1-digit numbers, 92
  by 2-digit numbers, 94
  by 3-digit numbers, 98
  by 10, 100, or 1000, 106
  calculator and, 97, 233, 447, 467
  check by multiplying, 92
  estimation and, 18, 93, 95, 98, 109, 111, 226
  inverse operations and, 240
  mental math and, 11, 97, 445
  of customary units, 174
  of decimals, 102, 104
  of fractions, 222, 233
  of integers, 444
  of whole numbers, 104
  solving equations with, 466
  understanding, 14, 97, 107
  with mixed numbers, 224
  with zeros in the quotient, 96
Divisor, 93
Dodecagon, 342
Double bar graphs, 138
Double line graphs, 138

INDEX

Down payment, 220
Draw a picture or diagram, 22, 32, 60, 90, 146, 256, 332

Edges, 398, 410, 424
Elapsed time, 206
Equally likely, 370
Equations
    calculator and, 37, 65, 467
    mental math and, 465, 469
    solving multiple-step, 468
    solving with addition, 462
    solving with division, 466
    solving with multiplication, 466
    solving with subtraction, 464
Equilateral triangles, 344
Equivalent fractions, 186, 193, 246, 248
Estimation
    addition and, 12, 21, 40, 68, 127, 130, 254, 257
    angles and, 337
    area of a parallelogram and, 403
    area of a rectangle and, 403
    area of a square and, 403
    budgets and, 353
    calculator and, 63, 67, 95, 109
    circumference and, 351
    clustering, 68
    compatible numbers, 18, 130, 197, 226, 314
    deciding when to use, 20, 227
    division and, 18, 93, 95, 98, 109, 111, 226
    expectation and, 379
    fractions and, 195, 197, 199, 226, 254, 257
    front-end, 40
    gauge readings and, 196
    graphs and, 123, 127, 353
    integers and, 441, 447
    measurement and, 151, 153, 167
    multiplication and, 16, 63, 65, 67, 73, 111, 226
    number lines and, 195, 285
    percents and, 285, 314, 321
    products, 16
    proportions and, 275

    rounding, 12, 16, 254
    sales tax and, 321
    square roots and, 473
    subtraction and, 12, 13, 40, 111, 254, 257
    sums and differences, 12, 40, 254
    temperature and, 169
    volume of a cone and, 419
Evaluate
    expressions, 37, 39, 63, 75
    graphs, 132
Event, 370, 374
Exact answer, 20, 227
Expectation, 378
    estimation and, 379
Experimental probability, 372
Exponents, 78
Expressions, 37, 63, 460

Faces, 410, 424
Factor
    common, 188
    greatest common, 188
    prime, 82
    tree, 82
Factorials, 368
    calculator and, 369
Fahrenheit, 168, 471
Fathom, 149
Federal withholding taxes, 232
Finance charge, 220
Finance, personal. See *Personal finance*
Find a pattern, 22, 120, 146, 332
Flowchart, 25, 240
Fluid ounce, 152
Foot, 148
Formulas, 170
    area of a circle, 170, 406
    area of a parallelogram, 402
    area of a rectangle, 62, 170, 400
    area of a square, 412
    area of a triangle, 170, 404
    circumference of a circle, 350
    distance, 64
    perimeter of a rectangle, 348
    perimeter of a square, 348
    Pythagorean Theorem, 474

    simple interest, 170, 316
    solving problems with, 470
    table of, 495
    volume of a cone, 418
    volume of a cube, 170
    volume of a cylinder, 416
    volume of a pyramid, 418
    volume of a rectangular solid, 170
Fractions
    addition, 242, 246
    as decimals, 198
    as percents, 288
    as ratios, 270
    calculator and, 198, 233
    comparing, 194
    comparing with percents and decimals, 290
    decimals as, 200
    denominator, 184
    division of, 222, 233
    Egyptian, 247
    equivalent, 186, 193, 246, 248
    estimation, 195, 197, 199, 226, 254, 257
    improper, 190
    least common denominator, 193, 246, 248, 252
    lowest factors of, 186
    lowest terms, 188
    mental math, 195, 201, 203, 229, 243, 250, 253
    mixed numbers, 190, 194
    multiplication of, 216, 233
    numerator, 184
    on a number line, 191, 195, 221, 285
    percents as, 288
    probability and, 370, 375
    reciprocals, 221
    subtraction, 242, 246
Frequency table, 382
Front-end estimation, 40
Furlong, 149

Gallon, 152
Geometry
    angles, 336
    area, 400, 402, 404, 406

INDEX

base of a figure, 402, 404, 410
calculator and, 343, 345, 403, 407
circles, 350, 406
circumference, 350
congruent, 280
estimation and, 337, 351, 353, 403, 419
line segments, 334
lines, 334
mental math and, 339, 349, 415
perimeter, 348
pi, 350
points, 334
polygons, 342
polyhedrons, 424
rays, 336
segments, 334
space figures, 398, 410, 414, 416, 418
spatial visualization and, 424
surface area, 412
volume, 414, 416, 418
Gram, 164
Graphing ordered pairs, 476
Graphs
bar, 122
circle, 128, 139, 352
computer generated, 139
double bar, 138
double line, 138
estimation and, 123, 127, 353
evaluating, 132
histograms, 382
line, 126
mental math and, 129
of ordered pairs, 476
pictographs, 124
Greatest common factor, 188
Gross pay, 76
Group Activity, 21, 109, 127, 138, 151, 153, 155, 159, 185, 203, 281, 335, 343, 345, 347, 371, 373, 375, 377, 381, 409, 435, 467
Guess and check, 22, 90, 268, 458

Hectometer, 106, 158
Hexagon, 342
Histograms, 382
Horizontal scale, 122
Hypotenuse, 474

Improper fractions, 190
Inch, 148
Income tax, federal, 232
Independent events, 374
Installment plans, 220
Integers
addition of, 436
calculator and, 435, 439, 447, 451
comparing, 434
division of, 444
estimation and, 441, 447
mental math and, 443, 445
multiplication of, 442
on a number line, 434, 436, 438
operations with, 446
opposites, 436
subtraction of, 438
Interest, simple, 170, 316
Intersect, 340
Inverse operations, 113, 240, 464
calculator and, 113
solving equations with, 462, 466
Irregular area, 425
calculator and, 425
Isosceles triangles, 344

Kilogram, 164
Kiloliter, 162
Kilometer, 106, 158

League, 149
Least common denominator, 193, 246, 248, 252
Least common multiple, 192
Length
customary units of, 148
metric units of, 158
Line graphs, 126, 138, 139
computer generated, 139
double, 138
Line segments, 334
congruent, 342

Lines, 334, 340
parallel, 340
perpendicular, 340
skew, 340
Liter, 162
Logic, 420
Logical reasoning, 212, 398
Lowest terms, 188, 270

Make a table, 22, 120, 146, 332, 364
Make an organized list, 22, 60
Manipulatives, 35, 38, 47, 91, 95, 150, 167, 219, 247, 283, 297, 333, 336, 338, 343, 345, 347, 356, 372, 375, 377, 467
Maps, 33, 107, 151, 197, 254, 255, 260, 283, 476
Mass, metric units of, 164
Mean, 384
Measurement
accuracy of, 150
adding, 174
area, 62, 402, 404, 406
calculating, 152, 170, 172, 174
175
capacity, 152
changing units, 148
circumference, 350
customary system of, 148
dividing, 174
estimation and, 151, 155, 167
length, 148, 150
mass, 164, 166
mental math and, 153
metric system of, 158, 160
multiplying, 174
nonstandard units, 149
of an angle, 336, 357
perimeter, 348, 354
subtracting, 174
surface area, 412
to the nearest centimeter, 166
to the nearest quarter inch, 150
using, 408
volume, 414, 416, 418

INDEX

weight, 156
Median, 386
Mental math
   addition, 44, 79, 131, 243, 250
   angles and, 339
   as a calculation method, 46
   break apart numbers, 79, 229
   circle graphs and, 129
   compatible numbers, 250
   compensation, 131
   counting, 44
   division and, 97, 445
   fractions and, 195, 201, 203, 229, 243, 250, 253
   integers and, 443, 445
   measurement and, 153, 160
   multiplication and, 67, 79, 131
   percents and, 129, 312
   perimeter and, 349
   range and, 387
   solving equations and, 465, 469
   subtraction and, 44, 131, 187, 253
   unit prices and, 279
   unit rates and, 277
   volume of a cube and, 415
   volume of a rectangular prism and, 415
Meter, 106, 158
Metric scales, 166
Metric units
   estimation, 167
   measuring to the nearest centimeter, 166
   of capacity, 162
   of length, 158, 160
   of mass, 164
   place value and, 160, 162, 164
   reading metric scales, 166
   temperature, 168, 471
Mile, 148
Mileage chart, 490
Milligram, 164
Milliliter, 162
Millimeter, 158
Mixed Applications, 13, 17, 35, 45, 73, 75, 105, 109, 153, 155, 159, 161, 163, 165, 185, 191, 201, 217, 219, 223, 225, 250, 279, 281, 291, 307, 309, 349, 351, 377, 379, 405, 413, 417, 437, 439, 443, 463, 465, 469, 473, 475
Mixed numbers
   addition, 244
   as decimals, 200
   comparing, 194, 290–291
   decimals as, 200
   division with, 224
   improper fractions and, 190
   multiplication with, 218
   regrouping, 251–252
   subtraction, 244
Mixed Practice, 39, 69, 103, 313
Mixed Skills Review, 9, 19, 21, 49, 65, 75, 99, 105, 123, 127, 133, 137, 149, 163, 169, 191, 195, 203, 219, 223, 245, 253, 257, 271, 273, 277, 283, 311, 315, 319, 335, 337, 341, 369, 375, 381, 389, 405, 407, 409, 437, 445, 447, 463, 465, 471
   Extra, 497–512
Mode, 384
Multiple-step problems, 70
Multiplication. See also *Products*
   by 1-digit numbers, 62
   by 2-digit numbers, 64
   by 3-digit numbers, 66
   by 10, 100, or 1000, 74
   calculator and 65, 67, 75, 233, 447, 467
   estimation and, 16, 63, 65, 67, 73, 111, 226
   exponents, 78
   inverse operations and, 240
   mental math and, 67, 79, 131, 432
   of customary units, 174
   of decimals, 72, 74, 83
   of fractions, 216, 233
   of integers, 442
   of mixed numbers, 218
   of whole numbers, 62, 64, 66
   regrouping, 62
   solving equations with, 466
   understanding, 14
   with mixed numbers, 218

Negative numbers, 434
Net pay, 76
Nonstandard units, 149, 151
Number line
   decimals on a, 287
   fractions on a, 191, 195, 221, 285
   integers on a, 434, 436, 438
   percents on a, 285, 287
   probability on a, 371
   used in adding, 436
   used in subtracting, 438
Number Sense, 9, 13, 17, 19, 37, 67, 97, 99, 105, 129, 149, 157, 161, 163, 165, 187, 191, 192, 196, 197, 199, 201, 217, 221, 245, 247, 273, 285, 287, 291, 312, 371, 377, 379, 417, 435, 443, 447, 461, 467
Numbers. See also *Compatible numbers; Mixed numbers; Whole numbers*
   comparing and ordering, 6
   composite, 82
   finding given a percent, 310
   finding percent, one of another, 308
   finding percents of, 306
   negative, 434
   place value of, 4
   positive, 434
   prime, 82
   reading, 4
   standard, 4
   writing, 4
Numerator, 184
Numerical expressions, 37

Obtuse angles, 338
Octagons, 342
Odds, 376
Operations, 10, 14
   in BASIC, 261
   inverse, 240, 462, 464, 466
   order of, 112
Opposites, 436, 441
Order of operations, 112

INDEX

Ordered pairs, graphing, 476
Origin, 476
Ounce, 154
    fluid, 152
Outcomes, 366, 372, 376, 378
Overtime rates, 229

Pace, 151
Parallel lines, 340
Parallelogram, 346, 357
    calculator and, 403
    area of a, 402
Patterns, 75, 99, 173, 199, 245, 473
Paychecks, reading, 76
Pentagon, 342
Percents
    addition and, 128, 284
    as decimals, 286
    as fractions, 288
    calculator and, 287, 289, 307, 321
    circle graphs and, 128
    comparing with fractions and decimals, 290
    computer calculation of, 325
    decimals as, 286, 290
    estimation and, 285, 314, 321
    finding a number given, 310
    fractions as, 288
    mental math and, 129, 312
    of a number, 306
    of decrease, 318
    of increase, 318
    on a number line, 285, 287
    one number of another, 308
    ratios and, 284
    subtraction and, 128, 284
Perimeter, 348
Permutations, 368
Perpendicular lines, 340
Personal finance, 24, 42, 43, 52, 53, 76, 170, 204, 220, 229, 232, 316, 320, 352, 440, 462, 468
Pi, 350
Pictographs, 124
Pint, 152
Place value, 4, 5, 7, 160, 162, 164

Planes, 340
Points, 334
Polygons, 342, 343, 344, 345, 346, 347, 348, 356, 410
Polyhedrons, 410, 414, 418, 424
Positive numbers, 434
Pound, 154
Powers, 78, 483
Practice Through Problem Solving, 7, 35, 38, 51, 63, 75, 81, 111, 159, 173, 249, 255, 345, 475
Predictions, 380
Prime factorization, 82
Prime numbers, 82
Principal, 316
Prisms, 412, 414
    surface area of, 412
    triangular, 410
    volume of, 414
Probability
    compound, 374
    counting principle, 366
    estimation and, 379
    event, 370, 374
    expectation, 378
    experimental, 372
    factorials, 368
    independent events, 374
    odds, 376
    on a number line, 371
    outcomes, 366, 372, 376, 378
    permutations, 368
    predictions, 380
    samples, 380
    simple, 370
    tree diagrams, 366
Problem solving, 5, 22, 67, 73, 97, 107, 125, 149, 165, 193, 201, 223, 245, 247, 257, 271, 275, 277, 281, 291, 309, 311, 317, 321, 345, 347, 369, 373, 401, 409, 413, 445, 461, 465, 469, 473, 477
    Data Collection and Analysis, 80, 136, 294, 388, 480
    Determining Reasonable Answers, 48, 91, 130, 292, 354, 478

    Developing a Plan, 227
    Developing Thinking Skills, 50, 204, 230, 322, 422
    Evaluating Solutions, 171, 231, 275, 289, 311, 365
    5-Point Checklist, 22, 32, 42, 43, 48, 50, 60, 70, 76, 80, 90, 100, 102, 110, 120, 132, 134, 136, 146, 156, 170, 172, 182, 196, 202, 204, 214, 220, 227, 228, 230, 240, 256, 258, 259, 268, 278, 292, 294, 304, 316, 320, 322, 332, 352, 354, 355, 364, 380, 382, 388, 398, 408, 420, 422, 432, 440, 448, 449, 458, 470, 478, 480
    Group Decisions, 110, 172, 259, 355, 449
    Learning to Use Strategies, 32, 60, 90, 120, 146, 182, 214, 240, 268, 304, 332, 364, 398, 432, 458
    missing data, 448
    multiple-step problems, 70
    Practice Through Problem Solving, 7, 35, 38, 51, 63, 75, 81, 159, 173, 249, 255, 345, 475
    Strategy Practice, 258
    Understand the Situation, 8, 12, 16, 18, 34, 36, 62, 64, 72, 74, 92, 94, 104, 108, 112, 122, 124, 126, 128, 130, 131, 134, 158, 160, 168, 184, 194, 198, 202, 216, 218, 222, 226, 242, 244, 246, 252, 254, 270, 272, 274, 276, 278, 280, 282, 284, 288, 290, 306, 308, 310, 314, 316, 318, 324, 332, 334, 336, 338, 340, 342, 344, 346, 348, 352, 366, 368, 370, 374, 376, 378, 380, 382, 384, 386, 390, 402, 404, 406, 408, 412, 414, 416, 418, 424, 434, 436, 438, 440, 441, 442, 444, 462, 464, 468, 470, 472, 476, 478

INDEX

Using Data from a Chart, 228
Using Data from a Table, 202, 320
Using Data from an Advertisement, 102
Using Formulas, 170
Using Logic, 420
Working with Data, 382
Problem-solving strategies
 act it out, 22, 182
 choose an operation, 22, 70
 draw a picture or diagram, 22, 32, 60, 90, 146, 256, 332
 find a pattern, 22, 120, 146, 332
 guess and check, 22, 90, 268, 458
 make a table, 22, 120, 146, 332, 364
 make an organized list, 22, 60
 solve a simpler problem, 22, 304, 332
 use logical reasoning, 22, 214, 398
 use objects, 22, 32, 182
 work backward, 22, 240, 432
 write an equation, 22
Products, 62. See *Multiplication*
Proportions, 272
 estimation and, 275
 scale drawings, 282
 similar figures, 280
 solving, 274
Protractors, 336, 338, 343, 345, 347
Pyramids, 410
Pythagorean Theorem, 474
Pythagorean triple, 483

Quadrilaterals, 62, 346, 400, 402
Quart, 152
Quotients. See *Division*

Radius, 350
Ranges, 386, 387

Rates
 of interest, 316
 unit, 276
Ratios, 270, 272, 276, 280, 284
 Golden, 467
Rays, 336
Reciprocals, 221
Reconciling bank statements, 52
Rectangles, 346
 area of, 62, 170, 400
 perimeter of, 348
Rectangular prisms
 surface area of, 412
 volume of, 170, 414, 415
Regrouping, 34, 36, 174, 251, 252
Regular polygons, 342
Remainder, 92
Repeating decimals, 198
Rhombus, 346
Right angles, 338
Right triangles, 344, 474
Rod, 149
Rounding, 16
 decimals, 8, 202
 estimation using, 12, 16, 254
 money, 8
 whole numbers, 8
Running balance, 43

Sales tax, 320
 calculator and, 321
 estimating, 321
Samples, 380
Savings account, 42
Scale drawings, 282
 calculator and, 297
Scalene triangles, 344
Scales
 horizontal, 122
 of maps, 383
 reading customary, 156
 reading metric, 166
 temperature, 168
 vertical, 122
Segments, 334
Show Math, 35, 63, 95, 185, 196, 219, 221, 223, 225, 249, 251, 255, 271, 273, 283, 285, 347, 387, 399, 401, 403, 477

Similar figures, 280
Simple interest, 170, 316
Simple probability, 370
Skew lines, 340
Solutions, 462
Solve a simpler problem, 22, 165, 304, 332
Solving equations, 462, 468
Space figures, 398, 410, 416, 418
 surface area of, 412
 volume of, 414
Spatial visualization, 424
Sphere, 410
Square roots, 472
 calculator and, 473
 computer calculation of, 483
 estimation and, 473
 table of, 495
Square units, 400
Squares, 346
 area of, 400
 computer calculation of, 483
 of numbers, 78, 472
 perimeter of, 348
 table of, 495
State tax, 76
Statistics
 average, 384
 Data Collection and Analysis, 80, 136, 294, 388, 480
 frequency table, 382
 histogram, 382
 mean, 384
 median, 386
 mental math and, 387
 mode, 384
 range, 386
 stem-and-leaf diagrams, 390
Stem-and-leaf diagrams, 390
Subtraction. See also *Differences*
 across zero, 38
 calculator and, 37
 estimation and, 12, 13, 40, 111, 254, 257
 inverse operations and, 240
 mental math and, 11, 44, 131, 253
 of customary units, 174
 of decimals, 36

INDEX

   of fractions, 242, 246
   of integers, 438
   of mixed numbers, 244, 248, 252
   of percents, 128, 284
   of time, 206
   of whole numbers, 36
   regrouping and, 252
   solving equations with, 464
   understanding, 10
   using number lines, 438
Summary Questions, 23
Sums. See *Addition*
Suppose, 15, 50, 71, 73, 110, 121, 133, 137, 139, 147, 161, 172, 183, 189, 215, 217, 231, 241, 243, 259, 269, 293, 295, 305, 309, 355, 365, 383, 389, 401, 415, 419, 449, 469, 479, 481
Surface area, 412

Tables
   frequency, 382
   of formulas, 495
   of measurements, 494
   of squares and square roots, 495
   sales tax, 492
Take-home pay, 76
Talk Math, 19, 41, 45, 73, 127, 187, 275, 289, 307, 333, 339, 367, 369, 387, 401, 411, 473
Taxes, 76, 232, 320
Telephone bills, 98, 296
Temperature, 168, 169, 450, 471
Terminating decimals, 198
Thinking Actions, 50, 204, 230, 322, 422
Time
   elapsed, 206
   simple interest and, 316
   zones, 260
Ton
   customary, 154
   metric, 164
Trapezoids, 346
Tree diagrams, 366
Triangles, 344, 474
   area of, 170, 404
Triangular prism, 410

Unit price, 278
   calculator and, 279
   mental math and, 279
Unit rates, 276
   mental math and, 277
Use logical reasoning, 22, 214, 398
Use Objects, 22, 32, 38, 47, 91, 182, 247, 333

Variables, 7, 37, 63, 129, 225, 249, 460
Vertical scale, 120
Vertices, 336, 342, 410
Volume
   estimation and, 419
   mental math and, 415
   of a cone, 418
   of a cube, 170, 414
   of a cylinder, 416
   of a pyramid, 418
   of a rectangular prism, 170, 414

Wages, 66, 76, 77, 232, 324, 440, 449
Weight, customary units of, 154
Whole numbers
   addition of, 34
   comparing and ordering, 6
   division of, 92, 94, 96, 98
   division of decimals by, 104, 106
   multiplication of, 62, 64, 66
   place value, 4
   reading, 4
   regrouping, 251
   rounding, 8
   subtraction of, 36, 38
   writing, 4
Wind chill factor, 450
Withdrawals, 43
Work backward, 240, 432
Working with Variables, 7, 37, 63, 129, 225, 249
Write an equation, 22
Write Math, 17, 78, 93, 107, 153, 171, 187, 195, 225, 229, 291, 317, 443, 463
Write Your Own Problem, 11, 15, 33, 51, 71, 113, 121, 147, 183, 215, 227, 241, 269, 293, 305, 333, 391, 433, 459, 461, 479

Yard, 148

Zero, subtraction across, 38
Zeros, writing, with addition, 34
Zeros, writing in decimals, 7, 280